PARALLEL COMPUTATIONAL FLUID DYNAMICS

PRACTICE AND THEORY

J.M. Burgerscentrum

TU Delft

PARALLEL COMPUTATIONAL FLUID DYNAMICS

PRACTICE AND THEORY

Proceedings of the Parallel CFD 2001 Conference
Egmond aan Zee, The Netherlands (May 21-23, 2001)

Edited by

P. WILDERS
Delft University of Technology
Delft, The Netherlands

A. ECER
IUPUI, Indianapolis
Indiana, U.S.A.

J. PERIAUX
Dassault-Aviation
Saint-Cloud, France

N. SATOFUKA
Kyoto Institute of Technology
Kyoto, Japan

Assistant Editor
P. FOX
IUPUI, Indianapolis
Indiana, U.S.A

2002
ELSEVIER
Amsterdam – Boston – London – New York – Oxford – Paris – San Diego – San Francisco – Singapore – Sidney – Tokyo

ELSEVIER SCIENCE B.V.
Sara Burgerhartstraat 25
P.O. Box 211, 1000 AE Amsterdam, The Netherlands

First edition 2002

Library of Congress Cataloging in Publication Data
A catalog record from the Library of Congress has been applied for.

ISBN: 0-444-50672-1

∞ The paper used in this publication meets the requirements of ANSI/NISO Z39.48-1992 (Permanence of Paper).

Transferred to digital printing 2006

PREFACE

ParCFD 2001, the thirteenth international conference on Parallel Computational Fluid Dynamics took place in Egmond aan Zee, the Netherlands, from May 21-23, 2001. The specialized, high-level ParCFD conferences are organized yearly on traveling locations all over the world. A strong back-up is given by the central organization located in the USA (www.parcfd.org).

These proceedings of ParCFD 2001 represent 70% of the oral lectures presented at the meeting. All published papers were subjected to a refereeing process, which resulted in a uniformly high quality.

The papers cover not only the traditional areas of the ParCFD conferences, e.g. numerical schemes and algorithms, tools and environments, interdisciplinary topics, industrial applications, but, following local interests, also environmental and medical issues. These proceedings present an up-to-date overview of the state of the art in parallel computational fluid dynamics. We believe that on basis of these proceedings we may draw the conclusion that parallel CFD is on its way to become a basic engineering tool in design, engineering analysis and prediction. As such, we are facing a next step in the development of parallel CFD and we hope that the papers in this book will contribute to the inspiration needed for enabling this development.

P. Wilders

ACKNOWLEDGEMENTS

The local organizing committee of ParCFD 2001 received a lot of support, both financial and organizational. In particular, we want to thank the international scientific committee for its help in the refereeing process and for proposing excellent invited speakers. This enabled us to organize a high-level conference.

Financial support to ParCFD 2001 was obtained from:

- Delft University of Technology
- J.M. Burgers Centre
- Royal Dutch Academy of Sciences
- AMIF/ESF
- Eccomas
- Delft Hydraulics
- National Aerospace Laboratory NLR
- Platform computing
- Compaq
- Cray Netherlands

The financial support enabled us not only to organize an excellent scientific and social program, but also to set up an attractive junior researchers program and to grant some researchers from Russia.

The working group on "Affordable Computing" of the network of excellence MACSINET helped us to organize a very successful industrial day.

Finally, the main organizer, P. Wilders, wants to thank staff and colleagues of Delft University for their strong support from the early beginnings.

The local organizing committee,

A.W. Heemink (Delft University of Technology)
M.S. Vogels (National Aerospace Lab. NLR)
P. Wesseling (Delft University of Technology)
P. Wilders (Delft University of Technology)

TABLE OF CONTENTS

1. Opening Paper

Parallel Computational Fluid Dynamics – Practice and Theory
P. Wilders, A. Ecer, J. Periaux, N. Satofuka and P. Fox (Editors)

An Overview of ParCFD activities at Delft University of Technology

P. Wilders[a] *, B.J. Boersma[a] †, J.J. Derksen[a] ‡, A.W. Heemink[a] §, B. Ničeno[a] ¶, M. Pourquie[a] ||, C. Vuik[a] **

[a]Delft University of Technology, J.M. Burgers Centre
Leeghwaterstraat 21, 2628 CJ Delft, The Netherlands,
email: p.wilders@its.tudelft.nl

At Delft University of Technology much research is done in the area of computational fluid dynamics with underlying models ranging from simple desktop-engineering models to advanced research-oriented models. The advanced models have the tendency to grow beyond the limit of single-processor computing. In the last few years research groups, studying such models, have extended their activities towards parallel computational fluid dynamics on distributed memory machines. We present several examples of this, including fundamental studies in the field of turbulence, LES modelling with industrial background and environmental studies for civil engineering purposes. Of course, a profound mathematical back-up helps to support the more engineering oriented studies and we will also treat some aspects regarding this point.

1. Introduction

We present an overview of research, carried out at Delft University of Technology and involving parallel computational fluid dynamics. The overview will not present all activities carried out in this field at our University. We have chosen to present work of those groups, that are or have been active at the yearly ParCFD conferences, which indicates that these groups focus to some extent on purely parallel issues as well. This strategy for selecting contributing groups enabled the main author to work quite directly without extensive communication overhead and results in an overview presenting approximately 70% of the activities at our University in this field. We apologize on forehand if we have overseen major contributions from other groups.

At Delft University of Technology parallel computational fluid dynamics is an ongoing research activity within several research groups. Typically, this research is set up and hosted within departments. For this purpose they use centrally supported facilities, most often only operational facilities. In rare cases, central support is given as well for developing purposes. Central support is provided by HPαC, http://www.hpac.tudelft.nl/, an institution for high perfor-

*Dept. Applied Mathematical Analysis, Section Large Scale Systems
†Dept. Mechanical Engineering, Section Fluid Mechanics
‡Dept. Applied Physics, Kramers Laboratorium
§Dept. Applied Mathematical Analysis, Section Large Scale Systems
¶Dept. Applied Physics, Section Thermofluids
||Dept. Mechanical Engineering, Section Fluid Mechanics
**Dept. Applied Mathematical Analysis, Section Numerical Mathematics

mance computing splitted off from the general computing center in 1996. Their main platform is a Cray T3E with 128 DEC-Alpha processors, installed in 1997 and upgraded in 1999.

From the paralllel point of view most of the work is based upon explicit parallel programming using message passing interfaces. The usage of high-level parallel supporting tools is not very common at our university. Only time accurate codes are studied with time stepping procedures ranging from fully explicit to fully implicit. Typically, the explicit codes show a good parallel performance, are favorite in engineering applications and have been correlated with measurements using fine-grid 3D computations with millions of grid points. The more implicit oriented codes are still in the stage of development, can be classified as research-oriented codes using specialized computational linear algebra for medium size grids and show a reasonable parallel performance.

The physical background of the parallel CFD codes is related to the individual research themes. Traditionally, Delft University of Technology is most active in the incompressible or low-speed compressible flow regions. Typically, Delft University is also active in the field of civil engineering, including environmental questions. The present overview reflects both specialisms.

Of course, studying turbulence is an important issue. Direct numerical simulation (DNS) and large eddy simulation (LES) are used, based upon higher order difference methods or Lattice-Boltzmann methods, both to study fundamental questions as well as applied questions, such as mixing properties or sound generation. Parallel distributed computing enables to resolve the smallest turbulent scales with moderate turn-around times. In particular, the DNS codes are real number crunchers with excessive requirements.

A major task in many CFD codes is to solve large linear systems efficiently on parallel platforms. As an example, we mention the pressure correction equation in a non-Cartesian incompressible code. In Delft, Krylov subspace methods combined with domain decomposition are among the most popular methods for solving large linear systems. Besides applying these methods in our implicit codes, separate mathematical model studies are undertaken as well with the objective to improve robustness, convergence speed and parallel performance.

At the level of civil engineering, contaminant transport forms a source of inspiration. Both atmospheric transport as well as transport in surface and subsurface regions is studied. In the latter case the number of contaminants is in general low and there is a need to increase the geometrical flexibility and spatial resolution of the models. For this purpose parallel transport solvers based upon domain decomposition are studied. In the atmospheric transport models the number of contaminants is high and the grids are regular and of medium size. However, in this case a striking feature is the large uncertainty. One way to deal with this latter aspect is to explore the numerous measurements for improvement of the predictions. For this purpose parallel Kalman filtering techniques are used in combination with parallel transport solvers.

We will present various details encountered in the separate studies and discuss the role of parallel computing, quoting some typical parallel aspects and results. The emphasis will be more on showing where parallel CFD is used for and how this is done than on discussing parallel CFD as a research object on its own.

2. Turbulence

Turbulence research forms a major source of inspiration for parallel computing. Of all activities taking place at Delft University we want to mention two, both in the field of incompressible flow.

A research oriented code has been developed in [12], [13]. Both DNS and LES methods are investigated and compared. The code explores staggered second-order finite differencing on Cartesian grids and the pressure correction method with an explicit Adams-Bathford or Runge-Kutta method for time stepping. The pressure Poisson equation is solved directly using the Fast Fourier transform in two spatial directions, leaving a tridiagonal system in the third spatial direction. The parallel MPI-based implementation relies upon the usual ghost-cell type communication, enabling the computation of fluxes, etc., as well as upon a more global communication operation, supporting the Poisson solver. For a parallel implementation of the Fast Fourier transform it suffices to distribute the frequencies over the processors. However, when doing a transform along a grid line all data associated with this line must be present on the processor. This means that switching to the second spatial direction introduces the necessity of a global exchange of data. Of course, the final tridiagonal system is parallelized by distributing the lines in the associated spatial direction over the processors. Despite the need of global communication, the communication overhead remains in general below 10%. Figure 1 gives an example of the measured wall clock time. The speed-up is nearly linear.

In figure 2 a grid type configuration at inflow generates a number of turbulent jet flows in a channel (modelling wind tunnel turbulence). Due to the intensive interaction and mixing, the distribution of turbulence becomes very quickly homogeneous in the lateral direction. A way to access the numerical results, see figure 3, is to compute the Kolmogorov length scales

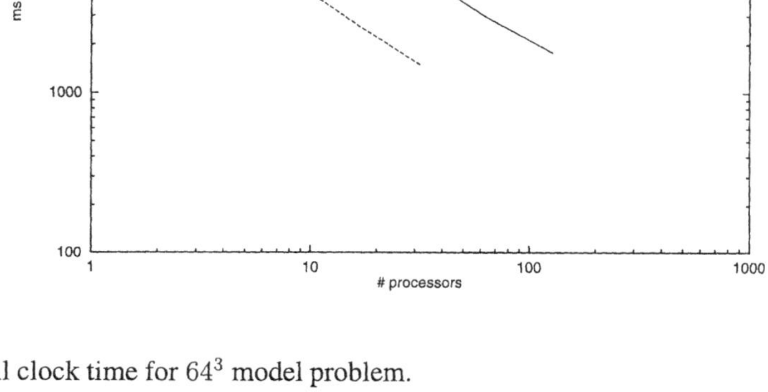

Figure 1. Wall clock time for 64^3 model problem.

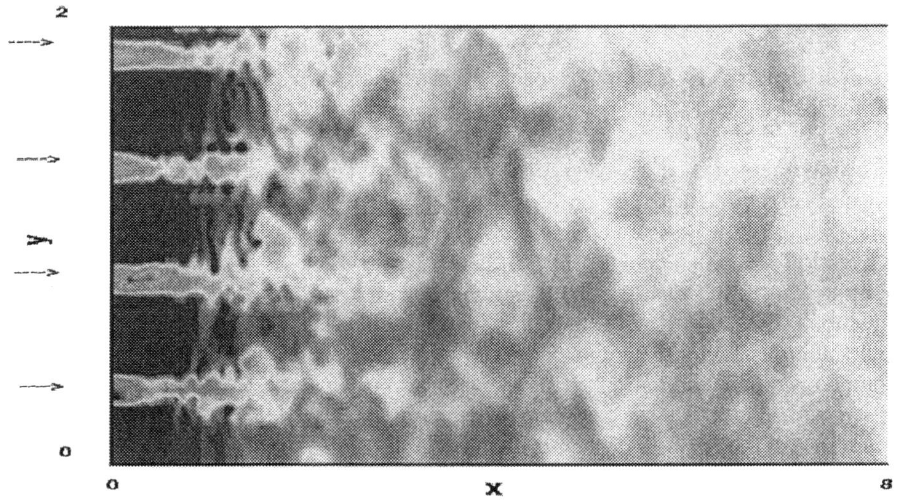

Figure 2. Contour plot of the instantaneous velocity for a flow behind a grid.

(involving sensitive derivatives of flow quantities). The grid size is 600 x 48 x 48 (1.5 million points), which is reported to be sufficient to resolve all scales in the mixing region with DNS for $R = 1000$. For $R = 4000$ subgrid LES modelling is needed: measured subgrid contributions are of the order of 10 %.

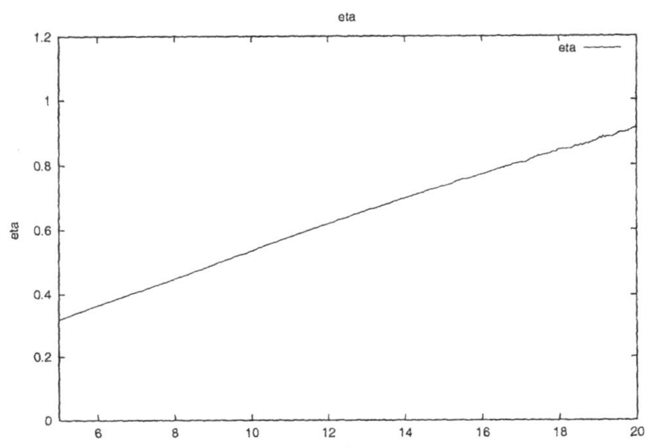

Figure 3. Kolmogorov scale.

A second example of turbulence modelling can be found in [11]. In this study the goals are directed towards industrial applications with complex geometries using LES. Unstructured co-located second-order finite volumes, slightly stabilized, are used in combination with the pressure correction method and implicit time stepping. Solving the linear systems is done with diagonally preconditioned Krylov methods, i.e. CGS for the pressure equation and BiCG for the momentum equations. The computational domain is split into subdomains, which are spread over the processors. Because diagonal preconditioning is used, it suffices to implement a parallel version of the Krylov method, which is done in a straightforward standard manner. As before, ghost-cell type communication enables the computation of fluxes, matrices, etc. As is well-known some global communication of inner products is necessary in a straightforward parallel implementation of Krylov methods.

Figure 4 presents some typical parallel performance results of this code. A nearly perfect speed-up is obtained. Here, the total number of grid points is around 400,000 and the number of subdomains is equal to the number of processors. Thus for 64 processors there are approximately 7000 grid points in each subdomain, being sufficient to keep the communication to computation ratio low. The code is memory intensive. In fact, on a 128 MB Cray T3E node the user has effectively access to 80 MB (50 MB is consumed by the system) and the maximal number of grid points in a subdomain is bounded by approximately 50,000. This explains why the graph in figure 4 starts off at 8 processors. LES results for the flow around a cube at $R = 13000$ are presented in figures 5. The results were obtained with 32 processors of the Cray T3E, running approximately 3 days doing 50,000 time steps.

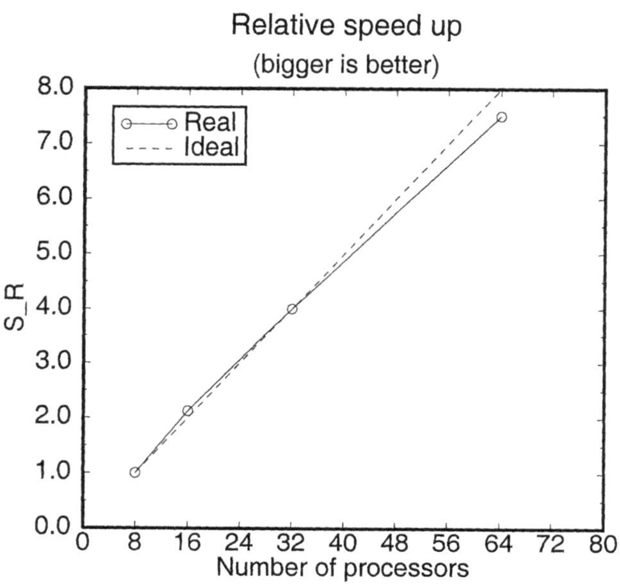

Figure 4. Speed-up on Cray T3E.

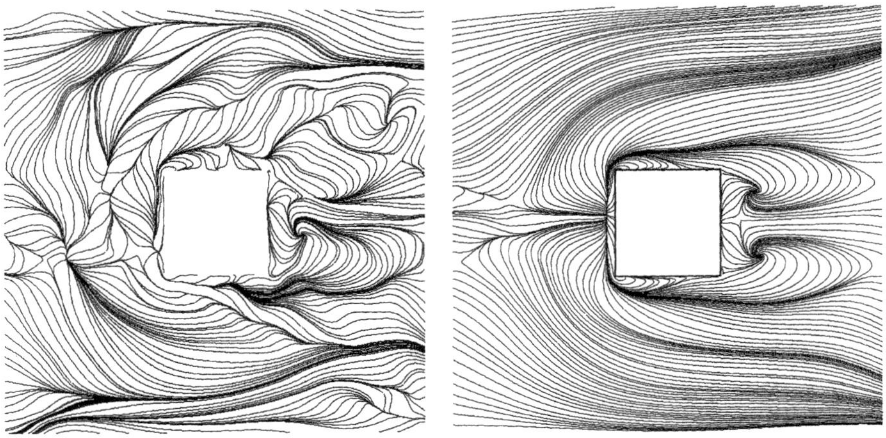

(a) instantaneous (b) averaged

Figure 5. Streamlines.

3. Sound generation by turbulent jets

It is well-known that turbulent jets may produce noise over long distances. Studying the flow properties of a round turbulent jet has been done in [1],[10]. As a follow-up the sound generation by a low Mach number round turbulent jet at $R = 5000$ has been investigated in [2]. For low Mach numbers the acoustic amplitudes are small and a reasonable approximation results from using Lighthill's perturbation equation for the acoustic density fluctuation $\varrho' = \varrho - \varrho_0$, which amounts to a second-order wave equation driven by the turbulent stresses via the source term $T_{ij,i,j}$, involving the Lighthill stress tensor T_{ij}. The equation is written as a system of two first-order equations and treated numerically by the same techniques used for predicting the jet. Typically, the acoustic disturbances propagate over longer distances than flow disturbances and the domain, on which Lighthill's equation is solved, is taken a factor of two larger in each spatial direction. Outside the flow domain Lighthill's equation reduces to an ordinary wave equation because the source term is set to zero. From figure 6 it can be seen that this is a suitable approach.

DNS computations are done using millions of grid point on non-uniform Cartesian grids. A sixth-order compact co-located differencing scheme is used in combination with fourth-order Runge-Kutta time stepping. In a compact differencing scheme not only the variables itself but also their derivatives are propagated. This introduces some specific parallel features with global communication patterns using the MPI routine MPI_ALLTOALL. With respect to communication protocols, this code behaves quite similar to the first code described in the previous section. Also here, communication overhead remains below 10%. Figures 7 and 8 present some results of the computation. Shortly after mixing the jet starts to decay. The quantity $Q = \partial\varrho'/\partial t$ is a

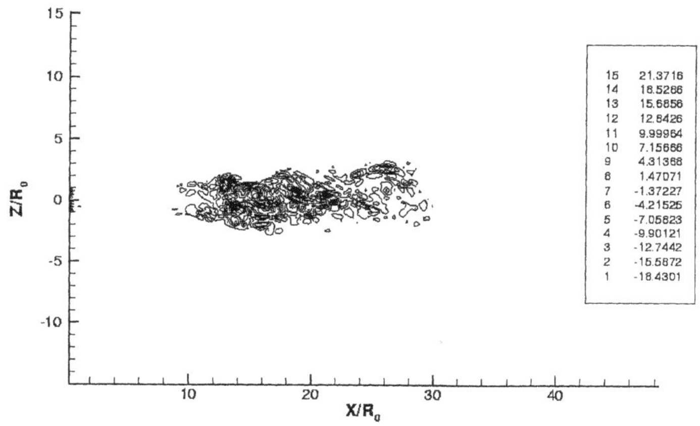

Figure 6. Magnitude of source term $T_{ij,i,j}$ in Lighthill's wave equation.

measure of the frequency of the sound. We can see distinct spherical waves originating from the point where the core of the jet collapses. The computations are quite intensive and the present results have been obtained on a SGI-Origin 3000 machine, located at the national computing center SARA (http://www.sara.nl/), using 16 processors.

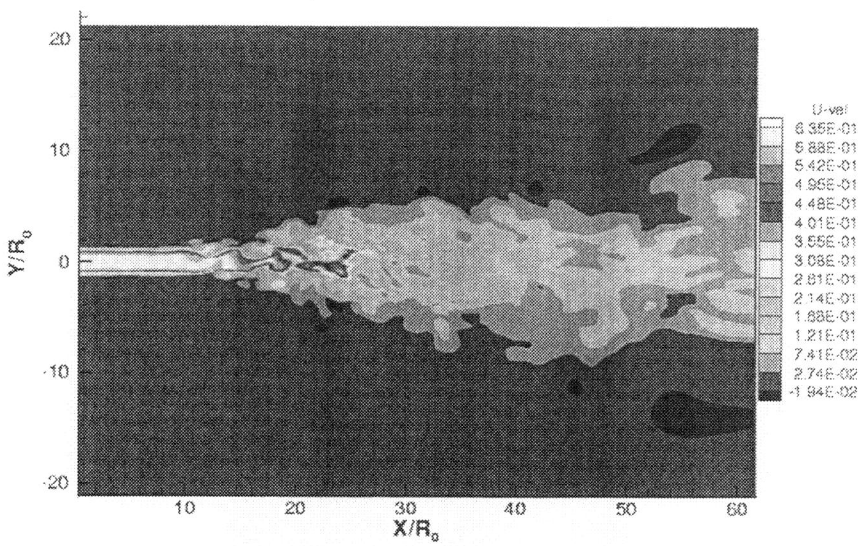

Figure 7. Contour plot of velocity.

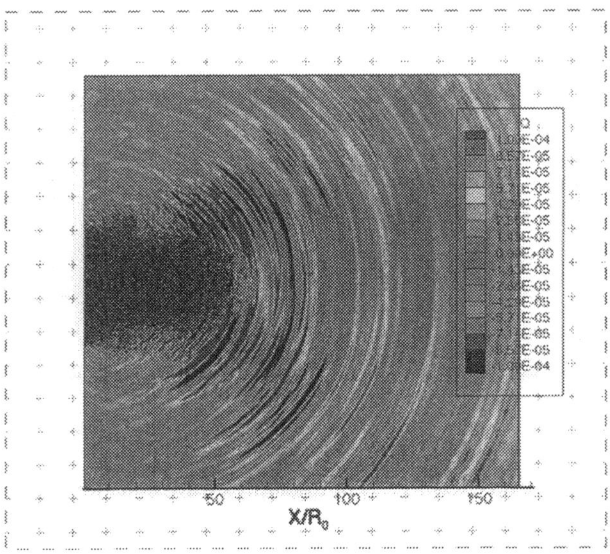

Figure 8. Contour plot of Q, measuring the frequency.

4. Stirring and mixing

Stirring in tanks is a basic operation in chemical process industries. Mixing properties depend strongly upon turbulence generated by the impeller. LES modelling of two basic configurations

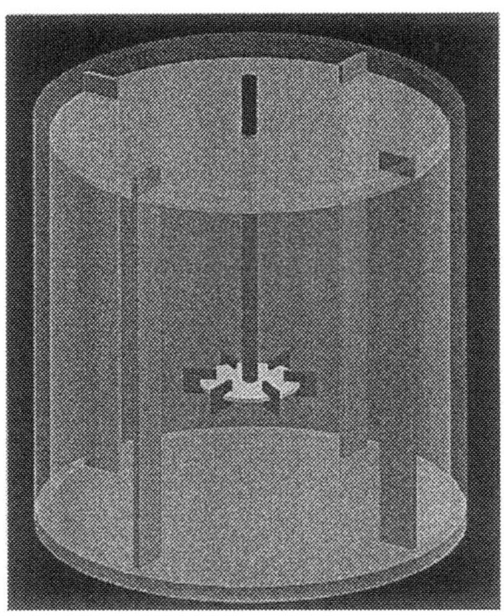

Figure 9. Disk turbine.

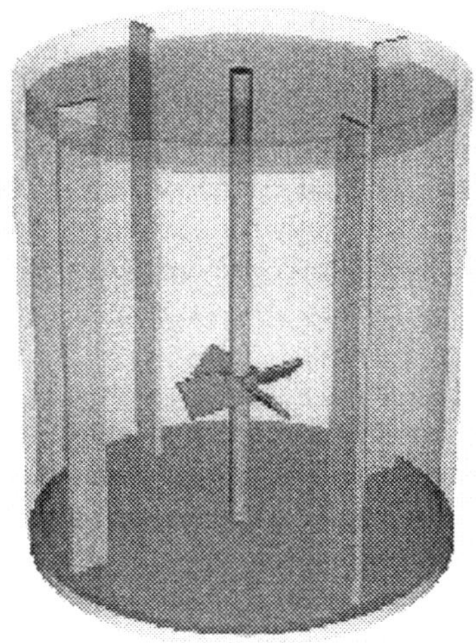

Figure 10. Pitched blade turbine.

at $R = 29000$, respectively $R = 7300$, see figures 9 and 10, has been done in [6], [5] using the Lattice-Boltzmann approach. This approach resembles an explicit time stepping approach

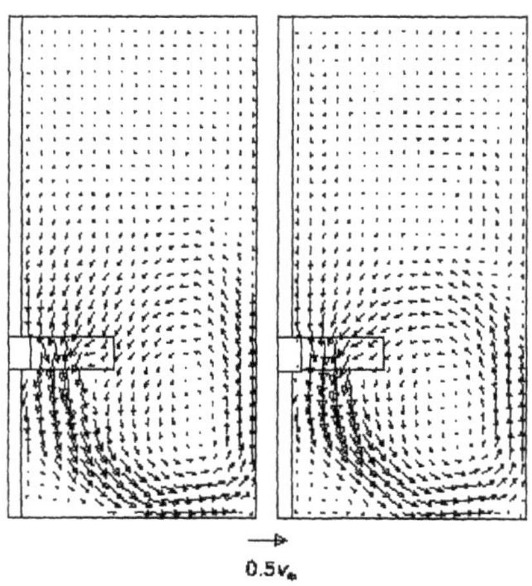

Figure 11. Pitched blade turbine. Average velocity. Left: LDA experiment. Right: LES on 360^3 grid.

in the sense that the total amount of work depends linearly upon the number of nodes of the lattice. In the specific scheme employed [7] the solution vector contains, apart from 18 velocity directions, also the stresses. This facilitates the incorporation of the subgrid-scale model.

Here, parallelization is rather straightforward. The nodes of the lattice are distributed over the processors and only nearest neighbour communication is necessary. In order to enhance possible usage by industry (affordable computing), a local Beowulf cluster of 12 processors and a 100Base TX fast Ethernet switch has been build with MPICH for message passing. On this cluster the code runs with almost linear speed-up, solving problems up to 50 million nodes, taking 2 days per single impeller revolution.

Here, it is worthwhile to notice that the impeller is viewed as a force-field acting on the fluid. Via a control algorithm the distribution of forces is iteratively led towards a flow field taking the prescribed velocity on the impeller, typically taking a few iterations per time step (< 5).

Most important for stirring operations are the average flow (average over impeller revolutions) and the turbulence generated on its way. It has been found that the average flow is well predicted, see figure 11. However, the turbulence is overpredicted, see figure 12. This is contributed to a lack of precision in the LES methodology.

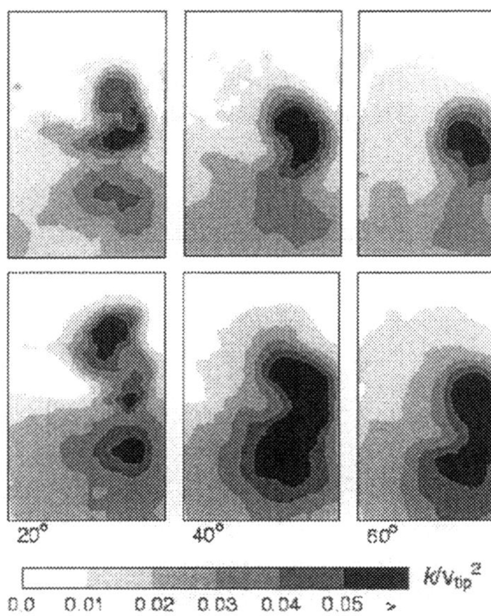

Figure 12. Pitched blade turbine. Contour plot of turbulent kinetic energy near the blade. Top row: experiment. Bottom row: LES on 240^3 grid.

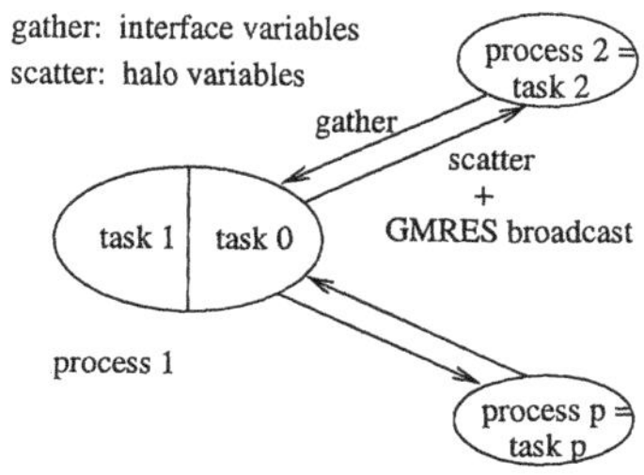

Figure 13. Communication patterns

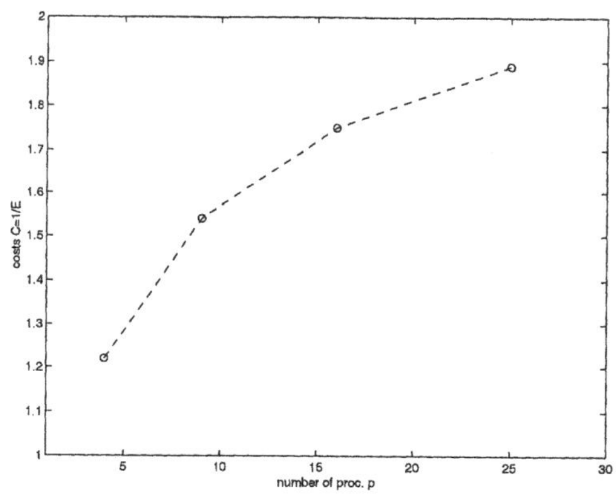

Figure 14. The costs C_p for a linearly growing problem size.

5. Tracer Transport

Tracer studies in surface and subsurface environmental problems form a basic ingredient in environmental modelling. From the application viewpoint, there is a need to resolve large scale computational models with fine grids, for example to study local behaviour in coastal areas with a complex bathymetry/geometry or to study fingering due to strong inhomogeneity of the porous medium. A research oriented code has been developed in [19], [17], [21]. Unstructured cell-centered finite volumes are implemented in combination with implicit time stepping and GMRES-accelerated domain decomposition. The parallel implementation is MPI-based and explores a master-slave communication protocol, see figure 13. The GMRES master process gathers the ghost cell variables, updates them, and scatters them back. Figure 14 presents the (relative) costs for a linearly growing problem size, such as measured on an IBM-SP2. It has been found that less than 15% of the overhead is due to communication and sequential operations in the master. The remaining overhead is caused by load imbalance as a consequence

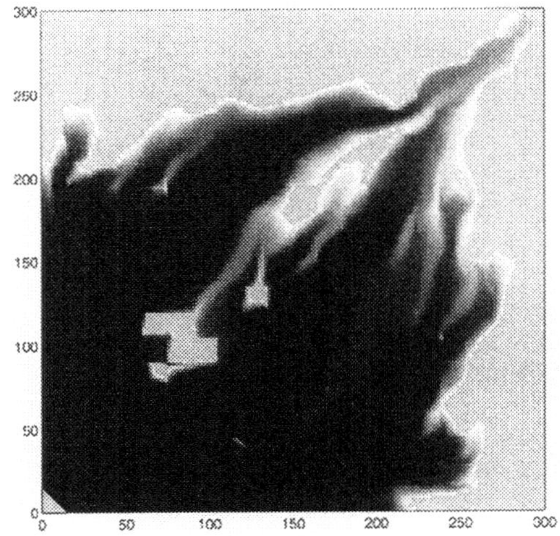

Figure 15. Concentration at 0.6 PVI.

of variations in the number of inner iterations (for subdomain inversion) over the subdomains. It is in particular this latter point that hinders full scalability, i.e. to speak in terms of [9], the synchronization costs in code with iterative procedures are difficult to control for large number of processors.

Typically, the code is applied off-line using precomputed and stored flow data in the area of surface and subsurface environmental engineering. Figure 15 presents an injection/production-type tracer flow in a strongly heterogenous porous medium. A large gradient profile is moving from the lower left corner to the upper right corner. Breakthrough - and arrival times are important civil parameters. It has been observed that arrival times in coastal applications are sometimes sensitive for numerical procedures.

6. Data assimilation

The idea behind data assimilation is to use observations to improve numerical predictions. Observations are fed on-line into a running simulation. First, a preliminary state is computed using the plain physical model. Next, this state is adapted for better matching the observations and for this purpose Kalman filtering techniques are often employed. This approach has been followed in [14] for the atmospheric transport model LOTOS (Long Term Ozone Simulation) for a region covering the main part of Europe. For the ozone concentration, figure 16 presents a contour plot of the deviations between a run of the plain physical model and a run with the same model with data assimilation. Figure 17 plots time series in measurement station Glazeburg, presenting measurements and results from both the plain physical model and the assimilated model. Figure 16 indicates that the adaptions by introducing data assimilation do not have a specific trend, that might be modeled by more simple strategies. Figure 17 shows the adaptions in more detail and it can be seen that they are significant.

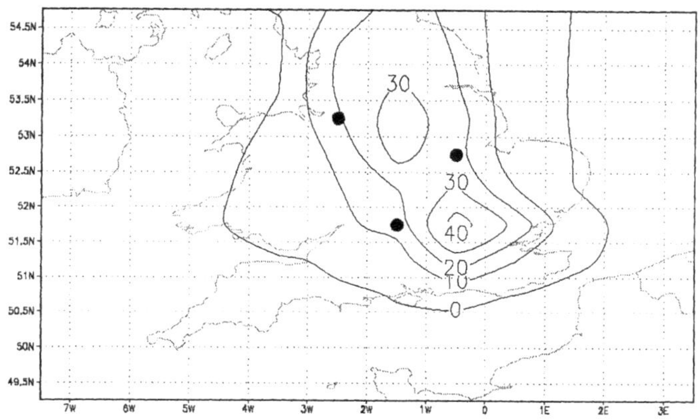

Figure 16. Adjustment of ozone concentration by assimilation.

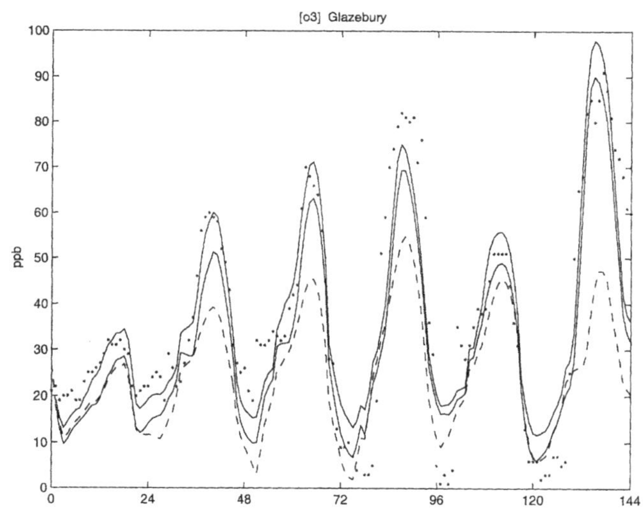

Figure 17. Ozone concentration at Glazeburg, dots:measurements, dashed: plain model, solid: assimilated.

Parallelization strategies have been investigated in [15] for a model with approximately $n = 160,000$ unknowns (26 species). The n x n covariance matrix P contains the covariance of uncertainties in the grid points and is a basic ingredient. Since P is too large to handle, approximations are introduced via a reduced rank formulation, in the present study the RRSQRT approximation (reduced rank square root) [16]. P is factorized ($P = SS'$), using the n x m

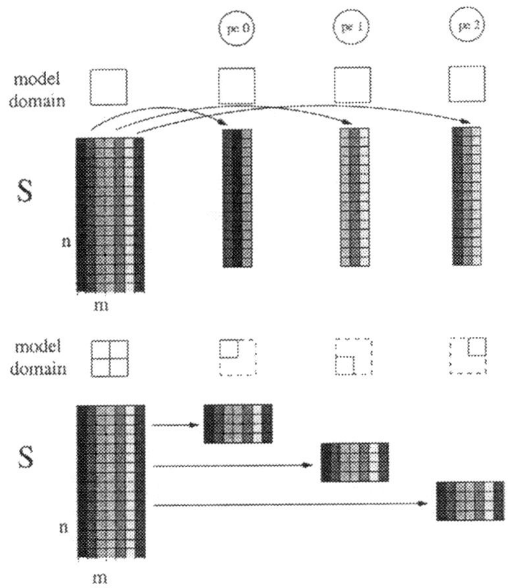

Figure 18. Decomposition: over the modes (columnwise) or over the domain (rowwise).

low-rank approximation S of its square root. For obtaining the entries of S the underlying LOTOS model has to be executed m times, computing the response for m different modes (called the forecast below). In the present study values of m up to 100 have been used.

An obvious way for parallelization is to spread the modes over the processors, running the full LOTOS model on each processor. In a second approach spatial domain decomposition is used to spread the LOTOS model over the processors, see figure 18. Figure 19 presents some performance results, such as obtained on the Cray T3E. Besides the forecast, several other small tasks (involving numerical linear algebra) have to be performed. However, their influence on the final parallel performance remains small, because only a fraction of the total computing time is spent here ($< 20\%$ in a serial run). For the problem under consideration the mode decomposition performs better. However, the problem size has been chosen in such a way that it fits into the memory of a single Cray T3E node (80 MB, see earlier), leading to a small problem size. For larger problem sizes it is expected that the situation turns in favour of the domain decomposed filter. Firstly, because the communication patterns show less global communication. Secondly, because of memory bounds. It shall be clear that scaling up with mode decomposition is more difficult in this respect, because the full physical model has to reside on each processor.

7. Domain decomposition methods

In most of the CFD codes one of the building blocks is to solve large sparse linear systems iteratively. A popular parallel method for engineering applications is non-overlapping additive

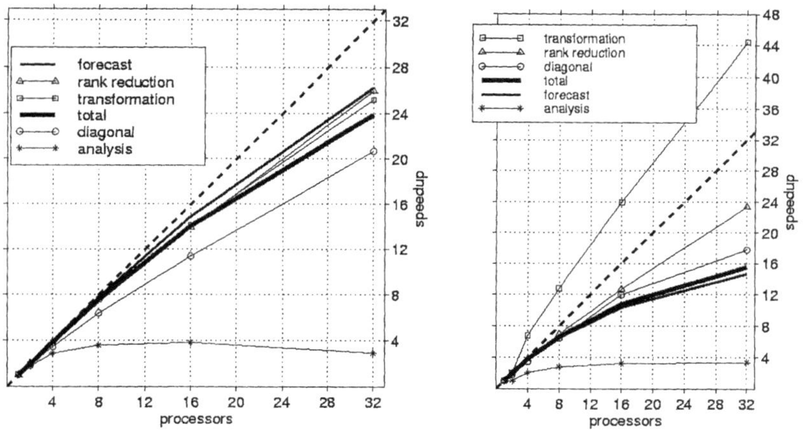

(a) mode decomposed filter

(b) domain decomposed filter

Figure 19. Speed-up for decomposed filter.

Schwarz (also called additive Schwarz with a minimum overlap). In Delft we prefer to use the equivalent, more algebraical, formulation, in which a Krylov method is combined with a block preconditioner of the Jacobi type. With respect to implementation this method is one of the easiest available. The method turns out to lead to an acceptable performance, in particular for time-dependent CFD problems with a strong hyperbolic character [4], [3], [20]. For problems with a strong elliptic character the situation is a little bit more complicated. A global mechanism for transfer of information is needed to enhance iterative properties.

From the mathematical point of view the key notion in obtaining a global mechanism is subspace projection. In dependence of the choice of the subspaces a diversity of methods can be generated, among which are the multilevel methods and methods of the multigrid type. In [18], [8] a deflation argument is used to construct suitable subspaces. Let us for simplicity consider the Poisson equation, discretized on a domain Ω, divided into p nonoverlapping subdomains. Let us denote the block-Jacobi preconditioned symmetrical linear system of n equations with $Au = f$. We use $u = Qu + (I - Q)u$ to split u into two components. Here, Q is a projection operator of (high) rank $(n-k)$ and $(I-Q)$ a projection operator of (low) rank k. The purpose of this splitting is to separate out some of the most 'nasty' components of u. We construct $(I - Q)$ by setting $(I - Q) = ZA_Z^{-1}Z^TA$ with $A_Z = Z^TAZ$ a coarse k x k matrix, being the restriction of A to the coarse space, and by choosing an appropriate n x k matrix Z of which the columns span the deflation subspace Z of dimension k.

It is easy to see that $(I - Q)u = ZA_Z^{-1}Z^Tf$, which can be executed at the cost of some matrix/vector multiplies and a coarse matrix inversion. For the parallel implementation a full copy of A_Z^{-1} in factorized form is stored on each processor. To obtain the final result, some nearest neighbor communication and a broadcast of length k are needed.

p	iterations	time	speedup	efficiency
1	485	710	-	-
4	322	120	5	1.2
9	352	59	12	1.3
16	379	36	20	1.2
25	317	20	36	1.4
36	410	18	39	1.1
64	318	8	89	1.4

Table 1
Speedup of the iterative method using a 480×480 grid

The remaining component Qu can be obtained from a deflated system, in which, so to speak, k coarse components have been taken out. Here, we use a Krylov method such as CG. As is well known, the convergence depends upon the distribution of the eigenvalues. Suppose that $0 < \lambda_1 < \lambda_2$ are the two smallest nonzero eigenvalues with eigenvectors $v_{1,2}$. Now, choose $Z = v_1$, i.e. deflate out the component in the v_1 direction. The remaining deflated system for obtaining Qu has λ_2 as the smallest nonzero eigenvalue, which allows the Krylov method to converge faster. Of course, the eigenvectors are not known and it is not possible to do this in practice. However, it has been found that a very suitable deflation of the domain decomposition type can be found by choosing $k = p$, with p the number of subdomains. Next, the vectors $z_q, q = 1, .., p$ of length n are formed with a zero entry at positions that are outside subdomain q and an entry equal to one at positions that are in subdomain q. Finally, the deflation space Z is defined as the span of these vectors. It is easy to verify that the coarse matrix A_Z resembles a coarse grid discretization of the original Poisson operator.

Table 1 presents some results for the Poisson equation on a Cray T3E. Most important, it can be seen that the number of iterations does not increase for larger p. Typically, the number of iterations increases with p for methods lacking a global transfer mechanism. Surprisingly, efficiencies larger than one have been measured. Further research is needed to reveal the reasons for this.

8. Conclusions and final remarks

Parallel computational fluid dynamics is on its way to become a basic tool in engineering sciences at least at Delft University of Technology. The broadness of the examples given by us illustrates this. We have also tried to outline the directions in which developments take place. Computations with millions of unknowns over moderate time intervals are nearly a day-to-day practice with some of the more explicit-oriented codes. Tools have been developed for post-processing the enormous amounts of data. For approaches, relying upon advanced numerical linear algebra and/or flexible finite volume methods, much remains to be done in order to scale up.

REFERENCES

1. B.J. Boersma, G. Brethouwer and F.T.M. Nieuwstadt, A numerical investigation on the effect of the inflow conditions on the self-similar region of a round jet, *Physics of Fluids*, **10**, pages 899-909, 1998.
2. B.J. Boersma, Direct numerical simulation of jet noise, In P. Wilders et al., editors, *Parallel Computational Fluid Dynamics 2001*, Egmond aan Zee, The Netherlands, May 21-23 2001, Elsevier 2002.
3. E. Brakkee and P. Wilders, The influence of interface conditions on convergence of Krylov-Schwarz domain decomposition for the advection-diffusion equation, *J. of Scientific Computing*, **12**, pages 11-30, 1997.
4. E. Brakkee, C. Vuik and P. Wesseling , Domain decomposition for the incompressible Navier-Stokes equations: solving subdomain problems accurately and inaccurately, *Int. J. for Num. Meth. Fluids*, **26**, pages 1217-1237, 1998
5. J. Derksen and H.E.A. van den Akker, Large eddy simulations on the flow driven by a Rushton turbine, *AIChE Journal*, **45**, pages 209-221, 1999.
6. J. Derksen, Large eddy simulation of agitated flow systems based on lattice-Boltzmann discretization, In C.B. Jenssen et al., editors, *Parallel Computational Fluid Dynamics 2000*, pages 425-432, Trondheim, Norway, May 22-25 2000, Elsevier 2001.
7. J.G.M. Eggels and J.A. Somers, Numerical simulation of free convective flow using the Lattice-Boltzmann scheme, *Int. J. Heat and Fluid Flow*, **16**, page 357, 1995.
8. J. Frank and C. Vuik, On the construction of deflation-based preconditioners, *Report MAS-R0009*, CWI, Amsterdam 2000, accepted for publication in SIAM J. Sci. Comput., available via http://ta.twi.tudelft.nl/nw/users/vuik/MAS-R0009.pdf
9. D. Keyes, private communication at *Parallel Computational Fluid Dynamics 2001*, Egmond aan Zee, The Netherlands, May 21-23 2001.
10. C.L. Lubbers, G. Brethouwer and B.J. Boersma, Simulation of the mixing of a passive scalar in a free round turbulent jet, *Fluid Dynamic Research*, **28**, pages 189-208, 2001.
11. B. Ničeno and K. Hanjalić, Large eddy simulation on distributed memory parallel computers using an unstrucured finite volume solver, In C.B. Jenssen et al., editors, *Parallel Computational Fluid Dynamics 2000*, pages 457-464, Trondheim, Norway, May 22-25 2000, Elsevier 2001.
12. M. Pourquie, B.J. Boersma and F.T.M. Nieuwstadt, About some performance issues that occur when porting LES/DNS codes from vector machines to parallel platforms, In D.R. Emerson et al., editors, *Parallel Computational Fluid Dynamics 1997*, pages 431-438, Manchester, UK, May 19-21 1997, Elsevier 1998.
13. M. Pourquie, C. Moulinec and A. van Dijk, A numerical wind tunnel experiment, In *LES of complex transitional and turbulent flows, EUROMECH Colloquium Nr. 412* , München, Germany, October 4-6 2000.
14. A.J. Segers, Data assimilation in atmospheric chemistry models using Kalman filtering, *PhD Thesis*, Delft University of Technology 2001, to be published.
15. A.J. Segers and A.W. Heemink, Parallization of a large scale Kalman filter: comparison between mode and domain decomposition, In P. Wilders et al. editors, *Parallel Computational Fluid Dynamics 2001*, Egmond aan Zee, The Netherlands, May 21-23 2001, Elsevier 2002.
16. M. Verlaan and A.W. Heemink, Tidal forecasting using reduced rank square root filters,

Stochastic Hydrology and Hydraulics, **11**, pages 349-368, 1997.

17. C. Vittoli, P. Wilders, M. Manzini and G. Fotia, Distributed parallel computation of 2D miscible transport with multi-domain implicit time integration, *J. Simulation Practice and Theory*, **6**, pages 71-88, 1998.
18. C. Vuik, J. Frank and F.J. Vermolen, Parallel deflated Krylov methods for incompressible flow, In P. Wilders et al., editors, *Parallel Computational Fluid Dynamics 2001*, Egmond aan Zee, The Netherlands, May 21-23 2001, Elsevier 2002.
19. P. Wilders, Parallel performance of domain decomposition based transport, In D.R. Emerson et al., editors, *Parallel Computational Fluid Dynamics 1997*, pages 447-456, Manchester, UK, May 19-21 1997, Elsevier 1998.
20. P. Wilders P. and G. Fotia, One level Krylov-Schwarz decomposition for finite volume advection-diffusion, In P.E. Bjorstad, M.S. Espedal and D.E. Keyes, editors, *Domain Decompostion Methods 1996*, Bergen, Norway, June 4-7 1996, Domain Decomposition Press 1998.
21. P. Wilders, Parallel performance of an implicit advection-diffusion solver, In D. Keyes et al., editors, *Parallel Computational Fluid Dynamics 1999*, pages 439-446, Williamsburg, Virginia, USA, May 23-26 1999, Elsevier 2000.

2. Invited and Contributed Papers

Parallel Computational Fluid Dynamics – Practice and Theory
P. Wilders, A. Ecer, J. Periaux, N. Satofuka and P. Fox (Editors)

Noise predictions for shear layers

A.V. Alexandrov [a],B.N. Chetverushkin [a] and T.K. Kozubskaya [a]

[a]Institute for Mathematical Modelling of Rus.Ac.Sci.,
4-A, Miusskaya Sq., Moscow 125047, Russia
e-mail: tata@imamod.ru

The paper contributes to the investigation of acoustic noise generation in shear layers with the use of advantages of parallel computing. Both noise propagation and generation are simulated by the linear acoustic equations with source terms which are derived, in its turn, on the base of complete Navier-Stokes equations system and triple flow decomposition. The mean flow parameters are predicted with the help of Reynolds averaged Navier-Stokes equations closed by $k - eps$ turbulence model. A semi-stochastic model developed in [1] and relative to SNGR model [2] is applied for describing the fields of turbulent velocity pulsation.

INTRODUCTION

As it is well known the acoustic noise arising within gas flows can significantly influence the whole gasdynamic process. For instance it may negatively affect the structure may cause a great discomfort both for the airplane or car passengers and the people around. So an adequate simulation of acoustic noise is a problem of high importance in engineering.

The difficulty in numerical prediction of aeroacoustics problems results in particular from a small scale of acoustic pulsation especially in comparison with large scale oscillations of gasdynamic parameters. This small scale places strict constrains on the numerical algorithms in use and requires powerful computer facilities due to the need of using higly refined computational meshes for resolving such small scale acoustic disturbances. In particular, to resolve high frequency perturbations under the requirement of 10 nodes per wave, it is necessary to use huge computational meshes. For instance, the resolution of frequency $200kHz$ even in a 2D domain of 1 square metre requires more than 1 million nodes.

The parallel computer systems with distributed memory architecture offer a robust and efficient tool to meet the requirement of large computational meshes. That's why the usage of parallel computer systems seems quite natural. All calculations in this paper were carried out on the parallel system MVS-1000.

1. MATHEMATICAL

Let us adapt the flow decomposition into mean and pulsation parameters. Then the dynamics of acoustic noise (both propagation and generation) can be described with the help of Linear Euler Equations with Source terms (LEE+S) [4], or Linear Disturbance (Acoustics) Equations with Sources (LDE+S or LAE+S) [1], which can be written in the following general form

$$\frac{\partial \mathbf{Q}'}{\partial t} + \mathbf{A}^x \frac{\partial \mathbf{Q}'}{\partial x} + \mathbf{A}^y \frac{\partial \mathbf{Q}'}{\partial y} = \mathbf{S}. \tag{1}$$

Here $\mathbf{Q}'$ is a conservative variables vector linearized on pulsation components which is defined as a vector consisting only of linear terms on physical pulsation variables

$$\mathbf{Q}' = \begin{bmatrix} \rho' \\ m' \\ n' \\ E' \end{bmatrix} = \begin{bmatrix} \rho' \\ \bar{u}\rho' + \bar{\rho}u' \\ \bar{v}\rho' + \bar{\rho}v' \\ \dfrac{\bar{u}^2+\bar{v}^2}{2}\rho' + \bar{\rho}\bar{u}u' + \bar{\rho}\bar{v}v' + \dfrac{1}{\gamma-1}p' \end{bmatrix}, \tag{2}$$

and $\mathbf{A}^x = \mathbf{A}^x(\bar{\rho},\bar{u},\bar{v},\bar{p})$ and $\mathbf{A}^y = \mathbf{A}^y(\bar{\rho},\bar{u},\bar{v},\bar{p})$ are the standard flux Jacobian matrices

$$\mathbf{A}^x = \begin{bmatrix} 0 & 1 & 0 & 0 \\ (\gamma-3)\dfrac{\bar{u}^2}{2} + (\gamma-1)\dfrac{\bar{v}^2}{2} & -(\gamma-3)\bar{u} & -(\gamma-1)\bar{v} & \gamma-1 \\ -\bar{u}\bar{v} & \bar{v} & \bar{u} & 0 \\ a^x_{14} & a^x_{24} & -(\gamma-1)\bar{u}\bar{v} & \gamma\bar{u} \end{bmatrix} \tag{3}$$

$$\begin{aligned} a^x_{14} &= (\gamma-2)\bar{u}\frac{\bar{u}^2+\bar{v}^2}{2} - \frac{\gamma}{\gamma-1}\frac{\bar{p}}{\bar{\rho}}\bar{u} \\ a^x_{24} &= \frac{\bar{u}^2+\bar{v}^2}{2} - (\gamma-1)\bar{u}^2 + \frac{\gamma}{\gamma-1}\frac{\bar{p}}{\bar{\rho}} \end{aligned} \tag{4}$$

$$\mathbf{A}^y = \begin{bmatrix} 0 & 0 & 1 & 0 \\ -\bar{u}\bar{v} & \bar{v} & \bar{u} & 0 \\ (\gamma-1)\dfrac{\bar{u}^2}{2} + (\gamma-3)\dfrac{\bar{v}^2}{2} & -(\gamma-1)\bar{u} & -(\gamma-3)\bar{v} & \gamma-1 \\ a^y_{14} & -(\gamma-1)\bar{u}\bar{v} & a^y_{34} & \gamma\bar{v} \end{bmatrix} \tag{5}$$

$$\begin{aligned} a^y_{14} &= (\gamma-2)\bar{v}\frac{\bar{u}^2+\bar{v}^2}{2} - \frac{\gamma}{\gamma-1}\frac{\bar{p}}{\bar{\rho}}\bar{v} \\ a^y_{24} &= \frac{\bar{u}^2+\bar{v}^2}{2} - (\gamma-1)\bar{v}^2 + \frac{\gamma}{\gamma-1}\frac{\bar{p}}{\bar{\rho}} \quad . \end{aligned} \tag{6}$$

The way of constructing of noise sources is a separate problem and more elaborately it's described in [1]. In brief the source term is approximated with help of semi-deterministic modeling of velocity turbulent fluctuations. One of the determining characteristics of sources is their frequencies. These frequencies are predicted with help of specially prearranged numerical experiment of flow field exposure to white noise irradiation from artificial sources. This technique allows to determine the most unstable frequencies. It's supposed that just these frequencies has a dominant role in the noise generation process.

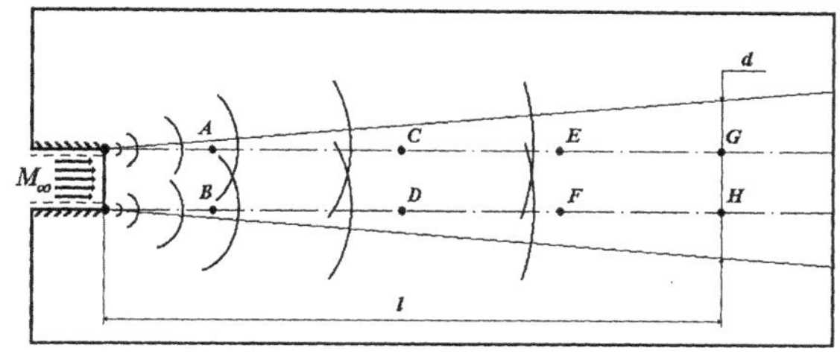

Figure 1. Scheme of mean flow exposure to noise radiation in jet

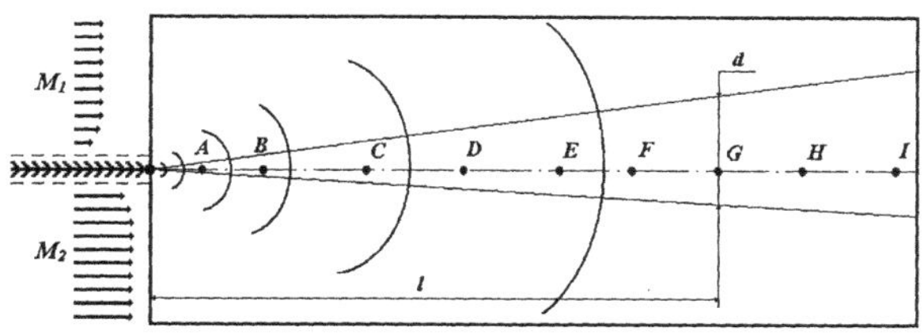

Figure 2. Scheme of mean flow exposure to noise radiation in mixing layer

The scheme of flow exposure to acoustic radiation is presented in Fig. 2 for plane mixing layers and in Fig. 1 for plane jets.

It has been discovered that the most amplified frequencies taken as characteristic well satisfies the following known relations for plane mixing layers.

$$f_0(x) = St\frac{u}{L}$$

Here L is a longitudinal distance that is a distance from a splitting plate tip to a point under consideration within the shear layer.

2. PARALLELIZATION AND NUMERICAL RESULTS

All the predictions have been performed on the base of explicit numerical algorithms. So the parallelization is based on geometrical domain partitioning in accordance with a number of processor available in a way that each subdomain is served by one processor unit. The computational domain is cut along one (transverse) direction, the data exchange is handled only along vertical splitting lines. The requirement of equal numbers of mesh nodes per processor is provided automatically. Such arrangement results in processor load balancing and, as a consequence, in reduction of idle time. This way of doing provides a good scalability and portability for an arbitrary number of processors. The corresponding codes are written in *C++* with the use of *MPI* – library.

The results on acoustics field modeling for free turbulent flows are demonstrated on the example of plane mixing layers.
As a plane mixing layer problem 3 test cases (for different Mach numbers) have been taken from [4]. In the paper presented, the mean flow components are predicted on the base of steady Reynolds Averaged Navier-Stokes equations closed by $k - eps$ turbulence model.
The growth of shear layer thickness of mean flow is represented in Fig.3. Here the value of local vorticity thickness δ_ω used is determined as

$$\delta_\omega = \frac{\Delta U}{|\partial\langle\tilde{u_1}\rangle/\partial y_2|_{max}}$$

Following [4] we replace $\langle u_1\rangle$ in expression for δ_ω on $\langle\tilde{u_1}\rangle$.
It is visible that the growth of δ_ω along the streamwise direction in the computations presented is practically the same as in [4].
In Fig. 3 vorticity the growth rate for case 1 and case 3 is demonstrated. One can see that the growth of the thickness of shear layer has a linear character. This fact is confirmed by numerous numerical and experimental data.

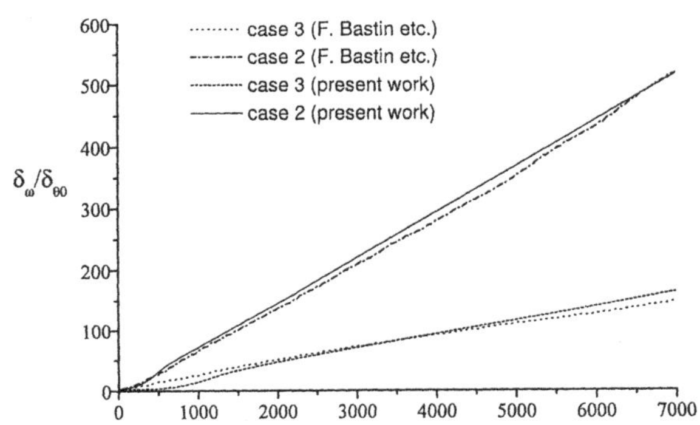

Figure 3. Vorticity thickness δ_ω growth for different shear layers (case 1 ($M_c = 0.19$) and case 2 ($M_c = 0.33$))

One can see that the results presented are close to those from [4]. The results presented here correspond to the subsonic test case with the following parameters: $M_a = 0.7$, $M_b = 0.3$, $M_c = 0.19$, $\delta_{\theta 0} = 0.107$ mm, $Re_{\delta_{\theta 0}} = U_a \delta_{\theta 0}/\mu_a = 2000$, where M_c is the convective Mach number, $\delta_{\theta 0}$ – the momentum thickness. A computational mesh of 547×687 has been used.

The instantaneous acoustic fields for pressure and transverse velocity pulsation are given in Figures 4 and 5.

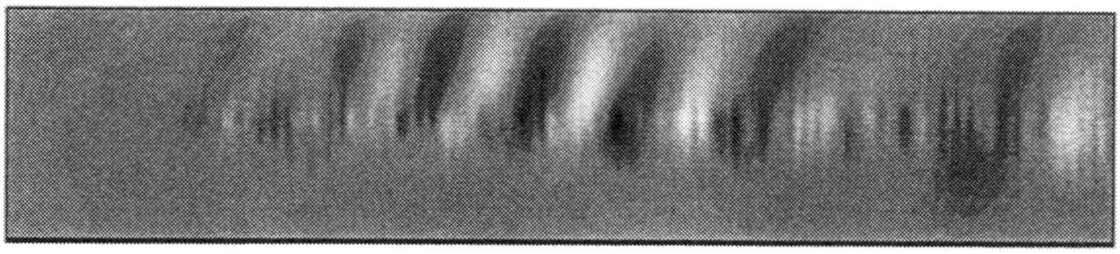

Figure 4. Pressure pulsation field in mixing layer

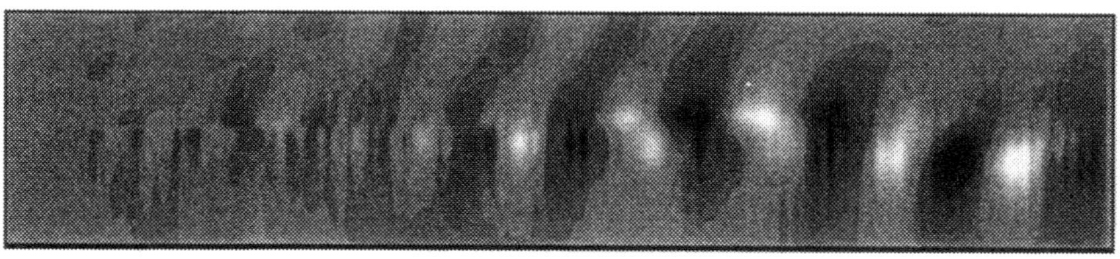

Figure 5. Transverse velocity pulsation field in mixing layer

The predictions have been performed on the multiprocessor system MVS-1000 consisting of several modules. There are sixteen computer nodes based on processors Alpha 21164 with operating memory 256 B. Each node has 4 links. The links connecting the processor units can be provided by the folowing communication processors: TMS320C44 made of *Texas Instruments* or SHARC of *Analog Devices.* The processors in a module the distance between which is the longest are connected by links as it shown in Fig 6. A total processor number in the system is 96 (may exceed 128).

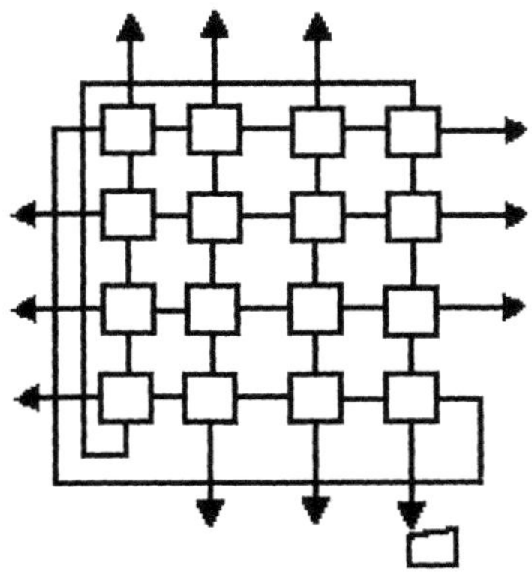

Figure 6. Scheme of a 16 processors module of MVS-1000

3. CONCLUSION

The usage of multiprocessor computer system allows to perform accurate predictions of noise propagation and generation processes for free turbulent gas flows. This has been demonstrated on the example of plane shear layers. To mathematically describe noise sources distributed within shear layers, a new numerical technique proposed in [1], [3] has been adopted to parallel computing. The technique has demonstrated its efficiency and readiness for further implementation to solving aeroacoustics problems in engineering.

ACKNOWLEDGEMENT

This work was supported in part by Russian Foundation for Basic Research (Grants No. 99-07-90388, No. 99-01-01215) and French-Russian A.M. Lyapunov Institute of Applied Mathematics and Informatics (Project 99-02).

REFERENCES

1. Kozubskaya, T.K. A Way of Acoustic Noise Modeling for Turbulent Gas Flows. — In CD ROM Proceedings of *European Congress on Computational Methods in Applied Sciences and Engineering*, 11-14 September 2000, Barcelona, Spain.

2. Christophe Bailly and Daniel Juve, "A Stochastic Approach To Compute Subsonic Noise Using Linearized Euler's Equations", *AIAA paper 99-1872.*
3. Alexandrov, A.V., Chetverushkin, B.N. and Kozubskaya, T.K., Numerical Investigation of Viscous Compressible Gas Flows by Means of Flow Field Exposure to Acoustic Radiation. — In Proceedings of *Parallel CFD 2000 conference*, Trondheim, Norway, May 22-24, 2000, North Holland, Elservier, (2000).
4. Bastin, F., Lafon, P. and Candel, S. "Computation of jet mixing noise due to coherent structures", *J. Fluid Mech.,* **335**, 261-304, (1997).

Parallel Computational Fluid Dynamics – Practice and Theory
P. Wilders, A. Ecer, J. Periaux, N. Satofuka and P. Fox (Editors)

Framework for Parallel Simulations in Air Pollution Modeling with Local Refinements

Anton Antonov

National Environmental Research Institute, Frederiksborkvej 399, P.O.Box 358, DK-4000 Roskilde;
e-mail: Anton.Antonov@uni-c.dk

Abstract

The Object-Oriented Paradigm (OOP) provides methodologies how to build flexible and reusable software. The OOP methodology of patterns and pattern languages was applied to construct the object oriented version of the large scale air pollution model known as the Danish Eulerian Model (DEM). In the paper will be described the general design of the object-oriented DEM, and some experiments with it run on parallel computers.

Key words: air pollution, design patterns;

Many difficulties must be overcome when large-scale air pollution models are treated numerically, because the physical and chemical processes are very fast. This is why it is necessary (i) to use large space domain in order to be able to study long-range transport pollutants in the atmosphere, (ii) to describe adequately all important processes when an air pollution model is developed, and (iii) to use fine grids in the discretization model. We will consider how the resulting from these conditions huge computational tasks can be approached with a program code for parallel simulations developed within the Object-Oriented (OO) paradigm. Our considerations are for the particular large-scale air pollution model, the Danish Eulerian Model (DEM, [12]).

Temporal and spatial variations of the concentrations and/or the depositions of various harmful air pollutants can be studied [12] by solving the system (1) of partial differential equations (PDE's):

$$\begin{aligned}\frac{\partial c_s}{\partial t} = &-\frac{\partial (uc_s)}{\partial x} - \frac{\partial (vc_s)}{\partial y} - \frac{\partial (wc_s)}{\partial z} \\ &+\frac{\partial}{\partial x}(K_x\frac{\partial c_s}{\partial x}) + \frac{\partial}{\partial y}(K_y\frac{\partial c_s}{\partial y}) + \frac{\partial}{\partial z}(K_z\frac{\partial c_s}{\partial z}) \qquad (1)\end{aligned}$$

$$+E_s + Q_s(c_1, c_2, ..., c_q) - (k_{1s} + k_{2s})c_s,$$
$$s = 1, 2, ..., q.$$

The different quantities that are involved in the mathematical model have the following meaning: (i) the concentrations are denoted by c_s; (ii) u, v and w are wind velocities; (iii) K_x, K_y and K_z are diffusion coefficients; (iv) the emission sources in the space domain are described by the functions E_s; (v) κ_{1s} and κ_{2s} are deposition coefficients; (vi) the chemical reactions used in the model are described by the non-linear functions $Q_s(c_1, c_2, \ldots, c_q)$. The number of equations q is equal to the number of species that are included in the model.

It is difficult to treat the system of PDE's (1) directly. This is the reason for using different kinds of splitting. A simple splitting procedure, based on ideas discussed in Marchuk[10] and McRae et al. [11], can be defined, for $s = 1, 2, \ldots, q$, by the following sub-models:

$$\frac{\partial c_s^{(1)}}{\partial t} = -\frac{\partial (u c_s^{(1)})}{\partial x} - \frac{\partial (v c_s^{(1)})}{\partial y} \tag{2}$$

$$\frac{\partial c_s^{(2)}}{\partial t} = \frac{\partial}{\partial x}\left(K_x \frac{\partial c_s^{(2)}}{\partial x}\right) + \frac{\partial}{\partial y}\left(K_y \frac{\partial c_s^{(2)}}{\partial y}\right) \tag{3}$$

$$\frac{dc_s^{(3)}}{dt} = E_s + Q_s(c_1^{(3)}, c_2^{(3)}, \ldots, c_q^{(3)}) \tag{4}$$

$$\frac{dc_s^{(4)}}{dt} = -(\kappa_{1s} + \kappa_{2s})c_s^{(4)} \tag{5}$$

$$\frac{\partial c_s^{(5)}}{\partial t} = -\frac{\partial (w c_s^{(5)})}{\partial z} + \frac{\partial}{\partial z}\left(K_z \frac{\partial c_s^{(5)}}{\partial z}\right) \tag{6}$$

The horizontal advection, the horizontal diffusion, the chemistry, the deposition and the vertical exchange are described with the system (2)-(6). This is not the only way to split the model defined by (1), but the particular splitting procedure (2)-(6) has three advantages: (i) the physical processes involved in the big model can be studied separately; (ii) it is easier to find optimal (or, at least, good) methods for the simpler systems (2)-(6) than for the big system (1); (iii) if the model is to be considered as a two-dimensional model (which often happens in practice), then one should just skip system (6).

The chemical sub-model reduces to large number of relatively small ODE systems, one such system per grid-point; therefore its parallelization is easy: the number of grid points (96×96, 480×480 in the two-dimensional DEM) is much bigger than the number of processors. From another hand, the two-dimensional Advection-Diffusion Submodel (ADS) that combines (2) and (3), poses the non-trivial question how it should be implemented for parallel computations, especially when higher resolution is required on specified regions i.e. locally uniform grids are used.

The development task considered is the implementation of an object-oriented framework for DEM. The framework should be amenable for simulations with local refinements, 3D simulations, and inclusion of new chemical schemes. The way this task is approached is to adopt the splitting procedure (2)-(6) and to build first a framework for the ADS. It was assumed that the ADS framework should be flexible on what parallel execution model is used. After scanning different books, articles, and opinions was decided to start a Conceptual Layering framework [5] with conceptual layer for Galerkin Finite Element Methods (GFEM), and building blocks layer comprising a mesh generator package, and a package of parallel solvers for linear systems. It was decided to develop a mesh generator for locally uniform grids, and to employ PETSc ([3],[4]) for the parallel solution of the linear systems. The initial Conceptual Layering framework mutated to a Multi-level (several conceptual layers) framework, because of the mesh generator. The OO construction of the framework employs design patterns [9]. The design of the GFEM layer is based on the Template Method Design Pattern (DP), which is combined with the Abstract Factory DP that provide consistency of the usage of a number of Strategies that provide different behavior flexibilities.

Since PETSc is based on the Message Passing Interface (MPI), and because the MPI model runs on all parallel architectures, MPI is reflected in the GFEM layer. The idea behind the way the parallelism is facilitated is similar to the ideas used in OpenMP, HPF and PETSc: the user achieves parallelism via domain decomposition, designing sequential code that is made parallel with minimal changes (comments in OpenMP and HPF, and name suffixes in PETSc). The Decorator DP for the data feeding was applied to the GFEM conceptual layer, and the Strategy DP for the parallel/sequential GFEM computations. Using these two patterns the sequential and the parallel code look in the same way: in order to add new parallel behavior, new classes are added, the existing code is not changed. Also, the use of the Strategy DP makes the GFEM classes independent of the linear solvers package.

The mesh generator provides description of the designed by the user grid. The grid nodes are ordered linearly according to their spatial coordinates. They are divided into classes: each class contains nodes with equal patches. The description is read by the GFEM layer and the nodes are distributed in equal portions among the parallel processes. GFEM classes know the lowest and the highest number of the nodes their processes are responsible for. For the machine presentation of the GFEM operators are used sparse matrices distributed over the processors; their handling is provided by PETSc.

The overall developed OO framework for DEM (OODEM) includes the ADS, the chemical sub-model, and a layer - the Data Handling (DH) layer - for the non-trivial issues of the data handling.

We would like to emphasize that the GFEM layer works in the GFEM Hilbert space. The Data Handling layer and the Mesh Generator are used to map the physical space of the simulation domain into a Hilbert space vector that is given to the GFEM layer. Because of this, and because OODEM uses static grids, it is easy to provide load balancing.

The Current Production Code of DEM has all the features of a "Big Ball of Mud"[8]. (Please note, this is not an antipattern: it is a common solution.) It can be seen that it has been developed via THROWNAWAY CODE, PIECEMEAL GROWTH, and KEEP IT WORKING ([8]). So, following [8] we first applied the pattern SWEEP IT UNDER THE RUG, and then applied SHEARING LAYERS. Currently in the top layer of OODEM "only the patterns that underlie the system remain, grinning like a Cheshire cat [8]".

Experiments with the framework was carried on, using data from the AUTO-OIL II project, [6] and [7]. Two types of locally refined grids were used, with several fine resolution regions. The finest resolution of the first grid was $10km \times 10km$; of the second, $2km \times 2km$. In both grids the refined regions was imbeded in a mother grid $50km \times 50km$; see Figure 1. The used GFEM scheme was with chapeau basis functions and the Crank-Nicholson time marching method. The number of nodes for the first grid is ≈ 16000; for the second ≈ 35000. The time step used is $900s$ for the advection, and $150s$ for the chemistry. We experimented with two different emission scenarios for selected months of year 1995. The experiments were carried on (Mips R12000, 300 MHz, 8 MB cache) of SGI Origin 2000 (a shared memory machine). Experiments for the framework scalability are shown on Figure 2. PETSc provides different types of preconditioners, and after trying several of them we chose the Additive Schwartz preconditioner for the sparse systems solution. After profiling we found that 70% of the ADS time is for solution of the sparse linear systems; 99% of the sparse systems solution is for matrix multiplication. Some scalability statistics for ADS (based on the PETSc profiling) are shown Figures 3 and 4.

It should be pointed out that the alternative in DEM to perform computations with resolution $10km \times 10km$, using the current production code, is to use total grid refinement with 230400 nodes and time step $150s$ for both the advection and chemical submodels. Clearly this task is much bigger than the one with local refinements, and the later should be preferred when we are interested in fine resolution results just on some parts of the modeling domain.

It is proven that the ADS framework is amenable for extensions to 3D simulations and schemes with higher order elements. The data feeding classes

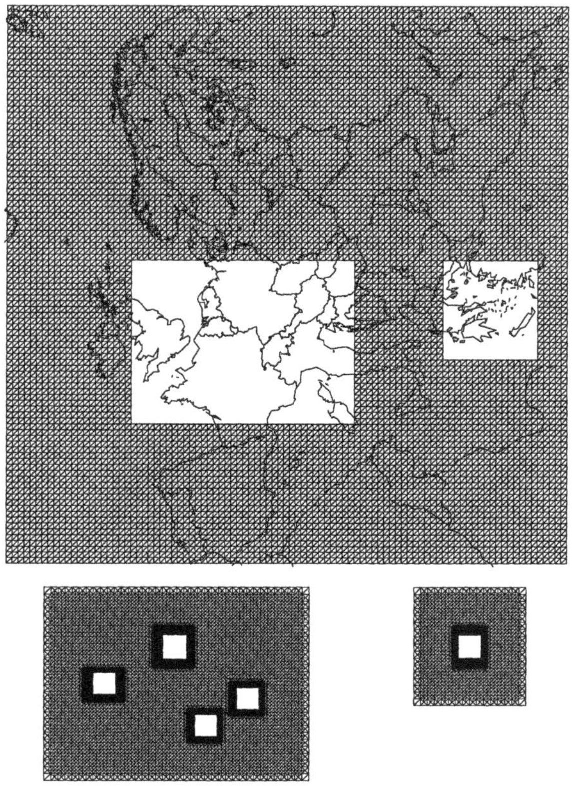

Fig. 1. A grid made with the OODEM Mesh Generator. Into the coarse grid above are imbeded the locally refined grid parts bellow.

for the GFEM layer can be extended/tuned for unstructured, automatically generated grids if it is desired.

We facilitated OODEM with the Observer DP to store the results. The objects from this pattern observe a class implementing the simulation clock. Observer DP can be used, together with the Decorator that provides parallelism, for status and performance monitoring. Also, it can utilize the PETSc performance monitoring tools and ALICE memory snooper, which belongs to the PETSc family. We consider implementing the Visitor DP for easier application of different bookkeeping tasks: Visitor's double dispatching provides easy way to traverse the class hierarchy with tasks like consistency checks or convergence monitoring.

OODEM is conceptually, class and example documented. The conceptual documentation can be found in [1]. The class documentation and the mesh generator documentation, can be found on [2].

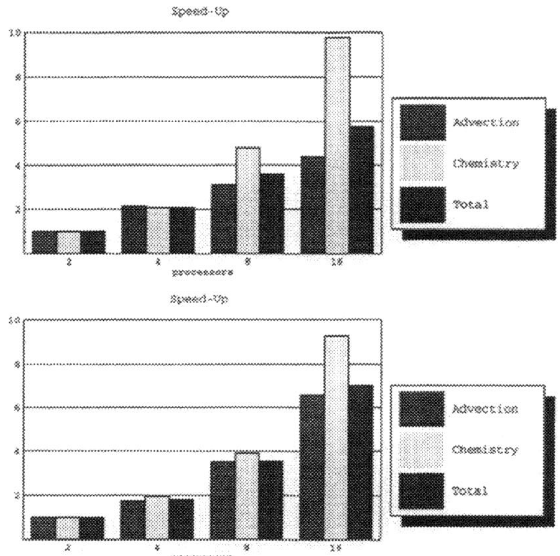

Fig. 2. OODEM scalability. Above: with grid nodes ≈ 16000, finest grid resolution $10km \times 10km$; $1.47 \div 1.57$ speed-up. Bellow: with grid nodes ≈ 35000, finest grid resolution $2km \times 2km$; $1.80 \div 1.90$ speed-up. All runs are for July'95, made in 3450 steps.

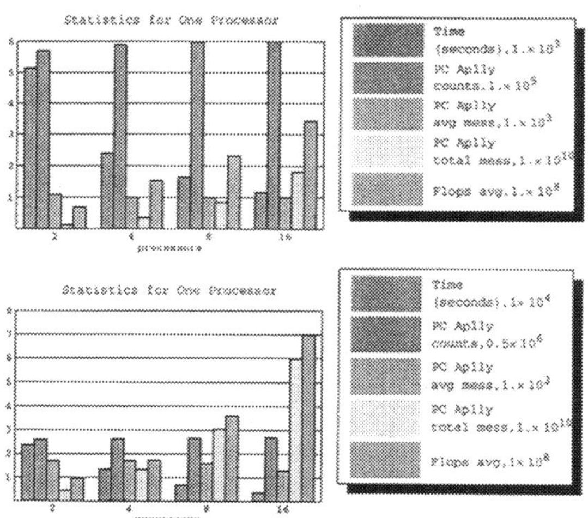

Fig. 3. Sparse system solution statistics (PETSc's MatMult). Above: with grid nodes ≈ 16000, finest resolution $10km \times 10km$; bellow: with grid nodes ≈ 35000, finest grid resolution $2km \times 2km$. All runs are for July'95, made in 3450 steps.

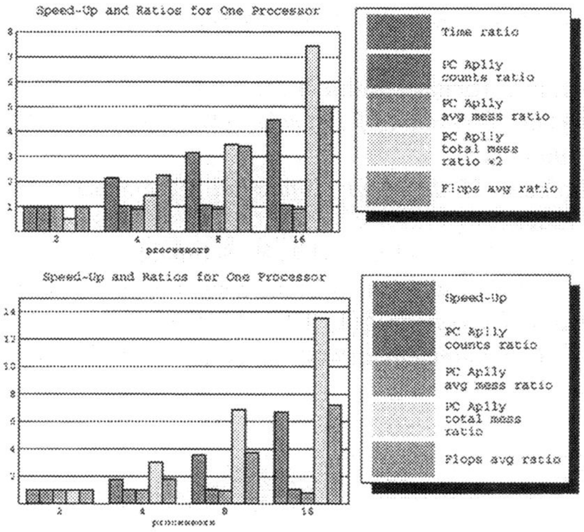

Fig. 4. Sparse system solution scalability.Above: with grid nodes ≈ 16000, finest grid resolution $10km \times 10km$; bellow: with grid nodes ≈ 35000, finest grid resolution $2km \times 2km$. All runs are for July'95, made in 3450 steps.

References

[1] Anton Antonov. *Object-Oriented Framework for Large Scale Air Pollution Models.* PhD thesis, Danish Technical University, April 2001.

[2] Anton Antonov. The Object-Oriented Danish Eulerian Model Homepage. http://www.imm.dtu.dk/ũniaaa/OODEM/, 2001.

[3] Satish Balay, William D. Gropp, Lois Curfman McInnes, and Barry F. Smith. Efficient management of parallelism in object oriented numerical software libraries. In E. Arge, A. M. Bruaset, and H. P. Langtangen, editors, *Modern Software Tools in Scientific Computing*, pages 163–202. Birkhauser Press, 1997.

[4] Satish Balay, William D. Gropp, Lois Curfman McInnes, and Barry F. Smith. PETSc home page. http://www.mcs.anl.gov/petsc, 2000.

[5] Shai Ben-Yehuda. Pattern language for framework construction. In Robert Hanmer, editor, *The 4th Pattern Languages of Programming Conference 1997*, number 97-34 in Technical Report. Washington University, Technischer Bericht, 1997. http://jerry.cs.uiuc.edu/p̃lop/plop97/Workshops.html.

[6] CONCAWE. *CONCAWE Review*, volume 9. Words and Publications, 10 2000. pages 10-12.

[7] European Commission for the Environment. Auto-Oil II Programme. http://europa.eu.int/comm/environment/autooil/, 2000.

[8] Brian Foote and Joseph Yoder. Big ball of mud. In Neil Harrison, Brian Foote, and Hans Rohnert, editors, *Pattern Languages of Program Design*

4, Addison-Wesley Software Patterns Series, chapter 29. Addison-Wesley, 1999.

[9] Erich Gamma, Richard Helm, Ralph Johnson, and John Vlissides. *Design Patterns. Elements of Reusable Object-Oriented Software.* Addison Wesley, 1995.

[10] G. I. Marchuk. *Methods for Numerical Mathematics.* Springer-Verlag, 2 edition, 1982.

[11] G. McRae, W. R. Goodin, and J. H. Seinfield. Numerical solution of the atnospheric diffusion equation for chemically reacting flows. *Journal of Computational Physics*, 45(1):356–396, 1982.

[12] Z. Zlatev. *Computer Treatment of Large Air Pollution Models.* Kluwer, 1995.

Parallel Computational Fluid Dynamics – Practice and Theory
P. Wilders, A. Ecer, J. Periaux, N. Satofuka and P. Fox (Editors)

Aerodynamic studies on a Beowulf Cluster

K.J.Badcock[a], M.A.Woodgate[a], K.Stevenson[a], B.E.Richards[a], M.Allan[a], G.S.L.Goura[a], R.Menzies[a].

[a] Computational Fluid Dynamics Group,Department of Aerospace Engineering, University of Glasgow, G12 8QQ, U.K.

The computational fluid dynamics research at Glasgow University has focussed for several years on developing algorithms for compressible flow solutions and low cost hardware options which will allow the level of performance necessary for systematic studies in aerodynamics. The latest numerical methods, together with modern commodity hardware, have reached the point where the promise of such studies is being realised. The current paper will briefly describe a new beowulf cluster installed at Glasgow University and the compressible flow solver which is executed on it before focussing on four case studies to illustrate the possibilities for aerodynamics research opened up by these developments. These case studies are for a moving delta wing, time marching aeroelatic simulation, surge wave propagation in an engine duct and cavity flow.

1. Introduction

Presently, any major computing carried out in aerodynamics is done using some level of parallelisation and a popular way of achieving this is using distributed memory architecture. This not only gains benefit from carrying out the task simultaneously on many processors but also enables a large amount of memory required to define problems of geometric complexity. Parallelism is achieved in CFD codes by domain decomposition with each processor calculating the solution on a partition of the whole grid. Then values need to be communicated to each processor corresponding to information at the boundaries of grid partitions. This message passing is implemented by a high level library such as MPI (message passing interface). The coding of partitioning and message passing does not constitute major problems if the basic flow algorithm is well conceived. This is all built into the PMB code, solving the Reynolds averaged Navier-Stokes (RANS) equations with turbulence modelled, providing ease of use to the user.

Small academic research groups can have difficulty getting access to national facilities. Generally resource on these facilities is awarded to a group of collaborators who use common software on fundamental problems of physical complexity that have huge run times e.g. study of turbulence, ocean currents, meteorology etc. RANS calculations are mostly directed towards industrial problems that are large, mainly because of their geometrical complexity, but the speed derived from fast computers is directed towards a fast turn round time needed for rapid design decisions. This along with the significant time in generating a grid to make changes of a geometric nature to the problem, means irregular access to a facility that is incompatible with the major users. Thus CFD users solving problems modelled using the RANS equations are forced to look for other solutions than using national facilities, such as the Beowulf facility.

The cost of even a powerful Beowulf facility is low enough that it is within the grasp of a small research group. Also when building such a facility the codes to be used on it can define its configuration. A small focussed group will have a minimum of compute intensive codes to implement on the facility, so that it can be configured simply. The CFD Group uses only the PMB code. This could also be said of teams within industry or small companies targeted at specialist areas in which the Beowulf would be an ideal facility. In other words small groups do not need the flexibility of a mainframe, and the purchase of one would not be economic. Also they would be free of restrictions imposed of corporate facilities.

One of the economies associated with the use of a Beowulf facility is the use of Linux as the operating system. This has been shown to provide a very robust operating system. It is a public domain software but which is now accepted as being reliable and also that commercial software is configured to run on it. For these reasons Linux and the Beowulf concept is becoming acceptable more widely presently.

For a more complete background to the subject, consult the articles [6] and [7]. This paper is concerned with operation and use of the new facility.

2. Numerical Formulation

The flow simulation is based on an implicit steady state solver on multiblock grids [6]. This solver uses approximate Jacobian matrices and a preconditioner which is decoupled between blocks to enhance parallel performance. Unsteady flows are calculated within this framework using Jameson's pseudo time method. Previous presentations by the authors at the parallel CFD conference series have discussed the development of this method, including (a) the performance of the method on a Pentium 200 cluster for the calculation of various steady wing flows [1] and (b) rolling delta wing flows [2], (c) performance on a workstation cluster for two dimensional test cases [4] and the influence of preconditioning strategies for parallel performance [3]. Recent enhancements to the solver have included the coupling with structural models for aeroelastic simulations [5].

3. Cluster

Previous papers [1] [7] described the 16 node Pentium 200 cluster which was installed at Glasgow University in 1997. The attractions of such a facility are well documented and revolve around price and performance. The system has now been upgraded with the development of a 32 node cluster. This new cluster will be referred to as simply "the cluster" in the following. The total cost of the cluster was 40,000 pounds sterling.

The cluster's internal connectivity is provided by a Cisco Catalyst 3548 fast Ethernet switch which provides 48 * 100BaseTX full duplex ports and two proprietary expansion slots one of which is populated with a 1000BaseSX module giving full duplex gigabit Ethernet connectivity to the cluster's main file server. The free expansion port can be utilised to chain additional Catalyst switches allowing the cluster to be scaled beyond it's current configuration.

The cluster's main file server is a dual 700MHz Pentium III box with 512Mb of RAM based around the ServerWorks III-LE chipset and is currently running Solaris 7 x86. This is accessed via NFS automounters on 7 different operating systems by client machines on the departmental LAN. The chipset features a 64-bit 66MHz PCI bus with a 200Gb filesystem used for data generated by processes on the back end nodes.

The backend nodes are made up of 32 * 750MHz AMD Athlon (Thunderbird) uniprocessor machines each with 512Mb of 100MHz DRAM. All the nodes have 100Mbps full duplex 3Com 3C905 network interface cards. The Athlon based machines are running Mandrake 7.2 with a newer 2.4.0 kernel primarily for it's NFSv3 enhancements.

The only software used on the backend nodes is the Glasgow flow solver. The lam mpi implementation is used for the message passing.

4. Applications

4.1. Delta Wing

Delta wing aerodynamics are characterised by two leading edge vortices. The response of these vortices during maneuvering flight is the dominant aspect of the flow. The leading edge vortices form due to flow separation from the leading edges of the wing (which are generally sharp on delta wings), with the resultant shear layer rolling up due to a spanwise pressure gradient along the surface of the wing. This pressure gradient also causes the outboard moving flow to separate from the wing surface, just outboard of the primary vortex core. This second flow separation results in secondary and tertiary vortices being formed. An important feature of such vortex dominated flows is the vortex burst. This is characterised by an abrupt deceleration of the flow in the vortex core, which causes the flow to stagnate and full scale turbulence to develop. The unsteady response of the vortex burst is a governing factor when considering delta wings in maneuvering flight since the forces generated on the wing are strongly influenced by the behaviour of the vortices.

A study was undertaken to assess the capabilities of an Euler code to predict the aerodynamic characteristics of maneuvering delta wings. Three types of motion were considered - pitching, rolling, and yawing. Calculations were performed at various mean incidence, amplitudes and frequencies of motion. The results were compared with wind tunnel test data. For the three types of motion considered, mean incidence was chosen such that the variation in size and intensity of the vortices can be examined (at the lower incidences), and that movement of the vortex burst location above the wing can be examined (at the higher incidences), figure 1.

The calculations in this case were carried out on grids with about 400k points. Typically three motion cycles with 50 (for pitching) or 200 (for yawing) time steps per cycle were required to obtain a time accurate and periodic flow. These cases were run on 8 processors with almost 100 per-cent efficiency. For a case at 27 degrees mean incidence yawing with an amplitude of 5 degrees and a frequency of 3 Hz the elapsed calculation time was 2.8 hours per cycle. For a case at 27 degrees mean incidence pitching with an amplitude of 5 degrees and a frequency of 3 Hz the elapsed calculation time was 48 minutes per cycle.

4.2. Surge Wave

Intakes are an important component of aircraft and the efficiency of such devices is crucial as they make major contributions to the performance and handling attributes.

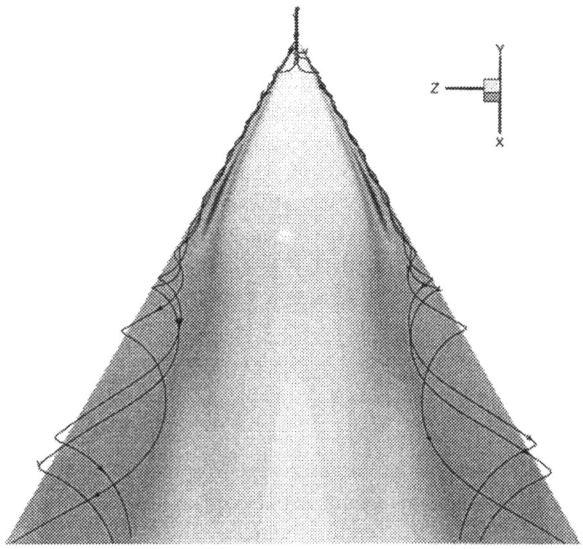

Figure 1. Surface pressure contours and vortex core streak lines near apex. Incidence = 27 degrees

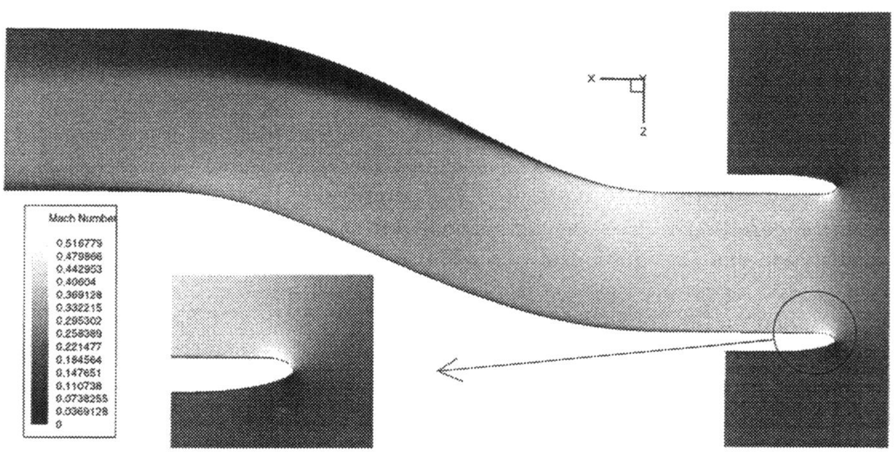

Figure 2. Symmetry plane Mach contours

Intake/airframe aerodynamic compatability is essential and the design must limit the possibility of compressor stall and engine surge among other things. Diffusing s-shaped intakes such as the RAE intake model 2129 are challenging to study due to the complex nature of the flow that develops as a result of the offset between intake cowl plane and the intake engine face plane. This offset generates flow separation at the first bend which can produce non-uniformities in total pressure acrosss the compressor face (distortion) that can induce compressor surge.

The Euler and Navier-Stokes equations have been usd to study the flow in S-shaped intake ducts. A boundary condition was implemented that allows for the modelling of engine demand. Calculations were then done for both a high and a low engine demand (mass flow demand) for which experimental and previous computational data is available. The computations were performed on the whole geometry including intake cowl and a large upstream region. The upstream region allows for the modelling of the flow from freestream conditions into the intake rather than just the flow in the intake itself. This negates the need to worry about complex intake entry conditions and allows for straighforward comparisons between flow solvers.

Grid converged solutions were obtained for the Euler and Navier-Stokes low mass flow calcualtions. Comparisons between previous computational results and experimental data showed an overall excellent agreement. The results of the Navier-Stokes calculation using the $k - \omega$ turbulence model on a grid of approximately 400,000 points are shown in figure 2. The calculations on processors required 1900 steady implicit steps at a CFL number of 30 to converge 8 orders of magnitude, taking around 3 hours of elapsed time to complete.

4.3. Cavity Flow

Cavities are of major interest as aircraft components in several situations, and in particular for internal weapons integration. The flow physics involved is very complex, resulting from motion of the shear layer, vortices inside the cavity and internal and external pressure waves. From the simulation point of view this is a very complex problem since there are a range of frequencies involved and large scale motions of the important flow features.

The Navier-Stokes equations were solved to simulate the three dimensional flow past a half and full span cavity. The grid contains 800 k points for the half span cavity and double this for the full span. About 1000 time steps were required to collect enough data for spectral analysis to determine frequencies and sound pressure levels. The half span case took an elasped time of 105 hours on 20 processors to complete this calculation.

4.4. Aeroelastic Response of a Wing

Flutter clearance is a serious issue before an aircraft can enter service. Flight testing increases the certification cost since each flight test hour costs about 50,000 pounds per hour for lighter aircraft. The increased use of computational methods to investigate flutter could reduce this time by improving the design and reducing the cost of these predictions. One possible area of advantage is in the transonic regime where current predictions are based on linear flow models, substantially decreasing confidence in the predictions.

Time marching simulations to determine aerostatic deformation and flutter boundaries have been undertaken. The aerostatic deformation of a transport type wing in transonic flow is shown in figure 3. The fluid grid in this case has 300k points and the structural modal consists of 18 modes. The calculation on only 2 processors took 2.4 hours to

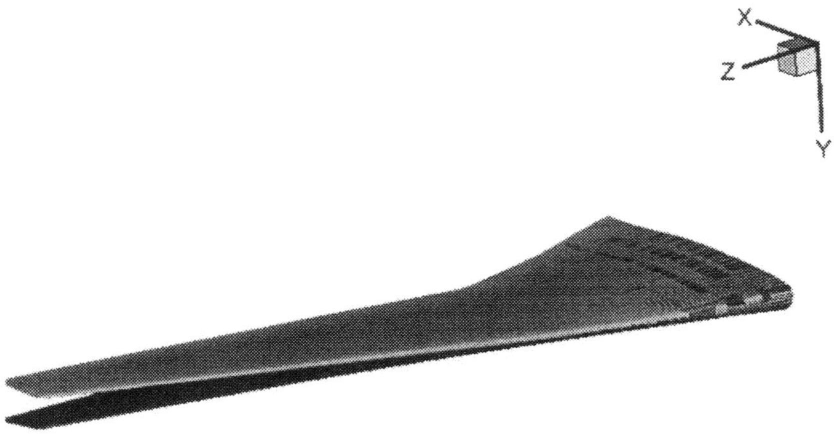

Figure 3. Aerostatic deflection for a transport type wing in a transonic flow.

converge 5 orders.

5. Evaluation and Perspective

The new cluster has made the routine, systematic and rigorous use of CFD for the analysis of aerodynamics a reality at the level of a small research group. The cost of the hardware is at a level to allow the facility to be dedicated to use by a group consisting of 10 researchers. This allows the scheduling of jobs to be done in an informal manner without the need for detailed load balancing software. The system management, based on LINUX, has proved straightforward and the system has operated robustly for nine months with no significant problems.

The pmb code which is executed on the cluster was ported from the old cluster without any program changes. Despite the faster processors in the new cluster no drop-off in efficiency was noted on the new cluster. However, it is estimated that if the next hardware upgrade delivers the same jump in processor performance whilst the networking remains at the same performance, then effort will need to be put into improving the parallel performance of the code.

The cluster has about 200 Gb of hard disk available. This capacity is quite limited for three-dimensional unsteady flow calculations. The post processing of the data, currently done sequentially by the commercial package tecplot, is a bottleneck in the simulation process. In addition, backing up data to tape is also slow and costly.

Current plans for aerodynamic applications to be carried out on the cluster are

- inviscid flutter analysis for a complete aircraft
- unsteady cavity simulation for cavity containing a store
- turbulent flow around manoeuvring delta wings.

Looking ahead to the next upgrade in 3-5 years, it is anticipated that for the same hardware cost it will be possible to increase the performance of the system by an order of magnitude but

- attention will need to be paid to improving the parallel performance of the code unless faster networks are also affordable
- improved mass storage devices will be needed within the price range of the other components of the cluster
- attention will need to be paid to parallel pre and post processing for the larger grids as these tasks will no longer be possible on sequentially.

REFERENCES

1. Woodgate, M. A., Badcock, K. J. and Richards, B. E. A parallel 3D fully implicit unsteady multiblock code implemented on a Beowulf cluster. Parallel Computational Fluid Dynamics - Towards Teraflops, Optimization and Novel Formulations. Ed. D. Keyes, A. Ecer, J. Periaux, N. Satofuka, and P. Fox. Elsevier Science B.V. October 2000 pp 447- 455.

2. Woodgate, M., Badcock, K.J., and Richards, B.E. The solution of pitching and rolling delta wings using a Beowulf cluster. Parallel CFD Conference. Trondheim. May 2000.
3. Badcock, K. J., McMillan, W., Woodgate, M. A., Gribben, B. J., Porter, S., Richards, B. E. Integration of an implicit multiblock code into a workstation cluster environment Parallel Computational Fluid Dynamics '96: Algorithms and Results using Advanced Computers, P. Schiano et al (Eds), Elsevier Science B.V. Amsterdam, pp. 408-415, 1997.
4. F. Cantariti, L. Dubuc, B. Gribben, M. Woodgate, K. Badcock, B. Richards and W. McMillan Integration of an implicit multiblock code into a workstation cluster environment Parallel Computational Fluid Dynamics '97: Recent Developments and Advances Using Parallel Computers, D.R. Emerson et al (Eds), Elsevier Science B.V. Amsterdam, pp. 169-175, 1998.
5. Goura, L., Badcock, K.J., Woodgate, M. and Richards, B.E. Implicit method for the time marching analysis of flutter. Aeronautical Journal. Volume 105, Number 1046, pp 199-214. April 2001.
6. Badcock, K.J., Richards, B.E. and Woodgate, M.A., Elements of computational fluid dynamics on block structured grids using implicit solvers, Progress in Aerospace Sciences, Vol 36, pp 351-392, September 2000.
7. McMillan, W.S., Woodgate, M.A., Richards, B.E., Gribben, B.J., Badcock, K.J. and Masson, C.A. and Cantariti, F., Demonstration of cluster computing for three dimensional CFD simulations, Aeronautical Journal, Vol 103, No 1027, pp 443-447, September 1999.

Parallel Computational Fluid Dynamics – Practice and Theory
P. Wilders, A. Ecer, J. Periaux, N. Satofuka and P. Fox (Editors)

Scalable numerical algorithms for efficient meta-computing of elliptic equations.

N. Barberou[a], M. Garbey[a], M. Hess[b], T. Rossi[c], M. Resh[b], J. Toivanen[c] and D. Tromeur-Dervout[a]

[a]Center for the Development of Scientific Parallel Computing, CDCSP/ISTIL - University Lyon 1, 69622 Villeurbanne cedex, France

[b]HLRS, University of Stuttgart, Germany

[c]Department of Mathematical Information Technology, University of Jyväskylä, Finland

This paper reports the recent development of the numerical Aitken-Schwarz domain decomposition method (M. Garbey and D. Tromeur-Dervout, proceedings of 12th International Conference on Domain Decomposition Methods, 2000) associated with the Partial Solution variant of Cyclic Reduction method (T. Rossi and J. Toivanen, SIAM J. Sci. Comput., 5, pp 1778-1796, 1999) on large scale parallel computers. The objective of this methodology is to obtain a numerically efficient algorithm for separable elliptic operator that is scalable on several hundred of processors. The use of PACX-MPI communication library software, developped by Resh & Al at HLRS [1,8], allows us to take advantage of the efficiency of vendor's MPI on each computer still with long distant connections. The results show the need of adapted numerical methods as the Aitken-Schwarz method to relax the communication constraint on a slow speed communication network that is inherent to metacomputing framework with no dedicated networks.

The plan of this article is as follows: Section 1 deals with the very parallel efficient solver PSCR and its underlying mathematical concepts. Section 2 describes the Aitken-Schwarz two levels domain decomposition methodology. Section 3 gives the preliminary results obtained by the combination of the two solvers in a European metacomputing framework. Section 4 will give the planed experiments and developments.

1. numerical methods

Partial Solution variant of Cyclic Reduction (PSCR) [12] was introduced by P.S. Vassilevski, and generalized by Yu.A. Kuznetsov [7]. This method is closely related with the partial fraction variant of classical cyclic reduction by R.A. Sweet [11].

The PSCR method is a multilevel hierarchical substructuring method with exact direct interface solvers. Only separable problems are taken in consideration in order to use special, very efficient techniques for the interface problems (optimized separation of variables method for problems with sparsity, the so-called partial solution technique).

Let us illustrate the substructuring with two subdomains, 1 and 2, separated by an interface γ:

$$\begin{pmatrix} A_{11} & A_{1\gamma} & 0 \\ A_{\gamma 1} & A_{\gamma\gamma} & A_{\gamma 2} \\ 0 & A_{2\gamma} & A_{22} \end{pmatrix} \begin{pmatrix} u_1 \\ u_\gamma \\ u_2 \end{pmatrix} = \begin{pmatrix} f_1 \\ f_\gamma \\ f_2 \end{pmatrix}$$

The principle consists to write u_1 and u_2 with respect to u_γ, solve for u_γ, recover u_1 and u_2.

- Step 1, compute : $\tilde{f}_\gamma = f_\gamma - A_{\gamma 1}A_{11}^{-1}f_1 - A_{\gamma 2}A_{22}^{-1}f_2$
- Step 2, solve : $S_\gamma u_\gamma = \tilde{f}_\gamma$ where

  $S_\gamma = A_{\gamma\gamma} - A_{\gamma 1}A_{11}^{-1}A_{1\gamma} - A_{\gamma 2}A_{22}^{-1}A_{2\gamma}$
- Step 3, recover u_1 and u_2 : $u_1 = A_{11}^{-1}(f_1 - A_{1\gamma}u_\gamma), \quad u_2 = A_{22}^{-1}(f_1 - A_{2\gamma}u_\gamma)$.
- Make steps 1 to 3 recursive, i.e. decompose the subdomains into smaller ones and apply the same process for solving systems with A_{11} and A_{22}, avoiding redundant work in practical implementation.

The interface problem: $u_\gamma = S^{-1}\tilde{f}_\gamma$ can be solved from

$$\begin{pmatrix} A_{11} & A_{1\gamma} & 0 \\ A_{\gamma 1} & A_{\gamma\gamma} & A_{\gamma 2} \\ 0 & A_{2\gamma} & A_{22} \end{pmatrix} \begin{pmatrix} * \\ u_\gamma \\ * \end{pmatrix} = \begin{pmatrix} 0 \\ \tilde{f}_\gamma \\ 0 \end{pmatrix}$$

using optimized separation of variables technique with $\mathcal{O}(N\log^{d-2}N)$ operations for $d = 2, 3$ -dimensional problems, provided that $\dim(u_\gamma) = \mathcal{O}(\dim(\tilde{f}_\gamma)) = \mathcal{O}(N^{\frac{d-1}{d}})$. This is the case since u_γ and f_γ live on a $d-1$-dimensional subdomain interface hyperplanes.

The total computational cost of this method is $\mathcal{O}(N\log^{d-1}N)$ for $d = 2, 3$ -dimensional problems.

The parallelism in the PSCR method comes from the following two key observations:

- Subdomain solves in steps 1 and 2 are completely independent
- Partial solution technique (optimized method of separation of variables) decomposes the d-dimensional problems into a set of fully independent $d-1$-dimensional problems

So, there will be collective communication only in the Fourier transforms of right-hand side vector block $\tilde{f}_\gamma$ and the solution block u_γ, but they are related to $d-1$-dimensional hyperplanes. All other communication is of point-to-point type.

The (recursive) subdivision of subdomains is performed as follows: The radix of the method p , call can be parametrized. For 2D-problems the original rectangle is divided recursively into p smaller rectangles in one direction. One can use up to $\mathcal{O}(N^{\frac{1}{2}})$ processors. For 3D-problems the original parallelepiped is divided recursively into p^2 smaller parallelepipeds in two coordinate directions. One can use up to $\mathcal{O}(N^{\frac{2}{3}})$ processors.

In the implementation $p = 4$, it is almost optimal in terms of required floating point operations.

The parallel PSCR solver PDC3D developed by T. Rossi and J. Toivanen [9,10] is a fast direct solver for 3D separable operators. This method is numerically stable and it reaches a good scalability on computer platform with a high performance communication network (high bandwidth and low latency). Table 1 gives the time in seconds to solve a problem of size $511 \times 511 \times 64$ on a T3E platform at CSC, Espoo, Finland. It shows the good efficiency of the method, and how well it scales. The PSCR methodology can be parallelized on 3D processor grid, however the implemented parallel solver PDC3D at the present time is profiled to use 2D processor grid.

This very efficient solver requires high performance communication network mainly due to global reduction operations, to gather the partial solution. On the other hand, M. Garbey and D.

size2/size1	1	2	4	8	16	32	64
1					25.842	12.706	6.625
2				25.638	13.587	6.662	3.50
4			27.047	13.001	6.517	3.380	
8		26.834	13.546	6.769	3.429		
16	28.876	14.977	7.176	3.621			
32	18.317	9.144	4.595				
64	13.352	6.665					

Table 1
Solution times in seconds for the PDC3D solver for an elliptic separable operator of size $511 \times 511 \times 64$ *distributed on a 2D topological network of size*1 $\times$ *size*2 *processors on a T3E platform.*

Tromeur-Dervout [2,4,6] have developed an Aitken-Schwarz algorithm (respectively Steffensen-Schwarz) for the Helmholtz operator, (respectively for general linear and non linear elliptic problems) that is highly tolerant to low bandwidth and high latency. We describe in the next section this methodology and how it can be combined to PSCR in order to perform efficient metacomputing.

2. Description of the new method

It is not an easy task to describe briefly the numerical ideas behind the Aitken Schwarz method. For simplicity, we will illustrate the concept with the discretized Helmholtz operator $L[u] = \Delta u - \lambda u,\ \lambda > 0$, with a grid that is a tensorial product of one dimensional grids, and a square domain decomposed into strip subdomains. Let us consider the homogeneous Dirichlet problem $L[U] = f\ in\ \Omega = (0,1),\ U_{|\partial\Omega} = 0$, in one space dimension. We restrict ourselves to a decomposition of Ω into two overlapping subdomains $\Omega_1 \bigcup \Omega_2$ and consider the additive Schwarz algorithm.

$$L[u_1^{n+1}] = f\ in\ \Omega_1,\ u_{1|\Gamma_1}^{n+1} = u_{2|\Gamma_1}^{n},\ L[u_2^{n+1}] = f\ in\ \Omega_2,\ u_{2|\Gamma_2}^{n+1} = u_{1|\Gamma_2}^{n}. \tag{1}$$

with given initial conditions $u_{1|\Gamma_2}^0, u_{2|\Gamma_1}^0$ to start this iterative process. To simplify the presentation, we assume implicitly in our notations that the homogeneous Dirichlet boundary condition are satisfied by all intermediate subproblems. This algorithm can be executed in parallel on two computers. At the end of each subdomain solve, the artificial interfaces $u_{2|\Gamma_1}^n$ and $u_{1|\Gamma_2}^n$ have to be exchanged between the two computers. In order to avoid as much as possible redundancy in the computation we fix once and for all the overlap between subdomains to be the minimum, i.e of size one mesh. This algorithm can be extended to an arbitrary number of subdomains and is nicely scalable, because the communications linked only subdomains that are neighbors. However it is one of the worst numerical algorithm to solve the problem, because the convergence is extremely slow. We introduce thereafter a modified version of this Schwarz algorithm so called Aitken-Schwarz that transforms this dead slow iterative solver into a direct fast solver while keeping the scalability of the Schwarz algorithm for moderate number of subdomains. The idea is as follows.

We observe that the interface operator T,

$$(u_{1|\Gamma_1}^n - U_{\Gamma_1}, u_{2|\Gamma_2}^n - U_{\Gamma_2})^t \rightarrow (u_{1|\Gamma_1}^{n+1} - U_{\Gamma_1}, u_{2|\Gamma_2}^{n+1} - U_{\Gamma_2})^t \tag{2}$$

is *linear.*

Therefore, the sequence $(u^n_{1|\Gamma_1}, u^n_{2|\Gamma_2})$ has pure linear convergence that is, it satisfies the identities:

$$u^{n+1}_{1|\Gamma_2} - U_{|\Gamma_2} = \delta_1(u^n_{2|\Gamma_1} - U_{|\Gamma_1}),\ u^{n+1}_{2|\Gamma_1} - U_{|\Gamma_1} = \delta_2(u^n_{1|\Gamma_2} - U_{|\Gamma_2}), \tag{3}$$

where δ_1 (respt δ_2) is the damping factor associated to the operator L in subdomain Ω_1 (respt Ω_2) [3]. Consequently

$$u^2_{1|\Gamma_2} - u^1_{1|\Gamma_2} = \delta_1(u^1_{2|\Gamma_1} - u^0_{2|\Gamma_1}),\ u^2_{2|\Gamma_1} - u^1_{2|\Gamma_1} = \delta_2(u^1_{1|\Gamma_2} - u^0_{1|\Gamma_2}), \tag{4}$$

So except if the initial boundary conditions matches with the exact solution U at the interfaces, the amplification factors can be computed from the linear system(4). Since $\delta_1\delta_2 \neq 1$ the limit $U_{|\Gamma_i}, i = 1,2$ is obtained as the solution of the linear system (3). Consequently, this generalized Aitken acceleration procedure gives the *exact* limit of the sequence on the interface Γ_i based on two successive Schwarz iterates $u^n_{i|\Gamma_i}$, $n = 1,2$, and the initial condition $u^0_{i|\Gamma_i}$. An additional solve of each subproblem (1) with boundary conditions $u^\infty_{\Gamma_i}$ gives the final solution of of the ODE problem. We can further improve this first algorithm as follows.

δ_1 and δ_2 can be computed before hand numerically or analytically as follows. Let (v_1, v_2) be the solution of

$$L[v_1] = 0\ in\ \Omega_1,\ v_{|\Gamma_1} = 1;\ L[v_2] = 0\ in\ \Omega_2,\ v_{|\Gamma_2} = 1; \tag{5}$$

We have then $\delta_1 = v_{|\Gamma_2}$, $\delta_2 = v_{|\Gamma_1}$.

Once (δ_1, δ_2) is known, we need only *one* Schwarz iterate to accelerate the interface and an additional solves for each subproblems. This is a total of two solves per subdomain. The Aitken acceleration thus transforms the additive Schwarz procedure into an *exact* solver regardless of the speed of convergence of the original Schwarz method, and in particular with minimum overlap.

This Aitken-Schwarz algorithm can be reproduced for multidimensional problems. As a matter of fact, it can be shown [6] that the coefficients of each wave number of the sine expansion of the trace of the solution generated by the Schwarz algorithm has its own rate of *exact* linear convergence. We can then generalize the one dimensional algorithm to two space dimensions as follows:

- step1 : compute analytically or numerically in parallel each damping factor δ^k_j for each wave number k from the two point one D boundary value problems analogues of (5).
- step2: apply one additive Schwarz iterate to the Helmholtz problem with subdomain solver of choice (multigrids, fast Fourier transform, PDC3D, etc...)
- step3:

  - compute the sine expansion $\hat{u}^n_{k|\Gamma_i}$, $n = 0,1,\ k = 1..N$ of the traces on the artificial interface $\Gamma_i, i = 1..2$ for the initial boundary condition $u^0_{|\Gamma_i}$ and the solution given by **one** Schwarz iterate $u^1_{|\Gamma_i}$, $i = 1,2$.

  - apply generalized Aitken acceleration *separately* to each wave coefficients in order to get $\hat{u}^\infty_{j|\Gamma_i}$.

  - recompose the trace $u^\infty_{j|\Gamma_i}$ in physical space.

- step4: compute in parallel the solution in each subdomains Ω_i, $i = 1,2$ with new inner BCs and subdomain solver of choice.

So far, we have restricted ourselves to domain decomposition with two subdomains. We show in [5], that a generalized Aitken acceleration technique can be applied to an arbitrary number $q > 2$ of subdomains with strip domain decomposition. Our main result is that no matter is the number of subdomains, the total number of subdomain solves required to produce the final solution is still **two**. However the generalized Aitken acceleration of the vectorial sequences of interface introduce a coupling between all interfaces. But we observe first that this generalized Aitken acceleration processes independently each waves coefficients of the sinus expansion of the interfaces. Second the highest is the frequency k the smallest is the damping factors δ_j^k, $j = 1..2q$. A careful stability analysis of the method shows that

- for low frequencies, we should use the generalized Aitken acceleration coupling all the subdomains.
- for intermediate frequencies, we can neglect this global coupling and implement only the local interaction between subdomains that overlap.
- for high frequencies, we do not use Aitken acceleration because one iteration of the Schwarz algorithm damps the high frequency error enough.

The algorithm has then the same structure than the two subdomains algorithm presented above. Step 1 and step 4 are fully parallel. Step 2 requires only local communication and scale well with the number of processors. Step 3 requires global communication of interfaces in Fourier space for low wave numbers, and local communications for intermediate frequencies. In addition for moderated number of subdomains, the arithmetic complexity of step3 that is the kernel of the method is negligible compared to step2.

Our algorithm can be extended successfully to 3d problems with multidimensional domain decomposition, grids that are tensorial product of one dimensional grids with arbitrary (irregular) space step, iterative domain decomposition method such as Dirichlet-Newman procedure with non-overlapping subdomains or red/black subdomains iterative procedure. For non linear elliptic problem, the Aitken acceleration is no longer exact. the so-called Steffensen-Schwarz variant is then a very efficient numerical method for low order perturbation of constant coefficient linear operators - see [6] and [5] for more details.

In the specific case of separable elliptic operator, the Aitken-Schwarz algorithm might be less efficient in terms of arithmetic complexity than PDC3D as the number of processors increases, but is is rather competitive with $O(10)$ subdomains.

The main trust of our paper is therefore to combine the two methods in order to have a highly efficient solver for the Helmholtz operator for metacomputing environments. This kind of solver can be use to solve the elliptic equations satisfy by the Velocity components of a incompressible NS code written in Velocity-Vorticity formulation. This elliptic part of this NS solver is usually the most time consuming part as theses equations must be solved very accurately to satisfy the Velocity divergence free constraint.

Our parallel implementation is then as follows: first one decomposes the domain of computation into a one-dimensional Domain Decomposition (DD) of $O(10)$ macro subdomains. This first level of DD uses Aitken Schwarz algorithm and the macro subdomains are distributed among clusters or distinct parallel computers. Secondly, each macro subdomain is decomposed into a two-dimensional DD: this level of DD use the PDC3D solver. Globally we have a three-dimensional DD and a two-level algorithm that matches the hierarchy of the network and access to memory.

	CrayS	CrayT
#proc	512	512
MHz	450	375
internal latency	12 μs	12 μs
internal bandwidth	320 MB/s	320 MB/s

Table 2
System configuration at Stuttgart and Helsinki.

3. The Results

We test the solvers in a metacomputing framework. In order to use the best communication software where it can be used, we link the numerical software with the PACX-MPI library [1,8]. This library allows to use MPI vendor implementation for inner communication between processor of a same supercomputer and TCP IP protocol for communication between processors that belong to different distant supercomputers. Two processors on each supercomputers must be added to manage distant communications. With this software we avoid firewall, and data representation problems. We are using the following hardware: Once for all we denote **CrayS** the Cray of HLRS in Stuttgart University and **CrayH** the Cray T3E of the National Scientific Computing Center of Finland at CSC. We make in this preliminary work three hypothesis:

- First, we restrict ourselves to the Poisson problem, i.e the Helmholtz operator with $\lambda = 0$. As a matter of fact, it is the worst situation for metacomputing because any perturbation at an artificial interface decreases linearly in space, instead of exponentially for the Helmholtz operator [3].
- Second, we neglect the load balancing that should be done on heterogeneous metacomputing Architecture. We verified that PDC3D solver is roughly 30% slower on CrayH than on CrayS for our test cases. However this will be done easily in our future experiment by adapting the number of grid points in each macro subdomains.
- Third we are running our metacomputing on the two supercomputers with the existing **ordinary ethernet network.** During all our experiments, the bandwidth fluctuated in the range $(1.6Mb/s - 2.1Mb/s)$ and the latency was about $30ms$.

First, let us show that the PDC3D solver cannot be used efficiently in metacomputing situation. Based on the performances -see Table 3- of the PDC3D on CrayS, we select the most efficient data distribution and run the same problem on the metacomputing architecture, i.e on CrayS and CrayH that share equally the total number of processors used. Table 4 gives a representative set of the performance of PDC3D on the metacomputing architecture (CrayS-CrayH). We conclude that no matter is the number of processors, most of the elapse time is spent in communications between the two computer sites. This conclusion holds for a problem of smaller size, that is 256^3: the elapsed time growths similarly from 0.76 s , up to 18.73s with 512 processors. In conclusion the PDC3D performance **degrades drastically** when one has to deal with a slow network.

Second we proceed with a preliminary performance evaluation of our two levels domain decomposition method combining Aitken-Schwarz and PDC3D (**AS**). We define the barrier between low and medium size frequencies in each space variable to be 1/4 of the number of waves; We do not accelerate the highest half of the frequencies. We checked that the impact on the numerical error against an exact polynomial solution is in the interval $[10^{-7}, 10^{-6}]$ for our test cases with

128 procs ($p_x \times p_y$)	256 procs ($p_x \times p_y$)	512 procs ($p_x \times p_y$)
25.9 s (4 x 32)	17.6s (4 x 64)	
22.0 s (16 x 8)	11.5s (16 x 16)	7.2 (16 x 32)
21.8 s (64 x 2)	11.2s (64 x 4)	**5.77 (64 x 8)**

Table 3

elapse time for PDC3D solver on CrayS with a problem of global size $511 \times 511 \times 512$

128 procs ($p_x \times p_y$)	256 procs ($p_x \times p_y$)	512 procs ($p_x \times p_y$)
72.0s (64 x 2)	77.2s (64 x 4)	75.1 s (64 x 8)

Table 4

elapse time for PDC3D solver on metacomputing architecture (CrayS, CrayH) with a problem of global size $511 \times 511 \times 512$

minimum overlap between macro subdomains. Tables 5 and 6 summarized our result and give average elapse time excluding the best runs. We observe from Table 6, for the smallest problem that AS is roughly 1.9 slower than PDC3D on a single parallel computer. However, as opposed to Table 2 result and considering the fact that we did not insure a proper load balancing, we obtain acceptable performances in the metacomputing configuration. Further, one can estimate the communication and waiting time lost because of the slow link in metacomputing configuration and we observe that its fraction reduced drastically from 45% to 25% as the problem size increases.

Further we checked that AS scales fairly well: Table 6 shows that when the size of the problems growths linearly with the number of processors, the elapsed time stays of the same order.

4. Developments

This work have been extended successfully toward two directions. First, experiments with more than two heterogeneous supercomputers. Second, developments of the numerical software to solve non linear non separable problems such Bratu problem. This claims to change the PSCR solver with a solver as multigrid solvers.

We will report on these later developments in a future paper.

Acknowledgement: We are grateful to the national computing centers CSC (Finland), Cines (France) and HLRS (Germany) which have been kind enough to give us access to their main computing ressources in interactive mode during our experiments. This work has been supported

global problem size $N_x \times N_y \times N_z$	Elapse time on CrayS with MPI		Elpase time in Metacomputing case
$512 \times 496 \times 496$	11.		25.2
$1024 \times 496 \times 496$	20.2		35

Table 5

elapsed time for ASreduced solver in second on CrayS and on metacomputing architecture (CrayS, CrayH)

Total number of processors	Global size of the problem	macro subdomains on CrayH	macro subdomains on CrayS	elapsed time
512	$342 \times 432 \times 432$	1	1	14.74
768	$513 \times 432 \times 432$	1	2	15.00

Table 6
elapsed time for AS solver in second on metacomputing architecture (CrayS, CrayH) with a problem of local size $171 \times 27 \times 27$ per processor

by ANVAR from France as well as grant 43066 and 66407 from the Academy of Finland.

REFERENCES

1. E. Gabriel, M. Resch, Th. Beisel, and R. Keller. Distributed Computing in a Heterogenous Computing Environment. In V. Alexandrov and J. Dongarra, editors, *Recent Advances in Parallel Virtual Machine and Message Passing Interface, Lecture Notes in Computer Science*, pages 180–188. Springer, 1998.
2. Marc Garbey. A Schwarz Alternating Procedure for Singular Perturbation Problems. *SIAM J. Sci. Comput.*, 17:1175–1201, 1996.
3. Marc Garbey and H.G. Kaper. Heterogeneous Domain Decomposition for Singularly Perturbed Elliptic Boundary Value Problems. *SIAM J. Numer. Anal.*, 34(4):1513–1544, 1997.
4. Marc Garbey and Damien Tromeur-Dervout. Operator Splitting and Domain Decomposition for Multicluster. In D. Keyes, A. Ecer, N. Satofuka, P. Fox, and J. Periaux, editors, *Proc. Int. Conf. Parallel CFD99*, pages 27–36, williamsburg, 1999. North-Holland. (Invited Lecture) ISBN 0-444-82851-6.
5. Marc Garbey and Damien Tromeur-Dervout. Aitken-Schwarz Method on Cartesian grids. In Marc Garbey, editor, *Proc. Int. Conf. on Domain Decomposition Methods DD13*. DDM org, 2001.
6. Marc Garbey and Damien Tromeur-Dervout. Two Level Domain Decomposition for Multicluster. In H. Kawarada T. Chan, T. Kako and O. Pironneau, editors, *Proc. Int. Conf. on Domain Decomposition Methods DD12*, pages 325–340. DDM org, 2001. (invited lecture), http://applmath.tg.chiba-u.ac.jp/dd12/proceedings/Garbey.ps.gz.
7. Yu. A. Kuznetsov and M. Matsokin. On Partial Solution of Systems of Linear Algebraic Equations. *Soviet J. Numer. Anal. Math. Modelling*, 4:453–468, 1989.
8. Michael Resch, Dirk Rantzau, and Robert Stoy. Metacomputing Experience in a Transatlantic Wide Area Application Testbed. *Future Generation Computer Systems*, (15)5-6:699–712, 1999.
9. Tuomo Rossi and Jari Toivanen. A Nonstandard Cyclic Reduction Method, its Variants and Stability. *SIAM J. Matrix Anal. Appl.*, 3:628–645, 1999.
10. Tuomo Rossi and Jari Toivanen. A Parallel Fast Direct Solver for Block Tridiagonal Systems with Separable Matrices of Arbitrary Dimension. *SIAM J. Sci. Comput.*, 5:1778–1796, 1999.
11. A. Sweet. A Parallel and Vector Variant of the Cyclic Reduction Algorithm. *SIAM J. Sci. Statist. Comput.*, 9:761–765, 1988.
12. P. S. Vassilevski. Fast Algorithm for Solving a Linear Algebraic Problem with Separable Variables. *C.R. Acad. Bulgare Sci.*, 37:305–308, 1984.

Parallel Computational Fluid Dynamics – Practice and Theory
P. Wilders, A. Ecer, J. Periaux, N. Satofuka and P. Fox (Editors)

DIRECT NUMERICAL SIMULATION OF JET NOISE

Bendiks Jan Boersma

J.M. Burgers Centre
Delft University of Technology
Mekelweg 2
2628 CD Delft
The Netherlands
email: b.j.boersma@wbmt.tudelft.nl

In this paper we will investigate the sound field of a round turbulent jet with a Mach number of 0.6 based on the jet centerline velocity and the ambient speed of sound. The flow field is obtained using Direct Numerical Simulation (DNS). The sound field is obtained by solving the Lighthill equation for the acoustic field. The simulation model is implemented on parallel computing platforms with help of the Message Passing Interface (MPI).

1. Introduction

A generic flow geometry of aeroacoustical sound production is a turbulent jet. Most people will be familiar with the sound of a jet engine of a commercial airliner. Stricter environmental measures around airports have put strong limitations on the sound that may produced by jets. Although significant sound reduction of these jet engines has been obtained over the last few decades, it is nevertheless required to reduce the sound of jet engines even more in view of the strong growth in air traffic foreseen in the future. The above mentioned jet engine is only one of the examples and other examples are aeroacoustical sound produced by high speed trains, wind noise around buildings, the sound comfort in cars but also ventilator noise in various household appliances. In this study we will focus on sound produced by turbulent jets because this flow is one of the benchmark flows for which a reasonable amount of experimental data is available.

Recently, with increasing computer power, it has become possible to calculate the acoustic field of simple flows using Direct Numerical Simulation (DNS), [1], [2]. Direct numerical simulations of high Mach number turbulent jets have been performed by [3]. In these simulations the sound is calculated with help of Kirchhoff surfaces. In low Mach number flows the acoustic amplitudes are very small and it is likely that acoustic equations like the one proposed by Lighthill [4] or Howe [5] will give more reliable results which are less contaminated by numerical errors. Furthermore, Kirchhoff methods can not predict sound emitted in the direction of the jet flow. While for low Mach number flows most of the sound is emitted in the forward direction.

In this paper we will describe a parallel computer model which solves the fully compressible Navier-Stokes equation with very accurate numerical methods. The Lighthill equation will be used to predict the far field sound of the jet.

2. Geometry and governing equations

In Figure 1 we show a sketch of the jet geometry. The jet is flowing form a circular hole into the ambient air. The velocity profile in the circular hole (jet orifice) is laminar. Downstream of the jet orifice the flow becomes gradually turbulent.

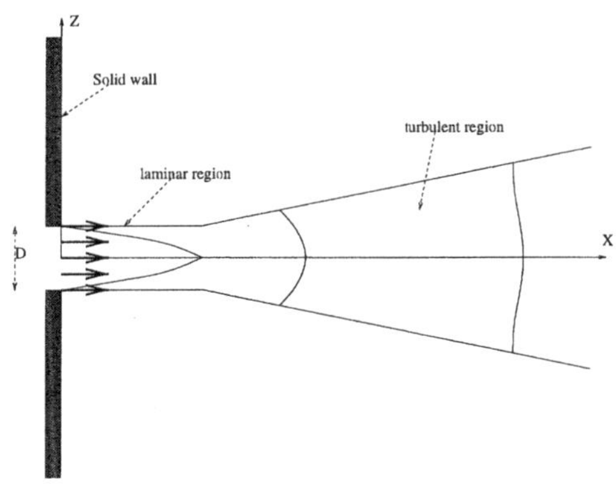

Figure 1. A sketch of the jet geometry. The transition from a laminar to a turbulent state occurs in general at a downstream location in between $7D$ and $13D$.

The jet flow is described by the well known compressible Navier-Stokes equations which can be found in various text books [6]. The equation for conservation of mass reads.

$$\frac{\partial \rho}{\partial t} + \frac{\partial \rho u_i}{\partial x_i} = 0 \tag{1}$$

where ρ is the fluids density and u_i the fluids velocity component in the ith coordinate direction. The equation which describes the conservation of momentum reads

$$\frac{\partial \rho u_i}{\partial t} + \frac{\partial \rho u_j u_i}{\partial x_j} = -\frac{\partial p}{\partial x_i} + \frac{\partial}{\partial x_i}\tau_{ij} \tag{2}$$

In which p is the pressure and τ_{ij} is the viscous stress given by:

$$\tau_{ij} = \mu S_{ij} = \mu \left(\frac{\partial u_i}{\partial x_j} + \frac{\partial u_j}{\partial x_i} - \frac{2}{3}\delta_{ij}\frac{\partial u_k}{\partial x_k} \right) \tag{3}$$

The dynamics viscosity μ is a weak function of the temperature in the gas. For the moment we will neglect this and assume that μ is constant.

For the energy equation in a compressible flow various formulations are possible. Here we choose for a formulation using the total energy, i.e. the sum of temperature and kinetic energy

$$E = \rho C_v T + \frac{1}{2}\rho u_i u_i \tag{4}$$

In which C_v is the specific heat at constant volume and T the temperature. The transport equation for the total energy E reads

$$\frac{\partial E}{\partial t} + \frac{\partial}{\partial x_i} u_i [E + p] = \frac{\partial}{\partial x_i} \kappa \left(\frac{\partial T}{\partial x_i} \right) + \frac{\partial}{\partial x_i} u_i S_{ij} \tag{5}$$

In which κ is the thermal diffusion coefficient, which is again a weak function of the fluids temperature. The formulation of the energy equation given above has the advantage that no source terms appear in the left hand side which would be the case in formulations using the temperature instead of the energy. The temperature T, the pressure p and the density ρ are related to each other by the equation of state

$$p = \rho R T \tag{6}$$

The important non-dimensional numbers for this flow are the Reynolds and Mach number. In this paper we will use the following definition for these numbers

$$Re = \frac{\rho_c U_c R_0}{\mu} \quad Ma = \frac{U_c}{c_\infty} \tag{7}$$

In which R_0 is the radius of the jet orifice, c_∞ the speed of sound and the subscript c denotes centerline quantities.

2.1. The acoustic field

The acoustic field of the jet can be calculated with help of acoustic analogons like the Lighthill equation:

$$\frac{\partial^2 \rho'}{\partial t^2} + c^2 \frac{\partial^2 \rho'}{\partial x_i^2} = \frac{\partial^2}{\partial x_i \partial x_j} T_{ij}. \tag{8}$$

In which ρ' is the acoustic density fluctuation of the gas, c the speed of sound and T_{ij} is the Lighthill stress tensor which is given by the following relation

$$T_{ij} = \rho u_i u_j + \mu \left(\frac{\partial u_i}{\partial x_j} + \frac{\partial u_j}{\partial x_i} - \frac{2}{3}\delta_{ij} \frac{\partial u_k}{\partial x_k} \right). \tag{9}$$

For turbulent flows the viscous term in the Lighthill stress tensor will be small and T_{ij} can be approximated by $T_{ij} \approx \rho u_i u_j$. Furthermore, if the Mach number is sufficiently small the density ρ can be replaced by the ambient value ρ_∞, resulting in the following equation for the acoustic density fluctuations

$$\frac{\partial^2 \rho'}{\partial t^2} + c^2 \frac{\partial^2 \rho'}{\partial x_i^2} = \rho_\infty \frac{\partial^2}{\partial x_i \partial x_j} u_i u_j. \tag{10}$$

3. Numerical method

In the previous section we have presented the governing equation for compressible flow. In this section we will describe how those equations are discretized.

A natural choice for the computation of a round jet would be to use a cylindrical coordinate system. In previous computational studies such systems have been used, [3], [7]. The problem when dealing with such a coordinate system is the treatment of the singularity at the centerline ($r = 0$) of the coordinate system. In the literature various methods are discussed, for a detailed overview we refer to [8]. None of these methods are able to retain a high order of numerical accuracy at the axis ($r = 0$) of the system. In physical space this axis will represent the jet centerline. An accurate simulation at the jet centerline is necessary because this is the area where most of the sound will be produced. In view of the problems mentioned above we have decided to use a Cartesian coordinate system for the complete flow domain.

The computational grid in the physical domain is non-uniform. Mapping functions $X_i = \eta_i(x_i)$, with $X_i = i\Delta X$ are used to map differential equation on a uniform grid in the computational domain, i.e.

$$\frac{\partial f}{\partial x} = \frac{\partial f}{\partial X}\frac{\partial X}{\partial x} \tag{11}$$

The mapping function $X_i = \eta_i(x_i)$ is chosen in such a way that $\partial X/\partial x$ can be integrated analytically to obtain the physical distribution of the gridpoints x_i. The derivative $\partial f/\partial X$ has been calculated with a 8th order compact finite difference scheme [9]:

$$\begin{aligned} \left.\frac{\partial f}{\partial X}\right|_i &= f_i' \\ \frac{3}{8}\left(f_{i-1}' + f_{i+1}'\right) + f_i' &= \frac{25}{32}\frac{f_{i+1} - f_{i-1}}{\Delta X} + \frac{3}{60}\frac{f_{i+2} - f_{i-2}}{\Delta X} - \frac{1}{480}\frac{f_{i+3} - f_{i-3}}{\Delta X} \end{aligned} \tag{12}$$

At the boundaries of the computational domain the accuracy of the compact scheme was reduced to third order, [9]. If we would have used a cylindrical system we would also have to reduce the order at the jet centerline to third order. Which on its turn would give an unreliable prediction of T_{ij}.

All the spatial derivatives in the continuity, momentum and energy equation are discretized with the 8th order approximation given above. The time integration has been performed with a standard 4th order Runga-Kutta method. The time step was fixed and the corresponding CFL number ($u_i\Delta t/\Delta x_i$)was approximately 1.0.

The Navier-Stokes equations are solved close to the jet orifice where there is a significant flow. Far away from the jet orifice there is no fluid motion and only the acoustic field, i.e equation (10) has to be solved. The source term in (10) is calculated on the Navier-Stokes grid. This is the only coupling between the two simulations, i.e. the acoustic waves do not influence the flow. This is a valid assumption if the Mach number of the flow is small.

3.1. Parallel implementation

The numerical method outlined above has been implemented with help of FORTRAN 77 and the Message Passing Interface (MPI). The computational domain with $N_x \times N_y \times N_z$ is in the x-direction distributed over the processors. If the number of processors is denoted by N_{proc} the number of grid points on each CPU is equal to $N_x/N_{proc} \times N_y \times N_z$,

Table 1
The wall-clock time of one timestep on a computational grid with 64^3 (Navier-Stokes) and 128^3 (Wave equation). The CRAY-T3E has 80 DEC-Alpha 300 Mhz processors (64 bit), the Beowulf cluster has 12 AMD-Athlon 900Mhz processors (32 bit), and the SGI-ORIGIN 3800 has 1024 R14000 CPU's (64 bit, 500Mhz). All the calculations are performed in FORTRAN using *real*8* as precision.

N_{proc}	CRAY-T3E	Beowulf Cluster	SGI-ORIGIN 3800
1	–	62.0 sec	46.7 sec
2	–	55.1 sec	20.5 sec
4	54.0 sec	41.5 sec	8.8 sec
8	29.0 sec	25.2 sec	4.2 sec
16	18.3 sec	–	2.8 sec
32	17.0 sec	–	1.9 sec

i.e. N_x/N_{proc} must be integer. On this data distribution all the derivatives in y and z direction in the governing equations are calculated. Once the derivatives are calculated the data is redistributed to a distribution $N_x \times N_y \times N_z/N_{proc}$, i.e. N_z/N_{proc} should be integer. On this distribution all the x-derivatives in the governing equations are calculated. The results obtained on the latter distribution are than transferred to the original distribution, and a Runga-Kutta (sub-)step is performed. The data is redistributed with help of the MPI routine MPI_ALLTOALL. Due to the computational intensity of the full compressible Navier-Stokes equations the ratio of computational time and communication time is reasonable large.

In Table 1 typical CPU times are shown for a computation on a grid of $64^3 + 128^3$ for various computer systems are shown. The scalability of the code is reasonable on the CRAY-T3E (due to limited amount of memory the minimum numbers of CPU's that could be used was 4). The Beowulf cluster does not scale well. Superior scalability is observed on the SGI-ORIGIN 3800. This is also the platform which is used to generate the results shown in the following sections.

4. Results

In this section we will present results obtained from the Direct Numerical Simulation of the jet and the sound field. The Reynolds and Mach numbers where equal $2.5 \cdot 10^3$ and 0.6 respectively. Two different computational domains are used, one with a small spatial size for the Navier-Stokes equations and one with a larger spatial size for the wave equations. The Navier-Stokes domain consisted of $160 \times 144 \times 144$ the x, y and z-direction respectively (x is streamwize direction). The wave domain consisted of $320 \times 272 \times 272$ point. For the jet-inflow profile a simple hyperbolic tangent profile of the following form is taken

$$U(r) = Ma\left(\frac{1}{2} - \frac{1}{2}\tanh[20(r - R_0)]\right) \tag{13}$$

In which R_0 is the radius of the jet and Ma the Mach number. The calculations have been continued until they reached a statistically steady state. After the calculations have reached this state they are continued for another 200 acoustic timescales R_0/c to obtain the statistics. In Figure 2 we show an instantaneous plot of the density field ρ. The figures show that the flow is laminar close to the jet nozzle and starts to become turbulent in the region $10 < x/R_0 < 15$ and becomes gradually fully turbulent farther downstream of the jet nozzle. In Figure 3 we show the mean velocity profile and mean axial flux along the

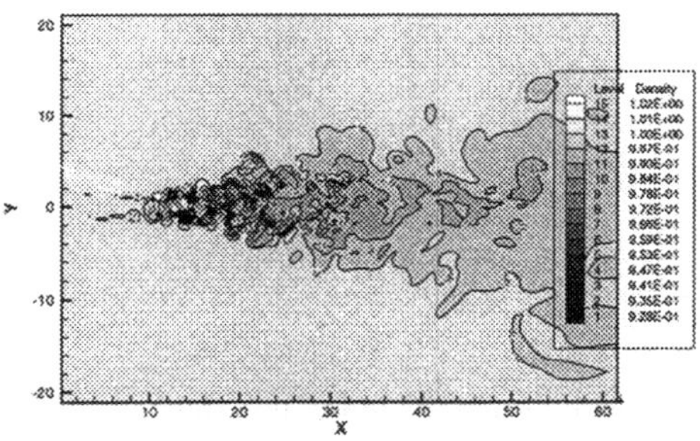

Figure 2. An instantaneous plot of the density field in the jet

jet centerline. In the region close to the jet orifice the centerline velocity is constant and then suddenly drops. The point were the centerline velocity suddenly drops is the point were most of the sound will be produced. The small difference between the profiles for u_x and ρu_x indicates that the compressibility of the flow is low.

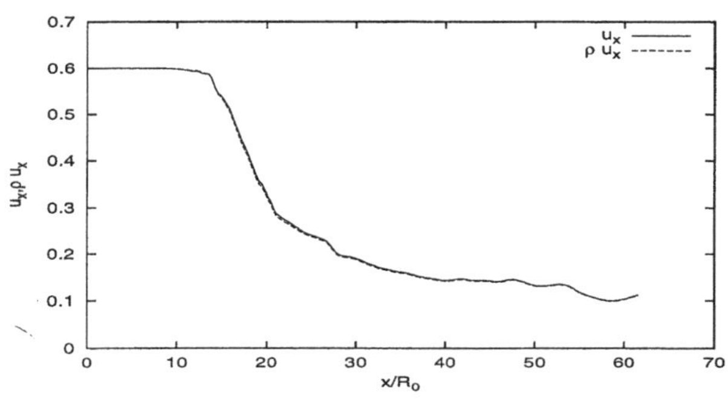

Figure 3. The mean axial velocity and axial flux at the jet centerline as a function of the downstream coordinate.

In Figure 4 an instantaneous plot of the right hand side of equation (10) is shown. The source term is large in the region $10 < x/R_0 < 30$, i.e the region in which the flow goes

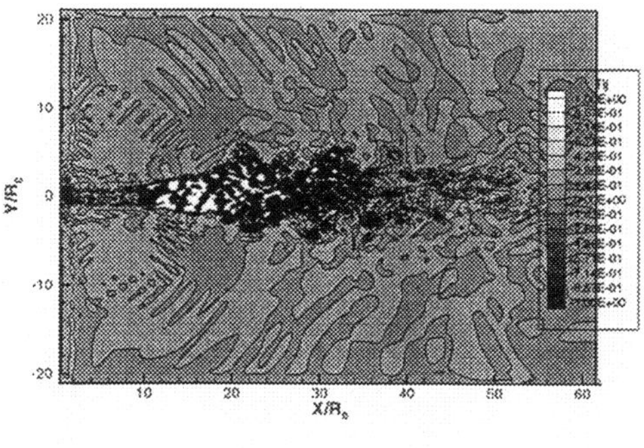

Figure 4. The acoustic source term obtained from the Navier-Stokes solution (right hand side of equation 9).

from a laminar to a turbulent state.

In Figure 5 the acoustic fields obtained with equations (10) is shown (the acoustic field is visualized with help of the dillatation $q = \partial\rho'/\partial t$). This sound field is very similar to the sound field observed in experiments. For instance, most of the sound is emitted under an angle of approximately 30° which is also found in the experiments by [10] and [11].

5. Conclusion

In this paper we have described a parallel computer model which is able to simulate compressible jet flows with a high numerical accuracy. Coupled with the flow a wave equation for the acoustic field is solved using the same numerical method. It has been shown that the code scales well on modern parallel computers like the ORIGIN-3800. The scalability on Beowulf clusters and on a CRAY-T3E is not very good which is caused by the rather slow communication between the nodes.

Acknowledgments

The author gratefully acknowledges the financial support form the Royal Dutch Academy of Science and Arts (KNAW). Computer time on the ORIGIN-3800 has been financed by the Dutch Supercomputing Foundation (NCF).

REFERENCES

1. Colonius, T., Lele, S.K., & Moin, P., 1997, Sound generation in a mixing layer, *J. Fluid Mech*, **330**, 375-409.
2. Mitchell, B.E., Lele, S.K., Moin, P., 1999, Direct computation of the sound generated by vortex pairing in an axissymmetric jet, *J. Fluid Mech*, **383**, 113-142.

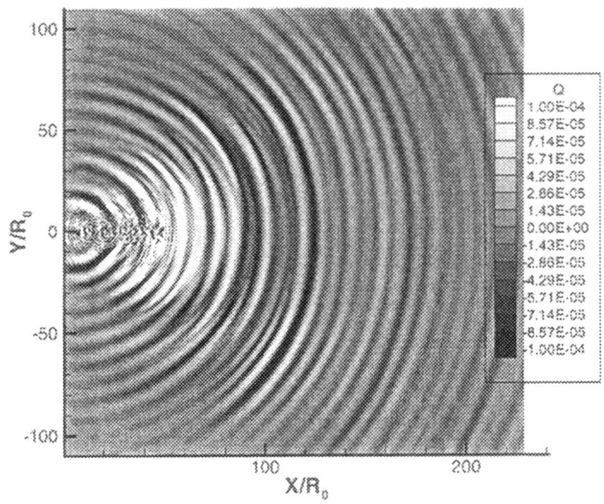

Figure 5. The acoustic field obtained with help of equation (10). The number of gridpoints for the acoustic field is equal to 320 × 272 × 272.

3. Freund, J.B., 2001, Noise sources in a low-Reynolds number turbulent jet flow at Mach 0.9, *J. Fluid Mech,*,**438**, 277-306.
4. M.J. Lighthill, On sound generated aerodynamically, *Proc. R. Soc. of London Ser. A,* 211, 564-587, 1952.
5. Howe, M.S., 1975, Contributions to the theory of aerodynamics sound, with application to excess jet noise and the theory of the flute, *J. Fluid Mech.*,**71**, 625-673.
6. Batchelor, G.K., (1967), *An introduction to fluid mechanics*, Cambridge University Press.
7. B.J. Boersma, G. Brethouwer and F.T.M. Nieuwstadt, A numerical investigation on the effect of the inflow conditions on the the self-similar region of a round jet, *Physics of Fluids* 10, 899-909, 1998.
8. Mohensi, K. & Colonius, T., 2000, Numerical Treatment of Polar coordinate singularities, *J. Comp. Phys.*, **157**, 787-795.
9. S.K. Lele, Compact finite difference schemes with spectral-like resolution, *J. Comp. Phys.* 103, 16-42, 1992.
10. P.A. Lush., 1971, Measurement of subsonic jet noise and comparison with theory, *J. Fluid Mech.*, **46**, 477-500.
11. E. Mollo-Christensen, 1967, Jet noise and shear flow instabilities seen from an experimenter's viewpoint, *J. Appl. Mech.*, **34**, 1-7.

Parallel Computational Fluid Dynamics – Practice and Theory
P. Wilders, A. Ecer, J. Periaux, N. Satofuka and P. Fox (Editors)

Migrating from a Parallel Single Block to a Parallel Multiblock Flow Solver

Thomas P. Bönisch[a], Roland Rühle[a]

[a] High Performance Computing Center Stuttgart
Allmandring 30, D-70550 Stuttgart, Germany

This paper describes the development of a multiblock structure to extend a simulation code simulating reentry flows on structured c-meshes. The new data structure, the load balancing approach with a block cutting algorithm and the handling of block sides at physical and inner boundaries are presented. Goal is the efficient calculation of reentry flows using multiblock meshes on current parallel supercomputer platforms.

1 INTRODUCTION

The flow simulation program URANUS [4] has been developed to calculate nonequilibrium flows around space vehicles reentering the earth's atmosphere. This program which is using single block c-meshes was parallelized within an earlier effort[5]. However, single block c-meshes contain a singularity in the mesh which is complicated to handle and which also limits the convergence speed. Moreover, there are topologies which cannot be meshed with one single block mesh, e.g. an X-38 with body flap.

One way to solve these problems is to use multiblock meshes, which consist out of structured blocks. These blocks can be combined in an unstructured way. Normally, we can assume continuity in the grid-line positions across the block boundaries. On one hand, such multiblock meshes need much more effort in generating than unstructured meshes do, but on the other hand the computation on the structured mesh of a block is much more efficient.

2 MULTIBLOCK APPROACH

Complementary to many other cases where flow simulation codes using multiblock meshes in the serial version were parallelized [1,2,3], we used a different approach. Since we have a quasi block structure within the mentioned parallel URANUS program already available, the idea is, to upgrade this flow simulation program in a way, that it is able to deal with multiblock meshes. For that extension of our parallel flow code, we have to consider the following properties of a multiblock mesh and its blocks:

- Each block within a multiblock mesh may have its own local coordinate system which is independent from the coordinate systems of its neighbours. The reason for different local coordinate systems is the irregular layout of the blocks in a multiblock mesh.
- At each of its six block sides a mesh block may have multiple neighbours and/or physical boundaries.
- Each block of a multiblock mesh may have a different size measured in number of mesh cells per block.
- There are irregular points, where more or less than eight blocks in 3D or three or five blocks instead of four in 2D respectively are connected to each other.

For the structure and the functionality of our parallel multiblock scheme this has several consequences.

2.1 Local coordinate system

The information of each block is stored according to its own local coordinate system. The halo information of the neighbours is stored on the local block according to its local coordinate system for efficiency reasons. This information in the overlapping regions cannot be used directly from the neighbouring block as a neighbour may have a different local coordinate system. The possibly necessary conversion in the storage sequence is automatically done during the data transfer. This means, the conversion is hidden from the flow calculation.

2.2 Physical boundaries

In the C-meshes used before, the occurrence of a physical boundary was bound to a specific index. At the lower end of the second dimension (with index j) for example, there was the boundary to the body's surface, on the upper end of the same index j was the inflow boundary. In a block of a multiblock mesh a physical boundary may occur at each block side or even only at a part of a block side. To calculate the values at physical boundaries efficiently, a specific data structure was created for each physical boundary type. There, all the data of one physical boundary type at a block are specified. It contains the subtype and the exact positions of these physical boundaries on the block. Using this data structure, there is no branching necessary for each of the six block sides where a physical boundary can reside. Additionally, all physical boundaries of one type at a block can be handled efficiently in a loop. Therefore, there is no code doubling necessary as it is, if you do branching. This prevents cut and paste errors which always happen during code doubling. Furthermore, there is only one code segment for each boundary type to be maintained and possibly updated.

2.3 Neighbour handling

The arrangement of the blocks within a multiblock mesh is unstructured and each block can have multiple neighbours on each of its six block sides. Therefore, the block number of the neighbours cannot be calculated out of the local block number and the block side as it was possible in the parallel single block code. Additionally, the number of the neighbour block and the logical number of the processor within the parallel execution environment where the neighbour resides is not necessarily equal.
Consequently, there is a new data structure implemented to store all the information about the relationship of the block and its particular neighbours. This information enfold the block number and the block side the local block is connected to, the neighbour blocks processor as far as the orientation of the neighbour block and the part of the local blocks side which is adjoined to the neighbour.
Furthermore, the programs communication structure has had to be changed, too. The different communication routines now have to deal with several possible communications on each block side. Due to performance reasons, the communication subroutines do polling on all block sides as soon as messages are expected. As soon as a message has been received, this message is processed. Then, the program checks for further messages as long as there is one to be received.

3 LOAD BALANCING

The possible difference in size of the mesh blocks in a multiblock mesh has a significant influence on the load balance when running a multiblock code in parallel. Putting each block on its own processor may lead to a substantial load imbalance and we are not longer free in choosing the number of processors to be attached to a simulation run. When using more processors as originally available blocks or when there is a large difference in size between the blocks, we have to split blocks which are too large to be efficiently calculated on one processor. For this, we implemented a cutting algorithm which cuts these blocks into as many equal sized pieces as necessary to make each of them fit onto one processor. In order to be able to cut these too large blocks in all of the three dimensions, only numbers of pieces which are multiples of 2 and 3 are allowed. Because, larger prime numbers as divider would lead to misshapen block sizes and possibly to complex interface conditions at boundaries when e.g. five new blocks have to be connected to seven new blocks on the neighbours side. The program is able to handle these complex neighbour connections, but more easier dependencies will result in less messages to be send.

3.1 Multiple blocks per processor

But we may not only have blocks which are too large for one processor, there may also be blocks which are too small to fully utilize the capacity of one processor.
In order to gain all the cycles of these processors, the new program is able to handle more than one block on each processor. For this, each block got its own data structure where all its values are stored, e.g. neighbours, physical boundaries, the local jacobian matrix, message handles, ... The data structure of the blocks with all their information are then organized in a linked list. The blocks are calculated one after another within each program part by just running through this linked list. There is no waste in memory or additional effort in memory control using this technology. Furthermore, the internal data structure of one block is flexible and easily extendable to meet future needs.
The communication between blocks on the same processor is also done using the implemented communication routines and MPI. Actually, a block does not know, whether its neighbour is located on the same processor as itself. It just sends a message to the processor with the number given in its local data structure. Thus, a processor may send a message to itself which is automatically handled by MPI. This does not lead to a dead lock as all point to point communications are using nonblocking communication routines.
The communication subroutines have also been adapted so that an incoming message is delivered to the block where the message belongs to.
For the calculation of global values where each block has a contribution, the results of one processor's blocks are calculated locally and then exchanged between the processors using the collective communication patterns of the programming model.
The number of blocks possible on each processor is only limited by the processors memory.

3.2 Load balancing approach

To obtain a good load balance a load balancing tool is essential to distribute the resulting blocks from the former steps to the available processors in an appropriate way. For the distribution itself, we added routines to transfer blocks between the processors. As load balancing tool, we are currently using parallel Jostle [6] even for the initial load balancing phase as the block information is distributed from the very beginning of the program. Accordingly, we will not loose to much performance due to the load balancing itself.
The load balancing tool works on graphs not on meshes. Therefore, we had to define a mapping between the multiblock mesh and a graph. The blocks are represented by the nodes of the graph, neighbour dependencies by the edges between the graph's nodes. The block size is represented by a weight given to the graph's nodes. With this information, the load balancing tool is able to calculate a distribution of the blocks which is near to the optimal block distribution. According to the suggestion of the load balancing tool, the blocks are redistributed.
Figure 1 to 3 show the load balancing procedure for a mesh with 68712 cells in 6 blocks. The largest block of the original mesh shown in figure 1 contains 32256 mesh cells, the smallest 1176 mesh cells. Figure 2 shows the mesh produced by the automatic block cut algorithm,

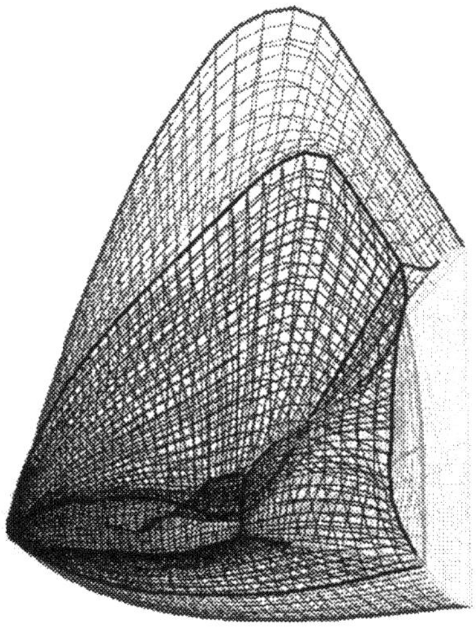

Figure 1. Original Multiblock Mesh

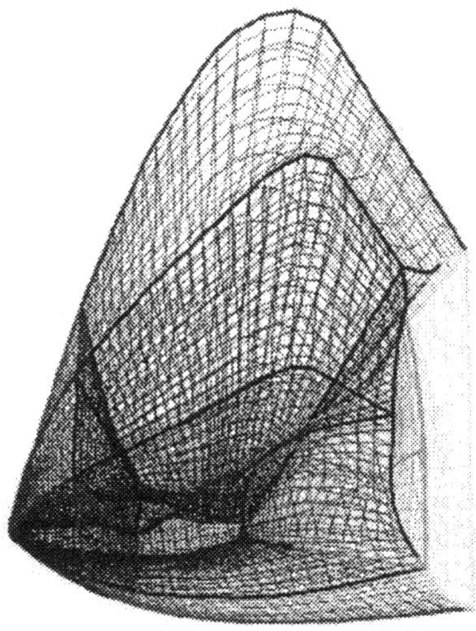

Figure 2. Mesh after blocks have been cut

which cuts the blocks which are too large for one processor. It is assumed that the calculation will be done on 9 processors. The resulting mesh now has 11 blocks. The largest block (black)

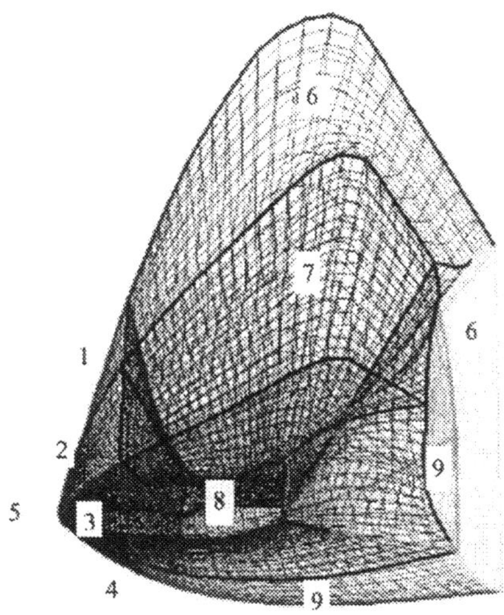

Figure 3. Mesh and block distribution after load balancer run

was cut into four pieces, the two blocks above and below the black (both dark grey) into two pieces each. The not visible block at the nose was not cut. In figure 3 the obtained distribution of the 11 blocks to the 9 processors is shown. Same numbers in the blocks mean same processor. The processor with the highest load has to calculate 8064 mesh cells, the processor with the lowest load has to calculate 6048 mesh cells. The load imbalance in this case is 18 %. The load imbalance could be less, if we would do an additional run cutting a small block into pieces, in order to fill the small gaps on the less loaded processors. But this is currently not implemented.

Due to the modular structure of the program, the replacement of the load balancing tool is easily possible.

4 RESULTS

With these adoptions we are now able to calculate reentry problems efficiently on parallel computers using multiblock meshes. Due to the usage of Fortran90 and MPI, the introduced multiblock flow simulation code is portable. It was tested on several parallel platforms including Cray T3E, Hitachi SR8000, IBM SP3, NEC SX-5, Compaq cluster and IA-64 architecture.

4.1 Speedup

For the Cray T3E a speedup measurement is shown in Figure 4. Here we used a mesh with 5 blocks and 192 000 mesh cells.

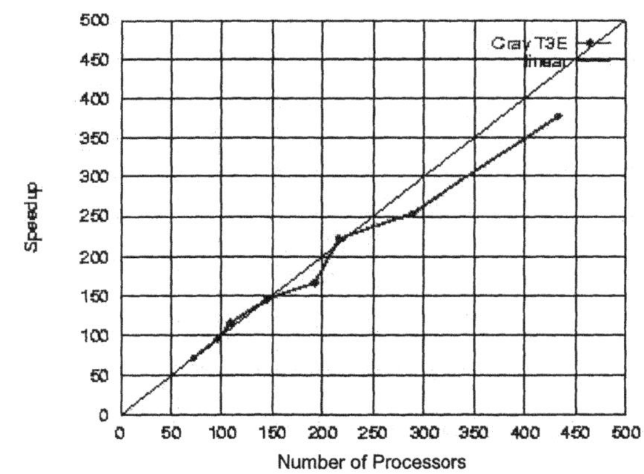

Figure 4. Speedup on Cray T3E for a 5 block 192 000 cell mesh

The processor numbers were chosen in a way to get perfect load balance. Nevertheless, there are slight fluctuations and superlinear speedup visible. The reason for these fluctuations is that for different processor numbers the blocks are cut in a different way. And the different proportion of mesh cells within the three dimensions of a block lead to a different more efficient or more inefficient cache and streams buffer usage. As one can see, the block size for 108 processors for example gives a better cache usage than the block size when running on 72 or 96 processors. The resulting speedup of 380 for 432 processors gives us an efficiency of 88% related to 72 processors which is the smallest number of CPU's where the used case is able to run.

4.2 Calculated result

Figure 5 shows the velocity calculated by solving the Euler equations on a multiblock mesh for X-38, the prototype of the planned Crew Rescue Vehicle (CRV) for the International Spacestation (ISS). The angle of attack is 40°, the velocity of the incoming flow is mach 6.

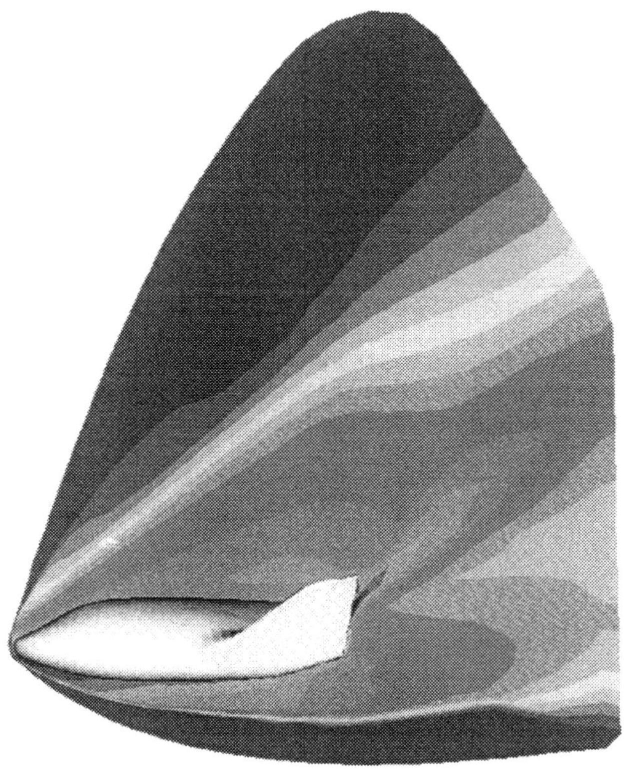

Figure 5. Calculated Euler solution on a multiblock mesh for X-38

5 ACKNOWLEDGEMENTS

This work was particularly supported by the Deutsche Forschungsgemeinschaft (DFG) within SFB259.

REFERENCES

[1] F.S. Lien, L. Chen and M.A. Leschziner, 'A Multiblock Implementation of a Non-Orthogonal, Collocated Finite Volume Algorithm for Complex Turbulent Flows', International Journal for Numerical Methods iin Fluids, Vol. 23, pp. 567-588, 1996

[2] M.A. Leschziner, F.S. Lien, 'Computation of Physically Complex Turbulent Flows on Parallel Computers with a Multiblock Algorithm' in Emerson et. al. (Eds.) 'Parallel Computational Fluid Dynamics' Recent Developments and Advances Using Parallel Computers, North-Holland, 1998, pp. 3-14

[3] N. Kroll, B. Eisfeld, H.M. Bleecke, 'FLOWer' in A. Schüller 'Portable Parallelization of Industrial Aerodynamic Applications (POPINDA)', Notes on Numerical Fluid Mechanics, Volume 71, Vieweg, 1999

[4] H.-H. Frühauf, M. Fertig, F. Olawsky, and Thomas Bönisch 'Upwind Relaxation Algorithm for Re-entry Nonequilibrium Flows', in E. Krause, W. Jäger 'High Performance Computing in Science and Engineering '99 ', Springer, Berlin 2000, pp. 365-378.

[5] T. Bönisch, R. Rühle, 'Portable Parallelization of a 3-D Flow-Solver' in Emerson et. al. (Eds.) 'Parallel Computational Fluid Dynamics' Recent Developments and Advances Using Parallel Computers, North-Holland, 1998, pp. 457-464

[6] C. Walshaw, M. Cross and M. Everett, 'Parallel Dynamic Graph Partitioning for Adaptive Unstructured Meshes', Journal of Parallel and Distributed Computing, 47(2), 1997, pp. 102-108

Parallel Computational Fluid Dynamics – Practice and Theory
P. Wilders, A. Ecer, J. Periaux, N. Satofuka and P. Fox (Editors)

Parallel Multidimensional Residual Distribution Solver for Turbulent Flow Simulations

D. Caraeni *, M.Caraeni † and L.Fuchs‡.
Division of Fluid Mechanics, Lund Institute of Technology, Sweden.

A new compact third-order Multidimensional Residual Distribution scheme for the solution of the unsteady Navier-Stokes equations on unstructured grids is proposed. This is a compact cell-based algorithm which uses a Finite-Element reconstruction over the cell to achieve its high-order of accuracy. The new compact high-order algorithm has an excellent parallel scalability, which makes it well suited for large scale computations on parallel-computers. Some results of Large Eddy Simulation of the fully developed turbulent channel flow are presented.

1. Introduction

Computational Fluid Dynamics requires new high-order algorithms which should combine the correct representation of the multidimensional physics with a good parallel scalability, for simulation of turbulent flows in complex geometries.

Residual Distribution (RD) schemes have been first proposed by Roe [9]. These schemes combine ideas both from Finite-Volumes (FVM) and Finite-Elements (FEM) methods. Basically, these algorithms can be written as loops through the cells in which the cell residual is computed and distributed to the cell-vertices accordingly with a multidimensional distribution scheme, followed by the nodal values update [11][6]. The approach has become popular in recent years due to some advantages it has as compared to classical finite-volumes methods: it is capable of better capturing the real multidimensional-flow physics [10] and more accurate [12]. Second order accuracy can be achieved by using a very compact stencil [6]. These schemes have been extended in [2] for second order accurate unsteady Navier-Stokes computations and applied for Large Eddy Simulation (LES) of turbulent compressible flows. The compactness of these schemes made them also highly suitable for parallelization [1]. Detailed informations about the Residual Distribution schemes approach can be found in [6][10][11][12].

In the present work we propose an extension of the Residual Distribution schemes from second- to third-order spatial accuracy, for unsteady Navier-Stokes computations. This is done while maintaining the same compact stencil as in the second order case, e.g. cell based computations. The idea is to use a high-order FEM procedure to compute the cell residual [4] together with a Linearity Preserving (LP) [11][6] distribution scheme. The proposed discretization requires a moderate increase in computational time (20%) and computer memory storage (45%) as compared with the second order algorithm. The accuracy of the new discretization has been

*Ph.D., LTH, Sweden dc@mail.vok.lth.se
†Ph.D. Student LTH, Sweden, mirela@mail.vok.lth.se
‡Professor, LTH, Sweden lf@mail.vok.lth.se

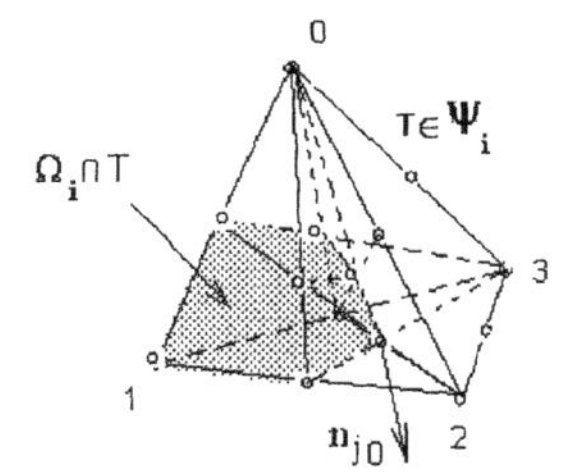

Figure 1. A sketch of the volumes Ω_i and Ψ_i around node i.

confirmed numerically [5]. Some details about the new discretization scheme are presented here. The third-order parallel algorithm shows the same super-linear parallel speed-up as the second order algorithm. Results of parallel LES simulations using the new compact third-order scheme are presented for the fully developed turbulent channel flow at Reynolds number of 5400.

2. Governing equations and solution algorithm

The Navier-Stokes equations, in tensorial conservative form, for a Cartesian coordinate system (x_1, x_2, x_3) can be written as:

$$\begin{cases} \rho_{,t} + (\rho u_j)_{,j} = 0 \\ (\rho u_i)_{,t} + (\rho u_i u_j)_{,j} = -p_{,i} + \sigma_{ij,j} \\ (\rho e)_{,t} + (\rho u_j h)_{,j} = (\sigma_{ij} u_i)_{,j} + (\lambda T_{,j})_{,j} \end{cases} \tag{1}$$

where σ_{ij}; $\sigma_{ij} = \mu\{(u_{i,j} + u_{j,i}) - \frac{2}{3}\delta_{ij}.u_{k,k}\}$, is the molecular stress tensor, δ_{ij} is the Kronecker delta function, μ is the dynamic molecular viscosity, λ is the conductive heat diffusivity coefficient, c_v and c_p are the specific heats at constant volume and constant pressure, respectively, $e = c_v T + \frac{u_i^2}{2}$ is the total energy per unit mass (specific energy), $h = c_p T + \frac{u_i^2}{2}$ is the total specific enthalpy and T is the temperature. The equation of state, $p = \rho R_{gas} T$, is used to close the system of equations.

To accurately simulate time dependent viscous flow situations, a Jameson-type dual time steps approach has been proposed in [2]. This procedure requires to perform subiterations in pseudo-time, for each real-time step, until a convergence is achieved. Denote by $U = (\rho, \rho u_i, \rho e)^T$, $i = [1..3]$, the vector of conservative variables in a Cartesian system of coordinates (x_1, x_2, x_3). The system of equations that has to be solved, when using the dual-time steps approach, can be written in a compact form as:

$$U_{,\tau} = -F^C_{j,j} + F^V_{j,j} - U_{,t} \tag{2}$$

Here τ and t are the pseudo- and the real-time, respectively. The convective flux vector $F_{j,j}^C$ and the diffusive flux vector $F_{j,j}^V$ have the following expressions:

$$F_j^C = \left\{ \begin{array}{c} \rho u_j \\ \rho u_k u_j + p\delta_{kj} \\ \rho u_j h \end{array} \right\} \quad F_j^V = \left\{ \begin{array}{c} 0 \\ \sigma_{kj} \\ u_k \sigma_{kj} + \lambda T_{,j} \end{array} \right\} \tag{3}$$

Denote by Ω_i the dual volume and by Ψ_i the reunion of all tetrahedral cells around node i, as shown in Figure 1. Denote also by T one of the tetrahedral cells which contain the node i ($\equiv$1 in this figure) and by n_{ji} the exterior normal of the face opposite to node i, with $j, k = [1..3]$. The shaded volume in Figure 1 represents the intersection between the dual-volume Ω_i and the tetrahedral cell T. Integrating the system of equations (2) over the control volume Ω one obtain the following integral form:

$$\iiint\limits_{\Omega} U_{,\tau}.dv = \iiint\limits_{\Omega} [(-F_j^C + F_j^V)_{,j} - U_{,t}].dv \tag{4}$$

Define $\Phi_T^C = \iiint\limits_T F_{j,j}^C.dv$, $\Phi_T^V = \iiint\limits_T F_{j,j}^V.dv$; $\Phi_T^{Uns} = \iiint\limits_T U_{,t}.dv$, the convective, diffusive and unsteady tetrahedral cell-residuals, respectively.

2.1. Pseudo-time term discretization

A first order discretization, both in space (using "mass-lumping") and in time, has been used in equation (4) for the pseudo-time derivative volume integral. $\iiint\limits_{\Omega_i} U_{,\tau}.dv = V_{\Omega_i}\left[\frac{U^{n+1,k+1} - U^{n+1,k}}{\Delta\tau}\right]$. This low-order accurate discretization is justified by the dumping properties required by the pseudo-time marching algorithm. V_{Ω_i} represents the volume of the dual volume Ω_i. The superscript n denotes the real time step, i.e. the real-time $t = n * \Delta t$, while k denotes the pseudo-time in the marching algorithm. The pseudo-time step $\Delta\tau$ is chosen to be the maximum local time step allowed by the stability requirements.

2.2. Update scheme

The update scheme we propose for unsteady Navier-Stokes computations, while using the dual-time step approach, corresponds to a "full-upwind" distribution scheme. Thus the convective, diffusive and the unsteady cell-residuals are computed with the required level of accuracy (third-order accuracy for the new proposed discretization) and an upwind Linearity Preserving distribution scheme (e.g. the distribution matrices remain bounded as the cell-residual goes to zero) is used to distribute these residuals towards the nodes. The update scheme can be written as:

$$\Delta U_i^{n+1,k+1} = U_i^{n+1,k+1} - U_i^{n+1,k} = -\frac{\Delta\tau}{V_{\Omega_i}} \sum_{T, i\in T} [\mathbf{B}_i^T(\Phi_T^C - \Phi_T^V + \Phi_T^{Uns})]^{n+1,k} \tag{5}$$

where $\mathbf{B}_i^T$ is the distribution matrix for node i. For conservation, the distribution matrices have to satisfy the relation $\sum\limits_{i\in T} \mathbf{B}_i^T = \mathbf{I}$ where the summation is taken over all vertices of the tetrahedral cell T. The properties of this numerical scheme depend on the definition of the distribution matrices $\mathbf{B}_i^T$ and on the computational accuracy of the cell-residuals. The cell-residuals and the distribution coefficients $\mathbf{B}_i^T$ are computed using the values $U^{n+1,k}$. The value

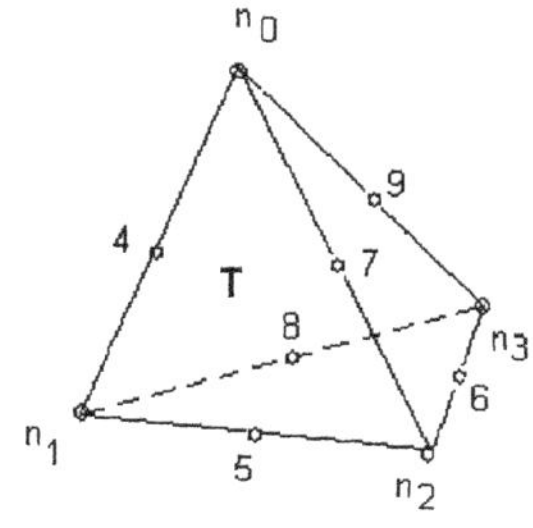

Figure 2. The definition of the nodes in a high order tetrahedral cell.

$U^{n+1,k}$ represents the approximation of the conservative variable vector U at the real-time step $(n+1)$, at the pseudo-time iteration k. Thus, this is an implicit scheme in real time which uses explicit pseudo-time marching iterations. A converged solution should be obtained at every new real-time step $(n+1)$. Full-Approximation-Storage (FAS) multi-grid iterations can be employed to accelerate the convergence. Details about the multigrid technique used can be found in [3]. Point implicit pseudo-time marching iterations can replace the explicit ones, to improve the smoothing properties of the update scheme. The Low Diffusion A (LDA) distribution scheme [11] has been used throughout the present work. The scheme described by equation (5) is a LP scheme. Numerical experiments showed that 'by using the update scheme (5) together with a LP distribution scheme, the accuracy of the numerical solution is determined by the accuracy of the cell-residual computation' [3][5]. Thus, by using a third-order FEM computation of the cell-residuals we obtained a third-order accurate scheme for unsteady viscous simulations.

2.3. Third-order compact convective term discretization

Computing the convective cell-residual Φ_T^C with higher-order of accuracy can be done if we assume that the field variables have a higher-order polynomial variation over the cell. In the classical RD discretization it is assumed that the parameter-variable $\mathbf{Z}$ has a linear variation over the cell. To obtain third-order of accuracy we will consider that $\mathbf{Z}$ has a quadratic variation over the cell. It is possible to do so if, before computing the cell-residual Φ_T^C, one computes and stores at the cell-vertices level the gradients of the $\mathbf{Z}$-variable, i.e. $\mathbf{Z}_{,j}$. Using the conservative variable U (or the parameter variable $\mathbf{Z}$) and the gradient $\mathbf{Z}_{,j}$ stored at the cell-vertices level, we can obtain a higher order reconstruction for the $\mathbf{Z}$-variable over the cell. Thus, the values of the $\mathbf{Z}$-variable in some virtual nodes, i.e. the nodes 4 to 9 in Figure 2, at the middle of the edges, are then defined using a formula, see[5], derived using a Least Square fitting technique. Using the values of the $\mathbf{Z}$-variable at the cell-vertices and in the virtual nodes (with indices between 4 and 9) one can compute the convective residual Φ_T^C with higher order accuracy by using a FEM approach. First, one use the Gauss theorem to transform the volume integral into a surface integral:

$$\Phi_T^C = \iiint\limits_T F_{j,j}^C . dv = \iint\limits_{\partial T} F_j^C . \overrightarrow{dS} \tag{6}$$

We observe that the convective-flux vector can be easily written in terms of the **Z**-variable as a quadratic function of **Z**. The surface integral in (6) can be decomposed (on tetrahedral cells):

$$\Phi_T^C = \iint\limits_{\partial T} F_j^C . \overrightarrow{dS} = \sum_{i=1}^{4} \left\{ \iint\limits_{face_i} \left(F_j^C \frac{n_{ji}}{|n_{ji}|} \right) d\sigma \right\} \tag{7}$$

Denote by $\overline{n}_{ji} = \frac{n_{ji}}{|n_{ji}|}$ the unit vector normal to the face i and $A_{face_i} = \iint\limits_{face_i} d\sigma$ the area of the face i. Now we compute the surface integrals on the tetrahedron faces, i.e. $\iint\limits_{face_i} \left(F_j^C \overline{n}_{ji} \right) d\sigma$, as:

$$\iint\limits_{face_i} \left(F_j^C \overline{n}_{ji} \right) d\sigma = \left\{ \begin{array}{c} \overline{n}_{ji} \iint\limits_{face_i} (Z_0 Z_j) d\sigma \\ \overline{n}_{ji} \iint\limits_{face_i} (Z_k Z_j + p\delta_{kj}) d\sigma \\ \overline{n}_{ji} \iint\limits_{face_i} (Z_4 Z_j) d\sigma \end{array} \right\} \tag{8}$$

with k=[1..3]. Finally, one has to compute the integrals $\mathbf{I}_{Z_k Z_j}^{face_i} = \iint\limits_{face_i} (Z_k Z_j) d\sigma$. By using the quadratic FEM shape functions on the face i, denoted here by $Q^{(l)}$ with $l = [1..6]$ (6 is the total number of nodes on a cell-face), the integral $\mathbf{I}_{Z_k Z_j}^{face_i}$ can be expressed as:

$$\mathbf{I}_{Z_k Z_j}^{face_i} = \iint\limits_{face_i} (Z_k Z_j) d\sigma = \sum_{i_0, i_1 = 1}^{6} \left\{ Z_k^{(i_0)} Z_j^{(i_1)} A_{face_i} \left[I_{QQ} \right]_{\alpha,\beta} \right\} \tag{9}$$

where the surface integral has been expressed in a simpler form. The matrix $[I_{QQ}]_{6,6}$ has been precomputed [4]. The same FEM approach, which has been used to compute the convective cell residual Φ_T^C with third-order accuracy, while assuming a quadratic variation of the parameter variable **Z** over the cell, can be applied in a similar way for the computation of the viscous and the unsteady cell residuals. Please note that in fact, at least in principle, by using this FEM approach one may obtain an even higher order expression for the cell-residual.

2.4. Third order diffusive term discretization

The extension of the order of accuracy for the diffusive cell-residual from second- to third-order can be done by exploiting the possibility of using the values the parameter variable **Z** and its gradients, $\mathbf{Z}_{,j}$, at the cell-vertices. Thus the diffusive cell-residual can be computed with the required accuracy $\Phi_T^V = \iint\limits_{\partial T} F_j^V . \overrightarrow{d\sigma} = \sum_{k=1}^{4} \iint\limits_{face_k} F_j^V . \overrightarrow{d\sigma}$ if the integrals $\iint\limits_{face_k} F_j^V . \overrightarrow{d\sigma}$ are computed with third-order accuracy using a FEM approach. Details about the third-order discretization for the diffusive term are omitted here for brevity.

2.5. Unsteady term discretization

For the time-discretization of the real-time term in (4) we propose a second order in time (implicit) finite-differences formulation:

$$U_{,t} = \frac{3U^{n+1,k} - 4U^{n} + U^{n-1}}{2\Delta t} \tag{10}$$

Again, here the superscript "n" represents the real time step and "k" the pseudo-time iteration. The conservative variable U at the real-time steps (n) and $(n-1)$ are also stored in nodes. Due to the fact that the algorithm is implicit with respect to the real time, the real time step Δt is free of stability restrictions and can be chosen on more physical grounds. For the spatial discretization of the real-time derivative integral in (4) we the same FEM approach. For a third-order spatial discretization:

$$\Phi_T^{uns} = \iiint_T U_{,t}.dv = \iiint_T \left[\sum_{\alpha=0}^{9} Q^{(\alpha)}.U_{,t}^{(\alpha)}\right].dv = \sum_{\alpha=0}^{9}\left[U_{,t}^{(\alpha)} \iiint_T Q^{(\alpha)}.dv\right] \tag{11}$$

which assumes a quadratic variation of the derivative $U_{,t}$ over the cell T and again $Q^{(\alpha)}$ is the quadratic shape function which corresponds to a node α, $\alpha = [0..9]$, see Figure 2. The values of $I_\alpha = \iiint_T Q^{(\alpha)}.dv$ are precomputed, see [5].

3. Parallel scalability

We tested the performance of the new discretization scheme, both for serial and parallel computations on Origin 2000 (R 12000 processors at 300MHz clock frequency). Thus, for a medium size problem (600,000 cells) the performances (per processor) of the algorithm are summarized in table below:

Scheme	2nd order scheme	3rd-order scheme
LDA	76 MFLOPS	97 MFLOPS

The new algorithm is computationally efficient due to the optimized mathematical kernel which computes the cell-residual. A practical measure of the efficiency of the parallel algorithm is to compare the parallel speed-up with the theoretical linear speed-up. The results for the parallel scalability are given in Figure 3. These results prove the excellent parallel scalability of the new algorithm (and of its second order counterpart).

4. LES results

The fully developed turbulent channel flow is a classic benchmark test case. Second order LES simulations of the turbulent channel flow using RD schemes have been reported in [2]. The third-order LES simulation presented here uses the same problem setup. The Reynolds number based on the bulk velocity and the channel height is $\mathrm{Re}_c = 5400$ and the bulk Mach number is $M_{bulk} = 0.15$. The "selective" Smagorinsky model [5] has been used to model the sub-grid scale effects. To obtain a statistically stationary turbulent channel flow, the flow has been simulated for enough flow-through times (i.e. the domain length in the streamwise direction divided by

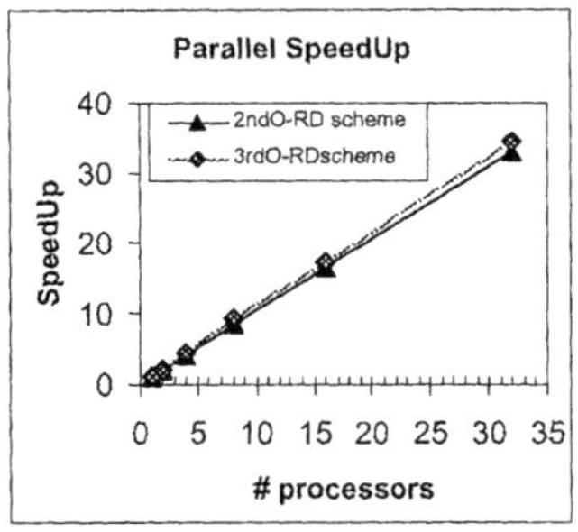

Figure 3. Parallel scalability of the 2nd and 3rd order **RD** discretizations.

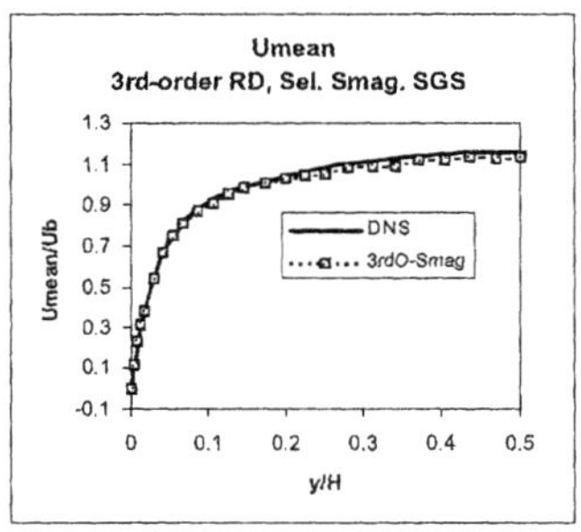

Figure 4. Turbulent channel flow results.

the bulk velocity). Then, the statistics have to be accumulated over at least $8 \div 10$ flow-through times. The results presented here are preliminary, as not enough time steps have been performed for obtaining meaningful statistics. The centerline velocity normalized by the bulk velocity, $\overline{U}_c$, has a value of 1.150 as compared with the experimental value of 1.162 [8]. Figure 4 depicts the planar averages of the time averaged normalized axial velocity, $\overline{U}_{mean}$. These results reproduce well the DNS results, although are not yet statistically converged. Statistically converged LES results will be compared with experiments by Kreplin [8] and with the DNS data of Kim [7]. These results will be presented elsewhere.

5. Conclusions

A new compact high-order Residual Distribution scheme has been proposed for Large-Eddy Simulations of turbulent compressible flows. Preliminary LES results for the fully developed turbulent channel flow simulation, while using the new algorithm are encouraging. The parallel scalability of the new algorithm is very good. The new algorithm will also be tested and validated against existing DNS and experimental results for a number of complex turbulent flow problems. All results obtained so far are very promising.

REFERENCES

[1] D. Caraeni, S. Conway and L. Fuchs. "About a parallel multidimensional upwind solver for LES". In R. Vielsmeier, F. Benkhaldoun and D. Hänel, eds., Finite volumes for complex applications II. Problems and perspectives, pp. 315-322, Hermes Science Publications, Paris 1999.

[2] D. Caraeni, L. Fuchs, "LES Using a Parallel Multidimensional Upwind Solver". First International Conference on Computational Fluid Dynamics, ICCFD-2000, Kyoto, Japan, 2000.

[3] D. Caraeni "Development of a Multidimensional Residual Distribution Solver for Large Eddy Simulation of Industrial Turbulent Flows", Ph.D. thesis, Lund Institute of Technology, ISBN-91-628-4280-3, September 2000.

[4] D. Caraeni, L. Fuchs. "A New Compact High Order Multidimensional Upwind Discretization". Proceedings of 4th World CSCC conference, Vouliagmeni, Greece, 2000.

[5] D. Caraeni, M. Caraeni and L. Fuchs. " A Parallel Multidimensional Upwind Algorithm for LES - A New Compact 3rd Order algorithm". AIAA 2001-2547, 15^{th}AIAA CFD conference, 2001.

[6] H. Deconinck and G. Degrez. "Multidimensional upwind residual distribution schemes and applications". In R. Vielsmeier, F. Benkhaldoun and D. Hänel, eds., Finite volumes for complex applications II. Problems and perspectives, pp. 27-40, Hermes Science Publications, Paris 1999.

[7] J. Kim, P. Moin and R. Moser. "Turbulence Statistics in Fully Developed Channel Flow at Low Reynolds Number". In J. Fluid Mech., vol. 177, pp. 133-166, 1987.

[8] H. Kreplin and H. Eckelmann. "Behavior of the Three Fluctuating Velocity Components in the Wall Region of a Turbulent Channel Flow". In Physics of Fluids, 22(7), pp. 1233-1239, 1979.

[9] P.L. Roe. "Fluctuations and signals, a framework for numerical evolution problems". In K.W. Morton and M.J. Baines, eds., Numerical Methods in Fluid Dynamics, pp.219-257, Academic Press, 1982.

[10] P.L. Roe. "Linear advection schemes on triangular meshes". CoA Report 8720, Cranfield Inst. Of Tech., 1987.

[11] E.van der Weide, H. Paillere, H. Deconinck. "Upwind Residual Distribution Methods for Compressible Flow. An alternative to finite volume and finite element methods". Presented at 28th CFD VKI Lecture Series, 1997.

[12] W.A. Wood and W.L. Kleb. "Diffusion characteristics of upwind schemes on unstructured triangulations". AIAA paper 98-2443, 1998.

Parallel Computational Fluid Dynamics – Practice and Theory
P. Wilders, A. Ecer, J. Periaux, N. Satofuka and P. Fox (Editors)

Parallel Implementation of a Line-Implicit Time-Stepping Algorithm

Lars Carlsson[a], Stefan Nilsson[a]

[a] Department of Naval Architecture and Ocean Engineering,
Chalmers University of Technology,
Chalmers tvärgata 8, SE-412 96 Göteborg, Sweden

A second-order accurate time-stepping algorithm is presented. It is based on a splitting of the discrete spatial operators and is applied to the Navier-Stokes equations for incompressible flows. A way to decompose the domain among the processors on a parallel computer is suggested by the operator splitting. Results from test runs are given, validating the implementation and demonstrating the parallel performance.

1. Introduction

We present a parallel version of a second-order accurate time-stepping method that is well suited for integrating the Navier-Stokes equations in boundary layers. In those regions the velocity varies little in the tangential direction but increases from zero at the boundary to the free-stream value in a very short distance in the normal direction. This calls for a very fine grid resolution near no-slip boundaries. We are interested in resolving the solution in time, but we cannot afford to use explicit methods in the boundary layer due to the severe restrictions inflicted on the time step. Instead we use a time-stepping algorithm which is based on operator splitting proposed in [1], where the discretized operators are divided into fast and slow components. The splitting makes the method fairly easy to parallelize.

In the next section we briefly describe the line-implicit time-stepping algorithm. In section 3, we show how the algorithm is implemented in parallel. In section 4 we show results validating the implementation and provide some results regarding parallel performance. Finally, in the last section we suggest some possible improvements.

2. Line-Implicit Time-Stepping

The purpose of the line-implicit time-stepping algorithm is to integrate Navier-Stokes equations in boundary layers with second-order temporal accuracy. For laminar flows, we observe different time scales in boundary layers. The solution changes more rapidly in the normal direction at no-slip boundaries (solid walls), than it does in the other directions.

2.1. Governing Equations

We use the velocity-pressure formulation of the Navier-Stokes equations. This formulation decouples the solving for the velocity from the pressure. We remark that solving for the pressure is another way to ensure that the solution for the velocity is divergence free. The momentum equations can be expressed as

$$\mathbf{u}_t = A(\mathbf{u})\mathbf{u} - \frac{1}{\rho}\nabla p + \frac{1}{\rho}\mathbf{f}. \tag{1}$$

In the remainder of this paper we will omit ρ, rescaling p and $\mathbf{f}$. In two space dimensions $\mathbf{u} = (u, v)^T$ and $\mathbf{f} = (f^{(x)}, f^{(y)})$. The spatial operators acting on the velocity are kept in $A(\mathbf{u})$, such that

$$A(\mathbf{u})\mathbf{u} = -(\mathbf{u} \cdot \nabla)\mathbf{u} + \nu\nabla^2\mathbf{u}, \tag{2}$$

where we assume a constant kinematic viscosity. We obtain Poisson's equation for the pressure by taking the divergence of the momentum equations and substituting the divergence-free condition, $\nabla \cdot \mathbf{u} = 0$. Again, in two space dimensions

$$\nabla^2 p = -\left(\frac{\partial u}{\partial x}\right)^2 - 2\frac{\partial v}{\partial x}\frac{\partial u}{\partial y} - \left(\frac{\partial v}{\partial y}\right)^2 - \nabla \cdot \mathbf{f}. \tag{3}$$

The boundary conditions for the velocity are shown in Table 1, for the pressure we use a Neumann-type boundary condition,

$$\mathbf{n} \cdot \nabla p = \mathbf{n} \cdot (-\mathbf{u}_t - (\mathbf{u} \cdot \nabla)\mathbf{u} - \nu\nabla \times \nabla \times \mathbf{u} + \mathbf{f}). \tag{4}$$

Note, that most of the terms disappear on a no-slip boundary. The reason for using the curl of the curl of the velocity is because of stability problems that occur if we keep the Laplacian in the pressure boundary condition. This is explained in more detail in [2].

2.2. Time-Discrete Formulation

We use a mixed Euler forward/backward algorithm to advance the solution for the velocity in time. Using this algorithm, we split the operators acting on the velocity into two parts, $A(\mathbf{u})\mathbf{u} = (A_I(\mathbf{u}) + A_E(\mathbf{u}))\mathbf{u}$. The indices I and E indicate whether we treat those operators in A implicitly or explicitly, respectively. At time-level k we know the solution for the velocity. It could be either the initial condition or the previously computed discretized solution. From here we compute the pressure. Then we advance two different solutions to the next time level, $k+1$. The time-discrete expression used to obtain the first solution looks like

$$(I_2 - (\Delta t)^k A_I(\mathbf{u}^*))\mathbf{u}^* = (I_2 + (\Delta t)^k A_E(\mathbf{u}^k))\mathbf{u}^k + (\Delta t)^k(-\nabla p^k + \mathbf{f}^k), \tag{5}$$

Table 1
The different boundary conditions for the velocity.

Boundary type	Mathematical formulation
inflow	$\mathbf{u} = \mathbf{g}$, where $\mathbf{g}$ is a velocity profile.
outflow	$(\mathbf{n} \cdot \nabla)\mathbf{u} = \mathbf{0}$
no slip	$\nabla \cdot \mathbf{u} = 0$ and $\mathbf{u} = 0$
slip	$(\mathbf{n} \cdot \nabla)(\mathbf{t} \cdot \mathbf{u}) = 0$ and $\mathbf{n} \cdot \mathbf{u} = 0$

The normal vector, $\mathbf{n} = (n^{(x)}, n^{(y)})$, and the tangential vector, $\mathbf{t} = (t^{(x)}, t^{(y)})$, both have length one.

where I_2 denotes the 2×2 identity matrix. We compute the second solution via an intermediate solution at time level $k + 1/2$

$$(I_2 - \frac{(\Delta t)^k}{2} A_I(\mathbf{u}^{k+1/2}))\mathbf{u}^{k+1/2} = (I_2 + \frac{(\Delta t)^k}{2} A_E(\mathbf{u}^k))\mathbf{u}^k + \frac{(\Delta t)^k}{2}(-\nabla p^k + \mathbf{f}^k). \quad (6)$$

At this point, we need to compute the pressure at time level $k + 1/2$ to get the second solution for the velocity

$$(I_2 - \frac{(\Delta t)^k}{2} A_I(\mathbf{u}^{**}))\mathbf{u}^{**} = (I_2 + \frac{(\Delta t)^k}{2} A_E(\mathbf{u}^{k+1/2}))\mathbf{u}^{k+1/2} + \frac{(\Delta t)^k}{2}(-\nabla p^{k+1/2} + \mathbf{f}^{k+1/2}). (7)$$

Finally, we do a Richardson extrapolation to get the desired second-order accurate solution at time level $k + 1$,

$$\mathbf{u}^{k+1} = 2\mathbf{u}^{**} - \mathbf{u}^{*}. \quad (8)$$

We estimate the time step $(\Delta t)^k = t^{k+1} - t^k$ with von Neumann analysis of the eigenvalues for the fully discretized problem. It is determined by the explicitly treated parts of the operators.

2.3. Spatial Discretization and Operator Splitting

We want solve the spatially discretized counterpart to (5)-(8). For this purpose we discretize on composite overlapping grids, using Xcog [3]. A composite overlapping grid consists of one or several component grids, and they are logically square in parameter space. All the x- and y-derivatives are expressed in second-order accurate centered differences in the r- and s-direction according to [4]. The r- and s-direction span parameter space. We use the line-implicit time-stepping algorithm in component grids with at least one no-slip boundary and a two-stage explicit Runge-Kutta algorithm in grids away from no-slip boundaries. The latter algorithm is also second-order accurate and needs the pressure to be computed at time level $k + 1/2$ as well. Here, we only give an example of how the x-derivative of a grid function, $h_{i,j}$, is expressed in terms of the grid spacings Δr, Δs and the components of the metric tensor of the mapping from parameter space to physical space (r_x, r_y, s_x and s_y). A more comprehensive description of the operator splitting and the line-implicit algorithm is given in [5]. The derivative

$$\left.\frac{\partial h}{\partial x}\right|_{i,j} = r_x|_{i,j} \frac{h_{i+1,j} - h_{i-1,j}}{2\Delta r} + s_x|_{i,j} \frac{h_{i,j-1} - h_{i,j-1}}{2\Delta s}. \quad (9)$$

We also assume that $i = 0, \ldots, M - 1$ and $j = 0, \ldots, N - 1$, where M and N are the number of grid points in the r- and s-direction, respectively. If we choose the s-direction to be treated implicitly, all the components of the velocity in the discretized counterpart to (2) go into the left hand side in (5)-(7). Thus, the example in (9) becomes

$$\left.\frac{\partial h}{\partial x}\right|_{i,j} = r_x|_{i,j} \frac{h^k_{i+1,j} - h^k_{i-1,j}}{2\Delta r} + s_x|_{i,j} \frac{h^{k+1}_{i,j-1} - h^{k+1}_{i,j-1}}{2\Delta s}. \quad (10)$$

This means that for each grid line in the r-direction we have to solve a non-linear algebraic system three times per time step. Those systems are in turn solved with Newton's method and for the linear systems of equations we use the `dgbsv` routine from LAPACK [6]. Since the coefficient matrices for the linear systems of equations are block tri-diagonal and due to the quadratical convergence of Newton's method, the amount of work for the line-implicit time-stepping algorithm is of the same order as for the explicit Runge-Kutta algorithm, which is of the order of the number of grid points.

3. Parallel Implementation

In this section we describe the implementation of the time-stepping algorithm on a distributed memory parallel computer. First the decomposition of the domain enforced by the spatial splitting of the momentum operators (2) and the consequences of this decomposition are illustrated. Then the distributed solution of the pressure equation is described.

All parallelization is done using MPI, although some of it is hidden in the data parallel libraries we use [7].

3.1. Domain Distribution

Due to the spatial splitting of the operators the different grid directions will be coupled with different strengths. The grid lines will be strongly coupled along the implicitly treated direction due to the non-linear algebraic system which needs to be solved for each such line. Along the explicitly treated direction we only need to build the right hand side of the non-linear systems, so the coupling is much weaker. It therefore makes sense to distribute the grid only along the explicitly treated direction. This will minimize the amount of communication between neighboring processors during each time-step. Each processor will solve the linear systems of equations pertaining to its local sub domain of the complete domain. Message passing is then only necessary once before these systems are solved. This is illustrated in Figure 1.

Although a one-dimensional decomposition of a two-dimensional domain is non-optimal if analyzed theoretically with regard to scalability, we do not expect this to have any negative impact on the parallel efficiency as the computation to communication ratio is very good for the line-implicit time-stepping algorithm.

3.2. Pressure Solution

The discretized pressure equation is distributed in the same manner as the momentum equations. Presently, the distributed system of equations is solved using preconditioned Krylov sub-space methods, taken from the Aztec numerical library [8].

The preconditioner used is of the domain decomposition type. It consists of an incomplete factorization of the local sub matrix augmented by some (possibly zero) elements from its neighboring sub domains [9].

4. Results

To validate the implementation and test the parallel scalability of the code we performed some numerical experiments.

4.1. Validation of Implementation

To check that our implemented algorithm really was second-order accurate, we ran a number of tests, using increasingly fine grids for the geometry shown in Figure 2(a). A

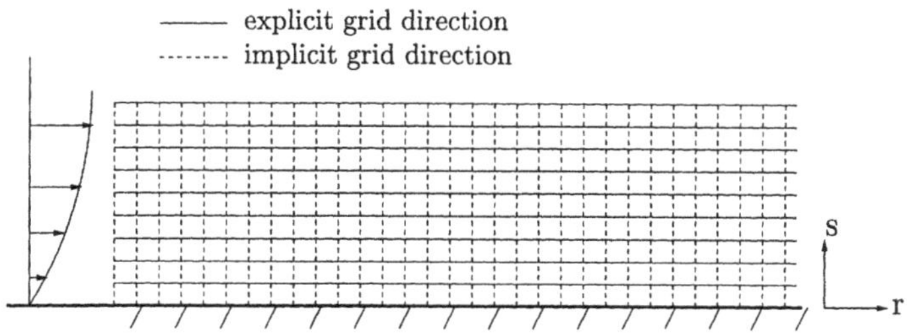

(a) Discretized boundary layer

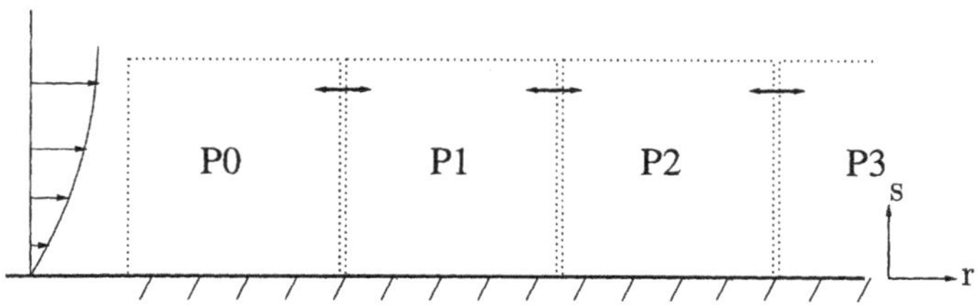

(b) Decomposition of boundary layer. ⟷ symbolizes necessary message passing between process $P0$ and $P1$, etc.

Figure 1. Illustration of distribution strategy.

fixed time interval was simulated, and the errors compared to a known analytic solution where then measured in the maximum norm for each grid. The results from these tests are shown in Figure 3.

4.2. Parallel Scalability

We also tested the scalability of the code as the number of processors used was increased. A channel flow problem, shown in Figure 2(b), discretized with approximately 0.5 million grid points was run with $1, 4, 8, 16, 64$, and 128 processors. A fixed time interval was simulated and the total time spent in the computational loop was measured for each run. The time to setup the problem was not taken into account. The tests where run on an IBM-SP computer, where each node consists of four PowerPC 604e processors sharing 1.6 GB of memory. The resulting speedup is shown in Table 2. The speedup is reasonable up until 64 processors and then decreases dramatically. This is most likely due to the problem size, which was to small to warrant the largest number of processors used in the scalability tests.

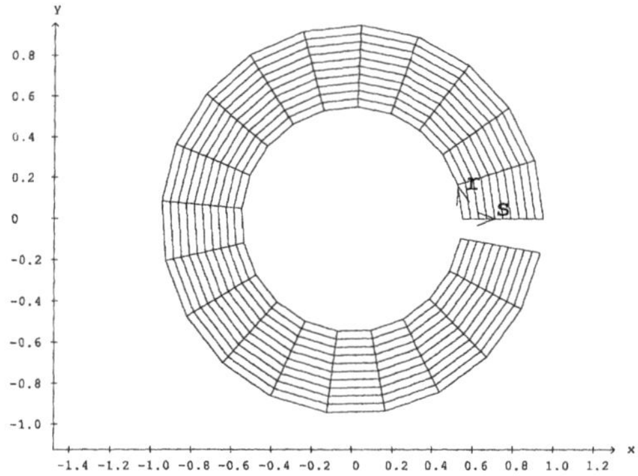

(a) Coarsest grid used for convergence tests.

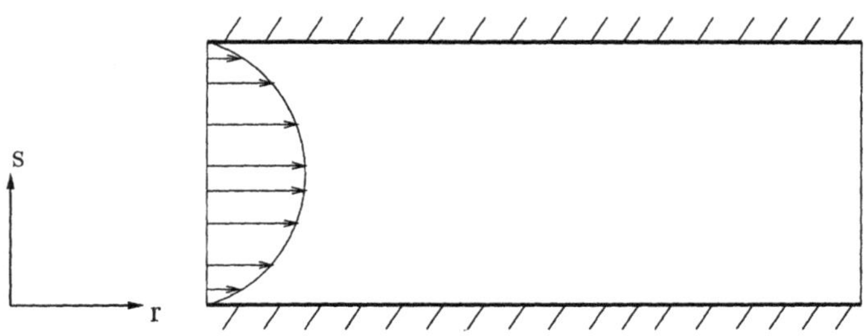

(b) Channel with parabolic velocity inflow profile, used for scalability tests.

Figure 2. Test problems.

5. Future Work

The computation to communication ratio for solving the pressure equation is worse than for the momentum equations. Furthermore, the iterative method combined with the preconditioners used will not scale well with increasing number of processors. This is something we will have to look deeper into, as the time to solve for the pressure probably will be the dominant part of the solution process.

One alternative is to use a multigrid algorithm. Then, we could use a line-relaxation scheme with the blocked lines lying in the implicitly treated direction. This might give a better ratio of computation to communication during the pressure solve. We also expect a multigrid algorithm to converge faster than the Krylov sub-space methods used now.

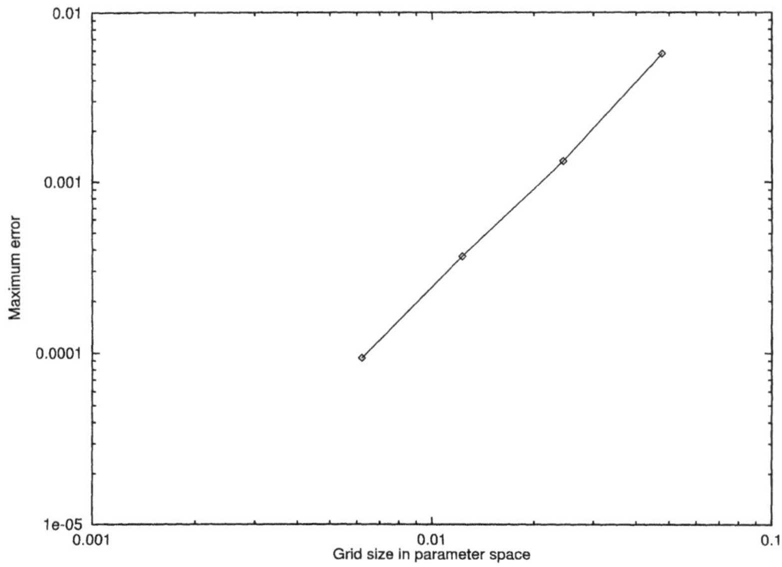

Figure 3. Maximum error of the computed u plotted against grid size. The convergence is approximately second order.

REFERENCES

1. Heinz-Otto Kreiss, Numerical solution of problems with different time scales II. Institute for Mathematics and its Applications, University of Minnesota, IMA Preprint Series #1294. (1995)
2. N. Anders Petersson, Stability of Pressure Boundary Conditions for Stokes and Navier-Stokes Equations. J. Comput. Phys. (2001) **172**
3. N. Anders Petersson, User's guide to Xcog, version 2.0. CHA/NAV/R-97/0048, Department of Naval Architecture and Ocean Engineering, Chalmers University of Technology. (1997)
4. J. F. Thompson, Z. U. A. Warsi and C. W. Mastin, Numerical Grid Generation. North-Holland, Amsterdam. (1985)
5. Lars Carlsson, A Line-Implicit Time-Stepping Algorithm. CHA/NAV/R-00/0071, Department of Naval Architecture and Ocean Engineering, Chalmers University of Technology. (2000)
6. E. Andersson *et al.*, LAPACK Users' Guide. ISBN 0-89871-294-7.
7. Daniel Quinlan, A++/P++ manual. Los Alamos National Laboratory, LANL Unclassified Report 96-3273. (1995)
8. Ray S. Tuminaro *et al.*, Official Aztec User's Guide Version 2.1. SAND99-8801J, Massively Parallel Computing Research Laboratory, Sandia National Laboratories, Albuquerque, NM 87185.

Table 2
Results from tests of parallel scalability.

# procs.	Speedup (all of code)	Speedup (only momentum eqs.)
1	1	1
4	3.4	4.1
8	6.8	8.6
16	14.6	18.0
64	52.6	51.4
128	37.9	39.8

The third column shows the speedup of the line-implicit algorithm alone.

9. Richard Barrett *et al.*, Templates for the Solution of Linear Systems: Building Blocks for Iterative Methods. SIAM. (1994)

Parallel Computational Fluid Dynamics – Practice and Theory
P. Wilders, A. Ecer, J. Periaux, N. Satofuka and P. Fox (Editors)

Parallel simulation of dense gas and liquid flows based on the quasi gas dynamic system *

B.N. Chetverushkin[a], N.G. Churbanova[a] and M.A. Trapeznikova[a]

[a]Institute for Mathematical Modeling, Russian Academy of Sciences,
4-A Miusskaya Square, 125047 Moscow, Russia

A new approach to simulation of dense gas and liquid flows is proposed. The model is based on the Enskog equation which is a generalization of the classical kinetic Boltzmann equation. A parallel semi-implicit algorithm is developed for the model implementation. The approach is validated by the 2D test problem of the isothermal lid-driven cavity flow at low Mach number. Then slightly compressible gas flows in a horizontal fast contact chemical reactor are predicted. Computations are performed on the multiprocessor Parsytec CC. The corresponding parallel C/MPI codes are scalable and portable. Efficiency of parallelization is very high.

1. INTRODUCTION

Slow motions of a viscous gas occur universally in industrial applications. If compressibility in these flows are connected only with the pressure but temperature variations are negligible then these gas flows can be treated as incompressible at Mach number $M \leq 0.1$. It is possible to simulate them using, for example, the Navier-Stokes equations for the incompressible liquid. However many practically important flows of this type are characterized by rather high density variations. Here compressibility appears from large temperature drops, variation of mass due to gas injection and so on. These are flows in turbo machinery cooling, in combustion chambers, in chemical reactors etc. Then simulation should be based on a mathematical model of the compressible gas and the employed model should take into account specific properties of low-Mach-number flows.

A model known as the quasi gas dynamic system (QGDS) of equations [5] was originally constructed for gases of usual and weak rarefied density. Computations of complex transonic and supersonic flows on the basis of QGDS were effective enough and showed merits of this approach. Generalization of QGDS to the case of subsonic flows seemed to be also very promising [6].

In the present paper a new quasi gas dynamic model for dense gases and liquids is proposed. This model is derived by averaging the kinetic Enskog equation. The model can be implemented by explicit or semi-implicit numerical methods and easily adapted to distributed memory multiprocessors. Computations predict that usage of the new QGDS instead of the previous one leads to more accurate results at low Mach numbers.

*This work is supported in part by ExxonMobil (Grant ISTC #1531p) and by Russian Foundation for Basic Research (Grants #99-01-01215, #00-01-00263, #00-01-00291 and #01-01-06202).

2. QUASI GAS DYNAMIC SYSTEM FOR DENSE GASES AND LIQUIDS

Kinetic schemes are advanced algorithms of computational fluid dynamics [1]-[4]. Derivation of kinetic schemes is based on discrete models for the one-partical distribution function of gas molecules. Averaging these models on molecular velocities with the collision vector components one can obtain difference schemes for gasdynamic parameters.

A model referred to as the quasi gas dynamic system (QGDS) of equations is closely connected with the kinetically consistent finite difference (KCFD) schemes [5]. KCFD schemes and QGDS were originally developed to describe flows of viscous compressible gases of usual and weak rarefied density. Such gases are characterized by large distance between molecules. Therefore KCFD schemes and QGDS are based on the classical kinetic Boltzmann equation for the one-partical distribution function: the Boltzmann equation is approximated by finite differences and then it is averaged on velocities of gas molecules. Additionally the next assumption is used: the distribution function is slightly varies near the equilibrium on the distance of the free path length of a molecule. Numerical solutions obtained on the base of this model are practically equivalent to results obtained by using the Navier-Stokes equations.

The original QGDS was formally employed outside of the domain of its physical adequacy to simulate flows in dense gases and liquids. Although these attempts were rather successful development of a kinetic model specially for gases of high density is very actual.

In dense gases the distance between two molecules has the same order of magnitude as a molecule itself. The distribution function for such gases complies with the kinetic Enskog equation which is a generalization of the Boltzmann equation:

$$\begin{aligned}\frac{\partial f}{\partial t}+\xi_i\frac{\partial f}{\partial x_i}+\frac{F_i}{m}\frac{\partial f}{\partial \xi_i} &= \int\int\int\Big[Y\left(\bar{x}+\frac{1}{2}\sigma\bar{k}\right)f(\bar{x},\bar{\xi}')f_1(\bar{x}+\sigma\bar{k},\bar{\xi}'_1)-\\ &\quad -Y\left(\bar{x}-\frac{1}{2}\sigma\bar{k}\right)f(\bar{x},\bar{\xi})f_1(\bar{x}-\sigma\bar{k},\bar{\xi}_1)\Big]\,g\;b\;db\;d\varepsilon\;d\xi_1\equiv J \qquad (1)\end{aligned}$$

The distinction between the Enskog and Boltzmann equations consists in the presence of the more complicated collision integral J. Here t is time, $\bar{x}$ is a spatial vector, $\bar{\xi}$ is a molecular velocity vector, f is a one-partical distribution density, $\bar{F}/m$ is an external force, σ is a molecule diameter, $\bar{k}$ is a unit vector of the direction between centers of molecules, Y is a coefficient of increase of the molecules' collision frequency. The rest notations are typical for the Boltzmann collision integral.

For further constructions it is convenient to expand the right hand side of (1) into the Taylor-series expansion with respect to σ up to the second order terms inclusively:

$$J\approx J_1+J_2+J_3+J_4+J_5+J_6 \qquad (2)$$

where

$$\begin{aligned}J_1 &= Y\int\int\int(f'f_1'-ff_1)g\;b\;db\;d\varepsilon\;d\xi_1\\ J_2 &= \sigma Y\int\int\int\bar{k}\cdot\left(f'\frac{\partial f_1'}{\partial\bar{x}}+f\frac{\partial f_1}{\partial\bar{x}}\right)g\;b\;db\;d\varepsilon\;d\xi_1\\ J_3 &= \frac{1}{2}\sigma\int\int\int\bar{k}\cdot\left(\frac{\partial Y}{\partial\bar{x}}\right)(f'f_1'+ff_1)g\;b\;db\;d\varepsilon\;d\xi_1\end{aligned}$$

$$J_4 = \frac{1}{2}\sigma^2 Y \int\int\int (\bar{k}\cdot\bar{k}: \left\{ f' \frac{\partial}{\partial \bar{x}}\frac{\partial f_1'}{\partial \bar{x}} - f \frac{\partial}{\partial \bar{x}}\frac{\partial f_1}{\partial \bar{x}} \right\} g\ b\ db\ d\varepsilon\ d\xi_1$$

$$J_5 = \frac{1}{2}\sigma^2 \int\int\int \left(\bar{k}\cdot\frac{\partial Y}{\partial \bar{x}}\right)\left(\bar{k}\cdot\left\{ f'\frac{\partial f_1'}{\partial \bar{x}} - f\frac{\partial f_1}{\partial \bar{x}}\right\}\right) g\ b\ db\ d\varepsilon\ d\xi_1$$

$$J_6 = \frac{1}{8}\sigma^2 \int\int\int \left(\bar{k}\cdot\bar{k}: \frac{\partial}{\partial \bar{x}}\frac{\partial Y}{\partial \bar{x}}\right)(f'f_1' - ff_1) g\ b\ db\ d\varepsilon\ d\xi_1$$

By the dot $(\cdot)$ the scalar product is denoted and the colon $(:)$ means the double contraction of tensors.

Integrating (1) over the velocity space with the collision vector components ψ:

$$\psi = m,\ m\bar{\xi},\ m\bar{\xi}^2/2$$

one can obtain the momentum equations:

$$\int \psi \left(\frac{\partial f}{\partial t} + \bar{\xi}\frac{\partial f}{\partial \bar{x}} + \frac{\bar{F}}{m}\frac{\partial f}{\partial \bar{\xi}}\right) d\xi = -\frac{\partial}{\partial \bar{x}}\Psi_\Phi \tag{3}$$

The term Ψ_Φ is responsible for the transport by collisions. It can be represented as follows:

$$\begin{aligned} \Psi_\Phi &= \frac{\sigma}{2}Y \int\int\int\int (\psi' - \psi) f f_1 \bar{k}\ g\ b\ db\ d\varepsilon\ d\xi\ d\xi_1 + \\ &+ \frac{\sigma^2}{4}Y \int\int\int\int (\psi' - \psi)\left(\bar{k}\cdot\left(f_1\frac{\partial f}{\partial \bar{x}} - f\frac{\partial f_1}{\partial \bar{x}}\right)\right)\bar{k}\ g\ b\ db\ d\varepsilon\ d\xi\ d\xi_1 \end{aligned} \tag{4}$$

Pay attention that in contrast to the Boltzmann equation momentum of the Enskog equation has a zero component in the right hand side (see (3), (4)): $\Psi_\Phi = 0$ only for $\psi = m$. The mass is not transported by collisions what is obvious from the physical standpoint. However molecular transport of the momentum and the energy is realized not only by direct movement of molecules but also by a series of sequential collisions of molecules.

Taking into account all the above formulas and remarks let us apply the previously developed technique of deriving KCFD schemes and QGDS (procedures of discretezation and averaging) [5] to the Enskog equation. As a result the quasi gas dynamic system (QGDS) of equations for gases of high density and liquids is derived — its general form is presented below:

$$\frac{\rho^{j+1} - \rho^j}{\tau} + \frac{\partial}{\partial x_k}(\rho u_k)^j = \frac{\partial}{\partial x_k}\frac{\tau}{2}\frac{\partial}{\partial x_l}(\rho u_k u_l + \sigma_{kl} p)^j - \frac{\partial}{\partial x_k}\left(\frac{\tau}{2}\frac{\rho F_k}{m}\right)^j \tag{5}$$

$$\begin{aligned} \frac{(\rho u_k)^{j+1} - (\rho u_k)^j}{\tau} + \frac{\partial}{\partial x_l}(\rho u_l u_k)^j + \frac{\partial p^j}{\partial x_k} - \left(\frac{\rho F_k}{m}\right)^j = \frac{\partial}{\partial x_l}\frac{\tau}{2}\frac{\partial}{\partial x_l}((\rho u_l^2 + p)u_k)^j + \\ + \sum_{r\neq l}\frac{\partial}{\partial x_r}\frac{\tau}{2}\frac{\partial}{\partial x_l}(\rho u_k u_r u_l)^j - \frac{\partial}{\partial x_l}\left(\frac{\tau}{2}\frac{F_l}{m}\rho u_k\right)^j - \frac{\partial}{\partial x_k}\left(\frac{\tau}{2}\frac{F_k}{m}\rho u_k\right)^j \end{aligned} \tag{6}$$

$$\frac{E^{j+1}-E^j}{\tau}+\frac{\partial}{\partial x_l}((E+p)u_l)^j-\left(\frac{\rho F_l u_l}{m}\right)^j=\frac{\partial}{\partial x_l}\frac{\tau}{2}\frac{\partial}{\partial x_l}((\rho u_l^2+p)E/\rho)^j+$$
$$+\sum_{k\neq l}\frac{\partial}{\partial x_k}\frac{\tau}{2}\frac{\partial}{\partial x_l}((E+2p)u_k u_l)^j+\left(\frac{3\tau}{2}\frac{\rho F_l^2}{m^2}\right)^j-\frac{\partial}{\partial x_k}\frac{\tau}{2}\left(\frac{F_l}{m}\rho u_k u_l+\frac{F_k}{2m}(\rho u_k^2+3p)\right)^j \quad (7)$$

where ρ is the density, $\bar{u}$ is the velocity, p is the pressure, $E=\rho(\varepsilon+u_k^2/2)$ is the total energy, ε is the inner energy, $\tau=2\mu/p$ is the kinetic time, μ is the viscosity coefficient.

It should be observed that for solving applied problems QGDS is usually normalized using reference values of all unknowns. One of the possible method of normalization at low Mach number leads to appearing the coefficient $1/Re$ (Re is Reynolds number) in front of the right hand sides of equations (5)-(7). Mach number M is included in boundary conditions for the temperature.

3. ALGORITHM OF COMPUTATIONS

For computations the quasi gas dynamic system can be simplified. Estimation of terms in the right hand sides of equations (5)-(7) allows to neglect terms with mixed derivatives. Let us also assume that there are no external forces.

Numerical implementation of QGDS is usually based on explicit or semi-implicit methods. In the explicit method the equations are discretized by explicit finite difference schemes with central differences. The semi-implicit method is more interesting: several iterations are performed at every time level. The half-sum of values from the previous time level and from the previous iteration of the current time level is used in approximations of spatial derivatives:

$$\frac{\rho^{j+1}-\rho^j}{\Delta t}+0.5*\left(\frac{\partial}{\partial x_k}(\rho u_k)^j+\frac{\partial}{\partial x_k}(\rho u_k)^n\right)=$$
$$=0.5*\frac{1}{Re}\left(\frac{\partial}{\partial x_l}\frac{\tau}{2}\frac{\partial}{\partial x_l}(\rho u_l^2+p)^j+\frac{\partial}{\partial x_l}\frac{\tau}{2}\frac{\partial}{\partial x_l}(\rho u_l^2+p)^n\right) \quad (8)$$

$$\frac{(\rho u_k)^{j+1}-(\rho u_k)^j}{\Delta t}+0.5*\left(\frac{\partial}{\partial x_l}(\rho u_l u_k)^j+\frac{\partial}{\partial x_l}(\rho u_l u_k)^n\right)+0.5*\left(\frac{\partial p^j}{\partial x_k}+\frac{\partial p^n}{\partial x_k}\right)=$$
$$=0.5\frac{1}{Re}*\left(\frac{\partial}{\partial x_l}\frac{\tau}{2}\frac{\partial}{\partial x_l}((\rho u_l^2+p)u_k)^j+\frac{\partial}{\partial x_l}\frac{\tau}{2}\frac{\partial}{\partial x_l}((\rho u_l^2+p)u_k)^n\right) \quad (9)$$

$$\frac{E^{j+1}-E^j}{\Delta t}+0.5*\left(\frac{\partial}{\partial x_l}((E+p)u_l)^j+\frac{\partial}{\partial x_l}((E+p)u_l)^n\right)=$$
$$=0.5\frac{1}{Re}*\left(\frac{\partial}{\partial x_l}\frac{\tau}{2}\frac{\partial}{\partial x_l}((E+2p)u_l^2+\frac{p}{\rho}(E+p))^j+\frac{\partial}{\partial x_l}\frac{\tau}{2}\frac{\partial}{\partial x_l}((E+2p)u_l^2+\frac{p}{\rho}(E+p))^n\right) \quad (10)$$

where j is the time level number, n is the iteration number. Spatial derivatives are approximated by central differences.

These algorithms are easily and efficiently adapted to parallel computers with distributed memory using the data parallelization (computational domain partitioning) principle and the model of message passing. It is a very important advantage of the proposed approach because detailed analysis of gas and liquid flows requires fine space-time grids and consequently leads to significant costs of computations. Parallel computers allow to solve such large-scale problems for acceptable time. The homogeneous variant of difference schemes is used and the domain is divided into subdomains containing the equal number of grid nodes — these two properties automatically provide a good load balancing of processors. In order to reduce time expenses on message passing the parallel algorithm implementing the semi-implicit method has no global communications on the processor network, all the data to be send/received by a processor during an iteration are combine in one large block, the number of data exchanges is minimized.

4. NUMERICAL RESULTS

4.1. Test Predictions

For validation of the new quasi gas dynamic model the well known 2D test problem of the isothermal lid-driven cavity flow was considered. Computations were performed for uniform square grids of 82×82, 162×162 and 322×322 nodes at Reynolds number Re $= 100$ and 400 and at Mach number in the range from 0.1 to 0.0.

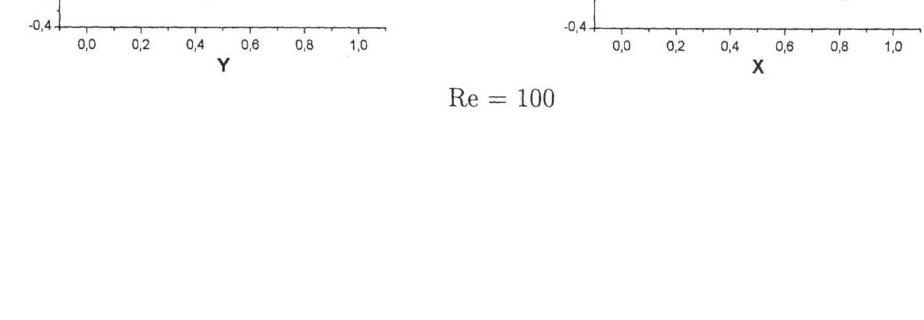

Re $= 100$

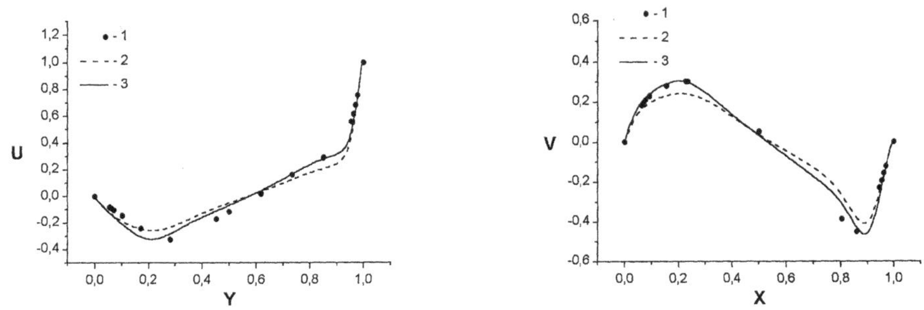

Re = 400

Figure 1. Velocity profiles ($M = 0.1$, the computational grid size is 162×162):
1 - the benchmark solutions (Ghia et al.)
2 - QGDS with the Boltzmann equation for the distribution function
3 - QGDS with the Enskog equation for the distribution function

Fig.1 shows the velocity profiles of vertical and horizontal midsections of the cavity. Solutions obtained on the basis of the Enskog equation were compared with solutions obtained by using the original QGDS based on the Boltzmann equation and with benchmarks from [7]. One can see a good agreement of all numerical results but the new approach indicates a higher accuracy.

4.2. Simulation of gas flows in chemical reactors

QGDS for dense gases and liquids is now used as a governing model for numerical simulation of gas flows in the horizontal fast contact chemical reactor. Calculations are performed in the 2D cartesian formulation, QGDS is normalized as it is described at the end of Sect.2.

The computational domain is a rectangular. The left boundary of the domain is the inlet with the Poisseilue velocity profile and a fixed temperature, the right boundary is the outlet with open boundary conditions (zero normal derivatives for velocity components and for the temperature), the upper and lower boundaries are thermally-insulated rigid walls (no-slip, no-permeability velocity conditions) with heated segments of a fixed temperature and of the unit length. These segments can be located at the unit distance from the inlet or at the middle of walls. Different Reynolds numbers, Mach number $M = 0.1$, different temperature regimes (different temperature drops between the inlet and the heated segments) are predicted.

The simulation is aimed at studying the structure of heat and fluid flows, the resulting temperature gradients and their dependence on boundary conditions. In Fig.2 the temperature field for one of the considered regimes is depicted. The temperature is minimal at the inlet ($T = 300$ K) and maximal at the heated segments ($T = 1200$ K).

The parallel semi-implicit algorithm implementing schemes (8)-(10) is adapted to the multiprocessor system Parsytec CC under the MPI communication library. The domain

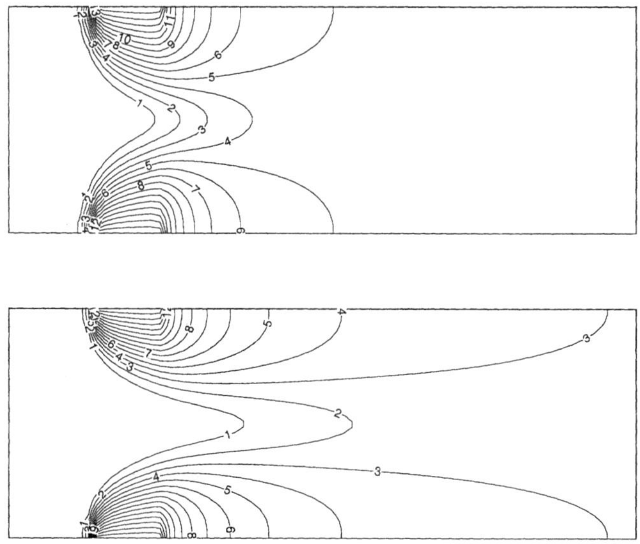

Figure 2. Temperature field for $Re = 500$ and $Re = 1000$

is divided into some quantity of equal subdomains in one direction according to the longer (horizontal) side. So the architecture of parallel processes is a pipeline. The developed parallel code is written in C, it is portable and scalable. The efficiency of parallelization depends on the computational grid size and the number of employed processors. Fig.3 shows the speed-up for grids of 324×44 and of 324×324 nodes on the relatively small number of processors. Thus if the number of processors are not more than 8 the efficiency for the considered real gas dynamics application is more than 90% what is a good result.

5. CONCLUSION

In the paper a new quasi gas dynamic model for describing flows of dense gases and liquids is stated. Numerical experiments demonstrate effectiveness of this approach for simulation of low-Mach-number flows on parallel computers with distributed memory. To increase the accuracy the model can be further explore considering a specific pressure behavior at low Mach number. It is reasonable to decompose the pressure into a sum of the volume-avaraged component and the dynamic component and to use different reference values of these components for normalization of the governing equations [6]. Then a new parallel implicit pressure-correction algorithm will be developed in order to decrease the total run time essentially.

Besides that in the future the authors intend to take into account chemical kinetics while simulating processes in chemical reactors.

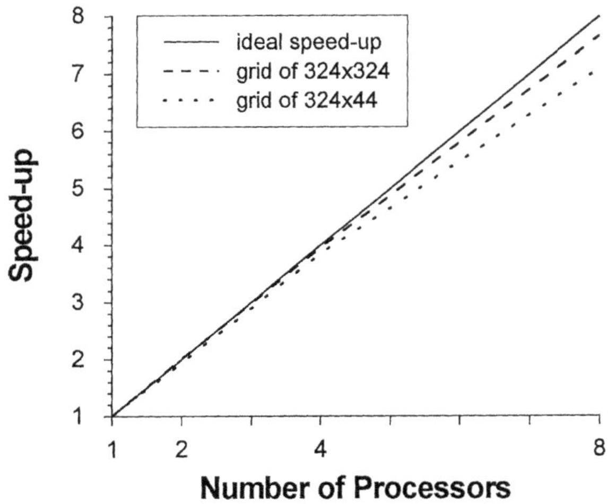

Figure 3. Speed-up of the parallel code for simulation of chemical reactors at different sizes of computational grids

REFERENCES

1. D.I. Pullin, Direct simulation methods for compressible gas flow, J. of Comp. Phys. 3 (1980) 231-244.
2. R.D.Reitz, One-dimensional compressible gasdynamic calculations using the Boltzmann equation, J. of Comp. Physics, 42(10) (1981) 103-105.
3. S.M. Deshpande, On the maxwellian distribution symmetric form and entropy conservation for Euler equations., NASA, Technical Paper 2583 (1986).
4. B. Perthame, The kinetic approach to the system of conservation laws. Recent advances in partial differential equations (El Esurial 1992), Res. Appl. Math., 30, Masson, Paris (1994).
5. T.G. Elizarova and B.N. Chetverushkin, Kinetically coordinated difference schemes for modeling flows of a viscous heat-conducting gas, J. Comp. Math. and Math. Phys. 11 (1988) 64-75.
6. M.A. Trapeznikova, N.G. Churbanova and B.N. Chetverushkin, Simulation of gas flows at low Mach number using parallel computers, In CD-Rom Proc. of ECCOMAS'2000 (2000).
7. V. Ghia, K.N. Ghia and C.T. Shin, High-Re solutions for incompressible flow using the Navier-Stokes equations and a multi-grid method, J. Comp. Phys. 48 (1982) 387-411.

Parallel Computational Fluid Dynamics – Practice and Theory
P. Wilders, A. Ecer, J. Periaux, N. Satofuka and P. Fox (Editors)

DLB 2.0 – A Distributed Environment Tool for Supporting Balanced Execution of Multiple Parallel Jobs on Networked Computers

Y.P. Chien, J.D. Chen, A. Ecer, H.U. Akay, and J. Zhou

Computational Fluid Dynamics Laboratory
Purdue School of Engineering and Technology
Indiana University-Purdue University Indianapolis (IUPUI)
Indianapolis, IN 46202, USA

1. INTRODUCTION

There are many under-used computing resources. However, these computers may have different owners and may run under different operating systems with different computing and storage resources. One may run programs on these computers but does not have exclusive use of the resources. One may also find it difficult to keep track of the availability of the resources varying in time. Everyone thinks that her/his job should have a high priority.

Load balanced execution of parallel programs on distributed systems is desired on the application level as well as on the system level. The perception of load balancing by end users is often very different from that of the system administrators. End users only consider how to make their parallel jobs finish fast and do not care how the other users are affected. The system administrators are more concerned about each user/process having a fair share of the CPU time. Since the load balancing of a tightly coupled parallel job often needs the information in the application level and computer system level, system schedulers such as LSF, Condor, etc., cannot provide a good load balanced distribution for the multiple parallel jobs on the same set of computers. We have previously published that the accuracy of the computer load is really misleading when an unbalanced tightly coupled parallel job is running on a computer [1]. The misleading computer load information can severely affect the quality of process scheduling. When multiple parallel jobs are run on the same set of computers, the load balancing of one parallel job may affect the load balancing of other parallel jobs if they are not cooperated. We have proposed an algorithm to provide correct load information and to enable multiple jobs to coordinate their load balancing effort so that the load balancing of one parallel job will not affect that of other parallel jobs [2].

In this paper, we describe a distributed system software tool, DLB 2.0, that implements the idea presented in [2]. This tool provides an environment that allows each parallel job to do its own load balancing while ensures that the load balancing of one parallel job does not affect the load balancing of other parallel jobs. DLB 2.0 allows many parallel jobs to be executed concurrently on the same set of computers. One user does not need to know the existence of other users.

DLB 2.0 takes advantage of computing resources that would otherwise be wasted and puts them in good use. DLB 2.0 can find idle machines as they become available and eliminate the machines from user suggested list if they are not available or heavily loaded. DLB 2.0 determines the load distribution for a parallel application based on the speed of the computers and speed of the network, and the computation to communication ratio of the parallel application program. In this way, tremendous amounts of computation can be done with very little intervention from the user.

2. DESCRIPTION OF DLB 2.0

DLB 2.0 is a system software tool that allows each parallel job to do its application level load balancing while ensuring that system is load balanced. DLB 2.0 is designed to run on a pool of machines that are distributively owned, in other words, not maintained or administered by a central authority. Any computer owner can join the pool without the permission of other machines in the pool. DLB 2.0 provides a distributed infrastructure to support load balancing. When a machine is in the pool, it allows any user who has account on this machine to do dynamic load balancing following the rules imposed by DLB 2.0. The rule is simple. a computer only allows one parallel job to use the computer in the load balancing calculation at any time. An analogy is that only one car can go through an all way stop intersection. DLB 2.0 enforces this rule and user does not need to consider this rule.

The following guiding principles are used for the development of DLB 2.0:

(1) The DLB environment is a community joined by all computers voluntarily.

(2) There is no central control (or master) to resolve the conflict of load balancing of multiple parallel jobs. All computers follow the same load balancing rules that are supported by the DLB 2.0 software. It is just like that if everyone follows the traffic rules, there will not be a traffic accident.

(3) There is no limit on how many computers can join the community.

(4) A computer does not need other computers to support the DLB environment.

(5) A parallel program does not have to use the DLB environment. However, this program will have disadvantage in getting its shares of CPU power compare with other jobs that use the DLB environment.

(6) A user does not need to know all the computers in DLB pool. The user only needs to know all the computers that he/she has accounts.

MDLB is a multi-agent based system that includes the following programs: Graphical User Interface (GUI), DLB Agent, User Agent, Job Agent, System Agent, Processes Tracker (PTrack), Communication Cost Tracker (CTrack), and Stamp Library. All these programs are stored in a system directory or in a specific DLB account so that all users can execute them. The files in this account are readable and executable but not writable. DLB 2.0 needs to be installed on every computer. The programs in DLB 2.0 are divided in three groups: system level, user level and application library.

The programs in the system level include System Agent, PTrack, and CTrack. These system level programs should be run on every computer as system processes or DLB account's processes. They are started and stopped by the system administrator (like the print demon on UNIX). The system level DLB programs are responsible for collecting system load information and network communication speed information, providing system information to the load balancer, and gather information from other computers.

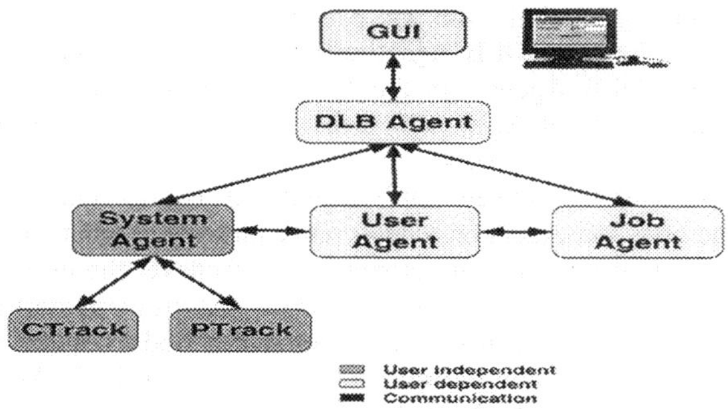

Figure 1. Software architecture of the DLB 2.0.

PTrack is a computer process-monitoring program. It runs on each computer and periodically (once every five elapsed minutes) reads the names and number of all the processes running on the computer. The names and number of running processes are passed to the System Agent periodically.

CTrack is a communication speed measurement program. It runs on every computer. It sends a round trip short message periodically between each pair of computers and finds the average time to send a message between every pair of computers. The communication speed information is passed to the System Agent periodically.

The System Agent is a program that runs on every computer. It accepts the registration of all jobs from all users through DLB agents, accepts the information from PTrack and CTrack periodically, determines the name and number of sequential and parallel processes running on the computer, keeps track on which parallel job is currently doing load balancing calculation, and keeps track on which job is idling for long time (possible hang up situation). System Agent provides all these information to whoever requests them.

The user level programs of DLB 2.0 include Graphical User Interface (GUI), DLB Agent, User Agent, and Job Agent. All user level programs are executed as user's processes. Each user needs only to run one User Agent and each parallel job needs to run one Graphical User Interface (GUI), one DLB Agent, and one Job Agent. The user level programs of DLB 2.0 do not need to be executed on the same computer. Anyone of them can be executed on any computer that the user has accounts on.

Graphical User Interface is developed for user to start and stop DLB Agent conveniently. Presently Graphical User Interface is running as a user process. User only needs to provide three information to start a parallel job through Graphical User Interface: (1) the list of all computers that the parallel program can be executed (this also means that the user has account on the computers), (2) the path of the parallel program on these computers, and (3) the desired number of parallel processes and desired number of computers to be used initially. In the future, Graphical User Interface will be a web service and be run as a system process.

DLB Agent is responsible for dynamic load balancing of parallel jobs. DLB Agent collects computer and application related information from System Agent and Job Agent for load balancing calculation. DLB Agent provides User Agent a suggested load distribution for executing a parallel job and executes the user's parallel application job through User Agent. Presently DLB Agent is running as a user process and each user needs to execute one DLB Agent. In the future, when Graphical User Interface is running as a web service, DLB Agent will be running as a system process so that only one DLB Agent is needed for a network.

User Agent is an interface between DLB Agent and parallel application jobs. Each user should execute one User Agent on at least one computer in order to use DLB 2.0. A User Agent can be started manually or automatically whenever the user is logged on the computer. User Agent is running as a user's process. At program start up, User Agent registers itself to System Agent so that the DLB Agent can find the where is a user's User Agent. Once found the connection information of User Agent, DLB Agent provides the load distribution information to the User Agent and asks the User Agent to start the Job Agent. The responsibility of the User Agent is only to start the user's Job Agent. It seems User Agent is not necessary since DLB Agent can start user's Job Agent directly. However, it will be essential when DLB Agent becomes a system process in the future.

Job Agent is responsible to start and stop user's parallel job and gather the execution timing information for the parallel job. The timing information is sent to the DLB Agent for load balancing. Since a user may run several parallel jobs concurrently, the User Agent starts and stops a Job Agent for each parallel job that the user wants to run.

The application library includes the Stamp library. The Stamp library is a set of functions that must be embedded into user's application program to get execution timing and communication information of a particular parallel job for future load balancing. Stamp library is written in C language and is callable from both FORTRAN and C programs. The timing information generated by the Stamp library is collected by the Job Agent and used by DLB Agent.

The DLB 2.0 supports load balancing of parallel jobs that run on computers of different architecture and different operating systems as long as the parallel job has an executable code on these machines. DLB 2.0 itself does not use PVM [3] or MPI. Parallel application programs use PVM or MPI. It just provides a suggestion of load distribution that is in the best interest of the user but does not enforce the user to run the job in the suggested load distribution. However, it is the disadvantage of the user if the suggested load distribution is not adopted. DLB 2.0 does not provide check pointing since a good check pointing for tightly coupled parallel program needs the knowledge of the application. We believe that it is the responsibility of the application program to provide check pointing. We are currently working on automatic restarting of the parallel application based on application provided check-pointing information if the program is stopped due to variety reasons. The execution of DLB Agent and System Agent is depicted in Figure 2.

3. EXPERIMENTS

DLB 2.0 has been successfully tested on 106 computers with 219 processors. The computers include PC, Sun workstations, IBM RS6000, and IBM PS running under different operating systems. The locations of these machines are shown in Figure 3.

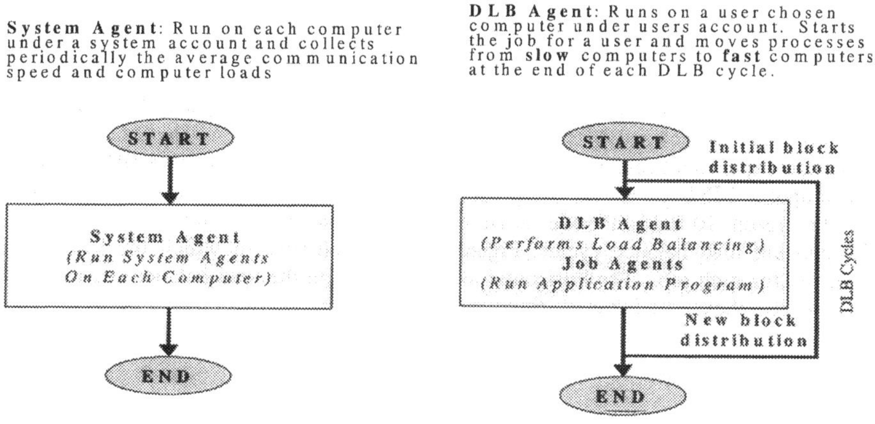

Figure 2. System Agent and DLB Agent are executed independently.

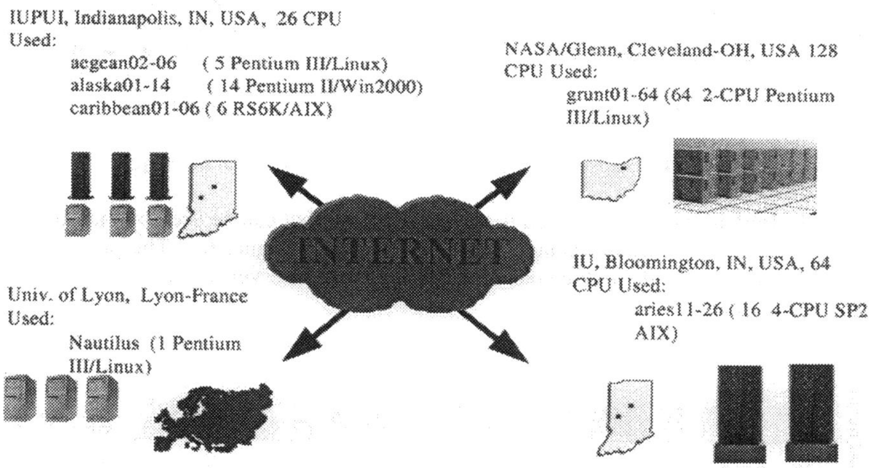

Figure 3. Computers used for testing DLB 2.0.

Conceptually, many parallel CFD programs can be load balanced concurrently on these computers. The following experiment used 18 machines that include 5 IBM RS6000 (IBM AIX) and 3 PC (Linux) at Indianapolis, Indiana, and 10 IBM SP (IBM AIX) nodes at Bloomington, Indiana. All RS6000s and PCs are single CPU machine. An RS6000 is 4 times faster than a PC. Each SP node has 4 CPUs. Each CPU of a SP node is two times faster than that of RS6000. The speed differences implies that if loads are balanced on those machines, for each block on PC, there should be 4 blocks on an RS6000, and 32 blocks on an SP node.

Two CFD jobs are executed concurrently. Job1 has 100 data blocks that are distributed to all 18 machines. Job2 has 50 data blocks that are distributed to 13 machines that include 5 IBM RS6000, 3 PC, and 5 IBM SP nodes. The following figures show the experiment result. In the figures, AIX1 to AIX5 represents 5 IBM RS6000 at Indianapolis. LINUX1 to LINUX3 represent 3 PC (Linux) at Indianapolis, and SP1 to SP10 represent 10 IBM SP nodes at Bloomington, Indiana. Each job had many mutually independent load balance cycles. Figure 4 shows the time at which the load balance occurred for each job. The horizontal axis shows when the load balance occurred. The vertical axis shows the job execution time per time step.

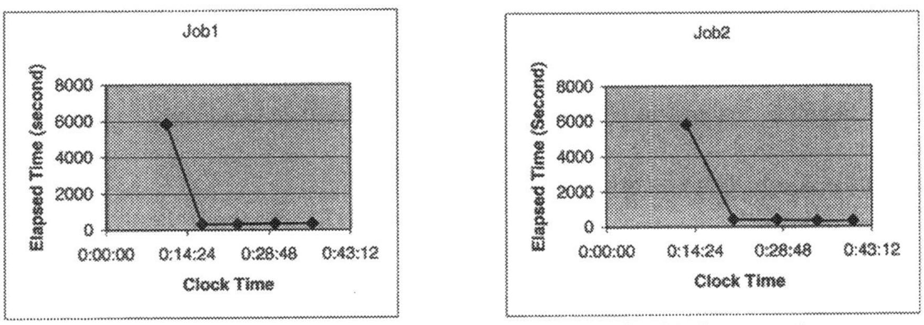

Figure 4. Elapsed time per time step in each load balance cycle.

Initially Job1 is evenly distributed to all the machines that can be used by that job (see Figure 5). Job2 is evenly distributed to 13 machines (see Figure 6). The processes of Job2 are considered as the extraneous load for Job1 and vise versa. A load balancer is used for each job.

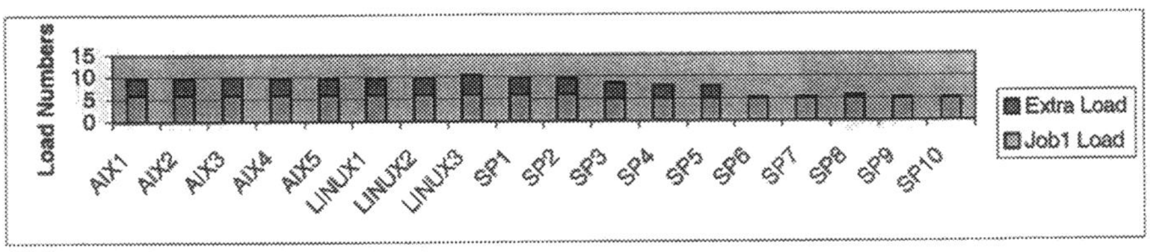

Figure 5. Initial load distribution for Job 1.

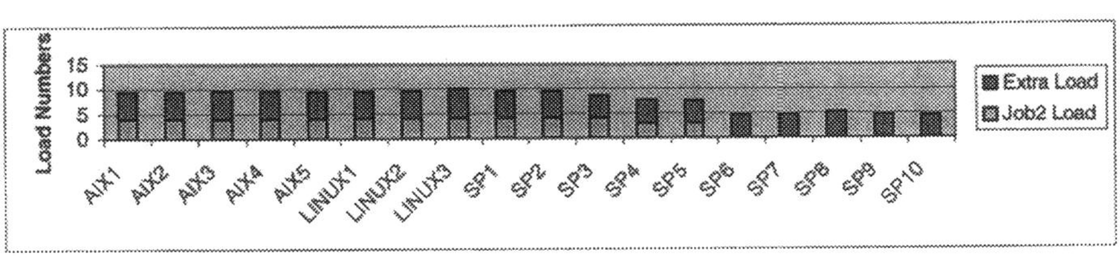

Figure 6. Initial load distribution for Job 2.

Job1 did first load balancing before Job2. After Job1's first load balancing, DLB moves all Job1 blocks from AIX and LINUX to SP except its first block (see Figure 7). After Job2 finishes its first cycle, Job2's load balancer moves all Job2's blocks to SP (see Figure 8After a while, Job1's load balancer moved some blocks to the machine that Job2 cannot use (see Figure 9). The machines are balanced at this time. Figure 10 shows that the load balancer for balance cycle 2 of jobs made no improvement.

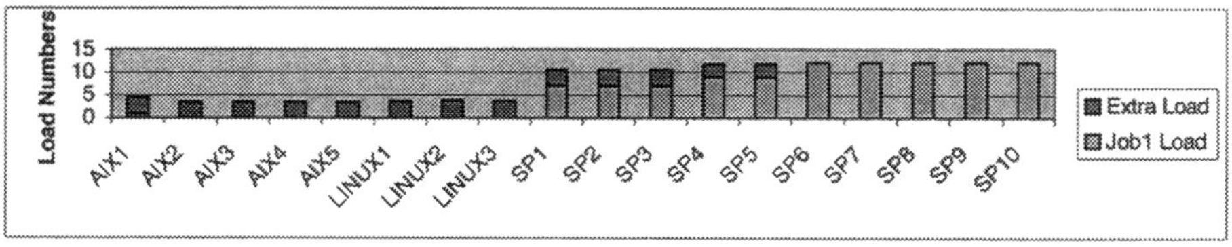

Figure 7. Load distribution after balance cycle 1 for Job1.

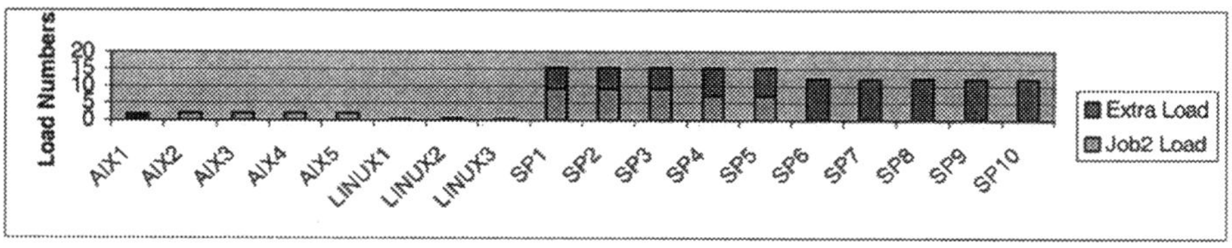

Figure 8. Load distribution after balance cycle 1 for Job 2.

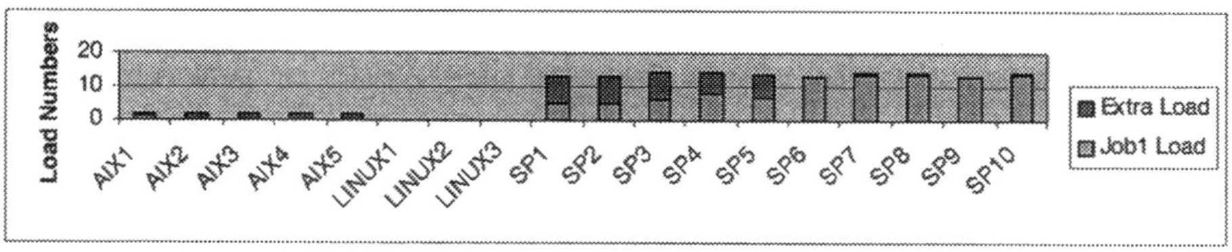

Figure 9. Load distribution after balance cycle 2 for Job 1.

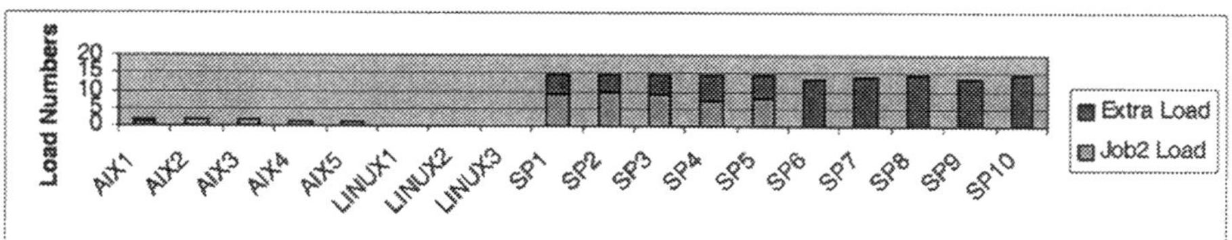

Figure 10. Load distribution after balance cycle 2 for Job 2.

4. DISCUSSION

Since DLB 2.0 is installed on each computer without the knowledge of other computers, a user cannot access other users' information through DLB 2.0. The information that the parallel job provides to the System Agent is only the name of the parallel job and connection port address of the User Agent. Since the User Agent can only execute a few predefined functions, there is no threat of system security even if a hacker has the access of these functions.

All programs except the DLB Agent in DLB 2.0 are written in Java. DLB Agent is written in a combination of Java and C++. The following software are required to run DLB 2.0: (1) Java run-time environment (JRE): JRE 1.2.2 or above, (2) Message passing package for parallel computing: PVM 3.4.4 or MPICH, and (3) a UNIX equivalent rsh capability. The rsh utility is required for running PVM and/or MPI on Windows NT/2000. There is commercial software that supports rsh on Windows based systems. DLB 2.0 currently supports the following computers:

- IBM RS6K with IBM-AIX 4.3 or above
- IBM SP with IBM-AIX 4.3 or above
- Sun Sparc with Sun OS
- Pentium PC with Windows NT 4.0 or Windows 2000
- Pentium PC with Linux OS

PTrack is the only system dependent program in DLB 2.0. Since PTrack is a very small program, DLB 2.0 can be quickly ported to other types of computers. For example, we only used a few days to port DLB 2.0 to Linux system.

5. CONCLUSIONS

The dynamic load balancer, DLB 2.0, is a software tool that provides support to many users who run parallel jobs concurrently on the distributed heterogeneous computers while maintaining the load balance from both user's point of view and system administrator's point of view. DLB 2.0 is a distributed system that has no master.

ACKNOWLEDGEMENT

The authors greatly appreciate NASA Glenn Research Center for the financial support of this research (Grant No. NAG3-2260). The authors are also grateful for the computer support provided by the IBM Research Center throughout this study.

REFERENCES

1. Chien, Y.P., Chen, J.D., Ecer, A., and Akay, H.U., "Tools to Support Load Balancing of Multiple Parallel Jobs," *Proceedings of Parallel Computational Fluid Dynamics,* 2000.
2. Ecer, A., Chien, Y.P., Akay, H.U., and Chen, J.D., "Load Balancing for Multiple Parallel Jobs," *Proceedings of 2000 European Congress on Computational Methods in Applied Sciences and Engineering*, September 2000.
3. Oak Ridge National Laboratory, "PVM, Parallel Virtual Machine," World Wide Web at http://www/epm.ornl.gov/pvm/.

Parallel Computational Fluid Dynamics – Practice and Theory
P. Wilders, A. Ecer, J. Periaux, N. Satofuka and P. Fox (Editors)

Parallel Computation of Thrust Reverser Flows for Subsonic Transport Aircraft

C. Chuck[1], S. Wirogo[2], D. R. McCarthy[3]
[1,3] Enabling Technology Research, Airplane Performance and Propulsion
The Boeing Company, P.O. Box 3707, MC 67-LF, Seattle, Washington 98124, USA
[1]Telephone: 425-237-3162 [3]425-234-8803 Fax: 425-237-8281
[2]Fluent Incorporated, 10 Cavendish Court, Centerra Resource Park
Lebanon, New Hampshire 03766-1442, USA
Telephone: 603-643-2600 Fax: 604-643-3976

KEYWORDS: Parallel unstructured solvers, industrial applications, domain decomposition

ABSTRACT

As our ability to compute complex fluid flows improves, the expectations of aircraft designers steadily keep pace. Today they rely increasingly on calculations which would not have been possible only a few years ago, using CFD not just to improve the performance of the usual wings and propulsion systems, but to avoid expensive downstream surprises in less familiar components.

1. INTRODUCTION

Modern jet transports employ a special system which reverses engine thrust during landing. When this system is activated, blocker doors drop into the fan stream closing off engine efflux to the rear; simultaneously, the rear portion of the engine cowling translates aft to expose an array of honeycomb structures which redirect the efflux partially forward. This system operates for only a few seconds in each flight, so its efficiency is not at issue. Nevertheless, there are critical design considerations. Since the reverser is in use precisely at the time when high lift devices, such as leading and trailing edge flaps and slats attached to the wing, are fully deployed, the plumes must be carefully directed so as not to impinge on these or other parts of the aircraft. Other effects to avoid are engine re-ingestion of the plume or of debris blown up from the runway, and envelopment of the vertical tail.

Because the placement and direction of the reverser affects the overall airplane configuration, these decisions must be made early in the program, often before any wind tunnel testing is done. Wind tunnel testing for these effects is very expensive and difficult in any event. Correcting any problem which might arise later on, say during flight test, would engender expensive retooling and unacceptable delays. Therefore, it is essential to verify the design computationally before anything is built.

In this work, we demonstrate a computational procedure for analyzing the flow around a typical jet transport. The aircraft model shown [Figure 1] is completely hypothetical, but representative of the cases of interest. We include the fuselage; horizontal and vertical stabilizers; wing with leading and trailing edge high lift devices deployed; engine pylon; and nacelle, with detailed inlet and multi-stream nozzle, all operating in reverse thrust on the ground. The procedure encompasses mesh generation, flow solution and post processing. Early calculations of simpler configurations using structured grid methods were reported in [1]. In this work, in order to accommodate increasing geometric complexity, fully unstructured meshes were use, obtained using the Tetra module of grid generation software from ICEM-CFD© Engineering. The flow fields were computed using Fluent V5©, an unstructured Navier-Stokes solver from Fluent Inc. Turbulence was modeled by using a two-equation k-ε model and wall functions near the solid surfaces. A more complete description of the physical models and assumptions used in the calculations can be found in [2].

2. COMPUTATIONS

The flow computations can require several days per case when run on a single processor. Thus, for the usual good reasons, we are driven to parallel processing, and, fortunately, the unstructured mesh approach is "embarrassingly" amenable to it. The calculations we show were done on the NAS SGI Origin© with a total of 512 R12000 CPUs. A typical surface mesh, together with the mesh on the ground and symmetry planes is shown in Figure 2. Figure 3 shows some mesh details in the vicinity of the nacelle, including several of the exposed cascade baskets. Mesh partitioning was done automatically by the Fluent© software. The solution was started with a first order scheme using initially small relaxation steps. Once transients exited the computational domain, the under-relaxation factors were ramped up and the solution procedure continued using second order scheme. As a stopping criterion, stable plume shape is preferable, though hard to monitor. Generally, the solution was assumed to have converged after the residuals dropped by 3 to 4 orders of magnitude. Depending on the number of CPUs, a final solution could generally be obtained in a few hours.

We undertook calculations at four different mesh densities from 1.9M to 6.28M cells and different numbers of CPUs from 4 to 480. A handful of cases at the extremes were impractical to run.

3. PERFORMANCE MEASURES

Of real importance to the analyst is information concerning the parallel performance of the procedure as a function of problem size and number of CPUs. It is by now standard to measure the efficiency of parallel scientific computing as the deviation of total computing time from linear speedup, presented as a function of the number of processors. However, from the point of view of an analyst operating on shared resources in an industrial environment, this is only part of the story. For one thing, the analyst often has a choice of software tools, and is interested to know if a particular tool is appropriate for parallel execution. This leads to the issue of algorithmic efficiency, i.e., a measure of whether the underlying solution algorithm itself deteriorates with the degree of parallelism. This is quite separate from the question of how long it takes to communicate boundary data between processors. In particular, one might expect that codes relying on strongly implicit methods would converge more slowly when extensively parallelized by domain decomposition.

The industrial practitioner generally agrees that the only reason to engage in parallel computing is to reduce wall clock time. Cost, however, is also important, and most computer centers have interesting schemes for levying it. Thus we adopt in what follows a slight generalization of the usual way of looking at parallel efficiencies. Without indulging in excessive rigor, we introduce a notation designed to expose the dependencies. Suppose we want to solve for a flowfield on grid G, using partitioning strategy Π. Then Π determines the number of processors, $p(\Pi)$. The time for one solver iteration on G using Π we call $T(G, \Pi)$, and the convergence rate on G using Π, that is, the residual reduction factor per iteration, we call $r(G, \Pi)$.

Assuming, rather generously, that the cost of computing per processor per unit time is a constant, C^0, we can easily write down the wall clock time and computing cost required to reduce the residual by a factor of q. These are, respectively,

$$W(G,\Pi,q) = \#\text{iterations} * \text{time/iteration}$$

$$= T(G,\Pi) \; * \; \frac{-q}{\log r(G,\Pi)}$$

and

$$C(G,\Pi,q) \approx \#\text{CPU's} * \text{wall clock time}$$

$$= C^0 * p(\Pi) * W(G,\Pi,q)$$

To express our results in terms of efficiencies, we normalize by introducing the quantities Π^0, the reference partition (often just a single domain); T^0, the time for one iteration on G using Π^0, and r^0, the convergence rate on on G using Π^0. Then putting it all together, we have

$$C(G,\Pi,q) = -\frac{C^0 T^0 q}{\log r^0} * \underbrace{\left[p(\Pi)\frac{T(G,\Pi)}{T^0} \right]}_{} * \underbrace{\left[\frac{\log r^0}{\log r(G,\Pi)} \right]}_{}$$

$$= \frac{\text{Charging, code, and problem characteri stics}}{\text{Computatio nal efficiency * Algorithmi c efficiency}}$$

Here the first factor is a function of the charging algorithm, the basic (single zone) algorithmic characteristics of the solver, the speed of a single CPU, and the required level of convergence. Except for the choice of solver, these items are usually not within the purview of the analyst. The second factor is (the reciprocal of) the computational efficiency: one hopes that the time, T, will decrease inversely with the number of processors, p. If it does not, this factor exceeds 1. The third factor is (again the reciprocal of) the algorithmic efficiency, the degree to which the solution procedure itself, as opposed to its implementation, tolerates domain decomposition.

The various components of these expressions were measured during our experiments. In our work, the partitioning strategy Π consisted of dividing the domain into roughly equal parts and assigning a single processor to each part. Thus $p(\Pi)$ was simply the number of subdomains.

We begin with the algorithmic efficiency. The solution algorithm underlying the solver is based on algebraic multigrid; when the problem is decomposed, the set of equations to be solved is decoupled into a number of subsets of fewer equations each. Since algebraic multigrid attempts to treat the set as a whole, one might expect it to exhibit some of the characteristics of an implicit solver. (See, e.g., [3].)

In figure 4 we superimpose a portion of the convergence histories for various numbers of processors (i.e., subdomains), finding that these histories are essentially identical, all with a convergence rate of about .992. We conclude that $r(G, \Pi) = r^0$, independent of Π, and the algorithmic efficiency is 1. This implies that the number of iterations required for convergence is also independent of Π, and so the overall efficiencies can be expressed through *per iteration* quantities.

The computational efficiency is shown in figure 5, and is a strong function of Π. To avoid claims of super-linear performance, presumably associated with the overhead of setting up parallel calculations for small numbers of sub-domains, we have normalized the curves to the partition Π^0 corresponding to 24 processors, as this was in fact the most computationally efficient choice for all 4 grid sizes. One sees the efficiency curve shifts to the right as the number of processors increases. This corresponds to the intuitive conclusion that more processors are useful as the problem size grows. It is apparent that for problems of the size undertaken here, processor choices in the range of 16 to 64 are generally appropriate. It is also clear that efficiency declines abruptly above this number.

Lastly, as outlined above, the quantities of immediate interest to the analyst running the code are the job cost and wall clock time. This may be viewed simply as a multiple objective optimization problem, in which both of these numbers are to be made small. Such problems seldom possess unique optima; more likely there will be a trade between speed and cost which can only be resolved once the budget and priorities are fixed. Since these cannot be known ahead, we represent these trades in a graph of the type shown in figure 6.

The axes in the figure represent the two quantities to be minimized. The curve is the locus of achievable results, parameterized by the number of processors. A point on the curve is a *Pareto* optimum if neither objective can be improved except at the expense of the other, that is, if there is no other point on the curve to its lower left. The Pareto boundary, then, is the southwest quadrant of the curve: there is no advantage to choices outside this region. In particular, we see from this type of analysis that there is more involved in the loss of computational efficiency above 128 processors than just the law of diminishing marginal utility. Beyond 128 processors, the marginal utility for both quantities is actually negative, since both the cost and the wall-clock time increase.

4. AERODYNAMIC RESULTS

The purpose of calculations of this kind in the aircraft industry is, of course, to investigate the flow field rather than the software. We therefore show some typical results. In this case, the nature of the calculations is more qualitative than quantitative: we are interested in where the flow goes, rather than detailed agreement with pressure measurements.

Calculations were made at a number of different flow conditions common during landing. We show results at Mach 0.076, or about 50 knots. This is a low speed achieved toward the end of reverser operation, and the plumes are well developed. Figure 7 shows a frontal view with particle traces. The plume from the lower inboard cascades can be seen to impinge on the runway at an angle, suggesting a possibility that debris could be blown up from the ground. The same plume rebounds from the runway and collides with plume from the other side of the airplane. This creates a high pressure region under the fuselage in the wing root area. In figure 8, closer to the engine, we can see evidence of re-ingestion of the reverser flow by the engine inlet. Based on these observations, the PD (preliminary design) engineer may choose to reorient the direction of some of the cascade baskets. The side view in figure 9, on the other hand, shows that the reverser efflux does not impinge on the wing leading edge slats. This information is far more detailed than any that could be obtained by wind tunnel or flight testing, and arbitrary particle traces can be created and examined from essentially any angle with standard flow visualization software

5. CONCLUSIONS

Parallel calculations of complex flows of practical interest in design are now not only commonplace, but is also indispensable. Here we have demonstrated the utility of a parallel, unstructured grid Navier-Stokes analysis of the extremely convoluted flows encountered in aircraft thrust reversers.

In addition, we have profiled the code's performance as a function of the number of processors and the grid density. We verified that the basic algorithmic behavior of the solver is insensitive to the degree of parallelism, and demonstrated that the effects of increasing the number of processors beyond a reasonable limit actually had a deleterious effect on both computing cost and wall clock time. Further, we devised a graphical tool to assist the analyst in determining the most efficacious number of processors to use given the relative priorities of speed and cost.

And, finally, we obtained computed flow fields which illustrate the power of these analyses to suggest improvements early on and decrease the risk of expensive downstream redesigns. Already, designers expect calculations of this kind to be routine. They ask for new configurations and new physics faster than we can respond. We see little remaining doubt that parallel computing represents the only near term hope of keeping pace with real applications.

6. ACKNOWLEDGEMENTS

We express our appreciation to the NAS computing center, and to Chuck Niggley and Cathy Schulbach of NASA, for their help in getting the Fluent code installed and for providing the computational resources; to S. Senthan and Anshul Gupta of ICEM CFD Engineering, for their assistance with grid generation; and to the Fluent corporation, for providing the software licenses and for assistance with the parallelization and running of the code.

REFERENCES

1. C. Chuck, E. Hsiao, J. Colehour, M. Su, and J. Jackson, Navier–Stokes Calculations of Under Wing Turbofan Nacelles, AIAA-98-2734, 1998
2. C. Chuck, Computational Procedures for Complex Three-Dimensional Geometries Including Thrust Reverser Effluxes and APUs, AIAA-2001-3747, 2001
3. McCarthy, D. R., Optimizing Compound Iterative Methods for Parallel Computation, Proceedings, Parallel CFD '96, May 20-23 1996, Capri, Italy.

PLOTS AND FIGURES

Figure 1 Generic Subsonic Transport Aircraft

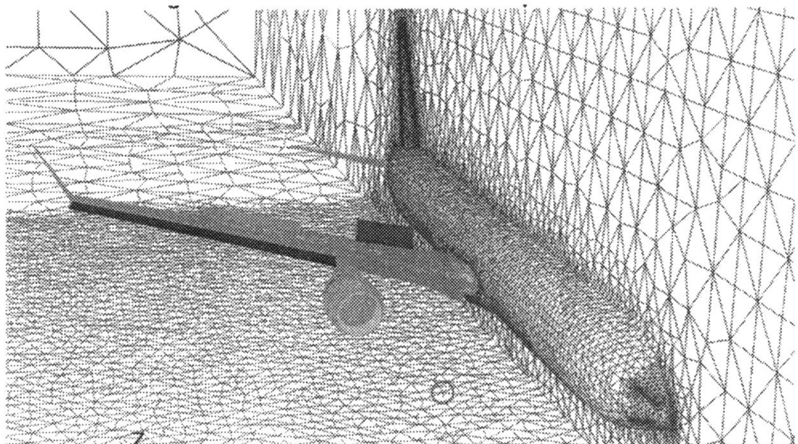

Figure 2 Surface Mesh of Airplane, Runway, Downstream and Symmetry Plane

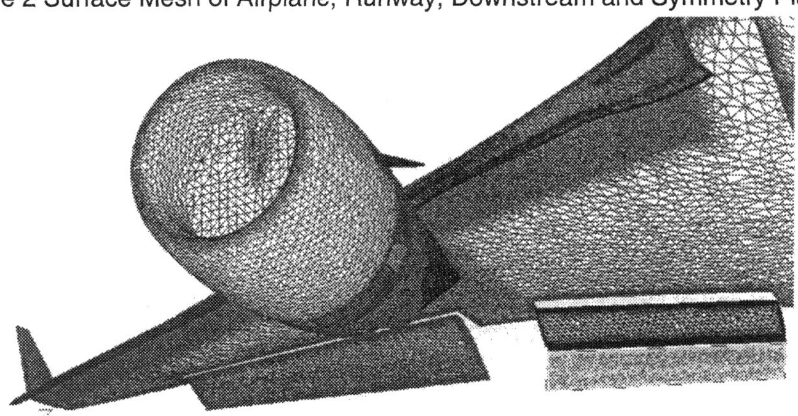

Figure 3 Surface Mesh near Nacelle Area

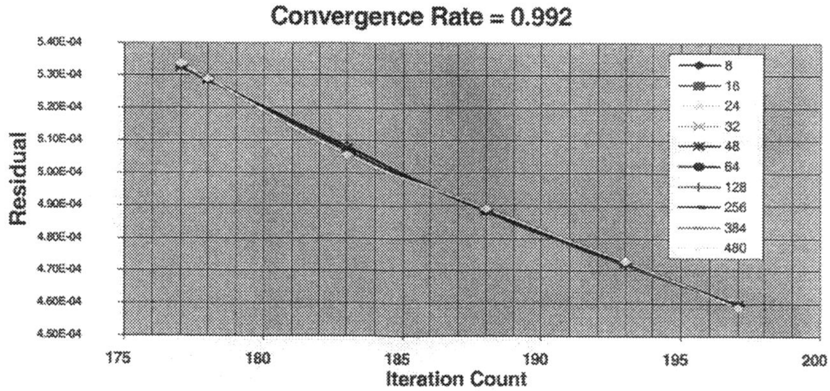

Figure 4 Convergence Rate

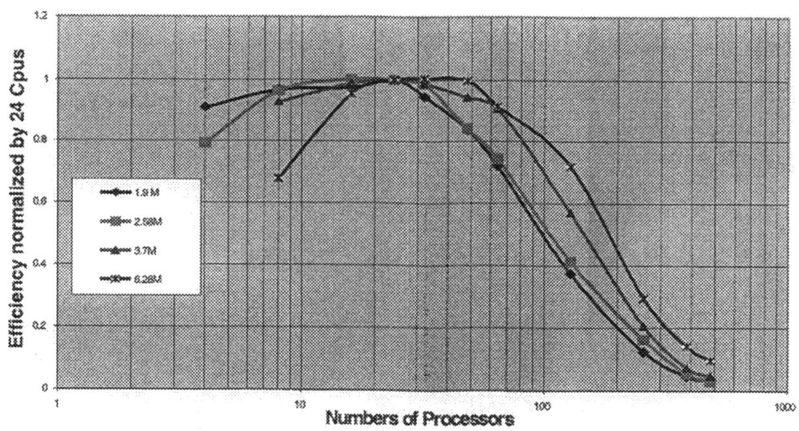

Figure 5 Computational Efficiency

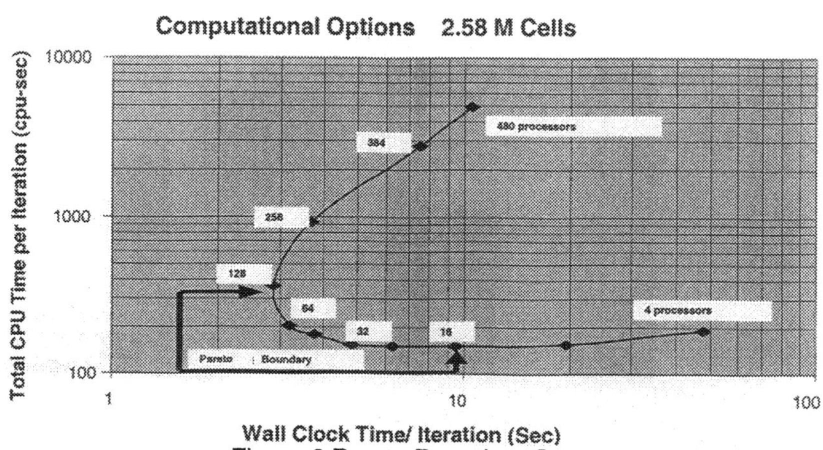

Figure 6 Pareto Boundary Curve

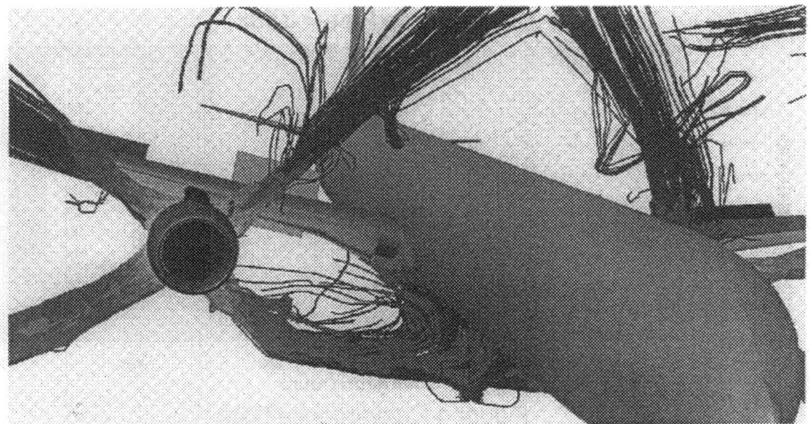

Figure 7 Particle Traces of Reverser Efflux Frontal View

Figure 8 Particle Traces of Reverser Efflux at Engine Inlet.

Figure 9 Particle Traces of Reverser Efflux from Wing Root Side View

Parallel Computational Fluid Dynamics – Practice and Theory
P. Wilders, A. Ecer, J. Periaux, N. Satofuka and P. Fox (Editors)

On a Fast Parallel Solver for Reaction-Diffusion Problems: Application to Air Quality Simulation *

W.E. Fitzgibbon †, M. Garbey †† and F. Dupros ††
† Dept. of Mathematics - University of Houston, USA
†† CDCSP- ISTIL - University Lyon 1, 69622 Villeurbanne France

In this paper, we consider reacting flow problems for which the main solver corresponds to reaction-diffusion-convection equation. Typical examples are large scale computing of air quality model but it can be applied also to heat transfer problems. We introduce a new familly of reaction-diffusion solvers based on a filtering technique that stabilizes the explicit treatment of the diffusion terms. We demonstrate that this method is numerically efficient with examples in air quality models that usually require the implicit treatment of diffusion terms. For general reaction-diffusion problems on tensorial product of one dimensionnal grids with regular space step, the filtering process can be applied as a black box post-processing procedure. Further, we show on critical components of the algorithm the high potential of parallelism of our method on medium scale parallel computers.

1. Introduction

We consider reacting flow problems for which the main solver corresponds to reaction-diffusion-convection equation:

$$\frac{\partial C}{\partial t} = \nabla.(K\nabla C) + (\vec{a}.\nabla)C + F(t,x,C), \tag{1}$$

with $C \equiv C(x,t) \in R^m$, $x \in \Omega \subset \mathbf{R}^3, t > 0$. A Typical example is an air pollution model where $\vec{a}$ is the given wind field, and F is the reaction term combined with source/sink terms. For such model m is usually very large, and the corresponding ODE system is stiff. The equation (1) can be rewritten as

$$\frac{DC}{Dt} = \nabla.(K\nabla C) + F(t,x,C), \tag{2}$$

where $\frac{D}{Dt}$ represents the total derivative.

The method of characteristics provides a good tool for the time discretization. The main problem remains the design of a fast solver for reaction diffusion who has good stability properties with respect to the time step but avoids the computation of the full Jacobian matrix. Usually one introduces an operator splitting combining a fast non linear ODE solver with an efficient linear solver for the heat operator. However the stiffness of the reaction terms induces some unusual missperformance problems for high order operator splitting. In fact, the classical splitting of Strang might perform less well than a first order source splitting [12]. Following the pionner work of A. Ecer published in this proceeding serie, we explore some alternative methodology

*This work was supported by the Région Rhône Alpes and the US Envir. Prot. Agency.

in this paper that consists of stabilizing with a posteriori filtering, the explicit treatment of the diffusion term. The diffusion term is then an additionnal term in the fast ODE solver and the problem is parametrized by space dependency. It is easy, at first sight, to have an efficient parallel algorithm due to the intense pointwise computation dominated by the time integration of the large system of ODEs. However load balancing is necessary and it should be dictated by the integration of the chemistry [5] [2] and therefore is strongly solution dependant.

The stabilizing technique based on filtering presented in this paper is limited to grid that can be mapped to regular space discretization or grid that can be decompose into overlapping subdomains with regular space discretization. We should point out that an alternative and possibly complementary methodology to our approach is the so called Tchebycheff acceleration [4] that allows so-called super time steps that decomposes into apropriate irregular time stepping.

The plan of this article is as follows. Section 2 presents the methodology for reaction diffusion problem first in one space dimension and second its generalisation to multidimensional problem with one dimensional domain decomposition. Section 3 gives examples of a computation of a simplified Ozone model. Section 4 comments on the parallel implementation of the method and first results on performance. Section 5 is our conclusion.

2. Method

2.1. Fundamental observations on the stabilization of explicit scheme

In this section, we restrict ourselves to the scalar equation

$$\partial_t u = \partial_x^2 u + f(u) \ , \ x \in (0, \pi), t > 0. \tag{3}$$

We consider the following second order accurate scheme in space and in time:

$$\frac{3u^{n+1} - 4u^n + u^{n-1}}{2\ dt} = 2\ D_{xx}u^n \ - \ D_{xx}u^{n-1} \ + \ f(u^{n+1}). \tag{4}$$

We recall that backward second order Euler is a standard scheme used for stiff ODEs [11], [13]. We restrict ourselves to finite difference discretization with second order approximation of the diffusion term. Extensions to finite volume will be reported elsewhere. The Fourier transform of (4) when neglecting the nonlinear term has the form:

$$\frac{3\ \hat{u}^{n+1} - 4\ \hat{u}^n + \hat{u}^{n-1}}{2\ dt} = \Lambda_k\ (2\ \hat{u}^n \ - \ \hat{u}^{n-1}), \tag{5}$$

where $\Lambda_k = \frac{2}{h^2}(cos(hk) - 1)$. The stability condition for wave number k has the form

$$2\ \frac{dt}{h^2}|cos(\frac{k\pi}{N}) - 1| \ < \ \frac{4}{3}, \tag{6}$$

with $h \ = \ \frac{\pi}{N}$. The maximum time step allowed is then

$$dt < \frac{1}{3}h^2. \tag{7}$$

However it is only the high frequencies that are responsible for such a time step constraint, and they are poorly handled by second order finite differences. Therefore the main idea is to construct a filtering technique that can remove the high frequencies in order to relax the constraint on the time step *while keeping second order accuracy in space.*

Let's $\sigma(\eta)$ be a filter function of order 8 [10]. We are going to filter the solution provided by the semi-implicit scheme (4) after each time step. We will neglect in the following notations the time step dependency of u and denote $u(0) = u_0$ and $u(\pi) = u_\pi$.

Because of the Gibbs phenomenon, the sine expansion of $u(x)$ is a very poor approximation of u. From [10], we observe that a discontnuity of $u(x)$ leads to a Fourier expansion with error $O(1)$ near the discontinuity and $O(\frac{1}{N})$ away from the discontinuity. We must apply a shift on $u(x)$ followed by a filter in such way that we preserve the accuracy on $u(x)$ and remove some of the high frequencies in order to lower the time step constraint on the explicit treatment of the diffusion term. We propose the following solution that is applied *at each time step*: first, we apply the low frequency shift:

$$v(x) = u(x) - (\alpha\ cos(x) + \beta),\ with\ \alpha = \frac{1}{2}(u_0 - u_\pi),\ \beta = \frac{1}{2}(u_0 + u_\pi). \tag{8}$$

Then, we extend v to $(0, 2\pi)$, with $v(2\pi - x) = -v(x),\ x \in (0, \pi)$. $v(x)$ is therefore a 2π periodic function that is $C^1(0, 2\pi)$.

Let $\hat{v}_k$ be the coefficients of the Fourier expansion of $v(x),\ x \in (0, 2\pi)$. The second step is to apply the filter:

$$\sigma_N v(x) = \Sigma_k \hat{v}_k \sigma(\kappa\frac{k}{N}) exp(ikx), \tag{9}$$

where $\kappa > 1$ is a stretching factor to be defined later on. The third step is to recover u from the inverse shift

$$u(x) = \sigma_N v(x) + \alpha\ cos(x) + \beta. \tag{10}$$

The correct choice for κ follows from the Fourier analysis with (6); we have

$$\kappa > \kappa_c = \frac{\pi}{acos(1 - \frac{2}{3}\frac{h^2}{dt})}. \tag{11}$$

In practice, because the filter damps some of the high frequencies less than $\frac{N}{\kappa}$, it can be suitable to take $\kappa = C_l\ \kappa_c$ with C_l that is less than 1. One can further compute optimum κ value for each time step by monitoring the growth of the highest waves that are not completely filtered out by $\sigma(\kappa\frac{k}{N})$. Further improvement consist to filter the residual in combination with a higher order shift in order to recover a solution with higher accuracy and lower κ [9].

2.2. Generalization to two space dimension

For simplicity, we restrict ourselves this presentation to 2 space dimensions, but the present method has been extended to 3 dimensional problems. Let us consider the problem

$$\partial_t u = \Delta u + f(u)\ ,\ (x, y) \in (0, \pi)^2, t > 0, \tag{12}$$

in two space dimension with Dirichlet boundary conditions

$$u(x, 0/\pi) = g_{0/\pi}(y),\ u(0/\pi, y) = h_{0/\pi}(x),\ x, y \in (0, \pi),$$

subject to compatibility conditions:

$$g_{0/\pi}(0) = h_0(0/\pi),\ g_{0/\pi}(\pi) = h_\pi(0/\pi).$$

Once again, we look at a scheme analogous to (4) with for example a five point scheme for the approximation of the diffusive term. The algorithm remains essentially the same, except the fact that one needs to construct an apropriate low frequency shift that allows the application of a filter to a smooth periodic function in *both* space directions. One first employs a shift to obtain homogeneous boundary condition in x direction

$$v(x,y) = u(x,y) - (\alpha cos(x) + \beta),\ with\ \alpha(y) = \frac{1}{2}(g_0 - g_\pi),\ \beta(y) = \frac{1}{2}(g_0 + g_\pi). \tag{13}$$

and then an additional shift in y direction as follows:

$$w(x,y) = v(x,y) - (\gamma cos(y) + \delta),\ with\ \gamma(x) = \frac{1}{2}(v(x,0) - v(x,\pi)),\ \delta(x) = \frac{1}{2}(v(x,0) + v(x,\pi \tag{14}$$

In order to guarantee that none of the possibly unstable high frequency will appear in the reconstruction step:

$$u(x) = \sigma_N w(x) + \alpha cos(x) + \beta + \gamma cos(y) + \delta, \tag{15}$$

high frequency components of the boundary conditions g must be filtered out as well.

The domain decomposition version of this algorithm with strip subdomain and adaptive overlap has been tested and gives similar results to the one dimensional case [7].

3. Application to a simplified Air Polution model

We have applied our filtering technique to air pollution models in situations where diffusion terms requires usually implicit solver in space. As a simple illustration, we consider the following reactions which constitute a basic air pollution model taken from [13]:

$$NO_2 + h\nu \longrightarrow^{k_1} NO + O(^3P)$$

$$O(^3P) + O_1 \longrightarrow^{k_2} O^3$$

$$NO + O_3 \longrightarrow^{k_3} O_2 + NO_2$$

We set $c_1 = [O(^3P)]$, $c_2 = [NO]$, $c_3 = [NO_2]$, $c_4 = [O_3]$. If one neglect viscosity, the model can be described by the ODE system:

$$c_1^\bullet = k_1 c_3 - k_2 c_1$$

$$c_2^\bullet = k_1 c_3 - k_3 c_2 c_4 + s_2$$

$$c_3^\bullet = k_3 c_2 - k_1 c_3$$

$$c_4^\bullet = k_2 c_1 - k_3 c_2 c_4$$

We take the chemical parameters and initial datas as in [13]. It can be shown that this problem is well posed, and that the vector function $c(t)$ is continuous [6]. At transition between day and night the discontinuity of $k_1(t)$ brings a discontinuity of the time derivative $c^\bullet$. This singularity is typical of air pollution problem. Nevertheless, this test case can be computed with 2nd Backward Euler (**BDF**) and constant time step for about four days, more precisely $t \in (0, 3.10^5)$ with $dt < 1200$. We use a Newton scheme to solve the nonlinear set of equations provided by BDF at each time step.

We recall that for air pollution, we look for numerically efficient scheme that deliver a solution with a 1 % error.

Introducing spatial dependancy with apropriate diffusion term in the horizontal direction and vertical transport, we have shown that our filtering technique produce accurate results [7].

We now are going to describe some critical elements of the parallel implementation of our method for multidimensional air pollution problems.

4. On the Structure and Performance of the Parallel Algorithm

In Air quality simulation, 90% of the elapsed time is usually spent in the computation of the chemistry. Using operator splitting or our filtering technique, this step of the computation is parametrized by space. Consequently, there are no communication between processors required and the parallelism of this step of the computation is (in principle) trivial. One however need to do the load balancing carefully, because the ODE integration of the chemistry is an iterative process that has a strong dependance on initial conditions.

In this paper, we restrict our performance analysis to the 10% of remained elapsed time spent to treat the diffusion term and possibly convective term that do require communication between processors.

For simplicity, We will restrict our System of reaction diffusion to two space dimensions. The performance analysis for the general case with three space dimensions give rise to analogous results.

The code has to process a 3 dimensional array $U(1 : N_c, 1 : Nx, 1 : Ny)$ where the first index corresponds to the chemical species, the second and third corresponds to space dependency. The method that we have presented in Sect 2 can be decomposed into two steps:

- Step1: Evaluation of a formula

$$U(:,i,j) := G(U(:,i,j), U(:,i+1,j), U(:,i-1,j), U(:,i,j+1), U(:,i,j-1)), \quad (16)$$

  at each grid points provided apropriate boundary conditions.

- Step 2: Shifted Filtering of U(:,i,j) with respect to i and j directions.

Step 1 corresponds to the semi-explicit time marching and is basically parametrized by space variables. The parallel implementation of Step 1 is straightforward and its efficiency analysis well known [3]. For intense point wise computation as in air pollution, provided apropriate load balancing and subdomain size that fit the cache memory, the speedup can be superlinear. The data structure is imposed by Step1 and we proceed with the analysis of the parallel implementation of Step2.

Step 2 introduces a global data dependencies across i and j. It is therefore more difficult to parallelize the filtering algorithm. The kernel of this algorithm is to construct the two dimensional sine expansion of $U(:,i,j)$ modulo a shift, and its inverse. One may use an off the shelf parallel FFT library that supports two dimension distribution of matrices -see for example http://www.fftw.org- In principle the arithmetic complexity of this algorithm is of order $N_c\, N^2 \log(N)$ if $Nx \sim N$, $Ny \sim N$. It is well known that the unefficiency of the parallel implementation of the FFTs comes from the global transpose of $U(:,i,j)$ across the two dimensional network of processors. Although for air pollution problems on medium scale parallel computers, we do not expect to have Nx and Ny much larger than 100 because of the intense pointwise computation induced by the chemistry. An alternative approach to FFTs that can use fully the vector data structure of $U(:,i,j,)$ is to write Step 2 in matrix multiply form:

$$\forall k = 1 :: N_c,\ U(k,:,:) := A_{x,sin}^{-1} \times (F_x \cdot A_{x,sin})\, U(k,:,:)\, (A_{y,sin}^{t} \cdot F_y) \times A_{y,sin}^{-t}, \qquad (17)$$

where $A_{x,sin}$ (respt $A_{y,sin}$) is the matrix corresponding to the sine expansion transform in x direction and F_x (respt F_y) is the matrix corresponding to the filtering process. In (17), $\cdot$ denotes the multiplication of matrices component by component. Let us define $A_{left} = A_{x,sin}^{-1} \times (F_x \cdot A_{x,sin})$ and $A_{right} = (A_{y,sin}^{t} \cdot F_y) \times A_{y,sin}^{-t}$. These two matrices A_{left} and A_{right} can be computed once for all and stored in the local memory of each processors. Since $U(:,i,j)$ is distributed on a two dimensional network of processors, one can use an approach very similar to the systolic algorithm [8] to realize in parallel the matrix multiply $A_{left} \times U(k,:,:) \times A_{right}$ for all $k = 1..N_c$.

Further we observe that the matrices can be approximated by sparses matrices while preserving the time accuracy of the overall scheme. The number of "non neglectable" coefficients growths with κ. Figure 1 gives the elapsed time on an EV6 processor at 500MHz obtained for the filtering procedure for various problem sizes, $\kappa = 2.$, and using or not the fact that the matrices A_{left} and A_{right} can be approximated by sparses matrices. This method should be competitive to a filtering process using FFT for large N_c and not so large N_x and N_y.

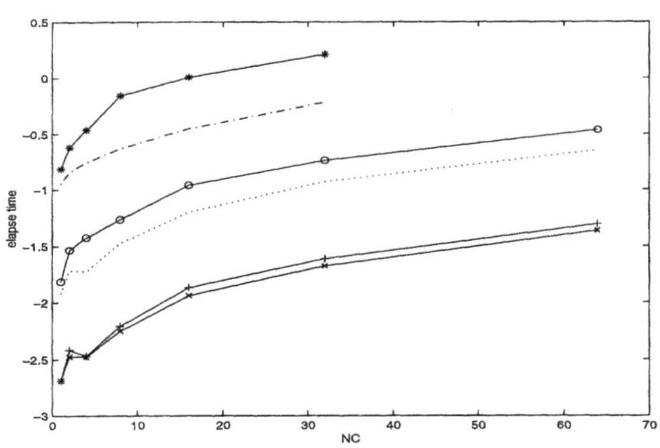

Figure 1. Elapse time of the matrix multiply form of the filtering processe as a function of N_c. With full matrices, '*' is for $N_x = N_y = 128$, 'o' is for $N_x = N_y = 64$, '+' is for $N_x = N_y = 32$. Neglecting matrix coefficients less than 1^{-5} in absolute value, '-.' is for $N_x = N_y = 128$, '.' is for $N_x = N_y = 64$, 'x' is for $N_x = N_y = 32$.

But the parallel efficiency of the algorithm as opposed to FFT on such small data sets is very high -see Table 1 to 2-

$px \times py$ proc.	$py = 1$	$py = 2$	$py = 4$	$py = 8$	$py = 16$
$px = 1$	100.00	98.0	90.9	84.2	70.0
$px = 2$	171.3	166.2	149.0	127.8	93.2
$px = 4$	158.1	151.9	130.1	100.0	60.4
$px = 8$	140.8	128.7	102.7	61.8	
$px = 16$	114.6	96.0	61.3		

Table 1: Efficiency on a Cray T3E with $N_c = 4$, Nx=Ny=128.

$px \times py$ proc.	$py = 1$	$py = 2$	$py = 4$	$py = 8$	$py = 16$
$px = 1$	100.00	97.3	88.1	79.9	66.2
$px = 2$	120.2	117.0	103.4	91.7	73.1
$px = 4$	110.9	106.6	94.8	81.7	60.5
$px = 8$	99.9	96.5	83.0	66.1	
$px = 16$	83.7	78.0	61.4		

Table 2: Efficiency on a Cray T3E with $N_c = 20$, Nx=Ny=128.

As a matter of fact, for $N_c = 4$, we benefit of the cache memory effect, and obtain perfect speedup with up to 32 processors. For larger number of species, $N_c = 20$ for example, we observe a deterioration of performance, and we should introduce a second level of parallelism with domain decomposition in order to lower the dimension of each subproblems and get data set that fits into the cache.

5. conclusion

In this paper, we have introduced a new familly of fast and numerically efficient reaction-diffusion solvers based on a filtering technique that stabilize the explicit treatment of the diffusion terms. We have shown the potential of this numerical scheme. Further, we have demonstrated on critical components of the algorithm the high potential of parallelism of our method on medium scale parallel computers. In order to obtain scalable performance of our solver on large parallel systems with O(1000) processors, we are currently introducing a second level of parallelsim with the overlapping domain decomposition algorithm described in [9].

thanks: we thanks Jeff Morgan for many interesting discussions. We thanks the Rechenzentrum Universität of Stuttgart for giving us a nice access on their computing resources.

REFERENCES

1. P.J.F.Berkvens, M.A.Botchev, J.G.Verwer, M.C.Krol and W.Peters, *Solving vertical transport and chemistry in air pollution models* MAS-R0023 August 31,2000.
2. D. Dabdub and J.H.Steinfeld, *Parallel Computation in Atmospheric Chemical Modeling,* Parallel Computing Vol22, 111-130, 1996.
3. A. Ecer et al, *Parallel CFD Test Case,* http://www.parcfd.org
4. V.I.Lebedev, Explicit Difference Schemes for Solving Stiff Problems with a Complex or Separable Spectrum, Computational Mathematics and Mathematical Physics, Vol.40., No 12, 1801-1812, 2000.
5. H. Elbern, *Parallelization and Load Balancing of a Comprehensive Atmospheric Chemistry Transport Model,* Atmospheric Environment, Vol31, No 21, 3561-3574, 1997.
6. W.E.Fitzgibbon, M. Garbey and J. Morgan, *Analysis of a Basic Chemical Reaction Diffusion Tropospheric Air Pollution Model,* Tech. Report Math deprt. of UH, March 2001.

7. W.E.Fitzgibbon, M. Garbey, *Fast solver for Reaction-Diffusion-Convection Systems: application to air quality models* Eccomas CFD 2001 Swansea Proceedings, September 2001.
8. I. Foster, *Designing and Building Parallel Programs,* Addison-Wesley Publishing Cie 94,
9. M. Garbey, H.G.Kaper and N.Romanyukha, On Some Fast Solver for Reaction-Diffusion Equations DD13 Lyon 2000 http://www.ddm.org, to appear.
10. D. Gottlieb and Chi-Wang Shu, *On the Gibbs Phenomenon and its Resolution,* SIAM review, Vol39, No 4, 644-668, 1997.
11. A. Sandu, J.G. Verwer, M. Van Loon, G.R. Carmichael, F.A. Potra, D. Dadbud and J.H.Seinfeld, *Benchmarking Stiff ODE Solvers for Atmospheric Chemistry Problems I: implicit versus explicit,* Atm. Env. 31, 3151-3166, 1997.
12. J.G.Verwer and B. Sportisse, *A Note on Operator Splitting in a Stiff Linear Case,* MAS-R9830, http://www.cwi.nl, Dec 98.
13. J.G.Verwer, W.H.Hundsdorfer and J.G.Blom, *Numerical Time Integration for Air Pollution Models,* MAS-R9825, http://www.cwi.nl, International Conference on Air Pollution Modelling and Simulation APMS'98.

Parallel Computational Fluid Dynamics – Practice and Theory
P. Wilders, A. Ecer, J. Periaux, N. Satofuka and P. Fox (Editors)

Algebraic Coarse Grid Operators for Domain Decomposition Based Preconditioners

L. Formaggia[a], M. Sala[b] *

[a]Département de Mathématiques, EPF-Lausanne, CH-1015 Lausanne, Switzerland

[b]Corresponding author. Département de Mathématiques, EPF-Lausanne, CH-1015 Lausanne, Switzerland. E-mail address: Marzio.Sala@epfl.ch

We investigate some domain decomposition techniques to solve large scale aerodynamics problems on unstructured grids. Where implicit time advancing scheme are used, a large sparse linear system have to be solved at each step. To obtain good scalability and CPU times, a good preconditioner is needed for the parallel iterative solution of these systems. For the widely-used Schwarz technique this can be achieved by a *coarse level* operator. Since many of the current coarse operators are difficult to implement on unstructured 2D and 3D meshes, we have developed a purely algebraic procedure, that requires the entries of the matrix only.
KEY WORDS: Compressible Euler Equations, Schwarz Preconditioners, Agglomeration Coarse Corrections.

1. INTRODUCTION

Modern supercomputers are often organised as a distributed environment and every efficient solver must account for their multiprocessor nature. Domain decomposition (DD) techniques provide a natural possibility to combine classical and well-tested single-processor algorithms with parallel new ones. The basic idea is to decompose the original computational domain Ω into M smaller parts, called subdomains $\Omega^{(i)}$, $i = 1, \dots, M$, such that $\cup_{i=1}^{N} \bar{\Omega}^{(i)} = \bar{\Omega}$. Each subdomain $\Omega^{(i)}$ can be extended to $\widetilde{\Omega}^{(i)}$ by adding an overlapping region. Then we replace the global problem on Ω with N problems on each $\widetilde{\Omega}^{(i)}$. Of course, additional interface conditions between subdomains must be provided.

DD methods can roughly be classified into two groups [2,4]. The former group may use non-overlapping subdomains and is based on the subdivision of the unknowns into two sets: those lying on the interface between subdomains, and those associated to nodes internal to a subdomain. One then generates a Schur complement (SC) matrix by "condensing" the unknowns in the second set. The system is then solved by first computing the interface unknowns and then solving M independent problems for the internal unknowns. In the latter, named after Schwarz, the computational domain is subdivided into overlapping subdomains, and local Dirichlet-type problems are then solved on each subdomain. In this case, the main problem is the degradation of the performance as the

*The authors acknowledge the support of the OFES under contract number BRPR-CT97-0591.

number of subdomains grow, and a suitable coarse level operator should be introduced to improve scalability [2].

This paper is organised as follows. Section 2 briefly describes the Schwarz preconditioner without coarse correction. Section 3 introduces the proposed agglomeration coarse correction. Section 4 reports some numerical results for real-life problems, while conclusions are drawn in Section 5.

2. THE SCHWARZ PRECONDITIONER

The Schwarz method is a well known parallel technique based on a domain decomposition strategy. It is in general a rather inefficient solver, however it is a quite popular parallel preconditioner. Its popularity derives from its generality and simplicity of implementation. The procedure is as follows. We decompose the computational domain Ω into M parts $\Omega^{(i)}, i = 1 \dots, M$, called subdomains, such that $\cup_{i=1}^{M} \bar{\Omega}^{(i)} = \Omega$ and $\Omega^{(i)} \cap \Omega^{(j)} = \emptyset$ for some i and j. To introduce a region of overlap, these subdomains are extended to $\widetilde{\Omega}^{(i)}$ by adding to $\Omega^{(i)}$ all the elements of Ω that have at least one node in $\bar{\Omega}^{(i)}$. In this case, the overlap is minimal. More overlap can be obtained by repeating this procedure. A parallel solution of the original system is then obtained by an iterative procedure involving local problems in each $\widetilde{\Omega}^{(i)}$, where on $\partial\widetilde{\Omega}^{i} \cap \widetilde{\Omega}^{(j)}$ we apply Dirichlet conditions by imposing the latest values available from the neighbouring sub-domains. The increase the amount of overlap among subdomains has a positive effect on the convergence history for the iterative procedure, but it may be result in a more computationally expensive method. Furthermore, the minimal overlap variant may exploit the same data structure used for the parallel matrix-vector product in the outer iterative solver, thus allowing a very efficient implementation with respect to memory requirements (this is usually not anymore true for wider overlaps). In the numerical results later presented we a used a minimal overlap, that is, an overlap of one element only. See [2,4,7] for more details.

3. THE AGGLOMERATION COARSE OPERATOR

The scalability of the Schwarz preconditioner is hindered by the weak coupling between far away sub-domains. A good scalability may be recovered by the addition of a coarse operator. Here we present a general algebraic setting to derive such an operator.

A possible technique to build the coarse operator matrix A_H for the system arising from a finite-element or finite volume scheme on unstructured grids consists in discretising the original differential problem on a coarse mesh, see for instance [7]. However the construction of a coarse grid and of the associated restriction and prolongation operators is a rather difficult task when dealing with a complex geometry. An alternative is to resort to algebraic procedures, such as the agglomeration technique which has been implemented in the context of multigrid [8]. The use of an agglomeration procedure to build the coarse operator for a Schwarz preconditioner have been investigated in [6,3] for elliptic problems. Here, we extend and generalise the technique and we will apply it also to non self-adjoint problems.

Consider that we have to solve a linear system of the form $A\mathbf{u} = \mathbf{f}$, which we suppose has been derived from the discretisation by a finite element procedure[2] of a differential problem posed on a domain Ω and whose variational formulation may be written in the general form

$$\begin{cases} \text{find } u \in V \text{ such that:} \\ a(u,v) = (f,v) \quad \text{for } \forall v \in V \, , \end{cases}$$

where $u, v, f : \Omega \to \mathbb{R}, \Omega \subset \mathbb{R}^d, d = 2, 3,\ a(\cdot,\cdot)$ is a bi-linear form and V is a Hilbert space of (possibly vector) functions in Ω. With (u,v) we denote the L_2 scalar product, i.e. $(u,v) = \int_\Omega uv d\Omega$. The corresponding finite element formulation reads

$$\begin{cases} \text{find } u_h \in V_h \text{ such that:} \\ (u_h, v_h) = (f, v_h) \quad \text{for } \forall v_h \in V_h \, , \end{cases}$$

where now V_h is a finite dimensional subspace of V generated from finite element basis functions. We can split the finite element function space V_h as

$$v_h = \sum_{i=1}^{M} V_h^{(i)}$$

where $V_h^{(i)}$ is set of finite element functions associated to the triangulation of $\Omega^{(i)}$, i.e. the finite element space spanned by the set $\{\phi_j^{(i)}, j = 1, \ldots, n^{(i)}\}$ of nodal basis function associated to vertices of $\mathcal{T}_h^{(i)}$, triangulation of Ω. Here we have indicated with $n^{(i)}$, the dimension of the space $V_h^{(i)}$. By construction, $n = \sum_{i=1}^{M} n^{(i)}$.

We build a **coarse space** as follows. For each sub-domain $\Omega^{(i)}$ we consider the set $\{\boldsymbol{\beta}_s^{(i)} \in \mathbb{R}^{n^{(i)}},\ s = 1, \cdots, l^{(i)}\}$ of linearly independent nodal weights $\boldsymbol{\beta}_s^{(i)} = \left(\beta_{s,1}^{(i)}, \ldots, \beta_{s,n^{(i)}}^{(i)}\right)$.

The value $l^{(i)}$ represents the (local) dimension of the coarse operator on sub-domain $\Omega^{(i)}$. Clearly we must have $l^{(i)} \le n^{(i)}$ and, in general $l^{(i)} << n^{(i)}$. We indicate with l the global dimension of the coarse space, $l = \sum_{i=1}^{M} l^{(i)}$.

With the help of the vectors $\boldsymbol{\beta}_s^{(i)}$, we define a set of local coarse space functions as linear combination of basis functions, i.e.

$$\mathcal{V}_H^{(i)} = \left\{ z_s^{(i)} : \Omega \to \mathbb{R} \mid z_s^{(i)} = \sum_{k=1}^{n^{(i)}} \beta_{s,k}^{(i)} \phi_k^{(i)},\ s = 1, \ldots, l^{(i)} \right\} .$$

It is immediate to verify that the functions in $\mathcal{V}_H^{(i)}$ are linearly independent. Finally, the set $\mathcal{V}_H = \bigcup_{i=1}^{M} \mathcal{V}_H^{(i)}$ is the base of our global coarse grid space V_H, i.e. we take $V_H = \text{span}\{\mathcal{V}_H\}$. By construction, $\dim(V_H) = \text{card}(\mathcal{V}_H) = l$.

Note that $V_H \subset V_h$ as it is built by linear combinations of function in V_h. Any function $W_H \in V_H$ may be written as

$$W_H = \sum_{i=1}^{M} \sum_{s=1}^{l^{(i)}} W_s^{(i)} z_s^{(i)}, \tag{1}$$

[2]the consideration in this Section may be extended to other type of discretisations as well, for instance finite volumes.

where the $W_s^{(i)}$ are the "coarse" degrees of freedom.

Finally, the coarse problem is built as

$$\begin{cases} \text{Find } U_H \in V_H : \\ a(U_H, W_H) = f(W_H) \ , \forall W_H \in V_H \ . \end{cases}$$

To complete the procedure we need a restriction operator $R_H : V_h \to V_H$ which maps a generic finite element function to a coarse grid function. We have used the following technique. Given $u \in V_h$, which may be written as

$$u = \sum_{i=1}^{M} \sum_{k=1}^{n^{(i)}} \phi_k^{(i)} u_k^{(i)},$$

where the $u_k^{(i)}$ are the degree of freedoms associated to the triangulation of $\Omega^{(i)}$, the restriction operator is defined by computing $U_H = R_H u$ as

$$U_H = \sum_{i=1}^{M} \sum_{s=1}^{l^{(i)}} U_s^{(i)} z_s^{(i)} \ , \ U_s^{(i)} = \sum_{k=1}^{n^{(i)}} \beta_{s,k}^{(i)} u_k^{(i)}, \ s = 1, \ldots l^{(i)}, \quad i = 1 \ldots, M.$$

At algebraic level we can consider a restriction matrix $R_H \in \mathbb{R}^{l \times n}$ and the relative prolongation operator R_H^T. The coarse matrix and right-hand side can be written as $A_H = R_H A R_H^T, \quad \mathbf{f}_H = R_H \mathbf{f}$.

Remark. The condition imposed on the $\boldsymbol{\beta}_s^{(i)}$ guarantees that R_H has full rank. Moreover, if A is non-singular, symmetric and positive definite, then also A_H is non singular, symmetric and positive definite.

The frame we have just presented is rather general. In the implementation of the Schwarz preconditioner carried out in this work we have made use of two decompositions. At the first level we have the standard decomposition used to build the basic Schwarz preconditioner. Each sub-domain Ω_i is assigned to a different processor. We have assumed that the number of sub-domains M is equal to the number of available processors. At the second level, we partition each sub-domain Ω_i into N_p connected parts $\omega_s^{(i)}, s = 1, N_p$. This decomposition will be used to build the agglomerated coarse matrix. In the following tables, N_p will be indicated as N_parts.

The coarse matrix is built by taking for all sub-domains $l^{(i)} = N_p$, while the element of $\boldsymbol{\beta}_s^{(i)}$ are build following the rule

$$\beta_{s,k} = \begin{cases} 1 & \text{if node } k \text{ belongs to } \omega_s^{(i)} \\ 0 & \text{otherwise.} \end{cases}$$

As already explained the coarse grid operator is used to ameliorate the scalability of a Schwarz-type parallel preconditioner P_S. We will indicate with P_{ACM} a preconditioner augmented by the application of the coarse operator (ACM stands for agglomeration coarse matrix) and we illustrate two possible strategies for its construction.

A one-step preconditioner, $P_{ACM,1}$ may be formally written as

$$P_{ACM,1}^{-1} = P_S^{-1} + R_H^T A_H^{-1} R_H$$

and it correspond to an additive application of the coarse operator.

An alternative formulation adopts the following preconditioner:

$$P_{ACM,2}^{-1} = P_S^{-1} + R_H^T A_{ACM}^{-1} R_H - P_S^{-1} A R_H^T A_{ACM}^{-1} R_H, \tag{2}$$

that can be obtained from a two-level Richardson method.

4. NUMERICAL RESULTS

Before presenting the numerical results we give some brief insight on the application problem we are considering, namely inviscid compressible flow around aeronautical configurations, and the numerical scheme adopted.

The Euler equations governs the dynamics of compressible inviscid flows and can be written in conservation form as

$$\frac{\partial \mathbf{U}}{\partial t} + \sum_{j=1}^{d} \frac{\partial \mathbf{F}_j}{\partial x_j} = 0 \quad \text{in } \Omega \subset \mathbb{R}^d \, , \, t > 0 \, , \tag{3}$$

with the addition of suitable boundary conditions on $\partial\Omega$ and initial conditions at $t = 0$. Here, $\mathbf{U}$ and $\mathbf{F_j}$ are the vector of conservative variables and the flux vector, respectively defined as

$$\mathbf{U} = \begin{pmatrix} \rho \\ \rho u_i \\ \rho E \end{pmatrix} \quad , \quad \mathbf{F}_j = \begin{pmatrix} \rho u_j \\ \rho u_i u_j + p\delta_{ij} \\ \rho H u_j \end{pmatrix} \, ,$$

with $i = 1, \ldots, d$. $\mathbf{u}$ is the velocity vector, ρ the density, p the pressure, E the specific total energy, H the specific total enthalpy and δ_{ij} the Kronecker symbol.

Any standard spatial discretisation applied to the Euler equations leads eventually to a system of ODE in time, which may be written as $dU/dt = R(U)$, where $U = (U_1, U_2, \ldots, U_n)^T$ is the vector of unknowns with $U_i = U_i(t)$ and $R(U)$ the result of the spatial discretisation of the Euler fluxes. An implicit two-step scheme, for instance a backward Euler method, yields

$$U^{n+1} - U^n = \Delta t R\left(U^{n+1}\right) \, , \tag{4}$$

where Δt is in general the time step but may also be a *diagonal matrix* of local time steps when the well known "local time stepping" technique is used to accelerate convergence to steady-state. The nonlinear problem (4) may be solved, for instance, by employing a Newton iterative procedure. In this case, a linear system has to solved at each Newton step.

Table 1 reports the main characteristics of the test cases used in this Section. At each time-step we have used one step of the Newton procedure. The starting CFL number is 10, and it has been multiplied at each time step by a factor of 2. The linear system

Table 1
Main characteristics of the test cases.

name	M_∞	α	N_nodes	N_cells
FALCON_45k	0.45	1.0	45387	255944
M6_23k	0.84	3.06	23008	125690
M6_42k	0.84	3.06	42305	232706
M6_94k	0.84	3.06	94493	666569
M6_316k	0.84	3.06	316275	1940182

has been solver with GMRES(60) up to a tolerance on the relative residual of 10^{-3}. For the Schwarz preconditioner, an incomplete LU decomposition with a fill-in factor of 0 has been used, with minimal overlap among subdomains. The coarse matrix problem has been solved using an incomplete LU decomposition to save computational time. Moreover, since the linear system associated with the coarse space is much smaller than the linear system A, we solve it (redundantly) on all processors.

For the numerical experiments at hand we have used the code THOR, developed at the von Karman Institute. This code uses for the spatial discretisation the multidimensional upwind finite element scheme [9]. The results have been obtained using a SGI Origin 3000 computer, with up to 32 MIPS14000/500Mhz processors with 512 Mbytes of RAM. The basic parallel linear solvers are those implemented in the the Aztec library [10], which we have extended to include the preconditioners previously described [11]; these extensions are freely available and can be downloaded.

Figure 1, left, shows the positive influence of the coarse operator. In particular, as the dimension of the coarse increases, we may notice positive effects on the number of iterations to converge. Moreover, the two-level coarse correction is substantially better than the one-level preconditioner, especially as the CFL number grows (that is, as the matrix becomes more non-symmetric). Figure 1, right, shows the convergence history for M6_316k at the 14th time step. We can notice that the coarse correction results in a more regular convergence. Figure 2 compares in more details P_S and $P_{ACM,2}$ for grids of different sizes and for different values of N_p. Finally, Table 2 reports the CPU time in seconds needed to solve the test case M6_94k using $P_{ACM,1}$ and $P_{ACM,2}$. In bold we have underlined the best result from the point of view of CPU time. Notice that, although the iterations to converge decreased as N_p grows, this value should not be too high to obtain good CPU timing. Moreover, $P_{ACM,2}$ outperform $P_{ACM,1}$, even if at each application of the preconditioner a matrix-vector product has to done.

5. CONCLUSIONS

A coarse correction operator based on an agglomeration procedure that requires the matrix entries only has been presented. This procedure does not require the construction of a coarse grid, step that can be difficult or expensive for real-life problems on unstructured grids. A single and a two-level preconditioner which adopts this coarse correction have been presented. The latter seems a better choice for the point of view of both iterations to converge and CPU time. Results have been presented for problems obtained from the

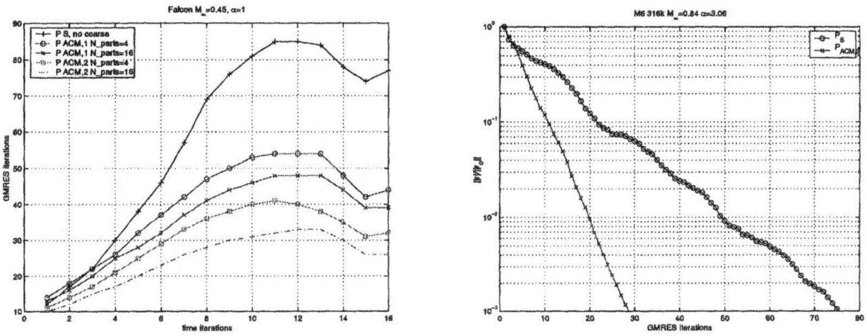

Figure 1. Comparison among different preconditioners for FALCON_45k (left) and convergence history at the 14th time step, using P_S and $P_{ACM,2}$ (right), using 16 SGI-Origin3000 processors.

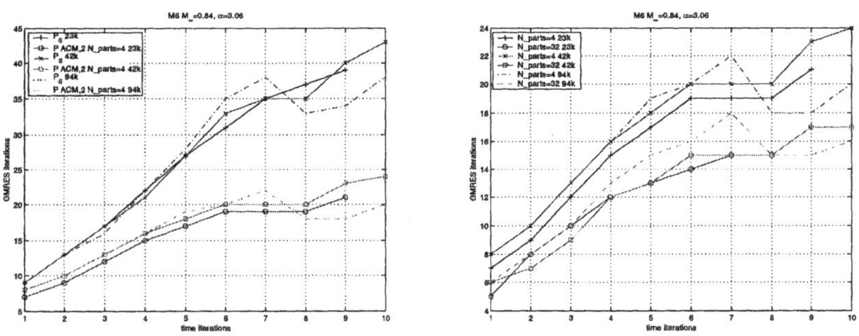

Figure 2. M6_94k. Iterations to converge with P_S and $P_{ACM,2}$ (left), and iterations to converge with $P_{ACM,2}$ using two different values of N_p (right), using 16 SGI-Origin3000 processors.

Table 2
M6_94k. SGI Origin-3000 processors, using $P_{ACM,1}$ and $P_{ACM,2}$.

	N_procs	N_p=4	N_p=8	N_p=16	N_p=32
$P_{ACM,1}$	8	+1.008e+03	**+9.784e+02**	+1.251e+03	+8.834e+02
$P_{ACM,1}$	16	**+5.025e+02**	+5.069e+02	+5.150e+02	+4.573e+02
$P_{ACM,1}$	32	**+2.080e+02**	+2.453e+02	+3.005e+02	+5.050e+02
$P_{ACM,2}$	8	+9.348e+02	+9.456e+02	**+9.093e+02**	+9.256e+02
$P_{ACM,2}$	16	+4.586e+02	**+4.052e+02**	+4.139e+02	+4.426e+02
$P_{ACM,2}$	32	+1.644e+02	**+1.647e+02**	+1.814e+02	+5.156e+02

3-dimensional compressible Euler equations. The proposed coarse operator is rather easy to build and may be applied to very general cases. The proposed technique to build the weights $\beta_s^{(i)}$ produces a coarse correction which is equivalent to a two-level agglomeration multigrid. However, other choices are possible and currently under investigation.

REFERENCES

1. A. Quarteroni, A. Valli. *Numerical Approximation of Partial Differential Equations.* Springer-Verlag, Berlin, 1994.
2. A. Quarteroni, A. Valli. *Domain Decomposition Methods for Partial Differential Equations.* Oxford University Press, Oxford, 1999.
3. L. Paglieri, D. Ambrosi, L. Formaggia, A. Quarteroni, A. L. Scheinine. Parallel Computations for shallow water flow: A domain decomposition approach. *Parallel Computing* 23 (1997), pp. 1261-1277.
4. B.F. Smith, P. Bjorstad and W. Gropp. *Domain Decomposition, Parallel Multilevel Methods for Elliptic Partial Differential Equations.* Cambridge University Press, New York, 1st edition, 1996.
5. Y. Saad. *Iterative Methods for Sparse Linear Systems.* Thompson, Boston, 1996.
6. L. Formaggia, A. Scheinine, A. Quarteroni. A Numerical Investigation of Schwarz Domain Decomposition Techniques for Elliptic Problems on Unstructured Grids, *Mathematics and Computer in Simulations*, 44(1007), 313-330.
7. T. Chan, T.P. Mathew. Domain Decomposition Algorithm, *Acta Numerica*, 61-163, 1993.
8. M. H. Lallemand, H. Steve, A. Derviuex. Unstructured multigridding by volume agglomeration: current status, *Comput. Fluids*, 32 (3), 1992, pp. 397-433.
9. H. Deconinck, H. Paillère, R. Struijs and P.L. Roe. Multidimensional upwind schemes based on fluctuaction splitting for systems of conservation laws. *J. Comput. Mech.*, 11 (1993) 215-222.
10. R. Tuminaro, J. Shadid, S. Hutchinson, L. Prevost, C. Tong. AZTEC - A massively Parallel Iterative Solver Library for Solving Sparse Linear Systems. `http://www.cs.sandia.gov/CRF/aztec1.html`.
11. M. Sala. An Extension to the AZTEC Library for Schur Complement Based Solvers and Preconditioner and for Agglomeration-type Coarse Operators. `http://dmawww.epfl.ch/~sala/MyAztec/`.

Parallel Computational Fluid Dynamics – Practice and Theory
P. Wilders, A. Ecer, J. Periaux, N. Satofuka and P. Fox (Editors)

Efficient Parallel Simulation of Disperse Gas–Particle Flows on Cluster Computers

Th. Frank[a*], K. Bernert[a], K. Pachler[a] and H. Schneider[b]

[a]Chemnitz University of Technology, Research Group on Multiphase Flows, Reichenhainer Straße 70, 09107 Chemnitz, Germany

[b]SIVUS gGmbH, Schulstraße 38, 09125 Chemnitz, Germany

The paper deals with different methods for the efficient parallelization of Eulerian–Lagrangian approach which is widely used for the prediction of disperse gas–particle and gas–droplet flows. Several aspects of parallelization like e.g. scalability, efficiency and dynamic load balancing are discussed for the different kinds of Domain Decomposition methods with or without dynamic load balancing applied to both the Eulerian and Lagrangian parts of the numerical prediction. The paper shows that remarkable speed-up's can be achieved on dedicated parallel computers and cluster computers (Beowulf systems) not only for idealized test cases but also for "real world" applications. Therefor the developed parallelization methods offer new perspectives for the computation of strongly coupled multiphase flows with complex phase interactions.

1. Motivation

Over the last decade the Eulerian–Lagrangian (PSI–Cell) simulation has become an efficient and widely used method for the calculation of various kinds of 2– and 3–dimensional disperse multiphase flows (e.g. gas–particle flows, gas–droplet flows) with a large variety of computational very intensive applications in mechanical and environmental engineering, process technology, power engineering (e.g. coal combustion) and in the design of internal combustion engines (e.g. fuel injection and combustion). Considering the field of computational fluid dynamics, the Eulerian–Lagrangian simulation of coupled multiphase flows with strong interaction between the continuous fluid phase and the disperse particle phase ranks among the applications with the highest demand on computational power and system recources. Massively parallel computers provide the capability for cost–effective calculations of multiphase flows. In order to use the architecture of parallel computers efficiently, new solution algorithms have to be developed. Difficulties arise from the complex data dependence between the fluid flow calculation and the prediction of particle motion, and from the generally non–homogeneous distribution of particle concentration in the flow field. Direct linkage between local particle concentration in the flow and the numerical

*Email & WWW : Frank@IMech.TU–Chemnitz.DE, http://www.imech.tu–chemnitz.de

work load distribution over the calculational domain often leads to very poor performance of parallel Lagrangian solvers operating with a Static Domain Decomposition method. Good work load balancing and high parallel efficiency for the Lagrangian approach can be established with the new Dynamic Domain Decomposition method presented in this paper.

2. The Eulerian–Lagrangian Approach

Due to the limited space it is not possible to give a full description of the fundamentals of the numerical approach. A detailed description can be found in [4] or in documents on [5]. The numerical approach consists of a multi–block Navier–Stokes solver for the solution of the fluids equations of motion [1] and a Lagrangian particle tracking algorithm (Particle-Source-In-cell method) for the prediction of the motion of the particulate phase in the fluid flow field (see eq. 1).

$$\frac{d}{dt}\vec{x}_P = \vec{u}_P \quad ; \quad m_P\frac{d}{dt}\vec{u}_P = \vec{F}_D + \vec{F}_M + \vec{F}_A + \vec{F}_G \quad ; \quad I_P\frac{d}{dt}\vec{\omega}_P = -\vec{T} \qquad (1)$$

A more detailed description of all particular models involved in the Lagrangian particle trajectory calculation can be found in [3–5]. The equations of fluid motion are solved on a blockstructured, boundary–fitted, non–orthogonal numerical grid by pressure correction technique of SIMPLE kind (**S**emi–**I**mplicite **P**ressure **L**inked **E**quations) with convergence acceleration by a full multigrid method [1]. Eq.'s (1) are solved in the Lagrangian part of the numerical simulation by using a standard 4th order Runge–Kutta scheme. Possible strong interactions between the two phases due to higher particle concentrations have to be considered by an alternating iterative solution of the fluid's and particles equations of motion taking into account special source terms in the transport equations for the fluid phase.

3. The Parallelization Methods

3.1. The Parallel Algorithm for Fluid Flow Calculation

The parallelization of the solution algorithm for the set of continuity, Navier–Stokes and turbulence model equations is carried out by parallelization in space, that means by application of the domain decomposition or grid partitioning method. Using the block structure of the numerical grid the flow domain is partitioned in a number of subdomains. Usually the number of grid blocks exceeds the number of processors, so that each processor of the PM has to handle a few blocks. If the number of grid blocks resulting from grid generation is too small for the designated PM or if this grid structure leads to larger imbalances in the PM due to large differences in the number of control volumes (CV's) per computing node a further preprocessing step enables the recursive division of largest grid blocks along the side of there largest expansion. The grid-block-to-processor assignment is given by a heuristicly determined block–processor allocation table and remains static and unchanged over the time of fluid flow calculation process.

Fluid flow calculation is then performed by individual processor nodes on the grid partitions stored in their local memory. Fluid flow characteristics along the grid block boundaries which are common to two different nodes have to be exchanged during the

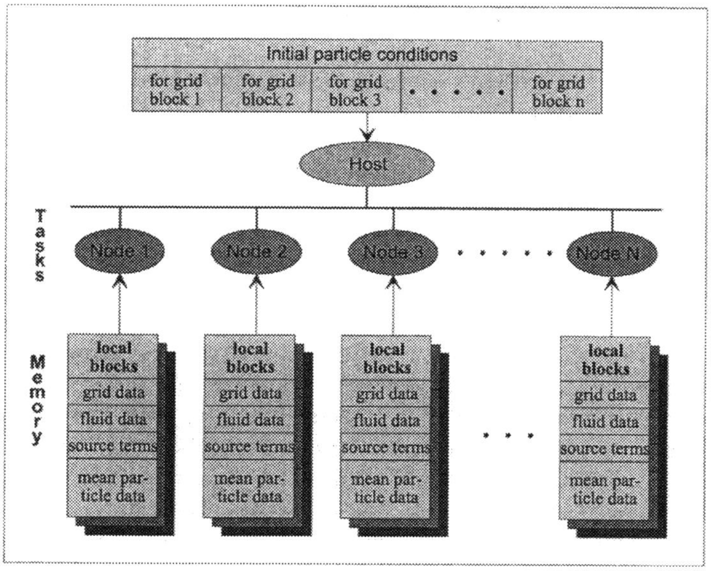

Figure 1. Static Domain Decomposition method for the Lagrangian solver.

solution process by inter–processor communication, while the data exchange on common faces of two neighbouring grid partitions assigned to the same processor node can be handled locally in memory. More details of the parallelization method and results for its application to the Multi–grid accelerated SIMPLE algorithm for turbulent fluid flow calculation can be found in [1].

3.2. Parallel Algorithms for the Lagrangian Approach

Considering the parallelization of the Lagrangian particle tracking algorithm there are two important issues. The first is that in general particle trajectories are not uniformly distributed in the flow domain even if there is a uniform distribution at the inflow cross–section. Therefore the distribution of the numerical work load in space is not known at the beginning of the computation. As a second characteristic parallel solution algorithms for the particle equations of motion have to deal with the global data dependence between the distributed storage of fluid flow data and the local data requirements for particle trajectory calculation. A parallel Lagrangian solution algorithm has either to provide all fluid flow data necessary for the calculation of a certain particle trajectory segment in the local memory of the processor node or the fluid flow data have to be delivered from other processor nodes at the moment when they are required. Considering these issues the following parallelization methods have been developed :

Method 1: Static Domain Decomposition (SDD) Method

The first approach in parallelization of Lagrangian particle trajectory calculations is the application of the same parallelization scheme as for the fluid flow calculation to the Lagrangian solver as well. That means a Static Domain Decomposition (SDD) method. In this approach geometry and fluid flow data are distributed over the processor nodes of the PM in accordance with the block–processor allocation table as already used in the fluid flow field calculation of the Navier–Stokes solver.

Furthermore an explicit host–node process scheme is established as illustrated in Figure 1. The trajectory calculation is done by the node processes whereas the host process carries out only management tasks. The node processes are identical to those that do the flow field calculation. Now the basic principle of the SDD method is that in a node process only those trajectory segments are calculated that cross the grid partition(s) assigned to this process. The particle state (location, velocity, diameter, ...) at the entry point to the current grid partition is sent by the host to the node process. The entry point can either be at an inflow cross section or at a common face/boundary to a neighbouring partition. After the computation of the trajectory segment on the current grid partition is finished, the particle state at the exit point (outlet cross section or partition boundary) is sent back to the host. If the exit point is located at the interface of two grid partitions, the host sends the particle state to the process related to the neighbouring grid partition for continuing trajectory computation. This redistribution of particle state conditions is repeatedly carried out by the host until all particle trajectories have satisfied certain break condition (e.g. an outlet cross section is reached). During the particle trajectory calculation process the source terms for momentum exchange between the two phases are calculated locally on the processor nodes $1, \ldots, N$ from where they can be passed to the Navier–Stokes solver without further processing.

An advantage of the domain decomposition approach is that it is easy to implement and uses the same data distribution over the processor nodes as the Navier–Stokes solver. But the resulting load balancing can be a serious disadvantage of this method as shown later for the presented test cases. Poor load balancing can be caused by different circumstances, as there are :

1. Unequal processing power of the calculating nodes, e.g. in a heterogenous workstation cluster.

2. Unequal size of the grid blocks of the numerical grid. This results in a different number of CV's per processor node and in unequal work load for the processors.

3. Differences in particle concentration distribution throughout the flow domain. Situations of poor load balancing can occur e.g. for flows around free jets/nozzles, in recirculating or highly separated flows where most of the numerical effort has to be performed by a small subset of all processor nodes used.

4. Multiple particle–wall collisions. Highly frequent particle–wall collisions occur especially on curved walls where the particles are brought in contact with the wall by the fluid flow multiple times. This results in a higher work load for the corresponding processor node due to the reduction of the integration time step and the extra effort for detection/calculation of the particle–wall collision itself.

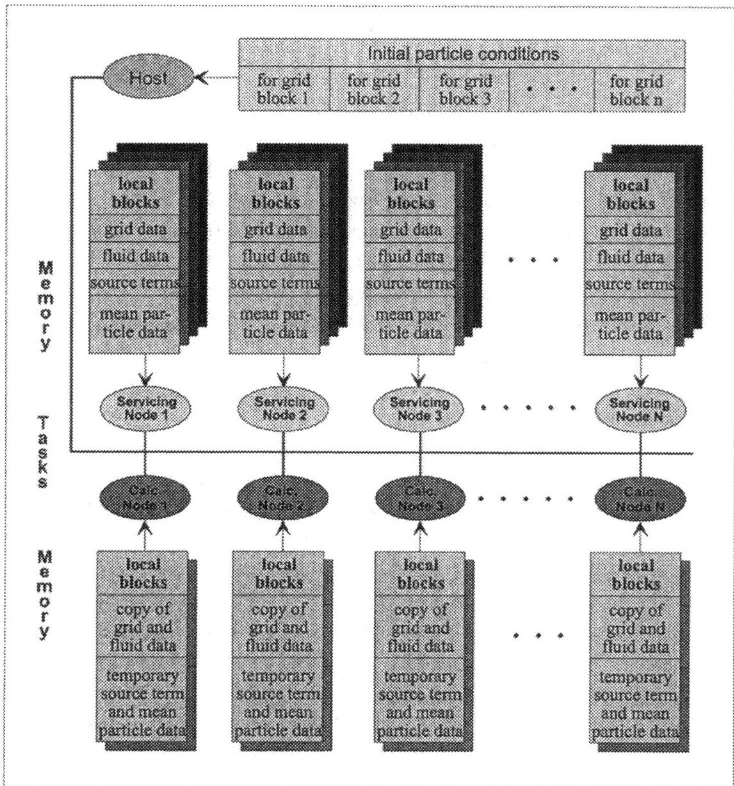

Figure 2. Dynamic Domain Decomposition (DDD) method for the Lagrangian solver introducing dynamic load balancing to particle simulation

5. Flow regions of high fluid velocity gradients/small fluid turbulence time scale. This leads to a reduction of the integration time step for the Lagrangian approach in order to preserve accuracy of the calculation and therefore to a higher work load for the corresponding processor node.

The reasons 1–2 for poor load balancing are common to all domain decomposition approaches and apply to the parallelization method for the Navier–Stokes solver as well. But most of the factors 3–5 leading to poor load balancing in the SDD method cannot be foreseen without prior knowledge about the flow regime inside the flow domain (e.g. from experimental investigations). Therefore an adjustment of the numerical grid or the block-processor assignment table to meet the load balancing requirements by a static redistribution of grid cells or grid partitions inside the PM is almost impossible. The second parallelization method shows how to overcome these limitations by introducing a dynamic load balancing algorithm which is effective during run time.

Method 2: Dynamic Domain Decomposition (DDD) Method

This method has been developed to overcome the disadvantages of the SDD method concerning the balancing of the computational work load. In the DDD method there exist three classes of processes : the host, the servicing nodes and the calculating nodes (Figure 2). Just as in the SDD method the host process distributes the particle initial conditions among the calculating nodes and collects the particle's state when the trajectory segment calculation has been finished. The new class of servicing nodes use the already known block-processor assignment table from the Navier–Stokes solver for storage of grid and fluid flow data. But in contrast to the SDD method they do not performe trajectory calculations but delegate that task to the class of calculating nodes. So the work of the servicing nodes is restricted to the management of the geometry, fluid flow and particle flow data in the data structure prescribed by the block-processor assignment table. On request a servicing node is able to retrieve or store data from/to the grid partition data structure stored in its local memory.

The calculating nodes are performing the real work on particle trajectory calculation. These nodes receive the particle initial conditions from the host and predict particle motion on an arbitrary grid partition. In contrast to the SDD method there is no fixed block-processor assignment table for the calculating nodes. Starting with an empty memory structure the calculating nodes are able to obtain dynamically geometry and fluid flow data for an arbitrary grid partition from the corresponding servicing node managing this part of the numerical grid. The correlation between the required data and the corresponding servicing node can be looked up from the block-processor assignment table. Once geometry and fluid flow data for a certain grid partition has been retrieved by the calculating node, this information is locally stored in a pipeline with a history of a certain depth. But since the amount of memory available to the calculating nodes can be rather limited, the amount of locally stored grid partition data can be limited by an adjustable parameter. So the concept of the DDD method makes it possible 1. to perform calculation of a certain trajectory segment on an arbitrary calculating node process and 2. to compute different trajectories on one grid partition at the same time by different calculating node processes.

4. Results and Discussion

Results for the parallel performance of the multigrid–accelerated Navier–Stokes solver MISTRAL-3D has been recently published.[1] So we will concentrate here on scalability and performance results for the Lagrangian particle tracking algorithms PartFlow–3D. Implementations of the SDD and DDD methods were based on the paradigm of a MIMD computer architecture with explicit message passing between the node processes of the PM using MPI. For performance evaluation we used the Chemnitz Linux Cluster (CLIC) with up to 528 Pentium-III nodes, 0.5 Gb memory per node and a FastEthernet interconnect. These data were compared with results obtained on a Cray T3E system with 64 DEC Alpha 21164 processors with 128 Mb node memory.

The first test case is a dilute gas–particle flow in a three times bended channel with square cross section of $0.2 \times 0.2\,m^2$ and inlet velocities $u_F = u_P = 10.0\ m/s$ ($Re = 156\,000$). In all three channel bends 4 corner vanes are installed, dividing the cross section

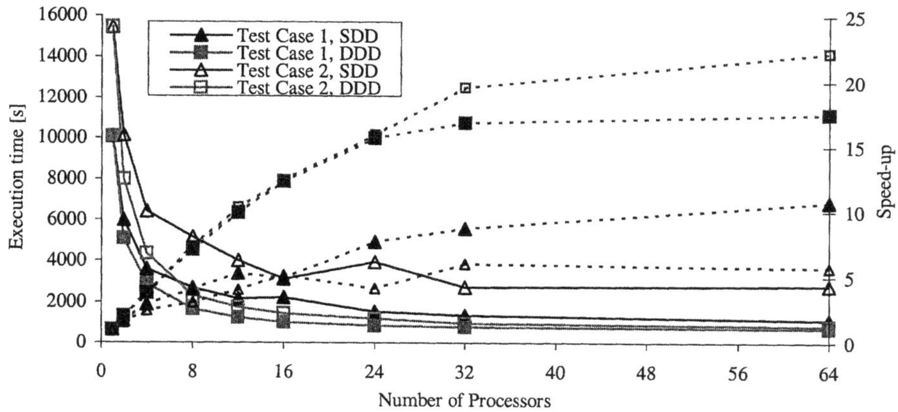

Figure 3. Execution time and speed-up vs. number of processor nodes; comparison of parallelization methods for both test cases.

of the bend in 5 separate corner sections and leading to a quite homogeneous particle concentration distribution. This corner vanes have been omitted for the second test case providing a typical strongly separated gas–particle flow. The numerical grid has been subdivided into 64 blocks, the number of finite volumes for the finest grid is $80 * 80 * 496 = 3\,174\,400$. For each of the test case calculations 5000 particle trajectories have been calculated by the Lagrangian solver.

Fig. 3 shows the total execution times, and the speed–up values for calculations on both test cases with SDD and DDD methods vs. the number of processor nodes. All test case calculations in this experiments had been carried out on the second finest grid level with 396.800 CV's. Fig. 3 shows the remarkable reduction in computation time with both parallelization methods. It can also be seen from the figure that in all cases the Dynamic Domain Decomposition (DDD) method has a clear advantage over SDD method.

Further the advantage for the DDD method for the first test case is not as remarkable as for the second test case. This is due to the fact, that the gas–particle flow in the first test case is quiet homogeneous in respect to particle concentration distribution which leads to a more balanced work load distribution in the SDD method. So the possible gain in performance with the DDD method is not as large as for the second test case, where the gas–particle flow is strongly separated and where we can observe particle roping and sliding of particles along the solid walls of the channel leading to a much higher amount of numerical work in certain regions of the flow. Consequently the SDD method shows a very poor parallel efficiency for the second test case due to poor load balancing between the processors of the PM (Fig. 3).

Figure 4 shows the comparison of test case calculations between the CLIC, an AMD–Athlon based workstation cluster and the Cray T3E. The impact of the Cray high-bandwith-low-latency interconnection network can clearly be seen from the figure. So the speed–up for the test case calculations on the Cray increases almost linearly with

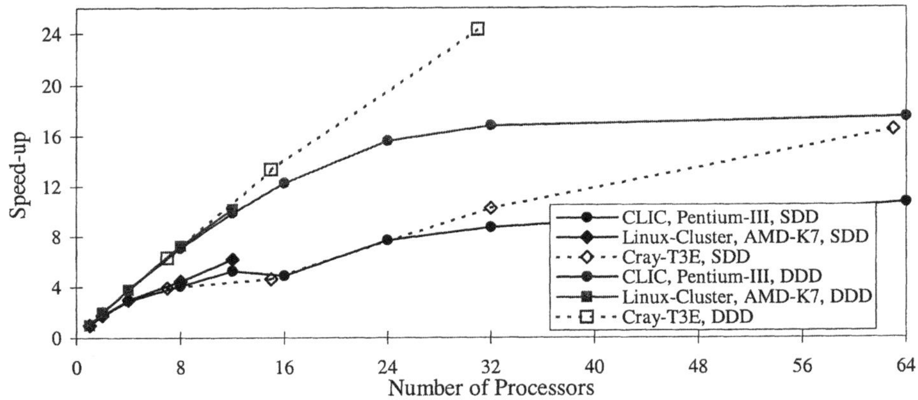

Figure 4. Comparison of parallel performace on Chemnitz Linux Cluster vs. Cray T3E.

increasing number of processors up to 32 nodes. On the CLIC we observe minor speed–up values and reach saturation for more than 32 processor nodes where a further substantial decrease of the total execution time for the Lagrangian solver could not be achieved. It was found from further investigations that this behaviour is mainly due to the limited communication characteristics of the Fast–Ethernet network used by the CLIC.

Acknowledgements

This work was supported by the German Research Foundation (Deutsche Forschungsgemeinschaft – DFG) in the framework of the Collaborative Research Centre SFB–393 under Contract No. SFB 393/D2.

REFERENCES

1. **Bernert K., Frank Th.** : "Multi–Grid Acceleration of a SIMPLE-Based CFD-Code and Aspects of Parallelization", IEEE Int. Conference on Cluster Computing — CLUSTER 2000, Nov. 28.-Dec. 2., 2000, Chemnitz, Germany.
2. **Crowe C.T., Sommerfeld M., Tsuji Y.** : "Multiphase Flows with Droplets and Particles", CRC Press, 1998.
3. **Frank Th., Wassen E.** : "Parallel Efficiency of PVM– and MPI–Implementations of two Algorithms for the Lagrangian Prediction of Disperse Multiphase Flows", JSME Centennial Grand Congress 1997, ISAC '97 Conference on Advanced Computing on Multiphase Flow, Tokyo, Japan, July 18–19, 1997.
4. **Frank Th.** : "Application of Eulerian–Lagrangian Prediction of Gas–Particle Flows to Cyclone Separators", VKI, Von Karman Institute for Fluid Dynamics, Lecture Series Programme 1999–2000, "Theoretical and Experimental Modeling of Particulate Flow", Bruessels, Belgium, 03.–07. April 2000.
5. Web site of the Research Group on Multiphase Flow, TUC, Germany. http://www.imech.tu-chemnitz.de/index.html – Index, List of Publications.

Parallel Computational Fluid Dynamics – Practice and Theory
P. Wilders, A. Ecer, J. Periaux, N. Satofuka and P. Fox (Editors)

Large Scale CFD Data Handling with Off-The-Shelf PC-Clusters in a VR-based Rhinological Operation Planning System

A. Gerndt[a], T. van Reimersdahl[a], T. Kuhlen[a] and C. Bischof[a]

[a]Center for Computing and Communication, Aachen University of Technology, Seffenter Weg 23, 52074 Aachen, Germany

The human nose can suffer from different complaints. However, many operations to eliminate respiration impairments fail. In order to improve the success rate it is important to recognize the responsiveness of the flow field within the nose's cavities. Therefore, we are developing an operation planning system that combines Computational Fluid Dynamics (CFD) and Virtual Reality (VR) technology. The primary prerequisite for VR-based applications is real-time interaction. A single graphics workstation is not capable of satisfying this condition and of simultaneously calculating flow features employing the huge CFD data set. In this paper we will present our approach of a distributed system that relieves the load on the graphics workstation and makes use of an "off-the-shelf" parallel Linux cluster in order to calculate streamlines. Moreover, we introduce first results and discuss remaining difficulties.

1. The Planning System

The human nose covers various functions like warming, moistening, and cleaning of inhaled air as well as the olfactory function. The conditions of the flow inside the nose are essential to these functionalities, which can be impeded by serious injury, disease, hereditary deformity or similar impairments. However, rhinological operations often do not lead to satisfactory results. We expect to improve this success rate considerably by investigating the airflow in the patient's nasal cavities by means of Computational Fluid Dynamics (CFD) simulation.

Therefore, we develop a VR-based rhinosurgical Computer Assisted Planning System to support the surgeon with recommendations from a set of possible operation techniques evaluated from the viewpoint of flow analysis [1]. For this the anatomy of nasal cavities extracted from computer tomography (CT) data will be displayed within a Virtual Environment. The geometry of the nose can be used to generate a grid, which the Navier-Stokes equations can calculate the flow simulation on. Nevertheless, the CFD simulation is an enormous time-consuming task and has to be calculated consequently as a pre-processing step. Afterwards flow features like streamlines can be extracted and visualized. During the virtual operation it is important to represent the pressure loss as a criterion of success. On completion of the virtual operation the surgeon can restart the flow simulation using the changed nasal geometry. This process can be reiterated until an optimum geometry is found.

2. The VR-Integration

A variety of commercial and academic visualization tools are available to represent flow fields, for instance as color coded cut planes. Also, streamlines or vector fields can be created. But due to the projection on a 2-dimensional display these visualization possibilities are often misinterpreted. This drawback can be avoided by integrating the computer assisted planning system into a Virtual Environment. Many people can profit directly from the Virtual Reality (VR) technology. On the one hand, aerodynamic scientists can inspect boundary conditions, the grid arrangement, and the convergence of the simulation model as well as the flow result. On the other hand, ear, nose, and throat specialists can consolidate their knowledge about the flow behavior inside the nose. Furthermore, before carrying out a real surgery it is possible to prepare and improve the operation within a virtual operation room.

The last aspect requires real-time interaction in the Virtual Environment with the huge time-varying data set, which resulted from the flow simulation of the total inspiration and expiration period. It is already difficult to represent the data, if the surgeon wants to explore the data set for one time level only. Head tracking and the stereoscopic projection, usually for even more than one projection plane, must not reduce the frame rate below a minimum limit. Including exploration of all time levels using additional interactive visualization techniques violates the frame-rate requirement and therefore prevents real-time interaction. In addition, if the surgeon wants to operate virtually, the planning system cannot furthermore be integrated in a usual stand-alone Virtual Reality system. Our Approach to handle such a complex system is a completely distributed VR system with units for the visualization and other units for the flow feature calculation and data management.

3. The Distributed VR-System

The foundation of the computer assisted planning system is the Virtual Reality toolkit ViSTA developed at the University of Technology Aachen, Germany [2]. Applications using ViSTA automatically run on different VR systems (e.g. the Holobench or the CAVE) as well as on a variety of OS platforms (e.g. IRIX, SUNOS, Win32 and Linux). ViSTA itself is based on the widely used World Toolkit (WTK). Moreover, we have implemented an interface into ViSTA in order to integrate further Open-GL-based toolkits like the Visualization Toolkit (VTK). VTK, an Open-Source project distributed by Kitware Inc., facilitates the development of scientific visualization applications [3]. Gathering all these components we could start to implement the planning system immediately without worrying about VR and CFD peculiarities.

In order to improve the performance of VR applications it may be possible to implement multi-processing functionalities. Multi-threading and multi-processing are convenient features to speed up a VR application on a stand-alone visualization workstation like our multi-processor shared-memory Onyx by SGI. However, extensive calculations and huge data sets can still slow down the whole system. Therefore, we have developed a scaleable parallelization concept as an extension of ViSTA, where it is possible to use the visualization workstation for the representation of graphical primitives only. The remaining time-consuming calculation tasks are processed on dedicated parallel machines. The raw

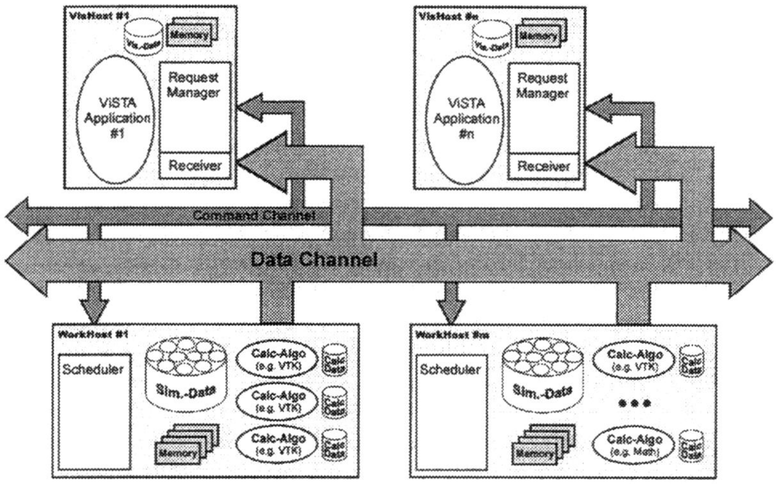

Figure 1. Important components and data flow

data sets, which for instance were yielded by a CFD simulation, are generally not needed on the visualization host anymore. Thus almost the whole memory and the power of all processors coupled with specialized graphics hardware are now available for the visualization and real-time interaction. In the next paragraphs a variety of additional design features are introduced in order to increase the performance even more.

3.1. The Design

Figure 1 shows the parallelization concept of ViSTA. On the upper row, the visualization hosts are shown. They can run independently or are connected to a distributed Virtual Environment. As a reaction to user commands, e.g. a request to compute and display streamlines at specified points, a request is created within ViSTA. Each of these requests is assigned a priority, which will actually determine how fast it is processed. Then it is passed on to a request manager, which chooses one of the work hosts (depicted on the lower row of figure 1) for completing the request. The request manager is an internal part of ViSTA. These request managers have to synchronize with each other to avoid a single work host being overloaded with requests, while other work hosts are idle.

Then the request is forwarded to the chosen work host, where a scheduler receives it. This scheduler selects a minimum and maximum number of nodes, which are to be utilized for the given request. These numbers depend on factors like computational speed, available memory, and the capacity of the network of the machine. Algorithms might actually slow down if too many nodes are used or if the network is too slow, so the number of nodes to use to fulfill a given request depends on the machine, on which the request is executed. The request is then added to the work queue of the scheduler, which is sorted with descending priorities. As soon as a sufficient number of nodes for the request with the highest priority are available, the scheduler selects the actual nodes (up to the maximum number assigned to the request), which will process the request. The selected nodes then start computing, and one of the nodes will send the result to the receiver on the visualization host, which sent the request. The receiver, which is the last part of ViSTA, is responsible for passing the result to the application.

This concept makes use of two different communication channels, as depicted in figure 1. On the one hand, the command channel, through which the requests are passed from the request manager to the work hosts, needs only very little bandwidth, since the messages sent along this channel are rather small (typically less than a hundred bytes). On the other hand, the results, which were computed on the work hosts, are quite large, up to several mega bytes. At this point, a network with a high bandwidth is necessary.

The concept offers one potential optimization feature: requests might be computed in advance on the actual user request. Even for large data sets and complex algorithms the work hosts will not be busy all the time, since the user will need some time to analyze the displayed data. During this time, requests can be computed in advance and then can be cached on the visualization hosts.

If a request manager receives a request, it first checks with its local cache, whether this request has already been computed. If so, the result is taken from the cache and immediately displayed. For this precomputation, requests of a very low priority can be generated. For this optimization to work, it is necessary to suspend any process working on any low priority request, when a request of a higher priority arrives.

3.2. Prototype Test Bed

A first prototype was implemented using the Message Passing Interface (MPI) as communication library [4]. This prototype supports only one visualization host and one work host. A kind of connection management dispatching the requests to systems, which are available and most suitable for this particular request, is still under construction. Thus, right now the user must determine the involved computer systems before starting the parallelized calculation task. This allowed us to quickly code and test the described concept.

MPI is speed optimized for each specific computer system. Therefore, it cannot be employed for heterogeneous systems. However, our prototype implements the data receiver of the visualization host by using of the MPI technology for the communication with the work host. In general (and this is just our goal), the visualization host and the work host are different systems. Fortunately, the Argonne National Laboratory (ANL) implemented a platform independent, system crossing and free available MPI version, called MPICH. The drawback of MPICH is the loss of some of possible speed, which is understandable because MPICH is based on a smallest common communication protocol, usually TCP/IP. Therefore we compared our MPICH based prototype with the MPI versions, which only works on heterogeneous platforms. For the final version of ViSTA, we consider using TCP/IP for the communication between the visualization hosts and the work hosts, thus we can profit by the faster MPI implementations for the calculations on high performance computer.

In order to assess our prototype we merely implemented a simple parallel function for

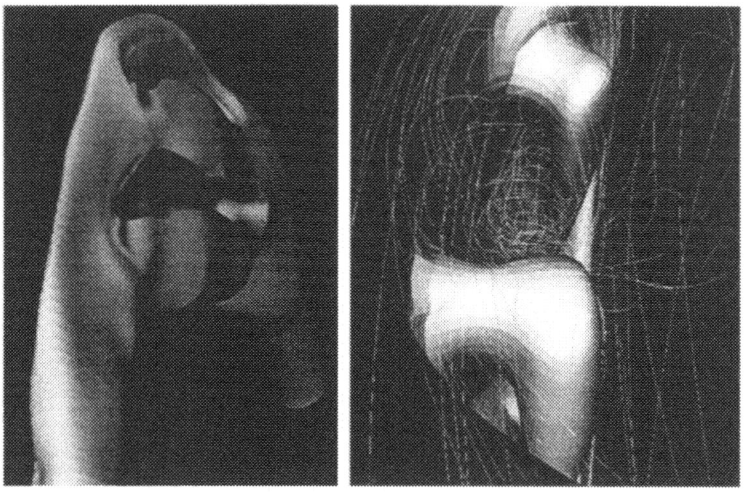

Figure 2. Outside view of one of the nose's cavities (pressure, color coded) (left), calculated streamlines inside of the nose (right)

computing streamlines. The complete data set of one time level, which is to be visualized, is read on each node of the parallel work hosts. The computation of streamlines is then split up equally on the available nodes, where the result is computed independently of the other nodes. Since the visualization host expects exactly one result for each request, the computed streamlines are combined on one node and then sent to the visualization hosts. Work in the area of meta-computing has shown that it might actually reduce communication time when messages over a slower network are combined [5].

The simulation of the airflow within a nose is a difficult and time-consuming process. Our first nose we have examined was not a human nose scanned by a CT, but an artificial nose, which was modeled as a "perfect" nose. More precisely, only one cavity was modeled. Flow experiments with this model resulted in first assumptions about the flow behavior during respiration. Right now we compare our simulation results with the results of these experiments. Furthermore, the current bounding conditions and multi-block arrangements are being adapted for a converging calculation. This adjustment is ongoing work, so that we took one time step of a preliminary multi-block arrangement for our parallelized prototype [1]. However, the final multi-block solution will also profit from the parallelization concept.

The used multi-block consists of 34 connected structured blocks, each with different dimensions, which yield into a total data set of 443.329 nodes. Moreover, for each node not only the velocity vector but also additional scalar values like density and energy are stored. Employing these informations more values, e.g. Mach number, temperature, and pressure, can be determined. In order to evaluate the parallelization approach a streamline source in form of a line was defined in the entry duct of the model nose. This resulted in streamlines flowing through the whole cavity. The model nose, property distribution, and calculated streamlines are depicted on figure 2.

3.3. The PC Cluster

The primary goal was to separate the system executing the visualization and the system, which is optimized for parallel computation. For our daily used standard stand-alone VR environment we use a high performance graphics workstation of SGI, the Onyx-2 Infinite Reality 2 (4 MIPS 10000, 195 MHz, 2 GByte memory, 1 graphics pipe, 2 raster managers), which should finally be used as our visualization host for the prototype test bed.

This system was coupled to the Siemen's hpcLine at the Computing Center of the university of Aachen. The hpcLine is a Linux cluster with 16 standard PC nodes (each consists of two Intel-PII processors, 400 MHz, 512 KByte level-2 cache, 512 MByte system memory), which are connected via a high performance network (SCI network, Scali Computer AS). This Linux cluster can achieve 12.8 Giga flops [6].

To determine the impact of the network bandwidth, we used different MPI implementations. On the one hand, we used the native MPI implementation on the SGI (SGI-MPI, MPI device: arrayd) and the hpcLine (ScaMPI 1.10.2, MPI device: sci), which offers a peak bandwidth of 95 and 80 MBytes/s, respectively. On the other hand, as already mentioned before, it is not possible to let these different libraries work together to couple an application on both platforms. Therefore, we used MPICH (version 1.1.2, MPI device: ch_p4), which is available for each of our target architectures, and which supports data conversion using XDR necessary for IRIX - Linux combinations. Since MPICH does not support the SCI network, the internal bandwidth of the hpcLine was reduced to about 10 MBytes/s. The Onyx and the hpcLine are connected with a 100 MBits/s Fast-Ethernet.

3.4. Results

Figure 3 shows the results of the nose application when computing 50 and 100 streamlines. The time needed for the calculation process is split up into the actual streamline calculation part, into a part needed for communication between all participating nodes, and into a part, which reorganizes the arising data structures into a unique data stream. The last step is needed, because MPI handles data streams of one data format only, e.g. only floating points, which we use for our implementation. The figure merely shows the time consumption of the worker nodes. The number of worker nodes does not include the scheduler, which is additionally running, but only plays a subordinary rule in this early prototype.

As a first result the Linux cluster is considerably faster in computing the results than the SGI. This supports the claim of the so-called meta-computing, where hosts of different architectures work together to solve one problem [5]. The hpcLine shows an acceptable speed-up mainly limited by the communication overhead. The floating-point conversion does not seem to have an essential impact on calculation time. Figure 4 shows all three parts without SGI results and as separate columns. Thus, they can be analyzed in more detail.

In contrast to earlier measurements where we used a merely simple CFD data set [7], the calculation load is not distributed equally on all nodes now. The distribution mainly depends on the start location of each streamline, which again controls its length and the

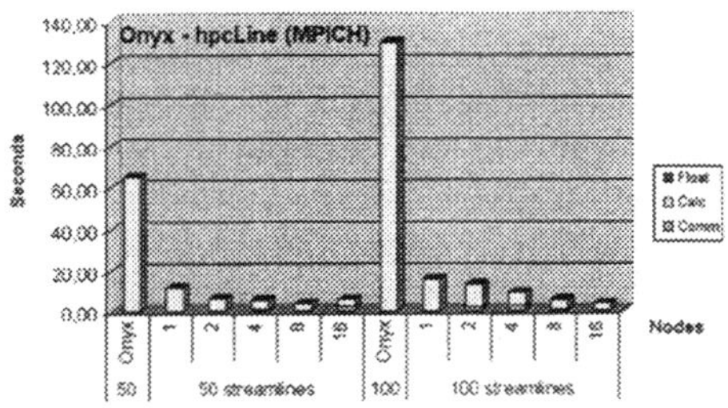

Figure 3. Calculation speed-up using different numbers of calculation nodes

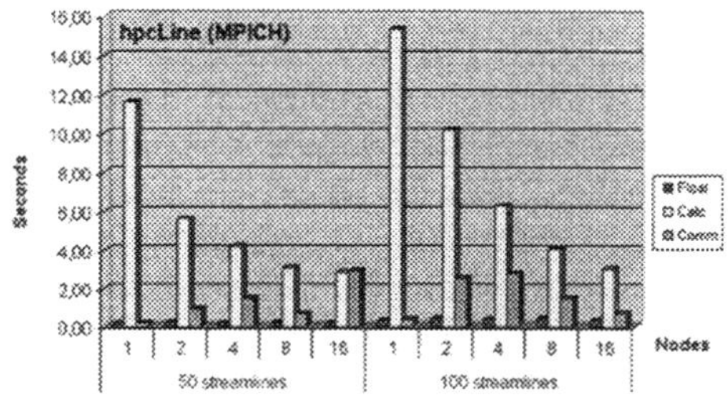

Figure 4. The behavior of the hpcLine in more detail

number of line segments. As we have no means of predicting the work load, each node calculates the same number of streamlines. Therefore, we could not achieve a speed-up by calculating 50 streamlines using sixteen nodes instead of eight (see figure 5). The slowest node determines the entire calculation time. In figure 5 we recognize a calculation peek at node 4 using 8 nodes and a peek at node 7 using 16 nodes, respectively. Both peeks are nearly equal which explains the missing speed-up. Moreover, strong varying calculation loads as well as a hanging or waiting result collecting node (our node 1) increase latency times, which again influence the measured communication part.

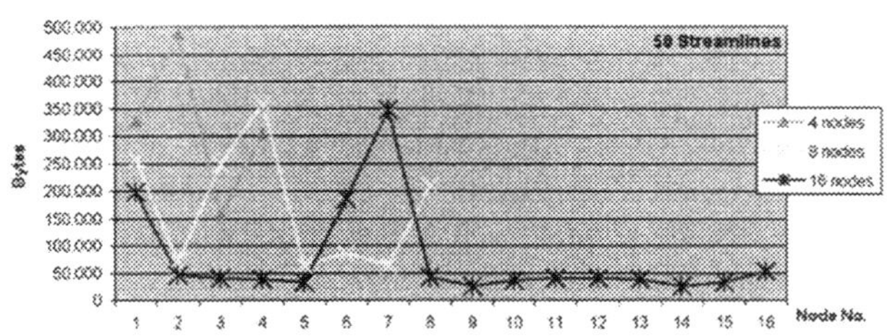

Figure 5. The resulting data size of calculated streamlines on each node

The significant influence of a fast network can be seen by comparing the result of the hpcLine alone and the coupled system SGI / hpcLine. On the coupled system, MPICH was employed, so that only the slower Fast-Ethernet was used to transmit the data on the work host. As a result, we finally achieve a maximum speed by utilizing 8 nodes to calculate 50 streamlines. This supports the importance of fast networks and the design issue that the scheduler on each work host decides on the number of nodes to use for a given request.

4. Conclusion and Future Work

Despite the achievement of real-time interaction on the visualization host it is conspicuous that the calculation expenditure of features within the flow field should be better balanced. This can already be achieved by a stronger integration of the scheduler, whose main job after all is the optimum distribution of incoming requests on the available nodes. This also includes the balanced distribution of one request on all nodes. Intelligent balancing strategies are going to be developed and will additionally speed-up the parallel calculation.

The simulation data was loaded in advance and was not shown in the measuring diagrams, because it is a part of the initialization of the whole VR system and therefore can be neglected. Otherwise loading the total data set into the memory of each node took approximately one minute. The Onyx has enough memory to accommodate the simulation data, however the nodes of the hpcLine already work at the limit. Larger data sets or unsteady flows definitely expect data loading on demand. Thus, we have started developing a data management system, where only the data package containing the currently needed data block of the whole multi-block grid is loaded into memory. Leaving the current block searching the next new flow particle position forces the worker node to load the appropriate neighboring block. Memory size and topological informations control the expelling from memory. Yet, extensive loading and removing data from harddisc to memory and vice versa is quite expensive and should be avoided. Probably prediction approaches can make use of a set of topologically and time linked blocks. Nevertheless, if one of the structured blocks are already be too large fitting in the memory of a node a splitting strategy (half-split, fourfold-split, eightfold-split) can be applied as a preprocessing step.

REFERENCES

1. T. van Reimersdahl, I. Hörschler, A. Gerndt, T. Kuhlen, M. Meinke, G. Schlöndorff, W. Schröder, C. Bischof, Airflow Simulation inside a Model of the Human Nasal Cavity in a Virtual Reality based Rhinological Operation Planning System, Proceedings of Computer Assisted Radiology and Surgery (CARS 2001), 15th International Congress and Exhibition, Berlin, Germany, 2001.
2. T. van Reimersdahl, T. Kuhlen, A. Gerndt, J. Henrichs, C. Bischof, ViSTA: A Multimodal, Platform-Independent VR-Toolkit Based on WTK, VTK, and MPI, Fourth International Immersive Projection Technology Workshop (IPT 2000), Ames, Iowa, 2000.
3. W. Schroeder, K. Martin, B. Lorensen, The Visualization Toolkit, Prentice Hall, New Jersey, 1998.
4. W. Gropp, E. Lusk, A. Skjellum, Using MPI - Portable Parallel Programming with the Massage-Passing Interface, Cambridge, MIT Press, Massachusetts, 1995.
5. J. Henrichs, Optimizing and Load Balancing Metacomputing Applications, In Proc. of the International Conference on Supercomputing (ICS-98), pp. 165–171, 1998.
6. http://www.rz.rwth-aachen.de/hpc/hpcLine
7. A. Gerndt, T. van Reimersdahl, T. Kuhlen, J. Henrichs, C. Bischof, A Parallel Approach for VR-based Visualization of CFD Data with PC Clusters, 16th IMACS world congress, Lausanne, Switzerland, 2000.

Parallel Computational Fluid Dynamics – Practice and Theory
P. Wilders, A. Ecer, J. Periaux, N. Satofuka and P. Fox (Editors)

An Optimised Recoupling Strategy for the Parallel Computation of Turbomachinery Flows with Domain Decomposition

Paolo Giangiacomo, Vittorio Michelassi, Giovanni Cerri

Dipartimento di Ingegneria Meccanica e Industriale, Università Roma Tre, Roma, Italy

The parallel simulation of two relevant classes of turbomachinery flow is presented. The parallel algorithm adopts a simple domain decomposition, which is particularly tailored to flow in turbomachines. The loss in implicitness brought by the decomposition is compensated by a sub-iterative procedure, which has been optimised to reduce the number of data exchanges and the time spent by MPI calls. The code has been applied to the simulation of an axial turbine stator and a centrifugal impeller. With 16 processors, speed-up factors of up to 14.7 for the stator and 13.2 for the impeller have been achieved at fixed residual level.

1. INTRODUCTION

Turbomachinery design is experiencing higher and higher benefits from the adoption of CFD techniques, in particular in the first stages of the design process. Massive tests of new design concepts over varying operating conditions require very fast computer codes for CFD to be competitive with experiments and to give results in a reasonable (from the industry point of view) time. In this scope, a great help may come from parallel computing techniques, which split the computational task among several processors [1]. In the search for higher code efficiency on distributed memory computers, the control and reduction of the time spent in exchanging data among processors is of fundamental importance. The parallel version of the time-marching implicit XFLOS code [2] adopts a simple domain decomposition which takes advantage of the peculiar features of turbomachinery flows, together with a sub-iterative procedure to restore the convergence rate of single-processor computations. The code has been applied to the simulation of the flow in an axial turbine stator and a centrifugal impeller, to test the effect on the speed-up and efficiency of an optimised data transfer strategy.

2. ALGORITHM

The XFLOS code solves the three-dimensional Navier-Stokes equations on structured grids, together with a two-equation turbulence model. For rotor flows, either absolute or relative variables may be chosen, and the present computations adopted absolute variables. The transport equations are written in unsteady conservative form and are solved by the diagonal alternate direction implicit (DADI) algorithm. The implicit system -in the unknown ΔQ- is thus split into the product of three simpler systems in the streamwise ξ, pitchwise η and spanwise ζ co-ordinates:

$$\left(L_\xi \times L_\eta \times L_\zeta\right)\Delta Q = RHS$$

The three operators are discretised in space by finite differences, and spectral-radius-weighed second-plus-forth-order artificial damping terms are added on both the explicit and the implicit size of the equations. This results in three scalar penta-diagonal linear systems to be solved in sequence.
Turbulence is accounted for by the k-ω model, together with a realisability constraint to limit overproduction of turbulent kinetic energy near stagnation points.

3. DOMAIN DECOMPOSITION

For multi-processor runs, a unique simply connected structured grid is decomposed into non-overlapping blocks in the spanwise direction only [2]. Each point of the overall grid belongs to one block only. To assemble fluxes at the interfaces, the solution in the outermost two layers is simply exchanged every time-step between neighbouring blocks, without interpolations. This procedure adds very little computational burden, and also ensures the identity of single and multiprocessor solutions at convergence. Figure 1 illustrates sample domain decomposition for the axial stator row and the centrifugal impeller.

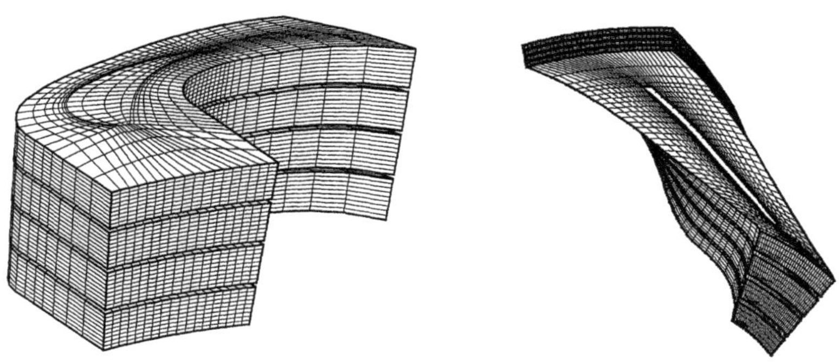

Figure 1 - Sample domain decompositions into non-overlapping blocks

This simple domain decomposition was deemed particularly advantageous for turbomachinery flows as long as the spanwise fluxes are much less than the dominant fluxes in the ξ and η directions. In fact, the decomposition does not alter either the RHS or the implicit operators L_ξ and L_η. Conversely, the explicit evaluation of spanwise fluxes and damping terms over the interfaces uncouples the overall operator L_ζ into one independent system per block, as visible in the left hand side of the following equation

$$
\underbrace{\left[\begin{array}{ccccc|ccccc}
\ddots & \ddots & \ddots & \ddots & & & & & & \\
\ddots & C & A & B & D & & & & & \\
 & E & C & A & B & (D) & & & & \\
 & & E & C & A & (B) & (D) & & & \\
\hline
 & & & (E) & (C) & A & B & D & & \\
 & & & & (E) & C & A & B & D & \\
 & & & & & E & C & A & B & \ddots \\
 & & & & & & \ddots & \ddots & \ddots & \ddots
\end{array}\right]}_{\text{block } n \qquad\qquad \text{block } n+1}
\left[\begin{array}{c}
\vdots \\
\Delta Q^{(m)}_{n,K-2} \\
\Delta Q^{(m)}_{n,K-1} \\
\Delta Q^{(m)}_{n,K} \\
\hline
\Delta Q^{(m)}_{n+1,1} \\
\Delta Q^{(m)}_{n+1,2} \\
\Delta Q^{(m)}_{n+1,3} \\
\vdots
\end{array}\right]
=
\left[\begin{array}{l}
\vdots \\
RHS_{n,K-2} \\
RHS_{n,K-1} - D\cdot\Delta Q^{(m-1)}_{n+1,1} \\
RHS_{n,K} - B\cdot\Delta Q^{(m-1)}_{n+1,1} - D\cdot\Delta Q^{(m-1)}_{n+1,2} \\
\hline
RHS_{n+1,1} - C\cdot\Delta Q^{(m-1)}_{n,K} - E\cdot\Delta Q^{(m-1)}_{n,K-1} \\
RHS_{n+1,2} - E\cdot\Delta Q^{(m-1)}_{n,K} \\
RHS_{n+1,3} \\
\vdots
\end{array}\right]
$$

in which the dropped-off terms have been put into brackets. Indices n,K and n+1,1 refer to the last node layer K of the n-th block and the first node layer of the (n+1)-th block, respectively. The reduced implicitness of the parallel DADI algorithm may be partly or fully restored by a sub-iterative procedure, which temptatively evaluates the neglected implicit terms, and reinserts them as a correction to the RHS. $\Delta Q^{(m)}$ is solved in function of $\Delta Q^{(m-1)}$ for a fixed number of sweeps, starting from $\Delta Q^{(0)} = 0$. Moreover, $\Delta Q^{(m-1)}$ in the outermost two layers must additionally be exchanged between neighbouring blocks. This correction is applied only to the five scalar equations of the mean flow.

4. RESULTS

The parallel XFLOS code has been applied to a research axial turbine stator and to a centrifugal impeller for industrial applications. The computational grids have I-type topology, and are shown in Figure 2. The number of grid points is 76×41×82 for the stator and 73×73×50 for the impeller, in the ξ, η and ζ directions, respectively.

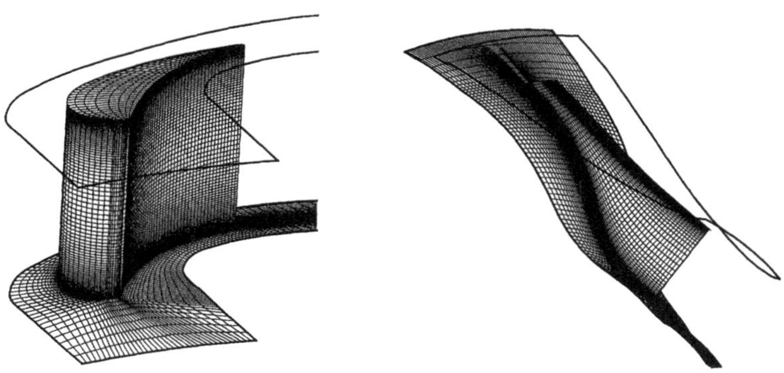

Figure 2. Computational grids for axial stator and centrifugal impeller.

The code has been run on a Cray T3E mainframe with MPI FORTRAN, and the execution times have been monitored by Cray's MPP Apprentice performance analysis tool.
Figure 3a shows the mean residual (average of residuals for the five mean flow equations) for the stator runs, without and with sub-iterative system re-coupling. With 8 processors the slow-down in convergence rate appears to be low, and single-processor convergence may be fully restored by 2 and 3 sweeps of the ζ-system with 8 and 16 processors respectively.

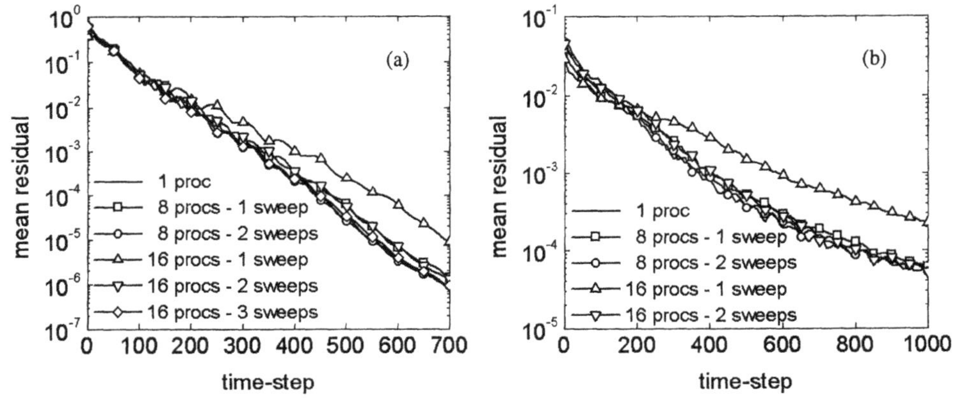

Figure 3. Convergence history for the stator (a) and the impeller (b)

The effect of domain decomposition and re-coupling on the centrifugal impeller convergence rate is shown in Figure 3b. Again, the slow-down in convergence rate is low with 8 processors. Single-processor convergence may be fully restored by 2 sweeps of the ζ-system with both 8 and 16 processors.
The sub-iterative recoupling of the L_ζ operator effectively restores single-processor convergence. However, the attractiveness of the sub-iterations closely depends on the required additional computational and MPI time. In particular, MPI time has a large incidence because of the large number of relatively small data packets to be exchanged during the sub-iterations. In fact, the coefficients of the L_ζ system do not change during the sweeps, and the system is solved by running all the sweeps on each η-constant layer before turning to the next. The coefficients can conveniently be stored in a two-dimensional array, but the temptative solution at and near the interfaces is exchanged once per sweep per layer. A considerable reduction in the MPI time and an improvement in the efficiency have been achieved with a more flexible solution of the L_ζ system, which can be performed on a variable number "idn" of η-layers simultaneously. The number and the size of packets to be exchanged are respectively divided and multiplied by idn, and the coefficients for all the idn layers have to be stored simultaneously in a larger three-dimensional array, the added dimension of which can be tuned as a parameter. However, the resulting increase in the memory size of the code should not be a problem in multi-processor computations, since the amount of RAM available to each CPU largely exceeds the needs of XFLOS. As an example, Table 1 summarised the memory

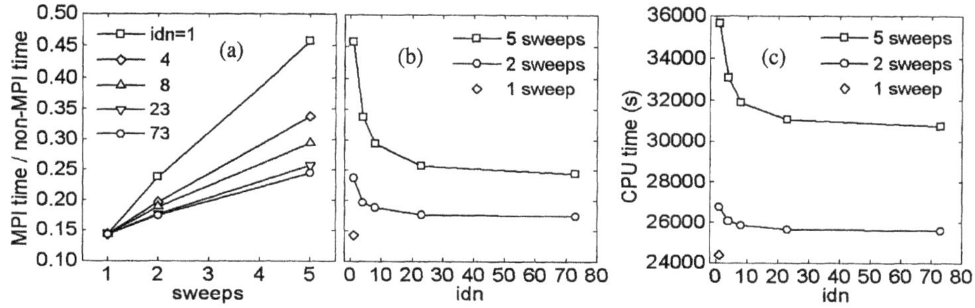

Fig. 4 - CPU time for the centrifugal impeller with 16 processors (a,b: MPI/non-MPI, c: total)

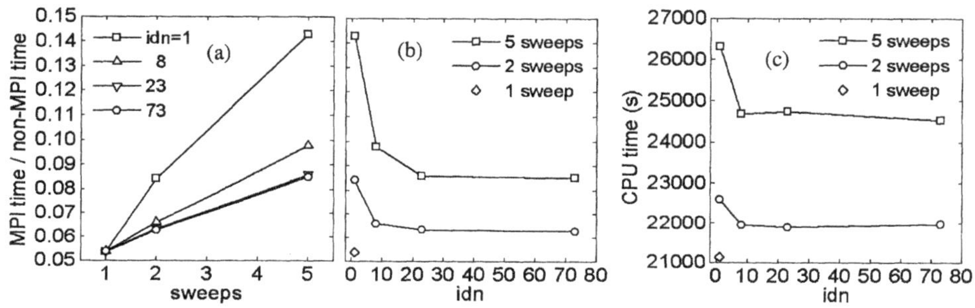

Fig. 5 - CPU time for the centrifugal impeller with 8 processors (a,b: MPI/non-MPI, c: total)

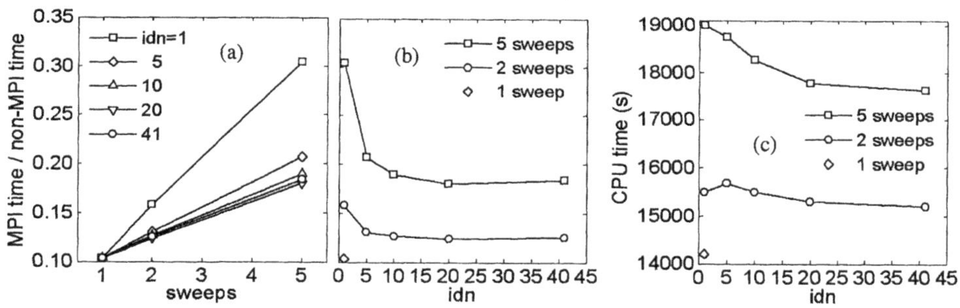

Fig. 6 - CPU time for the axial stator with 16 processors (a,b: MPI/non-MPI, c: total)

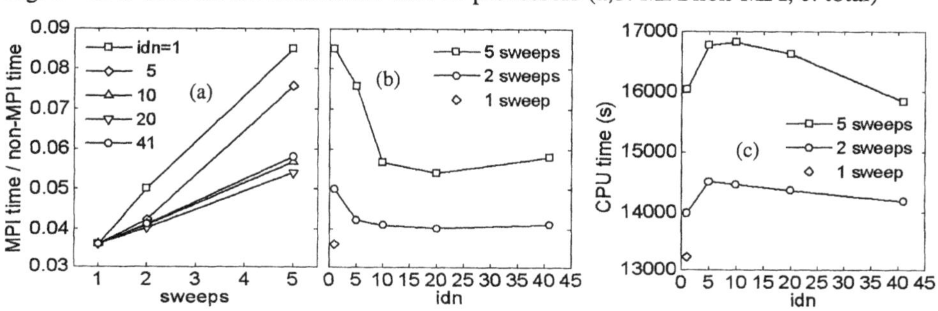

Fig. 7 - CPU time for the axial stator with 8 processors (a,b: MPI/non-MPI, c: total)

size of the code for the full grid of the centrifugal impeller. Observe that in parallel computations the minimum allowable spanwise dimension is greatly reduced.

Table 1
Memory size of XFLOS for 73×73×50 grid nodes (MB on an Alpha workstation)

idn	1	8	23	73
memory size	61.9	65.2	72.2	95.5

The effect of the parameter idn over the computation time of MPI procedures may be evaluated from Figures 4ab, 5ab, 6ab and 7ab, for the stator and for the impeller, with 8 and 16 processors. At fixed idn, the MPI time generally varies linearly with the number of sweeps, as expected. At a fixed number of sweeps, the reduced number of data exchanges results in a MPI time saving up to 47%. However, this does not necessarily turn into total CPU time saving, as shown in Figure 7c, referring to the stator with 8 processors, and in Figures 6c and 5c, although less evident. For the tests of Figure 7, a detailed analysis of the CPU time of individual subroutines has been carried out. Increasing idn generally results in a decrease in the MPI time, but it also alters the time spent in assembling and solving the ζ-system, probably because of caching and/or striding effects, due to the increasing size of coefficient arrays.

Table 2
Speed-up for the stator after 700 time-steps

idn	1	5	10	20	41
8 procs, 1 sweep	9.03	--	--	--	--
8 procs, 2 sweeps	8.25	--	--	--	--
16 procs, 1 sweep	16.44	--	--	--	--
16 procs, 2 sweeps	15.05	14.80	14.96	15.19	15.24
16 procs, 3 sweeps	13.92	13.78	13.92	14.19	14.39

Tables 2 and 3 compare the speed-up at fixed number of iterations for the impeller and for the stator respectively, for the different choices of the parameter idn. With one sweep, the efficiency for the stator is larger than one for both 8 and 16 processors, while the efficiency for the impeller is about 0.93 because of the very low spanwise dimension (only 3 node layers) of the blocks. The variation of the speed-up with idn reflects the variation of the CPU time illustrated in Figures 4 to 7. Increasing idn to the maximum allowable increased the speed-up by 3.4% for the stator with 16 processors and 3 sweeps, and by 4.7% for the impeller with 16 processors and 2 sweeps.

Table 3
Speed-up for the impeller after 1000 time-steps

idn	1	4	8	23	69
8 procs, 1 sweep	8.54	--	--	--	--
8 procs, 2 sweeps	7.99	--	8.22	8.24	8.21
16 procs, 1 sweep	14.81	--	--	--	--
16 procs, 2 sweeps	13.49	13.85	13.97	14.03	14.12

Table 4
Speed-up for the stator at $1 \cdot 10^{-5}$ mean residual level (560 time-steps with 1 processor)

Idn	time-steps	1	5	10	20	41
8 procs, 1 sweep	580	8.72	--	--	--	--
8 procs, 2 sweeps	550	8.40	--	--	--	--
16 procs, 1 sweep	690	13.35	--	--	--	--
16 procs, 2 sweeps	580	14.53	14.29	14.44	14.67	14.72
16 procs, 3 sweeps	560	13.92	13.78	13.92	14.19	14.39

Table 5
Speed-up for the impeller at $0.25 \cdot 10^{-3}$ mean residual level (590 time-steps with 1 processor)

Idn	time-steps	1	4	8	23	69
8 procs, 1 sweep	630	8.00	--	--	--	--
8 procs, 2 sweeps	590	7.99	--	8.22	8.24	8.21
16 procs, 1 sweep	970	9.01	--	--	--	--
16 procs, 2 sweeps	630	12.63	12.97	13.08	13.14	13.22

However, for the application point of view it is more meaningful to compare the efficiency at fixed final residual level (i.e. at the same quality of the solution), in order to take into account also the algorithmic efficiency, i.e. the decrease in the convergence rate brought by the domain decomposition. Tables 4 and 5 compare the efficiency for the impeller and for the stator respectively, together with the required number of time-steps to reach the pre-fixed residual level. For comparison, stator and impeller require 560 and 590 time-steps respectively with 1 processor.

Table 4 clearly shows the existence of a trade-off between the reduced time-step number to converge and the increased computational effort per time-step as the number of sweeps is increased. With both 8 and 16 processors it is not convenient to fully restore single-processor convergence rate, and the highest speed-up is found with 1 and 2 sweeps respectively. Increasing idn improves the speed-up by exactly the same extent as in Table 2. Similar comments may be drawn for the impeller (see Table 5). With 8 processors it is not convenient to adopt 2 sweeps, unless idn is set larger than one. The strong reduction in the convergence rate with 16 processors and 1 sweep requires 54% more time steps to reach the fixed residual level, and results in a very poor speed-up. However, the situation can be greatly improved by sub-iterating with a final speed-up factor of 13.22.

5. CONCLUSIONS

The presented domain decomposition is particularly simple, and allows easy code parallelisation without recalculations and interpolations. The slowdown in convergence rate with many processors may be compensated by a sub-iterative re-coupling procedure, and a trade-off between convergence improvement and increased computational effort can be found at an optimum number of sweeps. The variation of the parameter "idn", which controls number and size of data packets to be exchanged, allows a considerable reduction of the time spent by MPI procedures. However, this does not necessarily turn into reduced CPU time because of undesired effects on other subroutines, probably because of caching and/or striding effects, as the array dimensions increase with idn. Thus, the optimised sub-iterative procedure proved to be an efficient and effective mean to maintain the good convergence properties of the implicit method in the frame of a parallel code. The algorithm allows a speed-up factor of 14.7 for the stator and 13.2 for the impeller to be achieved at fixed residual level.

ACKNOWLEDGEMENTS

The Authors gratefully acknowledge CINECA computer centre for providing the technical support and CPU time.

REFERENCES

1 Schiano, P., Ecer, A., Periaux, J., Satofuka, N., (Ed.), "Parallel Computational Fluid Dynamics – Algorithms and Results Using Advanced Computers", 1997, Proceedings of the Parallel CFD '96 Conference, Capri, Italy, May 20-23 1996.

2 Michelassi, V., Giangiacomo, P., Simulation of Turbomachinery Flows by a Parallel Solver with Sub-iteration Recoupling. 1st International Conference on Computational Fluid Dynamics, July 10-14, 2000, Kyoto, Japan.

Parallel Computational Fluid Dynamics – Practice and Theory
P. Wilders, A. Ecer, J. Periaux, N. Satofuka and P. Fox (Editors)

Implementation of underexpanded jet problems on multiprocessor computer systems

I.A. Graur [a], T.G. Elizarova [a], T.A. Kudryashova [a], S.V. Polyakov [a], S. Montero [b]

[a] Institute for Mathematical Modeling, Russian Academy of Science, 125047 Moscow, Russia

[b] Instituto de Estructura de la Materia, CSIC, Serrano 121, 28006 Madrid, Spain

Parallel implementation for an axisymmetric supersonic jet simulation is presented. Numerical interpretation is based on the quasigasdynamic equation system. Parallel code is constructed using a domain decomposition technique. The Message Passing Interface standard have been used for organization of interprocessor data exchange. Calculations have been performed on cluster multiprocessor computer systems with distributed memory. Comparison of the numerical and experimental date based on Raman spectroscopy are fulfilled.

1. INTRODUCTION

The paper is devoted to the implementation of the numerical methods for the gasdynamic flows for the cluster multiprocessor computer systems with the Message Passing Interface (MPI) standard for organization of interprocessor data exchange. Underexpanded jet flow problem is used as an example of the technique applied. The calculations consist in solving quasigasdynamic (QGD) equations [1,2,3] under the flow conditions that allow the comparisons with the experimental results [4,5].

In the flow under consideration the pressure and density differ sharply from the nozzle exit section to the external parts on the jet. The geometrical configuration of the jet changes dramatically from several microns near the nozzle to several millimetres at the downstream section. These features and the complex shock configuration in the inner part of the jet require the implementation of fine computational grids, together with a large number of iterative steps for the solution convergence. So, the jet problem is time consuming and requires the implementation of a powerful computer system, namely parallel computer. The numerical method implemented here has inner structural parallelism and the use of parallel computers seems natural. The experience of using parallel transputer systems for implementation of QGD equations can be found in e.g. [6]. The use of modern parallel systems for semiconductor simulations is described in [7].

The calculations were carried out with flow conditions that allow comparison with the experimental results obtained at the Instituto de Estructura de la Materia, CSIC. The experimental part, based on high sensitivity Raman spectroscopy mapping, provides absolute density and rotational temperature maps covering the significant regions of the jet: zone of silence, barrel shock, Mach disk and subsonic region beyond the Mach disk [4,5]. The miniature jet diagnostic facility was used.

The comparison between numerical and experimental results shows the adequacy of the equation system and of the associated numerical procedure for treating the problem under consideration.

1. COMPUTATIONAL MODEL

The numerical interpretation is based on the QGD equations, constructed as an extension of the traditional Navier - Stokes (NS) equations. QGD equations reduce to NS ones for vanishing Knudsen numbers. For stationary flows, the dissipative terms in QGD equations are similar to NS ones with the additional contribution of order $O(Kn^2)$, where Kn is the Knudsen number [2,3].

The axisymmetric vector form of QGD equations in (r,z) co-ordinates reads:

$$\frac{\partial U}{\partial t}+\frac{\partial F}{\partial r}+\frac{F}{r}+\frac{\partial F_1}{\partial r}+\frac{\partial E}{\partial z}=\frac{\partial}{\partial z}\tau\left(\frac{\partial}{\partial z}V+p\frac{\partial Q_{p/\rho}}{\partial z}+\frac{p}{\rho}\frac{\partial Q_p}{\partial z}+\frac{1}{r}\frac{\partial rQ_1}{\partial r}+\frac{\partial Q_2}{\partial r}\right)+$$

$$\frac{1}{r}\frac{\partial}{\partial r}\tau\left(\frac{\partial rW_1}{\partial r}+r\frac{\partial rW_2}{\partial r}+p\frac{\partial Q_{p/\rho}}{\partial r}+r\frac{p}{\rho}\frac{\partial Q_p}{\partial r}+r\frac{\partial rQ_1}{\partial z}\right)+\frac{\partial}{\partial r}\tau\left(\frac{\partial rG}{\partial r}+\frac{\partial Q_2}{\partial z}\right)-\frac{2\tau}{r^2}G,$$

where:

$$U=\begin{pmatrix}\rho\\ \rho u_z\\ \rho u_r\\ E\end{pmatrix},\ E=\begin{pmatrix}\rho u_z\\ \rho u_z^2+p\\ \rho u_r u_z\\ u_z(E+p)\end{pmatrix},\ F=\begin{pmatrix}\rho u_r\\ \rho u_r u_z\\ \rho u_r^2\\ u_r(E+p)\end{pmatrix},\ F_1=\begin{pmatrix}0\\0\\p\\0\end{pmatrix},\ G=\begin{pmatrix}0\\0\\pu_r\\0\end{pmatrix},$$

$$V=\begin{pmatrix}\rho u_z^2+p\\ \rho u_z^3+3pu_z\\ \rho u_r u_z^2+pu_r\\ u_z^2(E+2.5p)+0.5pu_r^2\end{pmatrix},\ Q_p=\begin{pmatrix}0\\0\\0\\ \gamma p/(\gamma-1)\end{pmatrix},\ Q_{p/\rho}=\begin{pmatrix}0\\0\\0\\ \gamma p/(\rho(\gamma-1)\Pr)\end{pmatrix},$$

$$W_1=\begin{pmatrix}\rho u_r^2\\ \rho u_r^2 u_z\\ \rho u_r^3\\ u_r^2(E+2p)\end{pmatrix},\ W_2=\begin{pmatrix}p\\ pu_z\\ 2pu_r\\ 0.5p(u_r^2+u_z^2)\end{pmatrix},\ Q_1=\begin{pmatrix}\rho u_r u_z\\ u_r(\rho u_z^2+p)\\ \rho u_r^2 u_z\\ u_r u_z(E+2p)\end{pmatrix},\ Q_2=\begin{pmatrix}0\\ pu_z\\ 0\\ 0\end{pmatrix}.$$

Here $E=\rho(u_r^2+u_z^2)/2+p/(\gamma-1)$, $p=\rho(R/M)T$, $\tau=\mu/p$, μ- is the viscosity coefficient, Pr is the Prandtl number, γ is the specific heat ratio, M is the molar mass, R is the universal gas constant. The viscosity coefficient has been treated within the variable hard sphere (VHS) model, which leads to a thermal dependence in the form:

$\mu=\mu_e(\frac{T}{T_e})^{\omega}$, where $\mu_e=\mu_{ref}(\frac{T_e}{T_{ref}})^{\omega}$.

Table 1
Nozzle exit quantities

Stagnation temperature $T_0(K)$	295.
Stagnation pressure $p_0(kPa)$	100.
Temperature $T_e(K)$	249.2
Pressure $p_e(Pa)$	$3.824*10^4$
Number density $n_e(m^{-3})$	$1.1114*10^{25}$
Mean free path $\lambda_e(m)$	$11.73*10^{-8}$
Mach number Ma_e	1.01
Temperature of residual gas $T_\infty(K)$	200.
Knudsen number $Kn = \lambda_e/(2r_e)$	$3.747*10^{-4}$

The VHS molecular diameter $d_{ref} = d(T_{ref}) = 4.17\times10^{-10}m$, $\omega = 0.74$ and $\mu_{\mathrm{ref}} = \mu(\mathrm{T_{ref}}) = 1.656\times10^{-5}\mathrm{Nsm^{-2}}$ at $\mathrm{T_{ref}} = 273\mathrm{K}$ according to [8] have been used here for N_2. The nozzle exit quantities required for the calculation are reported in Table 1. They have been obtained from the conditions of the experiment by means of the isoentropic approximation assuming $\mathrm{Ma_e = u_e / a_e} = 1.01$ at the nozzle exit, where $\gamma = 1.4$, $\mathrm{r_e} = 156.5\mu\mathrm{m}$, $\mathrm{p_0} = 1\mathrm{bar}$, $\mathrm{T_0} = 295\mathrm{K}$.

Present computational efficiency was tested with respect to the experimental density and rotational temperature profiles of several shock waves, and corresponding wakes, recently reported [4]. These were generated by expansion of nitrogen through a nozzle of exit diameter D = 313μm, under nominal stagnation pressure ($\mathrm{p_0}$), and temperature ($\mathrm{T_0}$). The four shock waves used as a refence, henceforth referred to as A, B, C, D were located at distances to the nozzle $\tilde{z} = z/D$, of about 9, 18, 27, 36. The locations were fixed by the ratios of stagnation to residual pressure p_0 / p_∞, for $p_\infty = 4.2$, 1, 0.5,

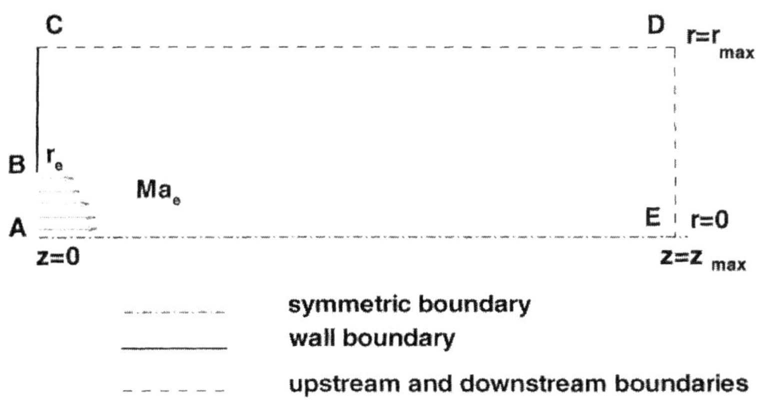

Figure1: Computational domain

0.28 mbar, respectively, controlling the residual pressure by means of an inlet needle valve in the expansion chamber.

The scheme of the computational domain is shown in Figure 1. At the nozzle exit we suppose a laminar boundary layer of the width $\delta = 0.18r_e$. The walls of the nozzle are considered to be adiabatic and Crokko's integral has been used for the temperature distribution near the wall [5].

2. NUMERICAL ALGORITHM AND PARALLEL REALIZATION

The computational domain is covered with a rectangular grid with space steps h_r and h_z. For $r < r_e$ the grid in the radial direction is uniform, with the smallest space step $h_r = h_{r,\min} = 0.1r_e$. For $r > r_e$, h_r increases between the next cells by a constant factor 1.05. The space step in the axial direction is uniform. . The limit of the computational domain in the radial direction is $r_{\max} = 100r_e$. The limit of the computational domain in the axial direction corresponds to the experimental conditions. The calculations were carried out for 3 computational grids given in Table 2.

For the solution of QGD equations the upwind – type splitting scheme with the space accuracy $O(h^2)$ was used based on the dissipative terms of QGD system [9]. The QGD equations are solved by means of an explicit algorithm where the steady-state solution is attained as the limit of a time evolving process. The computations stop when the steady-state solution is achieved.

The high performance parallel computer systems were used here in order to reproduce a detailed flow structure in a reasonable computation time. Present numerical method has been realised as a parallel program. For parallel realisation the geometrical parallelism has been implemented. This means that each processor provides calculations in its own subdomain. The whole computational domain is divided in z-direction. The number of subdomains is equal to the number of processors used.

Suppose that the computer system has p processors. Grid nodes set in z-direction $\Omega = \{0,1,...,nz\}$ is divided into subsets $\Omega^{(m)} = \{i_1^{(m)},...,i_2^{(m)}\}$, $m = 0,...,p-1$. Let nr is the number of points in r-axis, nz is the number of points in z-axis. That is, m-processor provides calculations of $(i_2^{(m)} - i_1^{(m)} + 1) * nr$ points. We form two-dimensional buffers for exchanges of inner boundary data. Data exchange among processors takes place after each time step, providing the synchronisation of the computations. One of the processors collects the results and saves them after each N-time step. If calculations are stopped, they can be continued with the saved files with another processors number (if need). The main sequential program can be included in the developed parallel program. Present software was written in FORTRAN.

Table 2
Computational grids

Type of grids	1	2	3
Number of points	141*91	281*91	561*91
h_z / r_e	1.0	0.5	0.25

All needed parallel functions are contained in MPI parallel libraries.

3. COMPUTATIONAL RESULTS

Two computer systems were used in present computational work:

1) MVS-1000M: 128 processors homogeneous parallel computer system equipped with Alpha-667 microprocessors. The total performance is over 130Gflops. Fast communication links give up to 250MB/sec data transmission rate.

2) Intel Pentium III: 24 processors cluster with distributed memory equipped with Intel Pentium III 600 processors. The total performance is over 14Gflops. Fast communication links give a 12MB/sec data transmission rate.

The speed-up for MVS-1000M and Intel-24 systems is shown in Fig. 2. The real efficiency of parallelisation for explicit schemes is close to 100% for sufficiently great number of computational nodes. Present parallel realisation is relatively simple and can be used for other algorithms based on iterative methods similar to the explicit finite-difference schemes.

According to the experimental data the computations were performed for 4 variants of the residual pressure (A, B, C, D). The grid convergence for variant A is plotted in Figures 3 and 4. The measured and calculated number density along the axis is normalized to the minimal value. For the temperature distributions the measured values are also shown. Note that in the experiment the rotational temperature was measured, but in numerical simulations the averaged temperature was calculated. The mentioned temperatures are close one to another except the vicinity of the shock-wave [5]. Decreasing the space steps leads to the convergence of the numerical results to the experimental values. Temperature is less affected to the grid step variations than the density.

In Figures 5, 6 show the density and temperature distributions for variant B (grid 3). For variants A and B calculated density and temperatures profiles nicely reproduce the

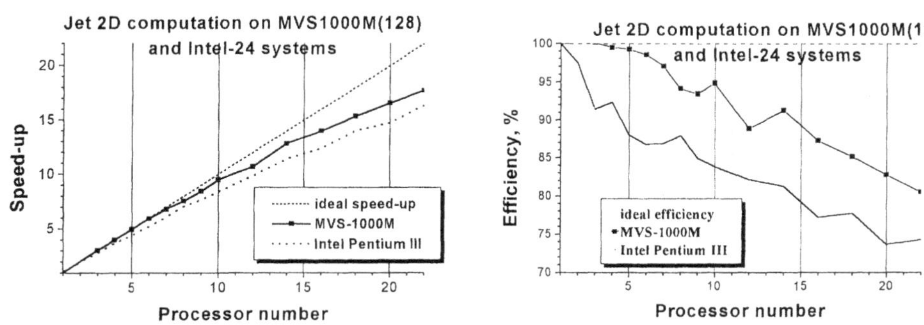

Figure 2: Comparison of parallel performance on MVS-1000M and Intel Pentium III

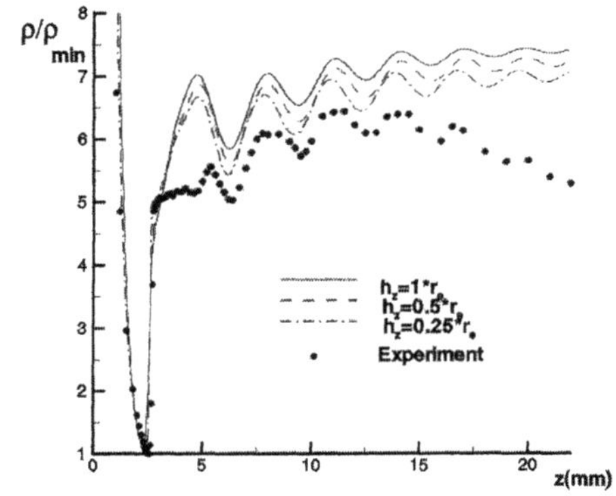

Fig. 3 Density distribution (variant A), grid convergence

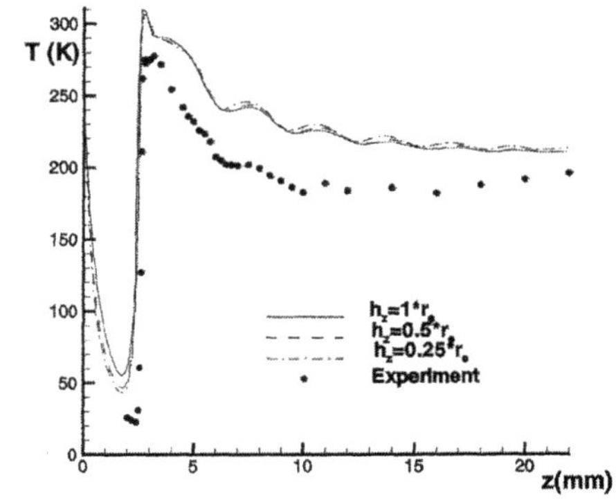

Fig. 4 Temperature distribution (variant A), grid convergence

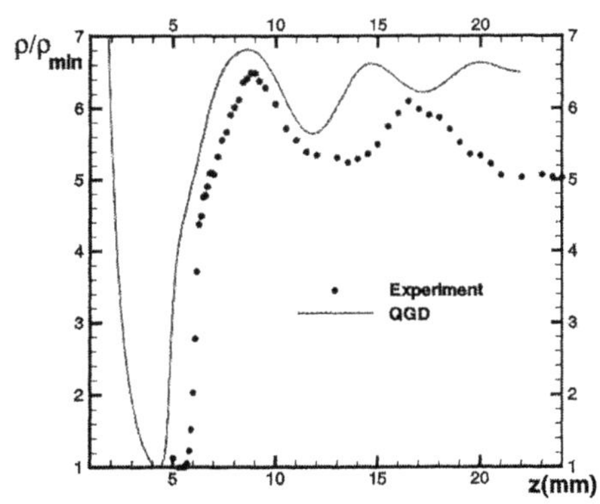

Fig. 5 Density distribution (variant B)

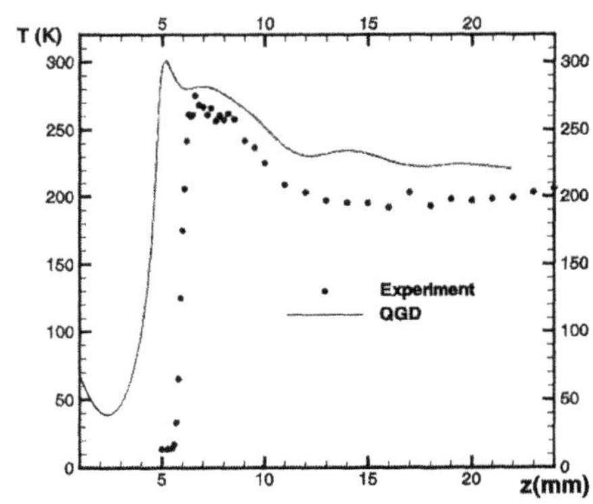

Fig. 6 Temperature distribution (variant B)

experimental data. For variants C and D (not shown here) corresponding to lower residual pressures and larger Kn numbers, the agreement between calculations and the experiment became poorer, especially for the temperature profiles prior to the shock.

Density and velocity maps (variant A, grid 3) presented in Figures 7 and 8, summarise the general features of an axisymmetric supersonic jet, namely zone of silence, Mach disk, barrel shock, slip region behind the Mach disk and the secondary shock waves, that are resolved with a qualitative agreement of the computations with the experiment. A trapped vortex (Fig.8) is formed beyond the Mach disk with a recirculation zone associated with a slow toroidal flow. In this structure the centreline velocity is reversed with respect to that in the zone of silence, differing qualitatively from the post-shock behaviour in one dimension

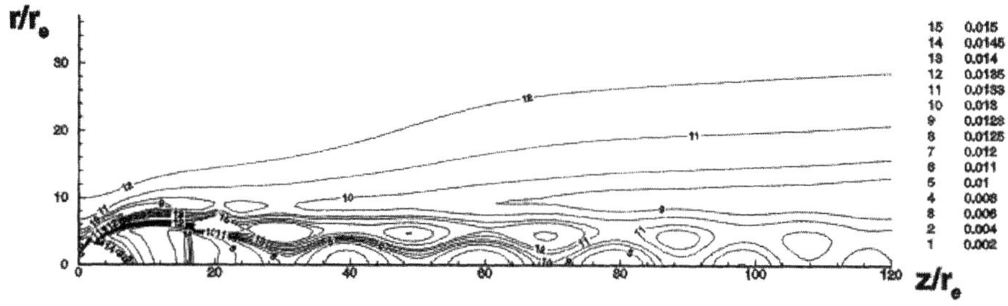

Figure 7: Isolines of density (variant A)

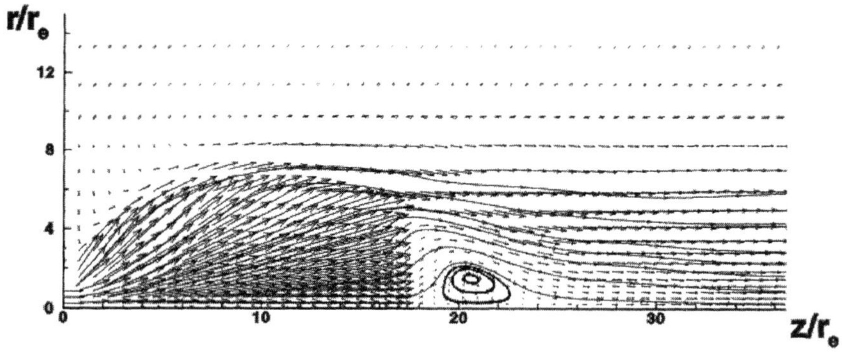

Figure 8: Flow field (variant A)

problem. The toroidal trapped vortex appears to be responsible for the collimation of the jet downstream from the shock wave.

4. CONCLUSIONS

The detailed description of an underexpanded jet flow field requires the fine time-space computational grids leading to the time-consuming computations, which naturally demand implementating high-performing computer systems. Present numerical results, and the efficiency estimations, show that the implemented numerical algorithm (explicit in time and homogeneous in space approximation of QGD equations) allows for an efficient use of the cluster multiprocessor computing systems described here.

The comparison of numerical and experimental results shows that the present method is adequate to simulate the underexpanded jet flow for sufficiently small Kn numbers.

REFERENCES

1. T.G. Elizarova, B.N. Chetverushkin. J. Comput. Math. Phys. Vol. 25, (1985) 164-169.
2. Yu.V Sheretov Quasihydrodynamic equations as a model for viscous compressible heat conductive flows, in book: Implementation of functional analysis in the theory of approaches, Tver University, (1997) 127 - 155 (in Russian).
3. T.G. Elizarova, Yu.V.Sheretov (2001) J. Comput. Math. Phys. Vol. 41, No 2, (2001) 219-234.
4. A. Ramos, B. Mate, G. Tejeda, J.M. Fernandez, S. Montero, (2000) Raman Spectroscopy of Hypersonic Shock Waves, Physical Rev E, October, 2000.
5. B. Mate, T.G. Elizarova, I.A. Graur, I. Chirokov, G. Tejeda, J.M. Fernandez, S. Montero, J. Fluid Mech., vol. 426, (2001) 177-197.
6. T.G. Elizarova, A.E. Dujsekulov, M. Aspnas, Computing and Control Engineering Journal, 1993, V.4, N 3, pp. 137-144.
7. T.A. Kudryashova, S.V. Polyakov, Simulation of 3D absorption optical bistability problems on multiprocessor computer systems, Proceedings of Moscow State Technological University "STANKIN", General physical and mathematical problems and modeling of technical and technological systems, ed. L.A. Uvarova, Moscow, (2001), 134-146.
8. G.A Bird, (1994), Molecular Gas Dynamics and the Direct Simulation of Gas Flows, Clarendon Press.
9. I.A. Graur (2001), J. Comput. Math. Phys. Vol. 41, No 11 (to be published).

Parallel Computational Fluid Dynamics – Practice and Theory
P. Wilders, A. Ecer, J. Periaux, N. Satofuka and P. Fox (Editors)

Numerical Simulation of Scramjet Engine Inlets on a Vector-Parallel Supercomputer

Susumu Hasegawa, Kouichiro Tani, Shigeru Sato

Kakuda Research Center, National Aerospace Laboratory
1 Koganezawa, Kimigaya, Kakuda, Miyagi, 981-1525, Japan

Flowfields in a scramjet inlet were numerically simulated by the Euler equations, and effect of the strut position in the scramjet inlet was also investigated. The computations were conducted for inflow Mach number of 5.3 on a vector-parallel supercomputer, i.e., NEC SX4/25CPU. The correlation was obtained between the inlet flowfields and strut positions.

1. Introduction

To accelerate research and development of advanced space engines, such as scramjet engines and reusable rocket engines, synergism of experiments and computation is indispensable. In addition to large-scale experimental facilities, such as RJTF (Ramjet Engine Test Facility)[1-6] and HIEST (High Enthalpy Shock Tunnel)[7], the Numerical Space Engine (NSE)[8-9], that is a numerical simulator for space engines, has been developed at the National Aerospace Laboratory, Kakuda Research Center. It is expected that the research and development using the experimental facilities will be facilitated by the use of the NSE. The main server of the NSE is a vector-parallel supercomputer, i.e., NEC SX4 / 25CPU. The NSE has been made use of for the elucidation of various phenomena inside the engines[10].

NAL conducted scramjet engine research at the flight conditions of Mach 6 by using the RJTF. It was found that engine performances depend on the strut positions in the inlet[4]. In order to clarify the dependence, flowfields in the scramjet engine inlet were numerically simulated by using the Euler equations, and effects of the strut position in the scramjet inlet were also investigated in our research.

2. Numerical Simulation Condition

In order to investigate influence of shock waves, expansion waves, and compression waves on flowfiled inside inlets, governing equations were assumed to Euler equations. Five processors on the NSE are used in parallel, and a turn around time of the run is 15 hours. In order to perform higher load computations such as combustion in a scramjet engine, it is necessary to develop more efficient and precise numerical techniques for vector parallel computers. The numerical method and inflow condition of flight Mach number 6 are stated in the following:

【Numerical Method】

Finite Volume Method

Multiblock

Implicit LU-SGS

Roe Scheme

3rd order Chakravarthy-Osher TVD

MPI Parallelization

Grid Size:100 x 100 x 50 (Symmetrical in the direction of Z)

【Inflow Condition】

Inlet Inflow Mach Number:5.3

Total Temperature :1450K

Total Pressure :4.8MPa

The inflow is uniform and contains no boundary layer thickness.

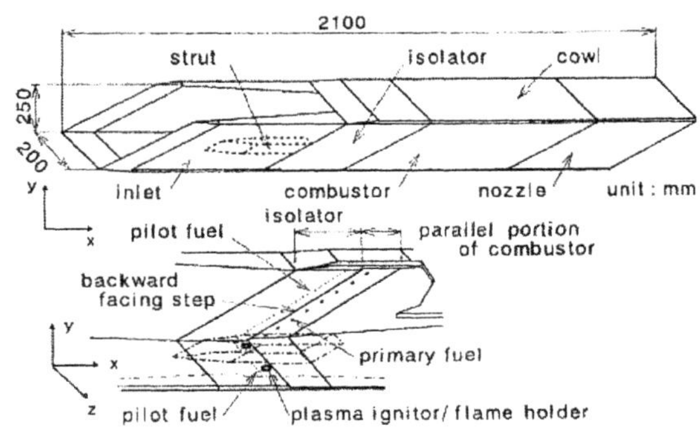

Figure1 Scramjet Engine Configuration

The engine configuration is showed in Fig.1. Computational region is limited to the inlet and the isolator. The length of the isolator is 200mm, and strut height is 1/5 of full-height strut size. The inlet is a sidewall compression type with a 6 degree-half angle. The leading edge is swept back at 45 degree. The geometrical contraction ratio is 2.86 without any strut. The influence of internal flow was investigated by changing distance between the strut leading edge and the isolator entrance. In the below, the difference of the flowfields about two cases are described from the numerical computational results. Schematic of the inlet and the isolator are depicted in Fig.2. In Case 1, the distance between the strut leading edge and the isolator entrance is 143mm, and in Case 2, the strut is shifted downstreams by 100mm, and the distance between the both is 43mm.

The RJTF experiments show the following results. In Case 1, when the fuel mass flow rate is small, the combustion is weak, and thrust is very low (called "weak combustion[1]"). In Case 2, while increasing the fuel flow rate, the combustion suddenly becomes intensive and the thrust increases (called "intensive combustion[1]") and keeps increasing with the increasing fuel flow rate[4]. If the fuel flow rate is increased too much, the engine goes into "unstart" and produces almost no thrust[1]. Engine performance depends on the strut positions.

Case 1: The distance between the strut leading edge and the isolator entrance is 143mm.

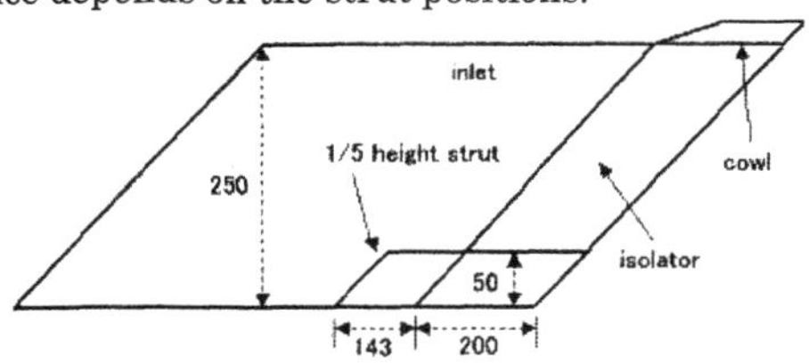

Case 2: The distance between the strut leading edge and the isolator entrance is 43mm

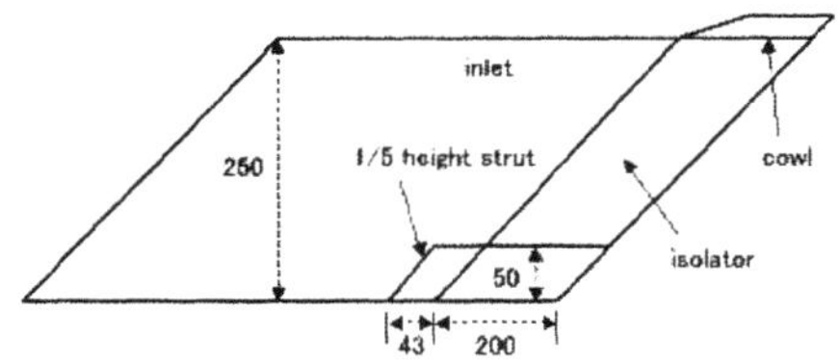

Fig.2 Schematic of the inlet and the isolator.

The grid in Fig.3 corresponds to the inlet and the isolator of Case 1. The total number of the grid points is 5 x 10^5 (I x J x K=100 x 100 x 50) and symmetrical calculation in the direction of K is performed. Since the grid size of the leading edges of the sidewall, the strut, and the cowl are made fine, the grid has sufficient resolution and gives no grid dependence. Minimum necessary block division is made, and the total number of blocks is 53.

3. Numerical Simulation Results

The numerical results based on the above conditions are described below. Comparison with experimental results and numerical results of Case 1 are shown in Fig.4. Figure 4(a) corresponds to the pressure on the sidewall versus the flow direction coordinate. Figure 4(b) corresponds to the pressure on the top wall versus the flow direction coordinate. In our calculation, it was assumed that the inflow is uniform and contains no boundary layer. The sidewall has its own boundary layer, but its thickness is not so much in the reality. The numerical results are mostly in agreement to the steep standup of the pressure and the position of maximum pressure obtained in the RJTF experiments[1] (see Fig.4 (a)). On the other hand, the pressure distribution on the top wall around 450mm is not in agreement with the RJTF experiments(see Fig.4 (b)). Since the inflow condition contains approximately 50mm boundary layer thickness in the RJTF nozzle flow, the region near the top wall is subsonic within the boundary layer. The pressure rise caused by the shock waves originated from the inlet leading edge is relaxed in the real flow, whereas observed in the computational result. But, the position of the maximum pressure, and the pressure distribution around 700mm are well in agreement. The same kind of result is also obtained in Case2. The CFD code used here has reasonableness.

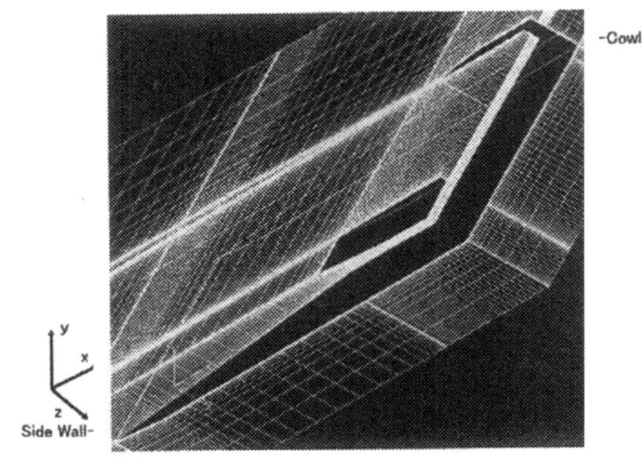

Figure 3 Grid of the inlet and the isolator of Case 1 (Size: 100 X 100 X 50).

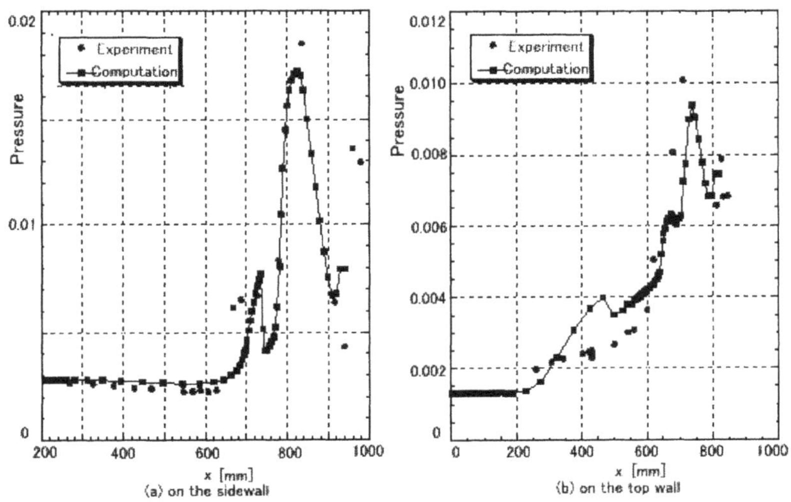

Figure 4 Comparison with experimental results[1] and numerical results of Case 1

(a) Pressure distribution on the sidewall, (b)Pressure distribution on the top wall.

The pressure is normalized by the total pressure of the inflow.

In order to clarify how the flow inside inlets changes with the positions of struts, the following Fig.5 through 8 show the numerically computed results. The distances between the strut leading edge and the isolator entrance are 143mm(Case 1) and 43mm(Case 2), respectively. Each pressure distribution on the top wall, on the cowl, in the symmetrical plane and in the isolator exit plane is displayed in Fig.5, Fig.6, Fig.7 and Fig.8, respectively. Since in the flow inside the inlets, not only shock waves from the inlet leading edge, from the cowl leading edge and from the strut leading edge but also expansion waves from the entrance of the isolator interfere one another, significantly different pressure patterns are made by the configuration of the inlet-strut system.

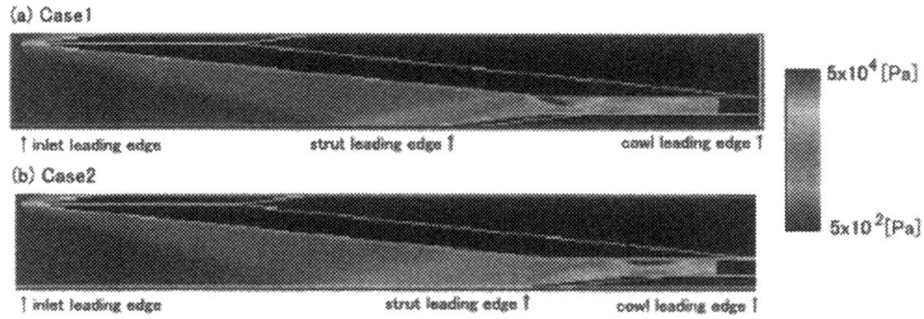

Figure 5 The pressure distribution on the top wall. The distances between strut leading edge and isolator entrance are (a)143mm(Case 1) and (b)43mm(Case 2), respectively.

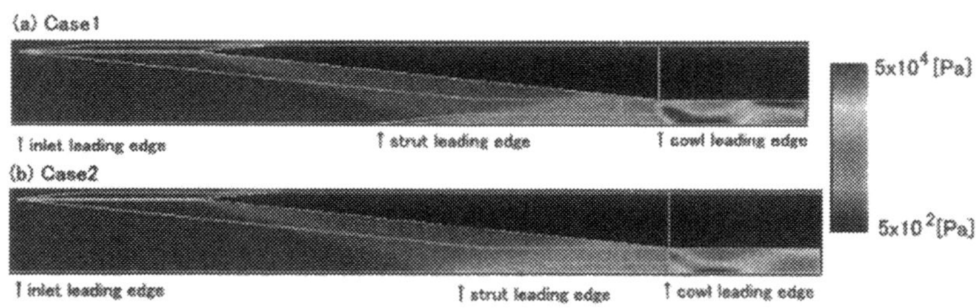

Figure 6 The pressure distribution on the cowl. The distances between strut leading edge and isolator entrance are (a)143mm(Case 1) and (b)43mm(Case 2), respectively.

Figure 5 (a) shows the pressure distribution on the top wall in Case 1. The shock wave from the inlet leading edge reflects in symmetrical plane, coalesces with the shock wave from the strut leading edge, and the coalescing shock waves collide with the sidewall. Thus, the first high pressure portion arises on the sidewall side. The coalescing shocks are weaken by the expansion waves generated by the shoulder of the strut, and the weaken coalescing shocks collide with the strut parallel part to produce the second high-pressure portion there. And the reflective waves from the strut parallel part collide with the sidewall and produce the third high pressure portion. Figure 5 (b) shows the pressure distribution on the sidewall in Case 2. The shock wave from the strut leading edge is shifted downstreams by 100mm in comparison with Case 1. Consequently, the high pressure portion in Case2 appears in the vicinity of the strut side near the isolator exit in comparison with Case1.

Figure 6 shows the pressure distributions on the cowl plane. The sidewall pressure on the cowl in the vicinity of the isolator exit in Case 1 is higher than that in Case 2

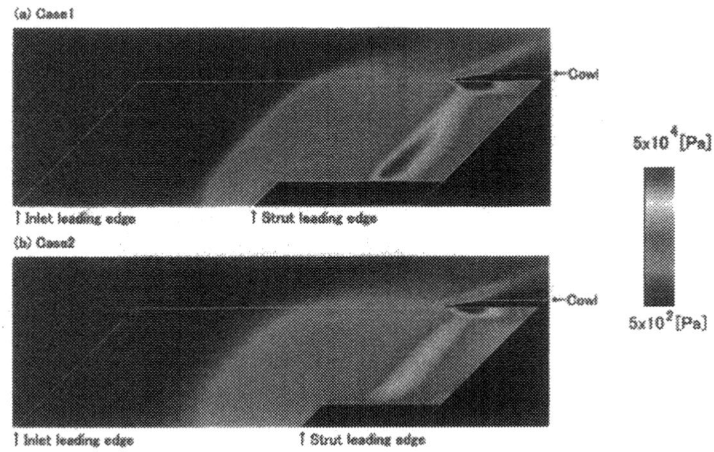

Fig.7 Pressure distribution in the symmetrical plane. The distances between strut leading Edge and isolator entrance are (a)143mm(Case 1) and (b)43mm(Case 2), respectively.

Figure 7 displays the pressure distributions in the symmetrical plane. The shock wave reflecting from the sidewall shown in Fig.5 collides with the strut and produces the high pressure portion above the strut in the symmetrical plane as well. The pressure level in the high pressure portion in Case1 is higher than in Case2, because the shock wave originated from the inlet leading edge merges with the shock wave originated from the strut leading edge. The influence of the pressure in the symmetrical plane is weaker as it is far from the strut.

Figure 8 gives the pressure distributions at the isolator exit of Case 1 and Case 2. The pressure in the vicinity of the sidewall on the cowl falls by shifting the strut leading edge downstreams by 100mm, whereas the pressure in the vicinity of the strut rises. It was found that it is possible to shift high pressure portion by moving the strut position.

4. Discussion

Although both of geometrically averaged pressures in the isolator exit cross section are the same at 0.7×10^{-2} (normalized by the total pressure in the inlet inflow), the maximum pressures are obtained 1.26×10^{-2} and 1.16×10^{-2} (normalized by the total pressure of the inlet inflow), respectively. By shifting the strut leading edge downstreams by 100mm, the maximum pressure falls by 7.78 %. The total pressure recoveries are obtained 84.6% and

85.1%, respectively, where the change of the total pressure recovery is at most 1% or less. For the given geometrical contraction ratio, the maximum pressure value is controllable by moving the strut position, although both geometrically averaged pressure at the isolator exit and the total pressure recovery hardly change.

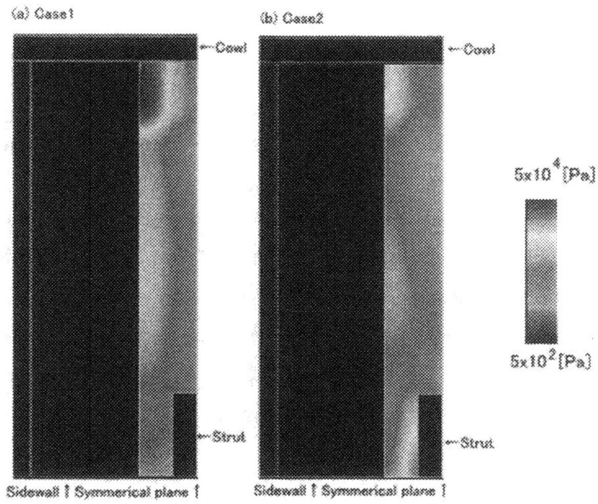

Fig.8 Pressure distribution at the isolator exit. The distances between strut leading edge and isolator entrance are (a)143mm(Case 1) and (b)43mm(Case 2), respectively.

We obtained very important phenomena[1] in the RJTF experiments: One is "weak combustion", and the other is "intensive combustion" in Case1, we obtained only the "weak combustion" when the fuel mass flow rate is increased, but. in Case2, we obtained the "intensive combustion" with the increasing fuel flow rate in the experiments[4]. It can be assumed that the transition to the intensive combustion can depends on the high pressure portion at the isolator exit. The shock wave that produces the high pressure portion near the sidewall in the vicinity of the cowl reflected with the sidewall, and the reflecting shock wave will produce the high pressure portion after the isolator exit in the symmetrical plane in Case 1. On the other hand, the shock wave that produces the high pressure portion near the strut reflected with the strut, and the reflecting shock wave will produce the high pressure portion on the sidewall near the fuel injector after the isolator exit in Case 2. Although the engine configuration in Case 2 may have the high pressure portion near the fuel injectors, the engine configuration in Case 1 can not have the high pressure portion near the fuel injectors. Therefore, in order to generate and maintain the intensive combustion, it is necessary to consider configurations of the inlet-strut system which generate the high pressure portion in the vicinity of the top wall near the strut.

It must be included in consideration that suitable pressure distribution for combustion can be obtained by controlling the strut position in the design of the inlet-strut system.

5. Conclusion

In this research, the numerical computation of the inlets under the Mach 6 test was performed. The flow generated as the result of interaction of shock waves and expansion waves was analyzed.

1) It is possible to shift high pressure portion in the inlet-strut system by moving the strut position.
2) For a fixed geometrical contraction ratio, the maximum pressure value at the isolator exit is controllable by moving the strut position. It is possible that the high pressure portion around fuel injectors can cause intensive combustion.
3) In order to generate the intensive combustion, it is necessary to take into consideration the strut position which produces high-pressure portion around fuel injectors when the inlet-strut system is designed.

Reference

1. T.Kanda et al,"Mach6 Testing of a Scramjet Engine Model", AIAA96-0380, Jan., 1996
2. T.Sunami et al, "Mach 4 Tests of a Scramjet Engine-Effects of Isolator," Proceedings of 13th International Symposium on Air Breathing Engine, Sep.1997,pp615-625
3. T.Kanda et al, "Mach 6 Testing of a Scramjet Engine Model," J.Propul.Power, 13.4.pp543-551.
4. S.Sato et al, "Scramjet Engine Test at the Mach 6 Flight Condition,"AIAA 97-3021, July 1997
5. T.Saito et al, "Mach 8 Testing of a Scramjet Engine Model," ISTS paper 96-a-2-11, 96
6. T.Kanda et. al, "Mach 8 Testing of a Scramjet Engine Model," AIAA99-0617,Jan. 1999
7. K.Itoh et al," Hypervelocity Aerothermodynamics and Propulsion Research using a High Enthalpy Shock Tunnel HIEST," AIAA 99-4960
8. S. Hasegawa et al, "The Virtual Test Bed Environment at NAL-Kakuda Research Center," Parallel Computational Fluid dynamics 1999、233-240
9. S.Hasegawa et al, " Development of the Numerical Space Engine at Kakuda Research Center, National Aerospace Laboratory," NAL-SP-41
10. S.Hasegawa et al, "Numerical Simulation of Scramjet Inlets – Estimation of Performance by Struts–," 50th Japan National Congress of Theoretical and Applied Mechanics,Tokyo, Japan, 2001

Parallel Computational Fluid Dynamics – Practice and Theory
P. Wilders, A. Ecer, J. Periaux, N. Satofuka and P. Fox (Editors)

Parallel Computation of Multigrid Method for Overset Grid

T. Hashimoto, K. Morinishi, and N. Satofuka

Department of Mechanical and System Engineering,
Kyoto Institute of Technology,
Matsugasaki, Sakyo-ku, Kyoto 606-8585, Japan

The purpose of this study is to develop an efficient parallel computational algorithm for complex geometries using overset grid technique. The procedure for parallel computation of multigrid method and the associated data strucure of one-dimensional listed type arrangement are described.

1. INTRODUCTION

At present, the analysis in the flowfield of the complex geometries has become practically important. Building the efficient computational algorithm which can analyze the flowfield around complicated shaped bodies brings a great contribution in engineering. Therefore, the purpose of this study is to develop an efficient parallel computational algorithm for analyzing the flowfield of complex geometries.

In this approach, grid generation and the load balancing in parallel computing are very important key factors.

First, for the flowfield where two or more complicated shaped bodies exist, overset grid technique [1] is used to avoid the difficulty of generating a single grid system over the complex field. For example, in a multi-element airfoil configuration, each grid of the elements is generated using a structured grid regardless of other elements. However, when the number of overlapping grids increases, an efficient procedure is needed for exchanging information between grids. In this study, an efficient treatment is proposed.

Next, in parallel computation, it is desirable to obtain the ideal parallel efficiency, but the efficiency may fall due to various factors. In this study, one-dimensional listed type data structure is introduced in order to divide the workload equally among PEs (Processor Element) and make computation efficiently for overset grid.

In addition, multigrid method is applied to parallel computation and its performance is investigated.

2. TREATMENT IN OVERSET GRID

The method of treating overset grid is represented here in the case of increasing overlapping grids. The overset grid technique in itself is referred to ref. [1]. Figure 1(a) shows the overset grid for a two-element airfoil. In this case, major grid is a Cartesian grid and covers the whole calculation domain. In addition, a minor grid is independently generated to each element using a structured grid regardless of other elements. However, when the number of overlapping grids increases, an efficient procedure is needed for exchanging information between grids. The efficient treatment adopted in this study is described as follows.

1. The grid points which exsit in and around the elements are excluded from the calculation. Figure 1(b)(c)(d) show overset grids after excluding the hole points.
2. Interpolation points are selected around the hole points and stored in a one-dimensional array in each grid successively. In each minor grid, the grid points at outer boundaries are added.
3. The cell surrounding each interpolation point is searched from the other grids. One of points consisting the cell is stored in another one-dimensional array successively corresponding to the interpolation point. In searching, the order of priority is given to each grid in advance.
4. The conservative variables are tranferred by a linear interpolation.

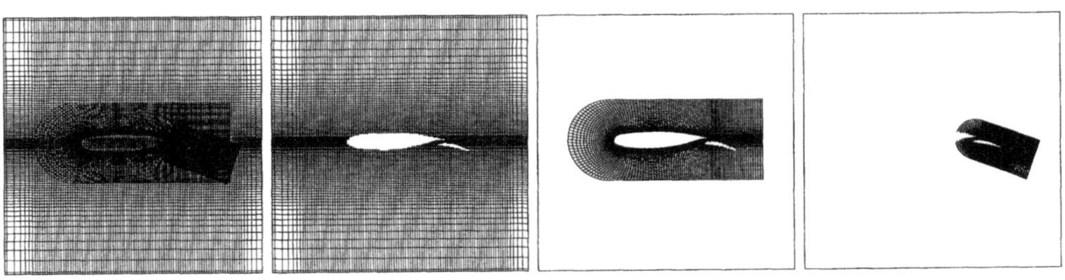

(a)overset grid (b)major grid (c)minor grid(main) (d)minor grid(flap)

Figure 1. Overset grid after excluding hole points.

3. NUMERICAL SCHEMES

The governing equations are the two-dimensional Reynolds-averaged Navier-Stokes equations. A central difference scheme with a Jameson's aritificial dissipation [2] is used for the spatial discretization. The time integration is lower-upper symmetric Gauss-Seidel implicit relaxation method [3]. The Baldwin-Lomax model [4] is used as a turbulence model. Moreover, for the acceleration of convergence to a steady-state solution, V-cycle multigrid method [3] is adopted.

At the far field one-dimensional Riemann invariants are used. At the wall, no slip conditions are specified and the flow is assumed to be adiabatic with zero temperature gradient in the normal direction. The boundary at the wake is defined by averaging the upper and lower points.

4. STRATEGY OF PARALLEL COMPUTATION

4.1. Domain decomposition

For the implementation on a parallel computer, a domain decomposition technique is adopted. In the domain decomposition, each subdomain should be equal size for load balancing, communication costs shold be minimaized and the decline of convergence due to deviding the calculation domain should be restrained. Therefore, it must be considered how those requests are attained. One of the best domain decomposition for overset grid selected from several cases of division is illustrated in Figure 2.

4.2. Data structure and data communication

The procedure for parallel computation is described here.

In order to make the domain decomposition suitable for overset grid, the data structure of one-dimensional listed type arrangement is introduced. The data is composed of a serial number and lists of information.

A serial number is given for each grid point from 1 to the total number of grid points. At first, the grid data are stored in a one-dimensional array deviding each grid into boundaries and inner fields as shown Figure 3 and Figure 4.

The difference between two type of data arrangements is shown in Figure 5. While each grid point in the conventional structured arrangement always knows the neighboring grid points, it is not so in the case of one-dimensional arrangement. Thus, lists for storing the serial number of the neighboring grid points at each grid point is prepared to keep the arrangements.

The advantage of one-dimensional listed type arrangement in comparison with the conventional structured arrangement is that data assignment in domain decomposition is very flexible. As a result, it is quite easy to make use of the best domain decomposition for overset grid. Moreover, in the numerical computation the conventional finite difference method can be applied because of retaining the structured arrangement through the serial number lists.

The data structure of each PE is allocated based on the initial data structure shown in Figure 4 when domain decompsition is carried out. A serial number is given once again in grid points distributed to each PE including grid points assigned for data communication. At this time, the grid points having the same part of calculation are gathered regardless of major grid and minor grid; outer boundary, solid boundary, interface boundary, inner field and so on are gathered in succession. Moreover, overlapping interface needed for data exchanges on each subdomain boundaries among PEs is added to this. In multigrid method, the grid points on coarse grid follow those of fine grid in each grid.

An appropriate data structure for data communication is also introduced so that each PE can communicate with all other PEs rapidly. In the data structure, lists for storing all necessary information in data communication are provided to each PE so that communication points of sending and those of receiving can be related. The procedure is explained using the illustration in data communication from PE1 to PE2 in Figure 6. In this Figure, PE1 gathers the data from its grid points which PE2 needs from PE1 using the list of PE1 and sends to PE2. Then, PE2 receives and distributes the data to the point using the corresponding list of PE2.

Message passing is handled with MPI (Message Passing Interface)

4.3. Procedure of multigrid method

Considering that conservative variables of all grid points on fine grid are modified by the correction, the data of two grid points outside along subdomain boundaries of each coarse grid is transferred by communications as shown Figure 7. The grid points on coarse grid are taken by those of fine grid distributed to each PE.

5. FLOW OVER A MULTI-ELEMENT AIRFOIL

This parallel code is applied to two test cases of compressible viscous turbulent flows over a multi-element airfoil and implemented on a Hitachi SR2201 parallel computer. The computer has 16PEs which are connected by a crossbar network. Each PE consist of a 150MHz PA-RISC chip, a 256 MB memory, and two cascade cashes.

The first test case is the flow over a two-element airfoil. Figure 8(a) shows the overset grid for a NLR7301 configuration. The number of grid points is that major grid is 129×129, minor grid of main and flap are 185×65 and 121×65 respectively. The total number of grid points is about 36500. Flow condition is a subsonic flow at a free stream Mach number of 0.185, an attack angle of 6.0°, and a Reynolds number of 2.51×10^6. The convergence criterion is 5 order magnitude reduction of L2-residual from its initial maximum value. Boldwin-Lomax turbulence model is applied to minor grids only. In multigrid method, 4 level of grid is applied.

Figure 8(a) shows Mach number contours. The interpolation between grids is very good. The computed surface pressure distributions are compared with corresponding experimental data in Figure 9(a). The comparison shows excellent agreement. Figure 10(a) shows convergence histories by 1PE without multigrid. The reduction is quite good.

In the performance of parallel computing, the speedup ratio is plotted in Figure 11(a). The speedup ratio is about 15.5 on 16PEs without multigrid. In detail, the speedup ratio and the efficiency are summarized in Table 1. In addition, the speedup of multigrid is presented in Table 2. The speedup and acceleration ratio in each level of grid are plotted in Figure 11(a). The efficiency is defined as

$$Efficiency = \frac{1}{N} \cdot \frac{T_1}{T_N} = \frac{1}{N} \cdot \frac{n_1}{n_N} \cdot \frac{t_1}{t_N} \tag{1}$$

where

N : Number of PEs
T : Total CPU time
t : CPU time per step
n : Number of time steps
n_N/n_1 : Number of time steps ratio
t_1/t_N : Speedup ratio per step

In Table 1, the number of time steps to a steady-state solution is almost same even if the number of PEs increases up to 16PEs, because the decline of convergence due to dividing the calculation domain is almost negligible. Therefore, the number of time steps ratio is constant mostly. Speedup ratio per time step linearlly increases up to 16PEs. Consequently, the total efficiency of 95% or more is achieved. In Table 2 and Figure 11(a), it is found that 2.6 times convergence acceleration is obtained by multigrid with 4 level of grid on 16PEs, and that overall speedup of 40.9 is achieved as the total improvement of the efficiency.

In the next test case, a three-element airfoil is carried out. Figure 8(b) shows the overset grid for a NHLP-2D configuration. The number of grid points is that major grid is 129×129, minor grid of main, slat and flap are 401×65, 177×65 and 193×65 respectively. The total number of grid points is about 66800. Figure 8(b) shows Mach number contours obtained at a free stream Mach number of 0.197, an attack angle of 4.01°, and a Reynolds number of 3.52×10^6. Figure 9(b) shows the computed surface pressure distributions compared to experimental data. The comparion is quite good. Figure 10(b) shows convergence histories by 1PE without multigrid. The reduction is more hard than the first test case. In the performance of parallel computing, it is recognized that the result of the same tendency compared to the first test case is obtained. In multigrid method, it is found that 3.4 times convergence acceleration is obtained by multigrid with 4 level of grid on 16PEs, and that overall speedup of 55.6 is achieved as the total improvement of the efficiency. This speedup is about 1.4 times as the first test case. That is why the second test case has about twice as many grid points as the first test case.

6. CONCLUSIONS

Parallel computation for overset grid is carried out to a multi-element airfoil by using one-dimensional listed type data structure. Following conclusions are obtained.

1. The computational result is in good agreement with experiment data.
2. The efficiency of 95% or more is attained up to 16PEs, because the decline of convergence due to dividing the calculation domain is almost negligible.
3. 3.4 times convergence acceleration in a three-element airfoil is obtained by using multigrid with 4 level of grid on 16PEs. Overall speedup of 55.6 is achieved as the total improvement of the efficiency.
4. It is found that the data structure introduced by this study is very effective in parallel computing.

REFERENCES

1. Steger, J. L. and Benek, J. A., On the Use of Compsite Grid Schemes in Computational Aerodynamics, Computer Methods in Applied Mechanics and Engeneering, 64,301-320,(1987).
2. Jameson, A. and Baker, T. J., Solution of the Euler Equations for Complex Configurations, AIAA Paper, 83-1929,(1983).
3. Yoon, S. and Kwak, D., Multigrid Convergence of an LU Scheme, Frontiers of Computational Fluid Dynamics, 319-338,(1994).
4. Baldwin, B. S., and Lomax, H., Thin Layer Approximation and Algebraic Model for Separated Turbulent Flows, AIAA Paper, 78-257,(1978).

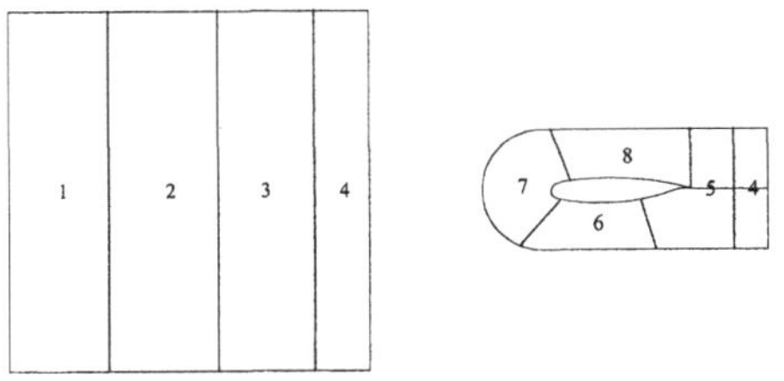

(a)major grid (b)minor grid

Figure 2. An example of the domain decomposition divided into 8 subdomains.

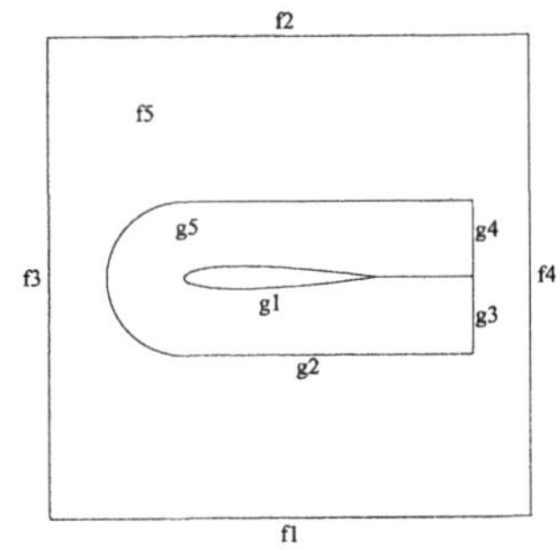

Figure 3. Boundaries and inner fields of overset grid.

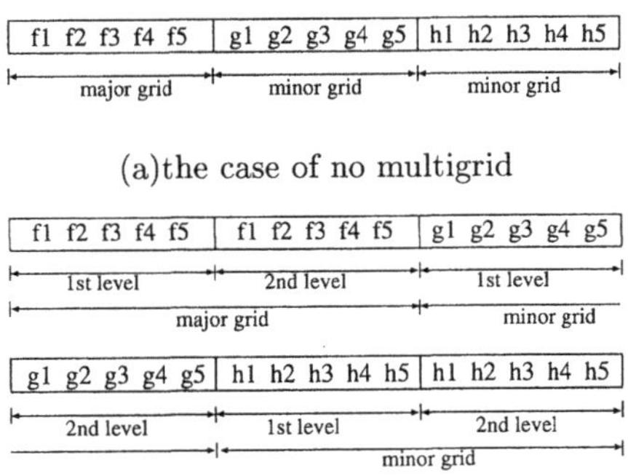

(b)the case of including 2 level of multigrid

Figure 4. Initial data structure.

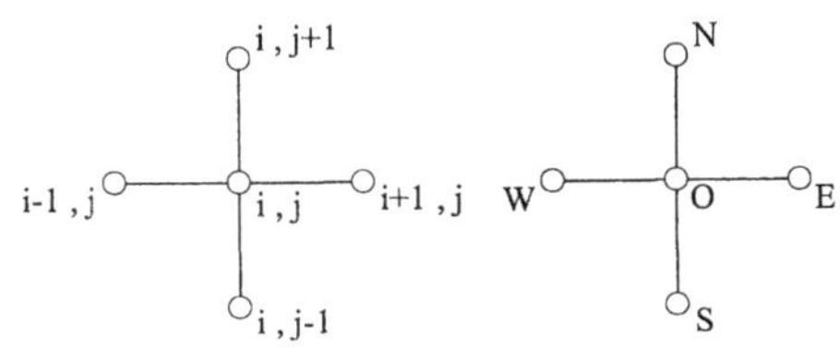

(a)2-D arrangement (b)1-D arrangement

Figure 5. Difference of data arrangement.

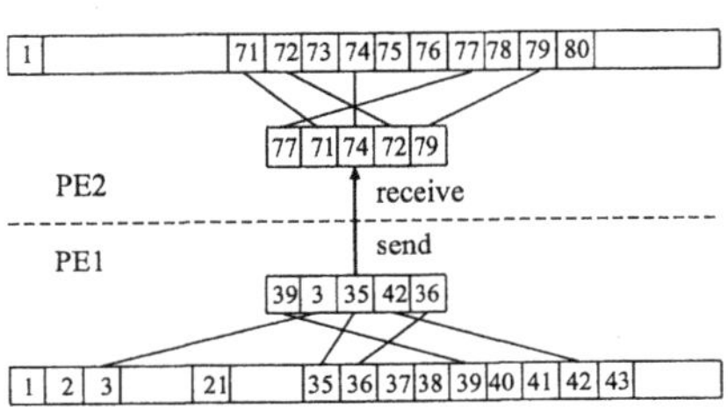

Figure 6. Data communication from PE1 to PE2.

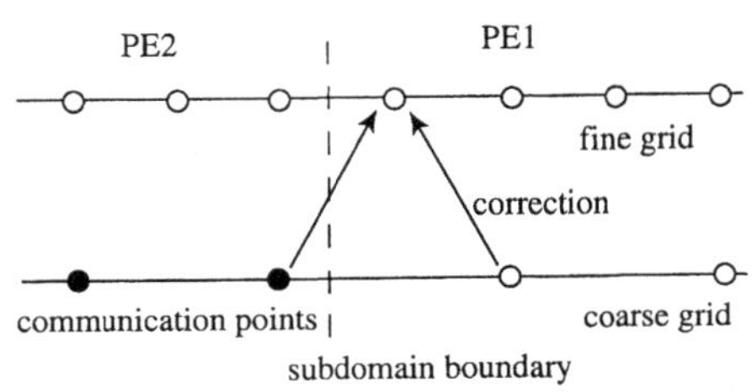

Figure 7. Communication points on coarse grid in PE1.

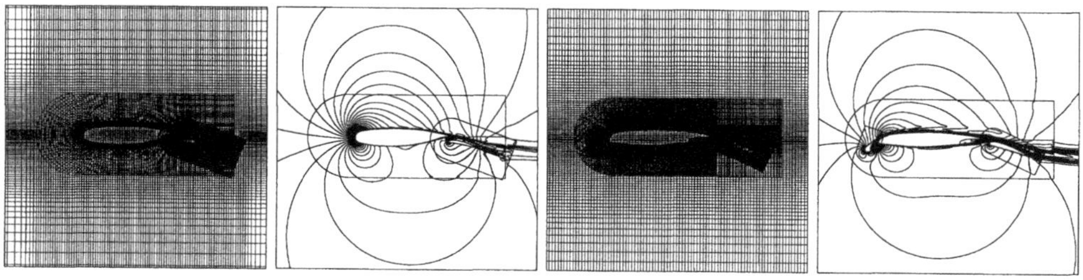

(a)two-element airfoil (b)three-element airfoil

Figure 8. Overset grid and Mach number contours.

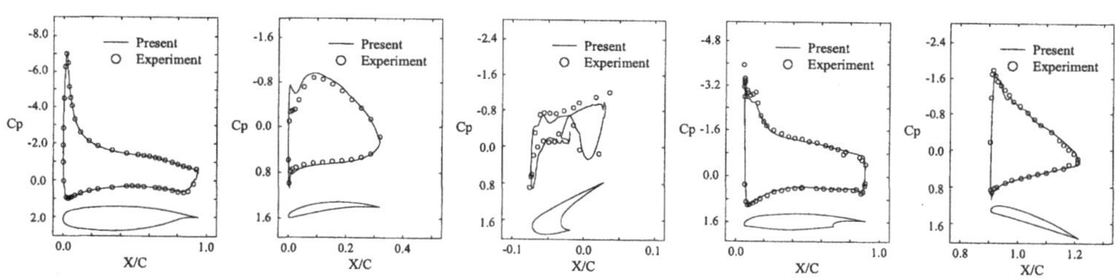

(a)two-element airfoil (b)three-element airfoil

Figure 9. Comparison of pressure distributions.

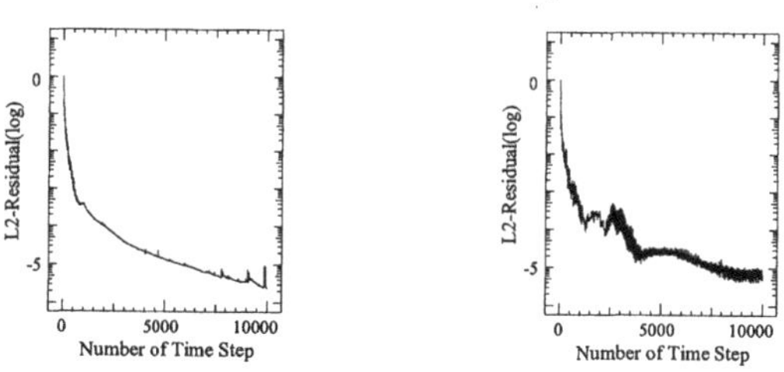

(a)two-element airfoil (b)three-element airfoil

Figure 10. Convergence histories.

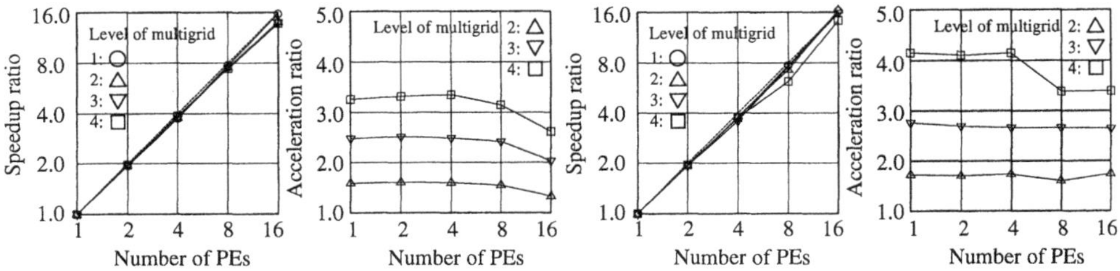

(a)two-element airfoil (b)three-element airfoil

Figure 11. Speedup and acceleration ratio of multigrid method.

Table 1
Speedup and efficiency

two-element airfoil/three-element airfoil					
Number of PEs	1	2	4	8	16
Number of time steps	5806	5804	5802	5831	5782
	7176	7134	7190	6960	7158
Number of time steps ratio	1.0	1.00	1.00	1.00	1.00
	1.0	0.99	1.00	0.97	1.00
Speedup ratio per step	1.0	1.91	3.84	7.97	15.54
	1.0	1.94	3.83	7.77	16.33
Efficiency	1.0	0.96	0.96	1.00	0.97
	1.0	0.98	0.96	1.00	1.02

Table 2
Speedup of multigrid method

two-element airfoil/three-element airfoil					
	Number of PEs				
Level of multigrid	1	2	4	8	16
1	1.0	1.91	3.81	7.82	15.58
	1.0	1.95	3.83	8.00	16.39
2	1.59	3.08	6.10	12.15	20.69
	1.72	3.31	6.62	12.82	28.57
3	2.48	4.82	9.49	18.92	31.70
	2.76	5.24	10.20	21.28	43.48
4	3.26	6.34	12.75	24.65	40.89
	4.15	8.00	15.87	27.03	55.56

Parallel Computational Fluid Dynamics – Practice and Theory
P. Wilders, A. Ecer, J. Periaux, N. Satofuka and P. Fox (Editors)

Parallel Computing of Transonic Cascade Flows Using the Lattice-Boltzmann Method

A. T. Hsu[a], C. Sun[a], C. Wang[a], A. Ecer[a], and I. Lopez[b]

[a] Department of Mechanical Engineering
Indiana University – Purdue University, Indianapolis, IN, 46202 USA

[b] NASA Glenn Research Center, Cleveland, OH, 44135 USA

The primary goal of the present work is to develop a lattice-Boltzmann (LB) model capable of simulating turbo-machinery flows. In the past, LB method has not been applied to turbo-machines because of its Mach number limitations and inability to handle complex geometries effectively. In our recent work, we have successfully developed a high Mach number LB model that is capable of shock capturing. In the current paper, we concentrated on establishing the capability of handling complex geometries. To demonstrate the new capability, a simulation of transonic cascade flows by the compressible LB model is performed and reported here. The parallelization and parallel efficiency of the scheme is also discussed in this paper.

1. INTRODUCTION

The lattice-Boltzmann method has been applied to many complex flow problems in the past[1-5]. However, its application to aeropropulsion applications has not been reported. To apply the lattice-Boltzmann method to aeropropulsion related flow problems, more specifically, to turbomachinery flow simulations, there are two major hurdles to be overcome. First of all, it has to be able to handle compressible flows, and secondly, it has to be able to handle complex geometry. In the past, the applications of the lattice-Boltzmann method are mostly restricted to incompressible flows. This restriction is caused by the very nature of the conventional lattice-Boltzmann method, where the density function is represented by particles that move one lattice length at every time step. The speed of sound calculated from the diffusion velocity of the particles therefore is always larger than the macroscopic velocity, resulting in the small Mach number restriction.

A new lattice-Boltzmann method was recently developed by the present authors that removed the low Mach number restriction [6]. In our previous work, an adaptive LB model for high-speed compressible flows has been developed and implemented on parallel computers. In contrast to standard LB models, this model can handle flows over a wide range of Mach numbers and capture strong shock waves. This compressible flow LB model is of same efficiency as standard LB models but consumes less computer memory. The total computation time is proportional to the total number of nodes. For the cases tested, high parallel efficiency is achieved. Results on a high Mach number shock wave reflection over a wedge were reported in our previous paper.

In the present work, we extended the ability of the high Mach number lattice-Boltzmann model to complex geometry, which include the ability to accurately simulate the boundary layers near curved boundaries. To this end, the boundary condition treatment for the high speed adaptive LB model is extensively studied. The model, together with the new boundary treatment, is tested on flat plate boundary layers with lattices are at an angle to the solid surface boundary. The model is then successfully applied to a two-dimensional supersonic cascade. The parallel implementation of the present LB model is also presented in this paper. The study shows that the cascade flow calculation can be successfully parallelized and that the parallel efficiency is superlinear for the number of computer nodes tested.

2. LB MODEL FOR COMPRESSIBLE FLOWS

In order to be able to treat supersonic flows, we introduce a larger particle velocity set, $S=\{\boldsymbol{r}\}$, into the present model, where $\boldsymbol{r}$ is the migrating velocity of the particles. We define $f(\boldsymbol{x}, \boldsymbol{r}, \boldsymbol{\xi}, \zeta, t)$ as the particle density distribution function for particles located at $\boldsymbol{x}$, with a continuous velocity $\boldsymbol{\xi}$, a discrete migrating velocity $\boldsymbol{r}$, and specific energy ζ. These particles will move to $\boldsymbol{x} + \boldsymbol{r}\Delta t$ after Δt, and transporting with them a momentum $m\boldsymbol{\xi}$ and energy $m\zeta$.

The macroscopic quantities, i.e., mass ρ, momentum $\rho\,\boldsymbol{v}$, and energy ρE, are defined as

$$\boldsymbol{Y} \equiv \sum_{r} \int_{D} \boldsymbol{\eta} f(\boldsymbol{x}, \boldsymbol{r}, \boldsymbol{\xi}, \zeta, t) d\boldsymbol{\xi} d\zeta , \tag{2.1}$$

where $D=D_1\times D_0$, $\boldsymbol{Y} \equiv (\rho, \rho\boldsymbol{v}, \rho E)$, and $\boldsymbol{\eta} \equiv (m, m\boldsymbol{\xi}, m\zeta)$.

In a BGK type of LB model, the Boltzmann equation is written as [1]

$$f(\boldsymbol{x}+\boldsymbol{r}\Delta t, \boldsymbol{r}, \boldsymbol{\xi}, \zeta, t+\Delta t) - f(\boldsymbol{x}, \boldsymbol{r}, \boldsymbol{\xi}, \zeta, t) = \Omega . \tag{2.2}$$

The collision operator is given as

$$\Omega = -\frac{1}{\tau} [f(\boldsymbol{x}, \boldsymbol{r}, \boldsymbol{\xi}, \zeta, t) - f^{eq}(\boldsymbol{x}, \boldsymbol{r}, \boldsymbol{\xi}, \zeta, t)],$$

where $f^{eq}(\boldsymbol{x}, \boldsymbol{r}, \boldsymbol{\xi}, \zeta, t)$ is the equilibrium distribution, which is completely determined by the macroscopic variables such as the fluid density, momentum, and energy.

On a uniform lattice, we consider the symmetric vector set $\{\boldsymbol{c}'_{j\nu}; j=1,\ldots, b_\nu\}$ connecting a node to its neighbors, where b_ν is the number of vector directions. If we take $\Delta t=1$, then these vectors are the particle velocities of the conventional LB model. In the following description, $\Delta t=1$ is implied, and velocities are used as distances without further explanation. For a hexagonal lattice we have $b_\nu=6$. The subscript ν represents the discrete velocity levels; here we choose two levels and $\nu = 1$ and 2. The module of $\boldsymbol{c}'_{j\nu}$ is c'_ν. Now let us consider the macroscopic fluid velocity vector $\boldsymbol{v}$ that starts from a node $\boldsymbol{x}$ and ends in a lattice triangle; let $\boldsymbol{v}_k$ (k=1,2,3) be the vectors from the node $\boldsymbol{x}$ to the three apexes of this triangle. We introduce the particle velocities $\boldsymbol{c}_{j\nu k}$, $\bar{\boldsymbol{c}}_{j\nu}$: $\boldsymbol{c}_{j\nu k} = \boldsymbol{v}_k + \boldsymbol{c}'_{j\nu}$, $\bar{\boldsymbol{c}}_{j\nu} = \boldsymbol{v} + \boldsymbol{c}'_{j\nu}$, and the fluctuating velocities $\boldsymbol{v}'_k$ (k=1,2,3): $\boldsymbol{v}'_k = \boldsymbol{v}_k - \boldsymbol{v}$. For high speed flow the fluctuating velocities $\boldsymbol{v}'_k$ is small.

For high-speed compressible flow, we define an equilibrium distribution as

$$f^{eq}(\boldsymbol{x}, \boldsymbol{r}, \boldsymbol{\xi}, \zeta, t) \equiv \begin{cases} d_{\nu k}\delta(\boldsymbol{\xi} - \bar{\boldsymbol{c}}_{j\nu})\delta(\zeta - \zeta_{j\nu}) & \text{for } \boldsymbol{r} = \boldsymbol{c}_{j\nu k} \\ 0 & \text{for other } \boldsymbol{r} \end{cases} \tag{2.3}$$

where $\delta(\boldsymbol{\xi})$ is a δ-function: $\delta(\boldsymbol{\xi}) = 0$ for $\boldsymbol{\xi} \neq 0$, and $\int g(\boldsymbol{\xi})\delta(\boldsymbol{\xi})\, d\boldsymbol{\xi} = g(\boldsymbol{0})$. In order to have an appropriate heat diffusion term in the energy equation, $\zeta_{j\nu}$ is calculated by $\zeta_{j\nu} = (1/2)(v^2 + 2\boldsymbol{c}'_{j\nu}\cdot\boldsymbol{v} + \bar{c}'^2) + \Phi$, where $\bar{c}'^2 = (1/\rho)\sum_{k,\nu} m b_\nu d_{\nu k} c'^2_\nu$, and a potential energy $\Phi = [1-(D/2)(\gamma-1)]e$ is introduced to obtain an arbitrary specific heat ratio; the internal energy e is given by $e = E - (1/2)v^2$. It is noted that $f^{eq}(\boldsymbol{x}, \boldsymbol{r}, \boldsymbol{\xi}, \zeta, t)$ is defined for all $(\boldsymbol{r}, \boldsymbol{\xi}, \zeta)$ in $S \times D_1 \times D_0$. However, $f^{eq}(\boldsymbol{x}, \boldsymbol{r}, \boldsymbol{\xi}, \zeta, t)$ is non-zero only for $(\boldsymbol{r}, \boldsymbol{\xi}, \zeta)$ in $\{\boldsymbol{c}_{j\nu k}\}\times\{\bar{\boldsymbol{c}}_{j\nu}\}\times\{\zeta_{j\nu}\}$, which is called support set of $f^{eq}(\boldsymbol{x}, \boldsymbol{r}, \boldsymbol{\xi}, \zeta, t)$. The support set is discrete and relatively small.

The coefficient $d_{\nu k}$ in equation (2.3) is defined as:

$$d_{\nu k} = \alpha_k d_\nu, \tag{2.4}$$

where $\alpha_k = \rho_k/\rho$; ρ_k is determined from the following equations

$$\sum_{k=1}^{3} \rho_k = \rho\,; \quad \sum_{k=1}^{3} \rho_k \boldsymbol{v}_k = \rho\boldsymbol{v}. \tag{2.5}$$

Where ρ and $\boldsymbol{v}$ are defined at node $\boldsymbol{x}$, and ρ_k's are the mass to be distributed to the 3 apexes of the destination lattice. For a two-dimensional model, Eq. (2.5) represent 3 scalar equations, and it can be proved that the system has unique non-negative solutions for ρ_k, k=1,2,3.

The coefficient d_ν in Eq. (2.4) is defined by

$$d_1 = \frac{\rho c'^2_2 - Dp}{b_1 m(c'^2_2 - c'^2_1)}$$

$$d_2 = \frac{Dp - \rho c'^2_1}{b_2 m(c'^2_2 - c'^2_1)}$$

where p is the macroscopic pressure and is given by

$$p = \sum_{v} mb_v d_v (1/D) c'^2_v$$

With these definition, the equilibrium distribution function $f^{eq}(x, r, \xi, \zeta, t)$ is now completely defined.

Using the Chapman-Enskog expansion, the following set of Navier-Stokes equations can be recovered from Eq. (2.2):

$$\frac{\partial \rho}{\partial t} + \mathrm{div}(\rho v) = 0, \tag{2.6}$$

$$\frac{\partial \rho v}{\partial t} + \mathrm{div}(\rho vv) + \nabla p = \mathrm{div}\{\mu[\nabla v + (\nabla v)^T - (\gamma-1)\mathrm{div} v I] + O(vv'_k v'_k)\}, \tag{2.7}$$

$$\frac{\partial \rho E}{\partial t} + \mathrm{div}(pv + \rho E v) = \mathrm{div}\{\mu v \cdot [\nabla v + (\nabla v)^T - (\gamma-1)\mathrm{div} v I]\} + \mathrm{div}\{\kappa \nabla e - (\gamma-1) e \nabla \kappa + O(v'_k v'_k)\}, \tag{2.8}$$

where

$$\mu = \kappa = \Delta t\,[\tau - (1/2)] \sum_{v} mb_v d_v (1/D) c'^2_v, \tag{2.9}$$

where I is a second-order unit tensor; D is space dimension; μ and κ are respectively viscosity and heat diffusivity. In Eq. (2.8) the first term and the second term of right-hand side correspond respectively to the dissipation and the heat diffusion.

3. BOUNDARY CONDITIONS

The conventional LB method is based on uniform lattices, and curved boundaries are either approximated by steps or treated with a particle bounce back boundary condition. Neither of the treatments is satisfactory for turbo-machine applications. In fact, we found that a bounced-back condition produces spurious pressure oscillations near the solid surface when applied to cascade flow simulations.

In order to successfully apply the LB model to turbomachinery calculations, we devised the following new treatment of the boundary condition. In order to maintain the uniformity of cells near the boundary, we introduce auxiliary nodes inside the solid wall.

The macroscopic variables at the auxiliary nodes are extrapolated from the values in the computational domain. The following conditions need to be satisfied in the extrapolation:

$$v = 0, \ \frac{\partial \rho}{\partial \vec{n}} = 0, \ \text{and} \ \frac{\partial e}{\partial \vec{n}} = 0.$$

4. VALIDATION STUDIES

(a) Flow over a flat plate. To validate the new procedure for viscous boundary conditions, a boundary layer flow over a flat plate was simulated using a mesh that does not coincide with the boundary (see Fig. 1). Figure 2 shows the non-dimensional velocity profiles versus η. The solid line represents the Blasius solution. The velocity profiles for various downstream locations match well with the Blasius solution.

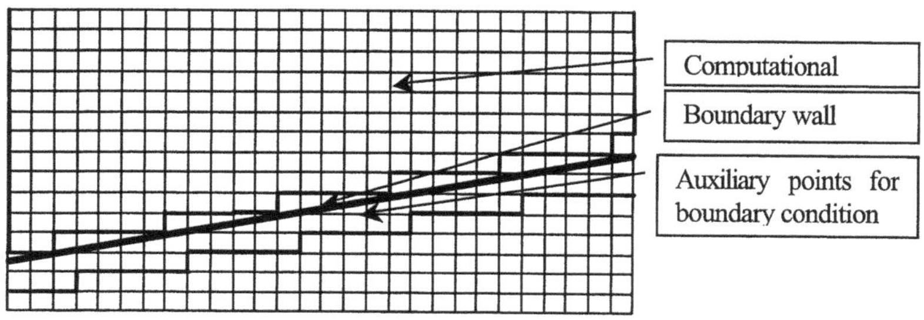

Figure 1. Boundary treatment for a flat plate boundary layer case.

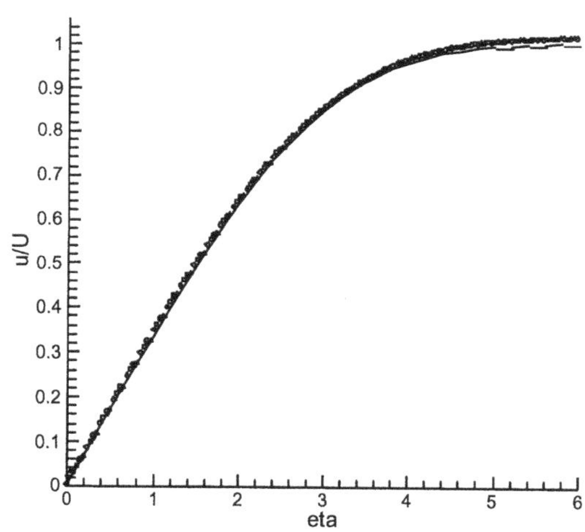

Figure 2. Flat-plate boundary layer solution using mesh not coinciding with the boundary: solid line is Blasius solution, and symbols are present numerical solutions at various down stream locations.

(b) Flow over a NACA 0012 airfoil. The new boundary procedure is applied to a NACA0012 airfoil. A comparison of pressure coefficients between solutions from the bounce-back boundary condition and present procedure is presented in Figure 3. The results clearly show an advantage for the new procedure.

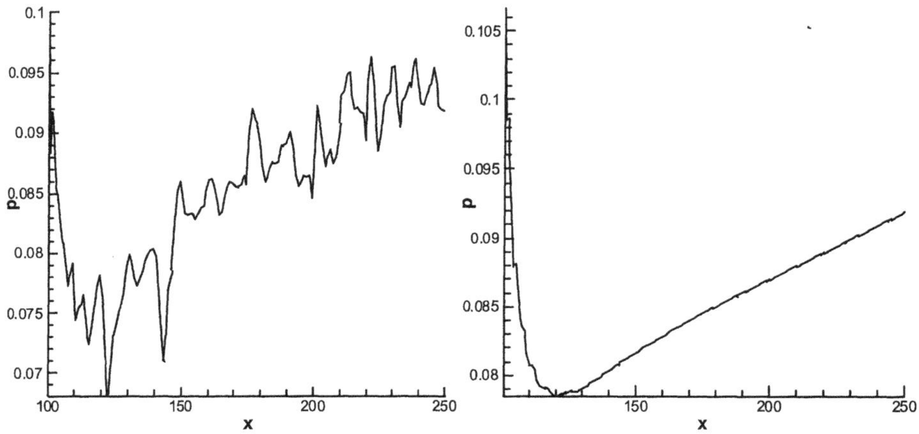

Fig. 3 Pressure distribution of a flow over NACA0012 obtained using a bounced-back wall condition (left) and the axiliary node method (right).

5. 2-D CASCADE FLOW

For the cascade simulation, the blade shape is taken from NACA0012 airfoil (Figure 4). At the upstream constant boundary conditions are imposed for density, pressure, and velocity $(\rho,\ p,\ u) = (1.0,\ 0.25/1.4,\ 1.5)$. The angle of attack is zero, and inflow Mach number is 3. Periodical boundary conditions are imposed on the upper and lower boundaries. Initial condition is set up to be the same as the upstream boundary condition, i.e. $(\rho,\ p,\ u) = (1.0,\ 0.25/1.4,\ 1.5)$.

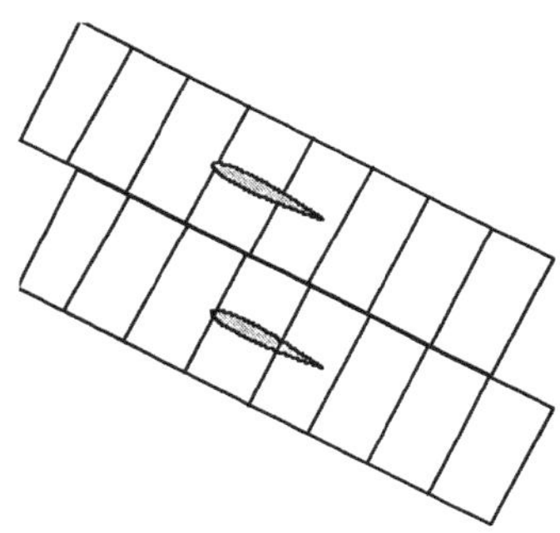

Figure 4. NACA 0012 cascade and solution domain decomposition

Figure 5 shows the Lattice Boltzmann solution for pressure through a cascade. Four blades are plotted. A lattice of 400X80 is used for each blade field.

Detached oblique shocks are formed in front of blades. Across these shocks the pressure, density, and internal energy increase, while the Mach number decreases. The flow changes direction crossing the shocks and forms a slight boundary-layer separation and reattachment at the lower surface of the blade, which can be observed from streamline plots.

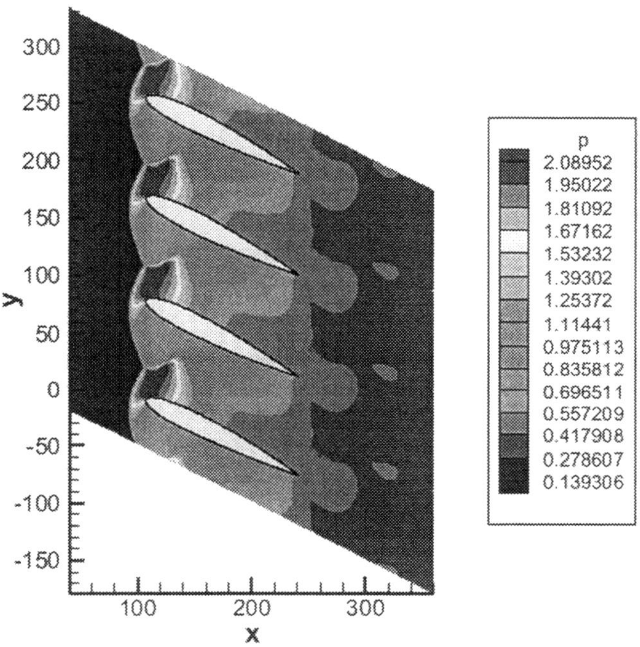

Figure 5. Pressure contours for transonic flow over a NACA0012 cascade.

5. PARALLEL EFFICIENCY

The cascade solution procedure is parallelized on a cluster of 64 Linux workstations. Parallelization is achieved through dividing the solution domain serially as shown in Fugure 6. The parallel efficiency of the parallel implementation is tested on 2, 4, 8, 16, 32, and 64 nodes. The efficiency based on wall clock time and the ideal efficiency based on CPU time (including data transfer time) are shown in Figure 6. Since the cluster is not dedicated, the wall clock time do not provide an accurate measure of the parallel efficiency. The CPU time based efficiency shows a superlinear efficiency up to 64 processors.

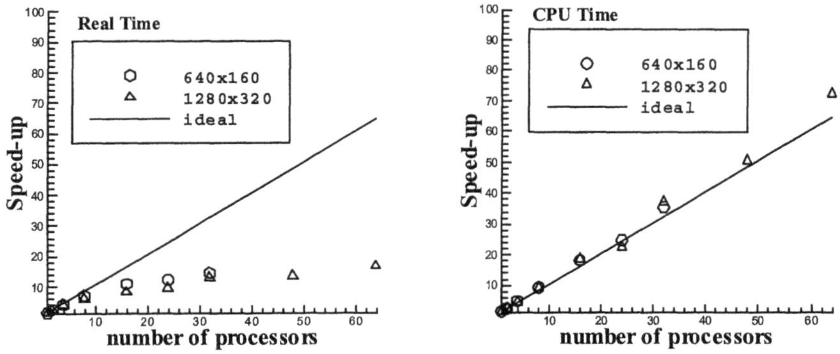

Figure 6. Parallel efficiency of the 2D cascade procedure for up to 64 processors based on wall clock (left) and CPU time plus data transfer time (right).

6. CONCLUSIONS

We have for the first time successfully simulated transonic cascade flows using a compressible lattice-Boltzmann model. A new boundary condition treatment is proposed for viscous flows near curved boundaries. Results on flatplate boundary layer and flows over a NACA0012 airfoil show that the new boundary condition produces accurate results. Preliminary results on supersonic cascade show that shocks, interactions between shocks, and boundary layer separation due to shock impingement are well captured. The parallel implementation of the scheme showed good parallel efficiency.

REFERENCES:

[1] S. Chen and G.D. Doolen, Annu. Rev. Fluid Mech., **30** (1998), 329.

[2] H. Chen, S. Chen, and W. Matthaeus, Phys. Rev. A, **45** (1992), R5339.

[3] Y. H. Qian, D. d'Humières, and P. Lallemand, Europhys. Lett., **17** (1992), 479.

[4] Y. H. Qian, S. Succi, and S. A. Orszag, Annu. Rev. of Comput. Phys. III, ed. Dietrich, W. S. (1995), 195.

[5] G. Amati, S. Succi, and R. Piva, Inter. J. of Modern Phys. C, **8**(4) (1997), 869.

[6] A.T. Hsu, C. Sun, and A. Ecer, , in: *Parallel Computational Fluid Dynamics*, C.B. Jenssen, etc., editors, 2001 Elsevier Science, p. 375

Parallel Computational Fluid Dynamics – Practice and Theory
P. Wilders, A. Ecer, J. Periaux, N. Satofuka and P. Fox (Editors)

Parallel Computation Of Multi-Species Flow Using a Lattice-Boltzmann Method

A. T. Hsu[a], C. Sun[a], T. Yang[a], A. Ecer[a], and I. Lopez[b]

[a] Department of Mechanical Engineering
Indiana University – Purdue University, Indianapolis, IN, 46202 USA

[b] NASA Glenn Research Center, Cleveland, OH, 44135 USA

As part of an effort to develop a lattice-Boltzmann parallel computing package for aeropropulsion applications, a lattice-Boltzmann (LB) model for combustion is being developed. In the present paper, as a first step to the extension of LB model to chemically reactive flows, we developed a parallel LB model for multi-species flows. The parallel computing efficiency of multi-species flow analysis using the LB Model is discussed in the present paper.

1. INTRODUCTION

A parallel computing lattice-Boltzmann module for aeropropulsion is being developed by the present authors. The objective of the research is to develop an inherently parallel solution methodology for CFD, to demonstrate applicability to aeropropulsion, and verify the scalability on massively parallel computing environment of the new method. Numerical simulation of aeropropulsion systems includes two major components; namely, turbomachinery simulations and combustion simulations. As a first step towards combustion simulations, the present paper reports the results on the extension of the lattice-Boltzmann method to multi-species flows and the parallelization of the scheme.

The lattice-Boltzmann (LB) method as applied to computational fluid dynamics was first introduced about ten years ago. Since then, significant progress has been made [1,2]. LB models have been successfully applied to various physical problems, such as single component hydrodynamics, magneto-hydrodynamics, flows through porous media, and other complex systems [3,4]. The LB method has demonstrated potentials in many areas with some computational advantages.

One attractive characteristics of the lattice-Boltzmann method is that it is naturally parallel. The computation of LB method consists of two alternant steps: particle collision and convection. The collision takes place at each node and is purely local, and is independent of information on the other nodes. The convection is a step in which particles move from a node to its neighbors according to their velocities. In terms of floating point operation, most of the computation for the LB method resides in the collision step and therefore is local. This feature makes the LB method particularly suited for parallel computations. As a result of the localized nature of computation, the scaling properties for parallel computing of the LB model are expected to be close to ideal. The scheme is expected to be fully scalable up to a large number of processors. Its successful application to parallel computing can supply industry with a tool for the simulation of realistic engineering problems with shorter turnaround time and higher fidelity. Parallel computing using the LB model has been pursued by many researchers [5-7]. Our research group recently developed an efficient parallel algorithm for a new high Mach number LB model [8].

The present paper presents the results of parallelization of a multi-species single phase flow LB model. The model is capable of treating species with different diffusion speeds. PVM library is applied to parallelize the solution procedure. The case of a single drop concentration, initially at rest, diffusing is a cross flow is used to test the parallel efficiency of the scheme. The multi-species LB model used in the present study is presented in Section 2, in Section 3, the test case and results for a 2-species flow is presented, and the parallel computing results and parallel efficiency are presented in Section 4.

2. MUTI-SPECIES LB MODEL

For flows with multi-species, we need to define a density distribution function for each species. Suppose that the fluid is composed of M species. Let c_v be the velocity set of the species v, $v=1,\ldots,M$; $\mathbf{c}_{jv}$ is the particle velocities of the species v in the direction j, $j=1, \ldots, b_v$; where b_v is the number of velocity directions. Let f_{jv} be the distribution function for the particle with velocity c_{jv}, the BGK type Boltzmann equation is written as:

$$f_{j\upsilon}(\vec{x}+\vec{c}_{j\upsilon}\Delta t, t+\Delta t) - f_{j\upsilon}(\vec{x},t) = \Omega_{j\upsilon} \tag{1}$$

where,

$$\Omega_{j\upsilon} = -\frac{1}{\tau}\left(f_{j\upsilon}(\vec{x},t) - f_{j\upsilon}^{eq}\right) \tag{2}$$

The macroscopic quantities, the partial mass densities and the momentum, are defined as follow:

$$\rho_\upsilon = \sum_j m_\upsilon f_{j\upsilon}\ ; \qquad \upsilon = 1,\cdots,M \tag{3}$$

$$\rho\vec{v} = \sum_{j,\upsilon} m_\upsilon f_{j\upsilon}\vec{c}_{j\upsilon} \tag{4}$$

where the total mass density is $\rho = \sum_\upsilon \rho_\upsilon$.

Following Bernardin et al [9] the equilibrium distributions are chosen as:

$$f_{j\upsilon}^{eq} = d_\upsilon\left\{1 + \frac{D}{c_\upsilon^2}\vec{c}_{j\upsilon}\cdot\vec{v} + \frac{D(D+2)}{2c_\upsilon^4}\left[\vec{c}_{j\upsilon}\vec{c}_{j\upsilon} : \vec{v}\vec{v} - \frac{c_\upsilon^2}{D}v^2\right]\right\} \tag{5}$$

where, $d_\upsilon = \frac{\rho_\upsilon}{bm_\upsilon}$; m_v and ρ_v are, respectively, the particle mass and the particle density of species v; $\mathbf{v}$ is the fluid velocity; and D is the number of spatial dimensions.

$\rho_1, \ldots, \rho_M, \rho\mathbf{v}$ are the macroscopic variables. Using the Chapman-Enskog expansion, we can derive the N-S equation from the Boltzmann equation (1):

$$\frac{\partial \rho \vec{v}}{\partial t} + div(\rho \vec{v}\vec{v}) = -\nabla p + div\left\{\nabla(\mu \vec{v}) + [\nabla(\mu \vec{v})]^T - \frac{2}{D} div(\mu \vec{v}) I_d\right\} \tag{6}$$

where the pressure p and viscosity μ is related to the microscopic quantities through the following realtions:

$$p = \frac{1}{D}\sum_{\upsilon} \rho_\upsilon c_\upsilon^2 - \frac{1}{D}\rho v^2 \tag{7}$$

$$\mu = \frac{\varepsilon T}{D+2}\left(\tau - \frac{1}{2}\right)\sum_{\upsilon} \rho_\upsilon c_\upsilon^2 \tag{8}$$

where T is time scale and e is a small number.

In the same way, we obtain the mass conservation equation for species ν:

$$\frac{\partial \rho_\upsilon}{\partial t} = -div(\rho_\upsilon \vec{v}) + \varepsilon T\left(\tau - \frac{1}{2}\right) div\left\{\nabla\left(\frac{\rho_\upsilon c_\upsilon^2}{D}\right) - \frac{\rho_\upsilon}{\rho}\sum_{\upsilon}\nabla\left(\frac{\rho_\upsilon c_\upsilon^2}{D}\right)\right\} \tag{9}$$

If this equation is summed up over ν the continuity equation $\frac{\partial \rho}{\partial t} + div(\rho \vec{v}) = 0$ is obtained. Let $Y_\upsilon = \frac{\rho_\upsilon}{\rho}$ to be the mass fraction of species ν. When the fluid is composed of two species we can write $Y_1 = Y$ and $Y_2 = 1 - Y$, and equation (9) is simplified:

$$\frac{\partial Y_\upsilon}{\partial t} + \vec{v} \cdot \nabla Y_\upsilon = \nabla \cdot (D_\upsilon \nabla Y_\upsilon) \tag{10}$$

where D_ν is the diffusion coefficient. For two species, the diffusion coefficient can be written as:

$$D_\upsilon = \frac{\varepsilon T}{D}\left(\tau - \frac{1}{2}\right)\left[c_1^2 + (c_2^2 - c_1^2)Y\right] \tag{11}$$

In this model the energy is conserved automatically. Since the magnitudes of the particle velocities of same species are the same (different in directions), the partial mass conservations ensure the energy conservation. Consequently, the energy equation is not an independent equation.

3. MASS CONVECTION-DIFFUSION SIMULATIONS

As a test case for parallel computing efficiencies, we applied the above-described model to a simple flow with two species. At the initial time a round gaseous droplet of species 2 of radius r (=16 nodes), is located at x=60, y=50, at rest, in a uniform flow of species 1, which has a mean velocity of V=0.5 along the horizontal axis. A schematic of the initial flow is shown in Figure 1. The particle velocities and the particle masses of the two species are respectively $c_1=1$, $c_2=\sqrt{3}$ (Fig. 2); $m_1=3$, $m_2=1$. The simulation is run on a 160X100 hexagonal lattice with τ set to be 1.0. A sample grid is shown in Figure 3.

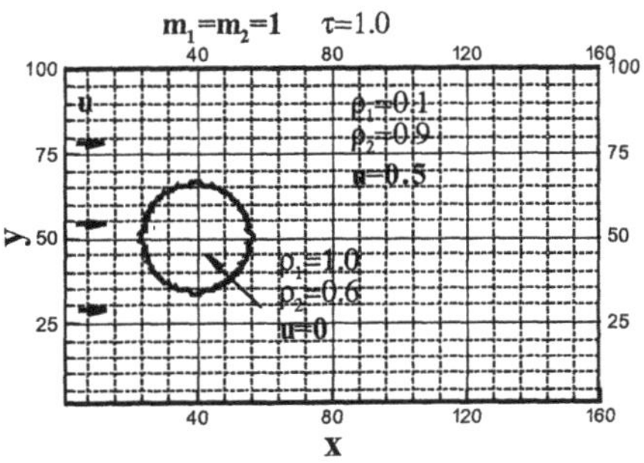

Figure 1. Intitial condition of the mass diffusion test case.

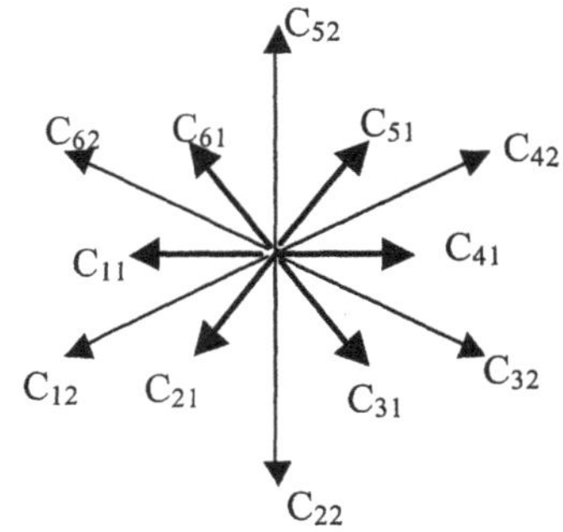

Figure 2. Particle velocities

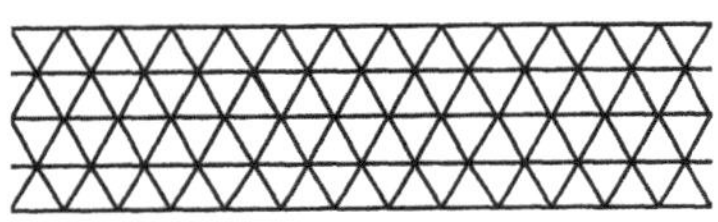

Figure 3. Schematics of grid used.

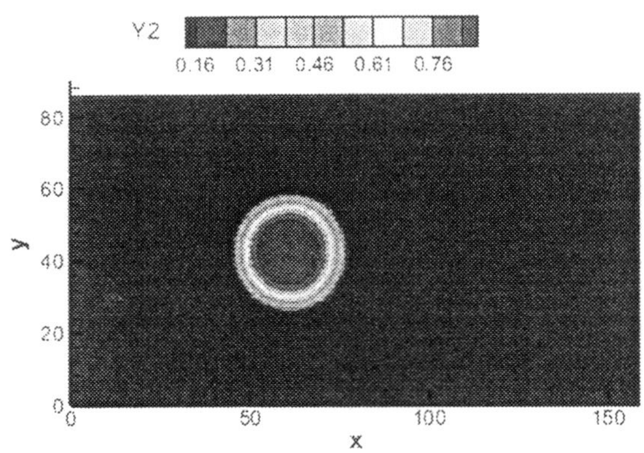

Figure 4: Y_2 at t=20. A rest droplet of species 2 in a flow of species 1 at a speed of 0.5

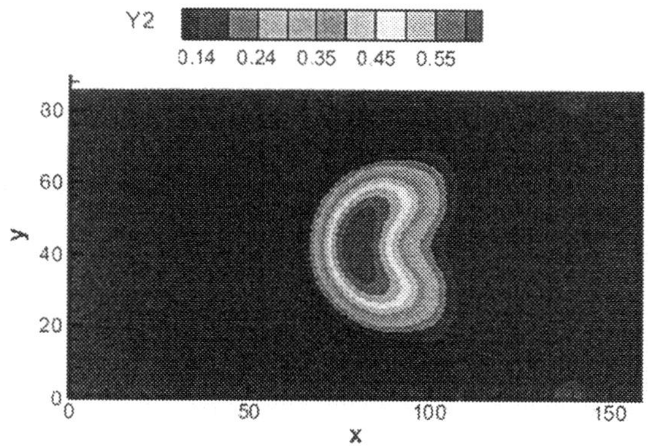

Figure 5: Y_2 at t=100. A rest droplet of species 2 in a flow of species 1 at a speed of 0.5

Figure 4 and Figure 5 show the distributions of Y_2 at instants t=20 and t=100. The initial round droplet is distorted by the freestream to a horseshoe shape. The diffusion effect is evident in that the concentration gradient continually to reduce as time increases.

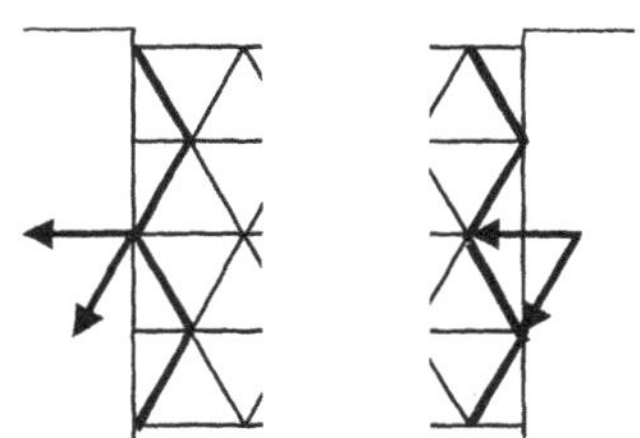

Figure 6 Buffer for message passing

4. PARALLEL COMPUTING

The LB model for multi-species is parallelized using the PVM routines. The solution block is divided into sub-domains. A buffer is created to store the particles that need to be transferred between sub-domains. As shown in Figure 6, the particles that leaves the right domain are collected into a buffer, and the buffer is send to the left domain through PVM. To test the parallel procedure and make sure that the solution does not deteriorate, we compared the solution from a single block calculation, a 2-block calculation, and a 32-blocks calculation. The result of this comparison is shown in Figure 7.

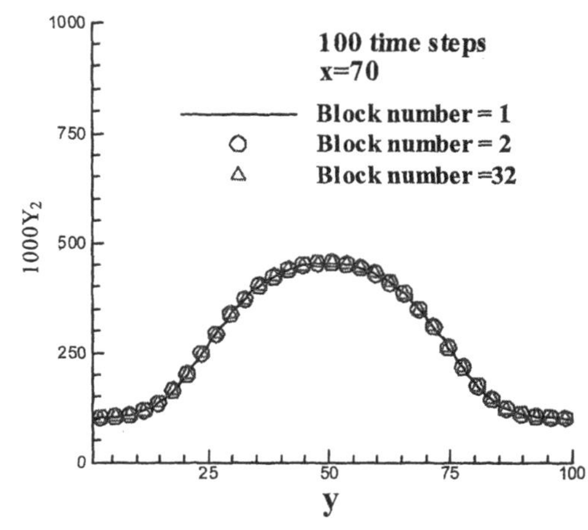

Figure 7 Verification of multi-block computations

The parallel performance of the algorithm is tested on a cluster of 32 Linux workstations. The rest results are listed in table 1, where the number of processors,

CPU time, clock time, and the performance as a percentage of the single processor time are listed in the table. The CPU time performance shows that under ideal situations, i.e., when waiting time is discounted, the parallel performance of the algorithm is super linear. The reason for the more than 100% performance is due to the fact that on a single processor, the memory requirement and paging can slow down the computation. Since the Linux cluster is not a dedicated system, the wall clock time includes waiting on other users, and does not reflect the ideal efficiency of the scheme.

Table 1. Parallel Performance

n: Num. of Proc.	T(n): CPU Time. s (500i)	R(n): Real Time s (500i)	T(1)/nT(n): %	R(1)/nR(n) %
1	332.60	409.66	100.00	100.00
2	146.11	189.56	113.82	108.06
4	63.83	87.38	130.27	117.21
8	30.74	46.46	135.25	110.22
16	15.26	33.72	136.22	75.93
24	9.19	24.07	150.80	70.91
32	7.66	24.12	135.69	53.08

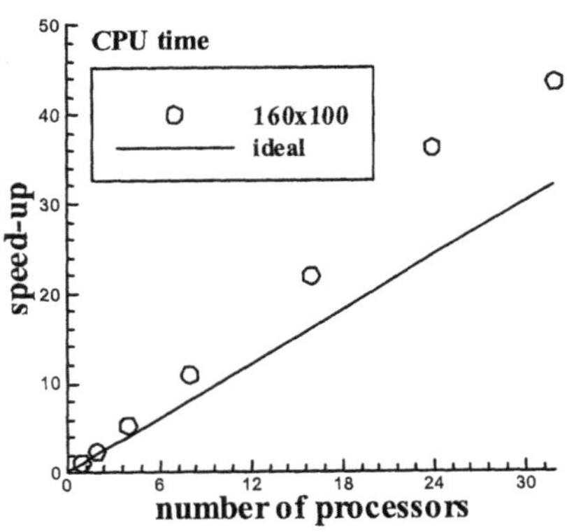

Figures 8. Ideal efficiency based on CPU time.

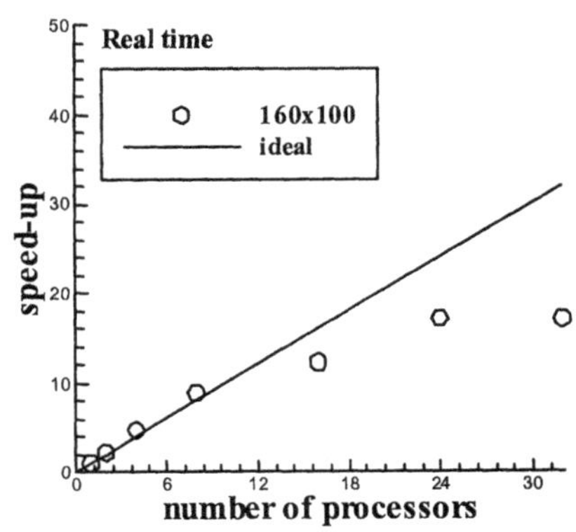

Figures 9. Efficiency based on wall clock time.

5. CONCLUSION

We have developed a parallel procedure for a multi-species, multi-speed, mass diffusion lattice Boltzmann model. Because of the multi-speed feature of the model, it is capable of treating preferential diffusion problems. Using the Chapman-Enskog method, we have derived from the BGK Boltzmann equation the macroscopic species transport equations. For low mean velocities (neglect convection effect in the equation) the partial mass conservation equations are then reduced to the Fick's law. The parallel efficiency of the solution module is tested on a 2-D convection-diffusion simulation. The ideal efficiency based on CPU shows superlinear behavior up to 32 processors.

6. REFERENCES

[1] H. Chen, S. Chen, and W. Matthaeus, Phys. Rev. A, **45** (1992), R5339.
[2] Y. H. Qian, D. d'Humières, and P. Lallemand, Europhys. Lett., **17** (1992), 479.
[3] S. Chen and G.D. Doolen, Annu. Rev. Fluid Mech., **30** (1998), 329.
[4] Y. H. Qian, S. Succi, and S. A. Orszag, Annu. Rev. of Comput. Phys. III, ed. Dietrich, W. S. (1995), 195.
[5] G. Amati, S. Succi, and R. Piva, Inter. J. of Modern Phys. C, **8**(4) (1997), 869.
[6] N. Satofuka, T. Nisihioka, and M. Obata, in: Parallel Computational Fluid Dynamics, Recent Development and Advences Using Parallel computers, D. R. Emerson, A. Ecer, J. Peraux, N. Satofuka and P. Fox, editors, 1998 Elsevier Science, 601.
[7] N. Satofuka and T. Nisihioka, in: *Parallel Computational Fluid Dynamics*, C. A. Lin, A. Ecer, J. Peraux, N. Satofuka and P. Fox, editors, 1999 Elsevier Science, p.171.
[8] A.T. Hsu, C. Sun, and A. Ecer, , in: *Parallel Computational Fluid Dynamics*, C.B. Jenssen, etc., editors, 2001 Elsevier Science, p. 375
[9] D. Bernardin, O. Sero-Guillaume, and C. H. Sun, Physica D, **47** (1991), 169.

A Weakly Overlapping Parallel Domain Decomposition Preconditioner for the Finite Element Solution of Convection-Dominated Problems in Three Dimensions

Peter K. Jimack[a*] Sarfraz A. Nadeem[a†]

[a]Computational PDEs Unit, School of Computing, University of Leeds, LS2 9JT, UK

In this paper we describe the parallel application of a novel two level additive Schwarz preconditioner to the stable finite element solution of convection-dominated problems in three dimensions. This is a generalization of earlier work, [2,6], in 2-d and 3-d respectively. An algebraic formulation of the preconditioner is presented and the key issues associated with its parallel implementation are discussed. Some computational results are also included which demonstrate empirically the optimality of the preconditioner and its potential for parallel implementation.

1. INTRODUCTION

Convection-diffusion equations play a significant role in the modeling of a wide variety of fluid flow problems. Of particular challenge to CFD practitioners is the important case where the convection term is dominant and so the resulting flow contains small regions of rapid change, such as shocks or boundary layers. This paper will build upon previous work of [1,2,6] to produce an efficient parallel domain decomposition (DD) preconditioner for the adaptive finite element (FE) solution of convection-dominated elliptic problems of the form

$$-\varepsilon\underline{\nabla}^2 u + \underline{b}\cdot\underline{\nabla}u = f \qquad \text{on } \Omega \subset \Re^3, \tag{1}$$

where $0 < \varepsilon << ||\underline{b}||$, subject to well-posed boundary conditions. An outline of the parallel solution strategy described in [1] is as follows.

1. Obtain a finite element solution of (1) on a coarse mesh of tetrahedra and obtain corresponding *a posteriori* error estimates on this mesh.

2. Partition Ω into p subdomains corresponding to subsets of the coarse mesh, each subset containing about the same total (approximate) error (hence some subdomains will contain many more coarse elements than others if the *a posteriori* error estimate varies significantly throughout the domain). Let processor i ($i = 1, ..., p$) have a copy of the entire coarse mesh and sequentially solve the *entire* problem using adaptive refinement only in subdomain i (and its immediate neighbourhood): the target number of elements on each processor being the same.

*Corresponding author: pkj@comp.leeds.ac.uk
†Funded by the Government of Pakistan through a Quaid-e-Azam scholarship.

3. A global fine mesh is defined to be the union of the refined subdomains (with possible minor modifications near subdomain interfaces, to ensure that it is conforming), although it is never explicitly assembled.

4. A parallel solver is now required to solve this distributed (well load-balanced) problem.

This paper will describe a solver of the form required for the final step above, although the solver may also be applied independently of this framework. The work is a generalization and extension of previous research in two dimensions, [2], and in three dimensions, [6]. In particular, for the case of interest here, where (1) is convection dominated, a stabilized FE method is required and we demonstrate that the technique introduced in [2,6] may still be applied successfully. The following section of this paper provides a brief introduction to this preconditioning technique, based upon what we call a *weakly overlapping* domain decomposition, and Section 3 presents a small number of typical computational results. The paper concludes with a brief discussion.

2. THE WEAKLY OVERLAPPING DOMAIN DECOMPOSITION PRECONDITIONER

The standard Galerkin FE discretization of (1) seeks an approximation u_h to u from a finite element space $\mathcal{S}_h$ such that

$$\varepsilon \int_\Omega \underline{\nabla} u_h \cdot \underline{\nabla} v \, d\underline{x} + \int_\Omega (\underline{b} \cdot \underline{\nabla} u_h) v \, d\underline{x} = \int_\Omega f v \, d\underline{x} \tag{2}$$

for all $v \in \mathcal{S}_h$ (disregarding boundary conditions for simplicity). Unless the mesh is sufficiently fine this is known to be unstable when $0 < \varepsilon << \|\underline{b}\|$ and so we apply a more stable FE method such as the streamline-diffusion algorithm (see, for example, [7] for details). This replaces v in (2) by $v + \alpha \underline{b} \cdot \underline{\nabla} v$ to yield the problem of finding $u_h \in \mathcal{S}_h$ such that

$$\varepsilon \int_\Omega \underline{\nabla} u_h \cdot \underline{\nabla}(v + \alpha \underline{b} \cdot \underline{\nabla} v) \, d\underline{x} + \int_\Omega (\underline{b} \cdot \underline{\nabla} u_h)(v + \alpha \underline{b} \cdot \underline{\nabla} v) \, d\underline{x} = \int_\Omega f (v + \alpha \underline{b} \cdot \underline{\nabla} v) \, d\underline{x} \tag{3}$$

for all $v \in \mathcal{S}_h$. In general α is chosen to be proportional to the mesh size h and so, as the mesh is refined, the problem (3) approaches the problem (2).

Once the usual local FE basis is defined for the space $\mathcal{S}_h$, the system (3) may be written in matrix notation as

$$A\underline{u} = \underline{b} \, . \tag{4}$$

If the domain Ω is partitioned into two subdomains (the generalization to p subdomains is considered below), using the approach described in Section 1 for example, then the system (4) may be written in block-matrix notation as

$$\begin{bmatrix} A_1 & 0 & B_1 \\ 0 & A_2 & B_2 \\ C_1 & C_2 & A_s \end{bmatrix} \begin{bmatrix} \underline{u}_1 \\ \underline{u}_2 \\ \underline{u}_s \end{bmatrix} = \begin{bmatrix} \underline{f}_1 \\ \underline{f}_2 \\ \underline{f}_s \end{bmatrix} \, . \tag{5}$$

Here $\underline{u}_i$ is the vector of unknown finite element solution values at the nodes strictly inside subdomain i ($i = 1, 2$) and $\underline{u}_s$ is the vector of unknown values at the nodes on the interface between subdomains. The blocks A_i, B_i, C_i and $\underline{f}_i$ represent the components of the FE system that may be assembled (and stored) independently on processor i ($i = 1, 2$). Furthermore, we may express

$$A_s = A_{s(1)} + A_{s(2)} \quad \text{and} \quad \underline{f}_s = \underline{f}_{s(1)} + \underline{f}_{s(2)} , \tag{6}$$

where $A_{s(i)}$ and $\underline{f}_{s(i)}$ are the components of A_s and $\underline{f}_s$ respectively that may be calculated (and stored) independently on processor i.

The system (5) may be solved using an iterative technique such as preconditioned GMRES (see [10] for example). Traditional parallel DD solvers typically take one of two forms: either applying block elimination to (5) to obtain a set of equations for the interface unknowns $\underline{u}_s$ (e.g. [5]), or solving the complete system (5) in parallel (e.g. [3]). The weakly overlapping approach that we take is of the latter form. Apart from the application of the preconditioner, the main computational steps required at each GMRES iteration are a matrix-vector multiplication and a number of inner products. Using the above partition of the matrix and vectors it is straightforward to perform both of these operations in parallel with a minimal amount of interprocessor communication (see [4] or [5] by way of two examples). The remainder of this section therefore concentrates on an explanation of our novel DD preconditioner.

Our starting point is to assume that we have two meshes of the same domain which are hierarchical refinements of the same coarse mesh. Mesh 1 has been refined heavily in subdomain 1 and in its immediate neighbourhood (any element which touches the boundary of a subdomain is defined to be in that subdomain's immediate neighbourhood), whilst mesh 2 has been refined heavily in subdomain 2 and its immediate neighbourhood. Hence, the overlap between the refined regions on each processor is restricted to a single layer at each level of the mesh hierarchy. Figure 1 shows an example coarse mesh, part of the final mesh and the corresponding meshes on processors 1 and 2 in the case where the final mesh is a uniform refinement (to 2 levels) of the initial mesh of 768 tetrahedral elements. Throughout this paper we refine a tetrahedron by bisecting each edge and producing 8 children. Special, temporary, transition elements are also used to avoid "hanging nodes" when neighbouring tetrahedra are at different levels of refinement. See [11] for full details of this procedure.

The DD preconditioner, P say, that we use with GMRES when solving (5) may be described in terms of the computation of the action of $\underline{z} = P^{-1}\underline{p}$. On processor 1 solve the system

$$\begin{bmatrix} A_1 & 0 & B_1 \\ 0 & \tilde{A}_2 & \tilde{B}_2 \\ C_1 & \tilde{C}_2 & A_s \end{bmatrix} \begin{bmatrix} \underline{z}_{1,1} \\ \underline{z}_{2,1} \\ \underline{z}_{s,1} \end{bmatrix} = \begin{bmatrix} \underline{p}_1 \\ M_2\underline{p}_2 \\ \underline{p}_s \end{bmatrix} , \tag{7}$$

and on processor 2 solve the system

$$\begin{bmatrix} \tilde{A}_1 & 0 & \tilde{B}_1 \\ 0 & A_2 & B_2 \\ \tilde{C}_1 & C_2 & A_s \end{bmatrix} \begin{bmatrix} \underline{z}_{1,2} \\ \underline{z}_{2,2} \\ \underline{z}_{s,2} \end{bmatrix} = \begin{bmatrix} M_1\underline{p}_1 \\ \underline{p}_2 \\ \underline{p}_s \end{bmatrix} , \tag{8}$$

Figure 1. An initial mesh of 768 tetrahedral elements (top left) refined uniformly into 49152 elements (top right) and the corresponding meshes on processor 1 (bottom left) and processor 2 (bottom right).

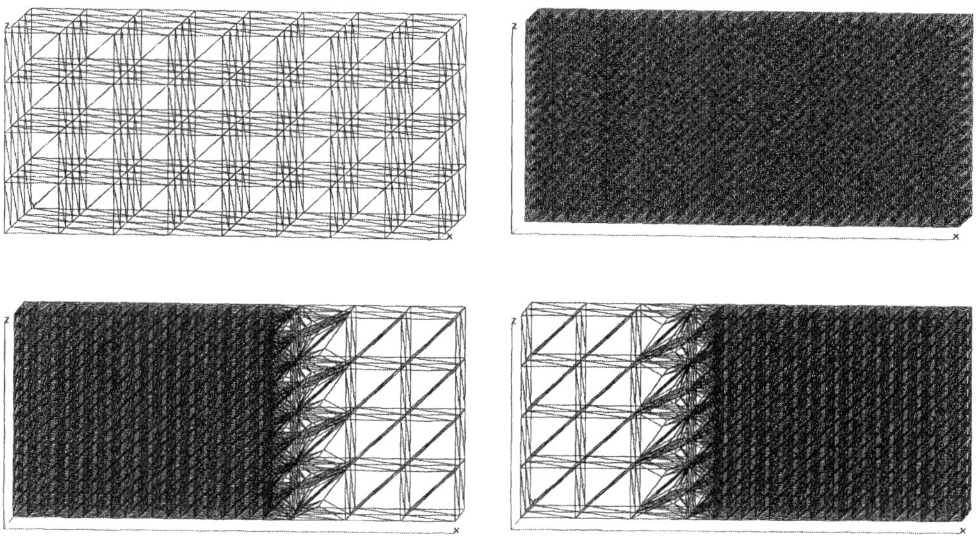

then set

$$\begin{bmatrix} \underset{\sim}{z}_1 \\ \underset{\sim}{z}_2 \\ \underset{\sim}{z}_s \end{bmatrix} = \begin{bmatrix} \underset{\sim}{z}_{1,1} \\ \underset{\sim}{z}_{2,2} \\ \frac{1}{2}(\underset{\sim}{z}_{s,1} + \underset{\sim}{z}_{s,2}) \end{bmatrix} . \tag{9}$$

In the above notation, the blocks $\tilde{A}_2$, $\tilde{B}_2$ and $\tilde{C}_2$ (resp. $\tilde{A}_1$, $\tilde{B}_1$ and $\tilde{C}_1$) are the assembled components of the stiffness matrix for the part of the mesh on processor 1 (resp. 2) that covers subdomain 2 (resp. 1). These may be computed and stored without communication. Moreover, because of the single layer of overlap in the refined regions of the meshes, A_s may be computed and stored on each processor without communication. Finally, the rectangular matrix M_1 (resp. M_2) represents the restriction operator from the fine mesh covering subdomain 1 (resp. 2) on processor 1 (resp. 2) to the coarser mesh covering subdomain 1 (resp. 2) on processor 2 (resp. 1). This is the usual hierarchical restriction operator that is used in most multigrid algorithms (see, for example [9]).

The generalization of this idea from 2 to p subdomains is straightforward. We will assume for simplicity that there is a one-to-one mapping between subdomains and processors. Each processor, i say, produces a mesh which covers the whole domain (the coarse mesh) but is refined only in subdomain i, Ω_i say, and its immediate neighbourhood. Again, this means that the overlapping regions of refinement consist of one layer of elements at each level of the mesh. For each processor i the global system (4) may be written as

$$\begin{bmatrix} A_i & 0 & B_i \\ 0 & \bar{A}_i & \bar{B}_i \\ C_i & \bar{C}_i & A_{i,s} \end{bmatrix} \begin{bmatrix} \underline{u}_i \\ \bar{\underline{u}}_i \\ \underline{u}_{i,s} \end{bmatrix} = \begin{bmatrix} \underline{f}_i \\ \bar{\underline{f}}_i \\ \underline{f}_{i,s} \end{bmatrix}, \tag{10}$$

where now $\underline{u}_i$ is the vector of finite element unknowns strictly inside Ω_i, $\underline{u}_{i,s}$ is the vector of unknowns on the interface of Ω_i and $\bar{\underline{u}}_i$ is the vector of unknowns (in the global fine mesh) outside of $\bar{\Omega}_i$. Similarly, the blocks A_i, B_i, C_i and $\underline{f}_i$ are all computed from the elements of the mesh inside subdomain i, etc.

The action of the preconditioner ($\underline{z} = P^{-1}\underline{p}$), in terms of the computations required on each processor i, is therefore as follows.
(i) Solve

$$\begin{bmatrix} A_i & 0 & B_i \\ 0 & \tilde{A}_i & \tilde{B}_i \\ C_i & \tilde{C}_i & A_{i,s} \end{bmatrix} \begin{bmatrix} \underline{z}_i \\ \tilde{\underline{z}}_i \\ \underline{z}_{i,s} \end{bmatrix} = \begin{bmatrix} \underline{p}_i \\ \bar{M}_i \bar{\underline{p}}_i \\ \underline{p}_{i,s} \end{bmatrix}. \tag{11}$$

(ii) Replace each entry of $\underline{z}_{i,s}$ with the average value over all corresponding entries of $\underline{z}_{j,s}$ on neighbouring processors j.
In (11) $\tilde{A}_i$, $\tilde{B}_i$ and $\tilde{C}_i$ are the components of the stiffness matrix for the mesh stored on processor i (this is not the global fine mesh but the mesh actually generated on processor i) which correspond to nodes outside of $\bar{\Omega}_i$. The rectangular matrix $\bar{M}_i$ represents the hierarchical restriction operator from the global fine mesh outside of $\bar{\Omega}_i$ to the mesh on processor i covering the region outside of $\bar{\Omega}_i$.

The main parallel implementation issue that now needs to be addressed is that of computing these hierarchical restrictions, $\bar{M}_i \bar{\underline{p}}_i$, efficiently at each iteration. Because each processor works with its own copy of the coarse mesh (which is locally refined) processor i must contribute to the restriction operation $\bar{M}_j \bar{\underline{p}}_j$ for each $j \neq i$, and processor j must contribute to the calculation of $\bar{M}_i \bar{\underline{p}}_i$ (for each $j \neq i$). To achieve this, processor i restricts its fine mesh vector $\underline{p}_i$ (covering Ω_i) to the part of the mesh on processor j which covers Ω_i (received initially from j in a setup phase) and sends this restriction to processor j (for each $j \neq i$). Processor i then receives from each other processor j the restriction of the fine mesh vector $\underline{p}_j$ (covering Ω_j on processor j) to the part of the mesh on processor i which covers Ω_j. These received vectors are then combined to form $\bar{M}_i \bar{\underline{p}}_i$ before (11) is solved. The averaging of the $\underline{z}_{i,s}$ in step (ii) above requires only local neighbour-to-neighbour communication.

3. COMPUTATIONAL RESULTS

All of the results presented in this section were computed with an ANSI C implementation of the above algorithm using the MPI communication library, [8], on a shared memory SG Origin2000 computer. The NUMA (non-uniform memory access) architecture of this machine means that timings for a given calculation may vary significantly between runs (depending on how the memory is allocated), hence all timings quoted represent the best time that was achieved over numerous repetitions of the same computation.

Table 1
The performance of the proposed DD algorithm using the stabilized FE discretization of the convection-diffusion test problem for two choices of ε: figures quoted represent the number of iterations required to reduce the initial residual by a factor of 10^5.

	$\varepsilon = 10^{-2}$				$\varepsilon = 10^{-3}$			
Elements/Procs.	2	4	8	16	2	4	8	16
6144	3	4	4	6	5	5	5	7
49152	3	4	4	6	4	5	5	7
393216	3	4	5	7	4	5	5	6
3145728	3	5	6	8	3	4	5	7

Table 2
Timings for the parallel solution using the stabilized FE discretization of the convection-diffusion test problem for two choices of ε: the solution times are quoted in seconds and the speed-ups are relative to the best sequential solution time.

	$\varepsilon = 10^{-2}$					$\varepsilon = 10^{-3}$				
Processors	1	2	4	8	16	1	2	4	8	16
Solution Time	770.65	484.53	347.61	228.39	136.79	688.12	442.44	277.78	187.16	108.75
Speed-Up	—	1.6	2.2	3.4	5.6	—	1.6	2.5	3.7	6.3

We begin with a demonstration of the quality of the weakly overlapping DD preconditioner when applied to a convection-dominated test problem of the form (1). Table 1 shows the number of preconditioned GMRES iterations that are required to solve this equation when $\underline{b}^T = (1, 0, 0)$ and f is chosen so as to permit the exact solution

$$u = \left(x - \frac{2(1 - e^{x/\varepsilon})}{(1 - e^{2/\varepsilon})} \right) y(1-y)z(1-z) \tag{12}$$

on the domain $\Omega = (0, 2) \times (0, 1) \times (0, 1)$. Two different values of ε are used, reflecting the width of the boundary layer in the solution in the region of $x = 2$. For these calculations the initial grid of 768 tetrahedral elements shown in Figure 1 (top left) is refined uniformly by up to four levels, to produce a sequence of meshes containing between 6144 and 3145782 elements.

It is clear that, as the finite element mesh is refined or the number of subdomains is increased, the number of iterations required grows extremely slowly. This is an essential property of an efficient preconditioner. In fact, the iteration counts of Table 1 suggest that the preconditioner may in fact be optimal (i.e. the condition number of the preconditioned system is bounded as the mesh is refined or the number of subdomains is increased), however we are currently unable to present any mathematical confirmation of this. In Table 2 we present timings for the complete FE calculations tabulated above on the finest mesh, with 3145728 tetrahedral elements.

Figure 2. An illustration of the partitioning strategy, based upon recursive coordinate bisection, used to obtain 2, 4, 8 and 16 subdomains in our test problem.

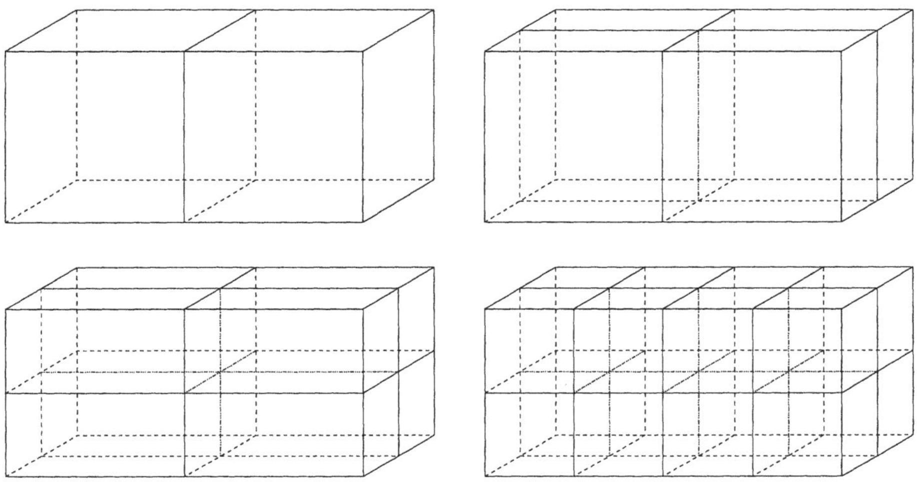

4. DISCUSSION

There are a number of features concerning the parallel timings in Table 2 that warrant further discussion. Perhaps the most important of these is that the algebraic action of the preconditioner that we have applied depends not only on the number of subdomains p, but also on the geometric properties of these subdomains. In each case the parallel implementation of these algorithms may be very efficient but if the algorithm itself is such that its sequential execution time is greater than that of the fastest available sequential algorithm (for this work we use [10]) then the speed-up will be adversely affected. In an effort to minimize this particular parallel overhead we have selected a simple partitioning strategy based upon recursive coordinate bisection. This strategy, illustrated in Figure 2, led to the best solution times that we were able to achieve in practice (from the small number of partitioning techniques so far considered). Furthermore, inexact solutions to the systems (11) have been used: reducing the residual by a factor of just 10 at each approximate solve (again using [10]). Whilst this can have the effect of slightly increasing the total number of iterations required for convergence it appears to yield the fastest overall solution time.

The main advantage of the partitions illustrated in Figure 2 is that the surface-area to volume ratio of the subdomains is small. This means that both the amount of additional refinement and the quantity of neighbour-to-neighbour communication is relatively small, making the cost of each iteration as low as possible. For convection-dominated problems however the number of preconditioned iterations required to converge to the solution may be decreased from those given in Table 1 by selecting a partition which contains long thin subdomains that are aligned with the convection direction. This trade-off that exists between minimizing the number of iterations required and minimizing the cost of each iteration is an important issue that is worthy of further investigation.

In addition to the particular question of subdomain shape and the more general issue of the overall partitioning strategy there are a number of further lines of research that need to be undertaken. When a nonlinear elliptic PDE is solved for example, the FE discretization leads to a nonlinear algebraic system which may be solved using a quasi-Newton method. At each nonlinear iteration a linear Jacobian system must be dealt with, and the Jacobian matrix itself may be partitioned and assembled in parallel using the block pattern of (10). These linear systems may then be solved using the weakly overlapping preconditioner to obtain a parallel nonlinear DD algorithm. The extension to linear and nonlinear systems of PDEs may then be undertaken and assessed.

REFERENCES

1. Bank, R.E., Holst, M.: A New Paradigm for Parallel Adaptive Meshing Algorithms. SIAM J. on Sci. Comp. **22** (2000) 1411–1443.
2. Bank, R.E., Jimack, P.K.: A New Parallel Domain Decomposition Method for the Adaptive Finite Element Solution of Elliptic Partial Differential Equations. Concurrency and Computation: Practice and Experience, **13** (2001) 327–350.
3. Chan, T., Mathew, T.: Domain Decomposition Algorithms. Acta Numerica **3** (1994) 61–143.
4. Gropp, W.D., Keyes, D.E.: Parallel Performance of Domain-Decomposed Preconditioned Krylov Methods for PDEs with Locally Uniform Refinement. SIAM J. on Sci. Comp. **13** (1992) 128–145.
5. Hodgson, D.C., Jimack, P.K.: A Domain Decomposition Preconditioner for a Parallel Finite Element Solver on Distributed Unstructured Grids. Parallel Computing **23** (1997) 1157–1181.
6. Jimack, P.K., Nadeem, S.A.: A Weakly Overlapping Parallel Domain Decomposition Preconditioner for the Finite Element Solution of Elliptic Problems in Three Dimensions. In Arabnia, H.R. (ed.): Proceedings of the 2000 International Conference on Parallel and Distributed Processing Techniques and Applications (PDPTA'2000), Volume III, CSREA Press, USA (2000) 1517–1523.
7. Johnson, C.: Numerical Solutions of Partial Differential Equations by the Finite Element Method. Cambridge University Press (1987).
8. Message Passing Interface Forum: MPI: A Message-Passing Interface Standard. Int. J. Supercomputer Appl. **8** (1994) no. 3/4.
9. Oswald, P.: Multilevel Finite Element Approximation: Theory and Applications. Teubner Skripten zur Numerik, B.G. Teubner (1994).
10. Saad, Y.: SPARSKIT: A Basic Tool Kit for Sparse Matrix Computations, Version 2. Technical Report, Center for Supercomputing Research and Development, University of Illinois at Urbana-Champaign, Urbana, IL, USA (1994).
11. Speares, W., Berzins, M.: A 3-D Unstructured Mesh Adaptation Algorithm for Time-Dependent Shock Dominated Problems. Int. J. for Numer. Meth. in Fluids **25** (1997) 81–104.

Parallel Computational Fluid Dynamics – Practice and Theory
P. Wilders, A. Ecer, J. Periaux, N. Satofuka and P. Fox (Editors)

Lattice-Boltzmann simulations of inter-phase momentum transfer in gas-solid flows

D. Kandhai, J.J. Derksen and H.E.A. Van den Akker

Kramers Laboratorium voor Fysische Technologie
Faculty of Applied Sciences, Delft University of Technology
Prins Bernhardlaan 6 2628 BW Delft, The Netherlands
email:B.D.Kandhai@tnw.tudelft.nl

Numerical simulations of dense gas-solid flows are generally based on the so-called two fluid equations. An important term in these equations is the momentum transfer of the solid phase on the gas phase and vice versa. This closure term is commonly modeled using empirical correlations. In this paper we consider direct numerical simulations of the inter-phase momentum transfer coefficient as a function of the Reynolds number and solid volume fraction for a periodic packing by means of a lattice-Boltzmann method. The use of parallel computing is important due to the large computational requirements of the simulations. Our preliminary results are in good agreement with previous findings reported in the literature.

1. Introduction

The progress in multi-phase reactor engineering is handicapped by a lack of understanding of the fluid dynamics involved. It is well known that hydrodynamics plays a crucial role in the dynamic behavior of for instance fluidized beds [1], [2]. Currently computational fluid dynamics methods are widely used to explore these effects.

Numerical simulations of hydrodynamics in gas-solid fluidized beds are generally based on the so-called two-fluid models [1]. In these models both phases are considered as inter-penetrating continua and mass and momentum balances are derived using volume averaging techniques. The momentum equations for the gas phase for example is given by,

$$\frac{\partial}{\partial t}(\epsilon\rho_f\mathbf{U}) + (\nabla\cdot\epsilon\rho_f\mathbf{U}\mathbf{U}) = -\epsilon\nabla p - \beta(\mathbf{U}-\mathbf{V}) - \nabla\cdot\epsilon\tau + \epsilon\rho_f\mathbf{g}, \tag{1}$$

where $\mathbf{U}$ and $\mathbf{V}$ are the velocity of the gas and solid phase respectively, ϵ is the gas volume fraction, ρ_f is the density of the gas, p is the pressure, τ is the stress tensor, $\mathbf{g}$ is gravity and β is inter-phase momentum transfer coefficient. The inter-phase momentum transfer coefficient is regularly modeled using semi-empirical relations obtained from experimental data on pressure drops of dense packed beds, the so-called Ergun relation [3]. Although this relation is based on fixed systems it may be applied to dynamic systems in the case of large density differences between the two phases [4]. Moreover for relatively dilute systems the well-known Richardson and Zaki [5] correlation, derived from sedimentation experiments, is often used. For more details see also Refs. [2] and [6].

Recently, Hill et al. presented a rigorous study concerning the dependence of the drag-force on the Reynolds number and solid volume fraction [10]. Using a lattice-Boltzmann method the drag force of static (dis)ordered mono disperse bead packings was computed. Their results clearly suggest a discrepancy between the Ergun equation and the lattice-Boltzmann simulations for volume fractions between 0.2 and 0.4. For systems of freely moving particles Wachmann et al. addressed the drag-closure problem in the regime of relatively small Reynolds numbers and high volume fractions by means of a finite-difference method [9]. Their numerical results are in good agreement with the Richardson-Zaki expression.

Our main interest is to study the validity of existing correlations for describing the inter-phase momentum transfer in the case of gas flows with static and freely moving solid particles for moderate Reynolds numbers in the range 1 to 50 and volume fractions between 0.1 and 0.6. The combined dynamical behavior of the solid particles interacting with the fluid and vice versa is modeled using a lattice-Boltzmann method (LBM) [7], [8]. We emphasize that for full 3D simulations the computational requirements for these simulations are large. Moreover, several parameter studies including those to explore finite-size effects, require a vast amount of simulations to be carried out. Obviously the use of parallel computing is of substantial importance for this work.

2. Simulation method

2.1. Lattice-Boltzmann method

LBM have been used in the past to simulate a wide variety of fluid flow applications. They originated from the lattice gas automata which are discrete models for the simulation of transport phenomena [7]. In these models the computational grid consists of a number of lattice points that are connected with some of their neighboring sites (depending on the model) by a bond or link. At each time step particles move synchronously along the bonds of the lattice and interact locally subject to physical conservation laws. The inherent spatial and temporal locality of the update rules makes this method ideal for parallel computation. The simplest model in the hierarchy of lattice-Boltzmann methods is the so-called Lattice-BGK model[7]. Several models for capturing the physics of moving solid-fluid interfaces within the lattice-Boltzmann framework exist [11]. Here we use the model proposed by O. Behrend [8].

2.2. Parallel computing

Parallelization of the lattice-Boltzmann algorithm is done using domain decomposition. The computational box is divided in sub domains of more or less equal volume. Each processor performs the computations on a specific box and information exchange between the neighboring processors is required for the update of the interprocessor boundary sites. It is by now well known that the lattice-Boltzmann scheme has perfect scalability characteristics with respect to the number of processors for typical computational problems (see e.g. [12]).

In the case of suspension flows parallelization is less trivial. The parallel performance of the solver is then affected by the dynamic workload distribution due to irregular particle motion and sophisticated dynamic load balancing methods might be usefull for obtaining efficient parallel simulations. For the current purpose we chose for a rather simple approach based on a master/slave model of parallelization. Although on the first glance simple this approach appears to be reasonable efficient, because in the simulations we consider a small number of particles and the computational time spend in the updating of the particle positions is only a small fraction of the total time (on the order of 10%).

3. Two-dimensional simulation results

3.1. Single cylinder

Before discussing the results obtained for the disordered arrays of particles, we first present two benchmark cases as a validation of our simulation program. The reference values are taken form Ref. [13]. In the first test-case we consider flow around a cylinder placed at the middle of a channel. The channel width is 128 lattice units, periodic boundaries are imposed at the ends of the channel and the upper and lower walls move with velocity u_w. We consider the case where the walls move and the cylinder is fixed and the case where the walls are fixed and the cylinder moves. Notice that these cases are physically equivalent due to galilean invariance. The Reynolds number, defined as $Re = \frac{u_w 2R}{\mu}$, is equal to 1 in all test cases. The results are shown in Table 1. There is clearly a good agreement with the corresponding results obtained with the finite-element solver.

Our second test case is the computation of the hydrodynamic forces acting on a fixed cylinder placed to a moving wall and forced to rotate counterclockwise. We considered two different particle radii and two different gaps between the center of the particle and the moving wall. The wall velocities were $u_w = 0.05$ and $u_w = 0.1$ for the bigger and smaller cylinder, respectively. The results obtained for the drag and lift forces on the cylinder are shown in table 2. We see that the agreement between the lattice-Boltzmann method and the Fluent simulations is quite good.

R		u_w	u_d	f_d
5.4	FEM	-0.04	0.0	0.966
	LBM	-0.04	0.0	0.956
	LBM	0.0	0.04	0.961
10.4	FEM	-0.02	0.0	1.158
	LBM	-0.02	0.0	1.125
	LBM	0.0	0.02	1.148

Table 1
The dimensionless force on the particle (f_d) as computed by the lattice-Boltzmann method (fixed and moving object/walls) and the finite-element method. The force acting on the cylinder is scaled by $\rho_f u_w^2$, and by the area of the cylinder.

Re_{rot}	f_l^*	f_l	f_d^*	f_d
0.1	-0.53	-0.6	4.02	4.1
0.25	-0.64	-0.72	4.02	4.14
1	-1.31	-1.43	4.04	4.17
2.5	-2.80	-3.11	4.09	4.22

Table 2
The dimensionless lift (f_l) and drag (f_d) forces on a cylinder with forced rotation computed by the lattice-Boltzmann method and a commercial finite-volume solver* (Fluent). The radius of the cylinder is 11.55 lattice units (l.u.) and the distance between the moving wall and the center point of the cylinder is 23.5 l.u.

3.2. Many cylinders

We studied a 2D system of freely moving particles in a gas flow. Initially the particles are distributed randomly in a box. The gas flow is driven upwards by applying a uniform body-force on the fluid in the vertical direction. As a consequence the hydrodynamic force acting on the particles increases and at some point balances the gravity force acting on the particles. Notice that in a periodic box (i.e. in the absence of periodic walls) a force balance is only obtained when the body-force acting on the fluid is equal to the total gravitational force acting on the particles. The input parameters in these simulations are the particle Reynolds number, Re_p, the density ratio between solid and gas phase, $\rho^* = \frac{\rho_s}{\rho_g}$, and the simulation cell dimensions. The output parameters of interest are the slip velocity, U_s, and the friction coefficient, f_c. The parameters are defined as follows:

- The particle Reynolds number, $Re_p = \frac{U_s D_p}{\nu}$, where D_p is the particle diameter and ν is the kinematic viscosity;
- The slip-velocity, $\mathbf{U_s} = <\mathbf{U}> - <\mathbf{V_p}>$, where $\mathbf{U}$ is the superficial average gas velocity and V_p is the average velocity of the particles;
- The friction coefficient, $f_c = \frac{<F_{drag}>}{\rho_g U_s^2 * A}$, with $< F_{drag} >$ the average drag force and A is the surface area of the particles.

In our simulations we consider a fixed density ratio between solid and gas phase ($\rho^* = 100$) and a volume fraction of 0.14 (a periodic box of dimension $7.5D_p \times 7.5D_p$ with $N = 10$ particles). As an illustration we show in Fig. 1, the particle arrangement after $t = 1000$, $t = 3000$ and $t = 6000$ time-steps, respectively. After $t = 1000$ time-steps the fluid flow is more or less developed and as time progresses the particles move due to the hydrodynamic interactions. Moreover channeling effects on the flow fields are observed due to heterogenities in the particle arrangement (data not shown). Initially the drag-force is zero as both particles and fluid are in rest. As the simulation progresses the drag force increases and balances the gravitational force. Due to the mobility of the particles, local heterogenities in the flow field are generated which in turn influence the dynamics of the particles. Thus we have a stationary state only in a statistical sense. The fluctuations

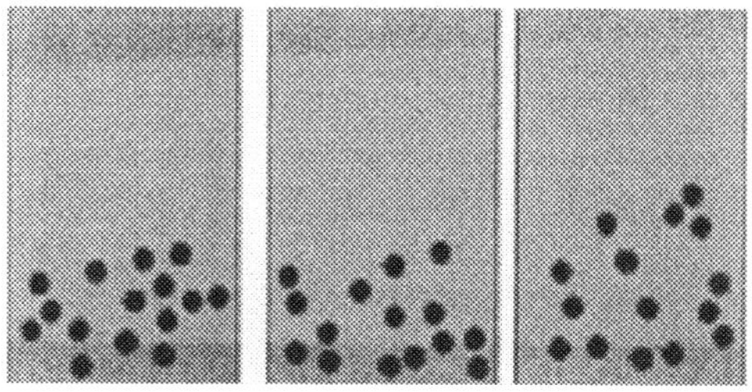

(a) (t=1000) (b) (t=3000) (c) (t=6000)

Figure 1. Particle arrangements for different times.

are also reflected in the slip velocity of the system (see Fig. 2). In Fig. 3 the drag coefficient as a function of the particle Reynolds number in the range 0.1 to 20 is shown. These values are on the order of 30 to 40% higher compared to that of an ordered array of cylinders. Moreover, a fit of the form $\frac{1}{Re^{0.75}}$ seems to describe the drag curve rather well. More detailed analysis of the intensity and typical duration of the fluctuations in the slip velocity and the drag force will be carried out in the future.

4. Three-dimensional simulation results

In 3D we considered the dimensionless drag force of a packed bed of spheres. The positions of the spheres are fixed and distributed randomly in a periodic box. The particle diameter is 20 lattice points and the gas-volume fraction (ϵ) is 0.8 and 0.9, respectively. The lattice dimensions are $100 \times 100 \times 100$ and the corresponding number of particles are 10 and 20, respectively. In these simulations the flow is driven by a constant body-force. The Reynolds number (based on the particle radius) is tuned by adjusting the kinematic viscosity of the fluid or the body-force. The results obtained for the drag force as a function of Re_p are shown in Fig. 4. Our results confirm the previous findings of Hill et al. [10] in that there is indeed a discrepancy between the lattice-Boltzmann simulations and the Ergun equation.

In the future we will extend this study to freely moving particles. Our initial guess is that lattices of $200 \times 200 \times 200$ might be appropriate for performing the simulations in the range of our interest. For such lattice dimensions the typical memory requirements are in the order of 500 MB. Moreover, the computation time depends strongly on the packing density; for dilute systems a large number of time steps is required to reach equilibrium, whereas for dense systems a stationary state is reached much faster. On the other hand larger particle radii are required for the simulation of the denser packings, because the gaps between the solid particles are then quite narrow.

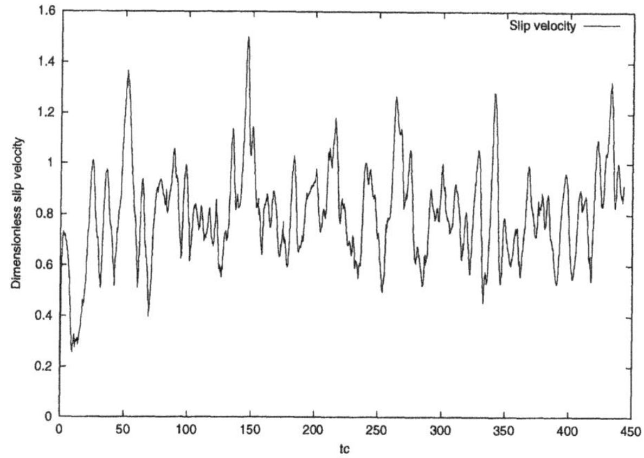

Figure 2. Slip velocity in time for $Re_p = 1.5$. A 2D system with 10 particles is considered. t_c is the typical time that is required to move a distance equal to the particle diameter in the case of a single-sphere sedimentation experiment. The slip velocity is made dimensionless by dividing with the slip velocity of a single sphere at $Re = 1.5$.

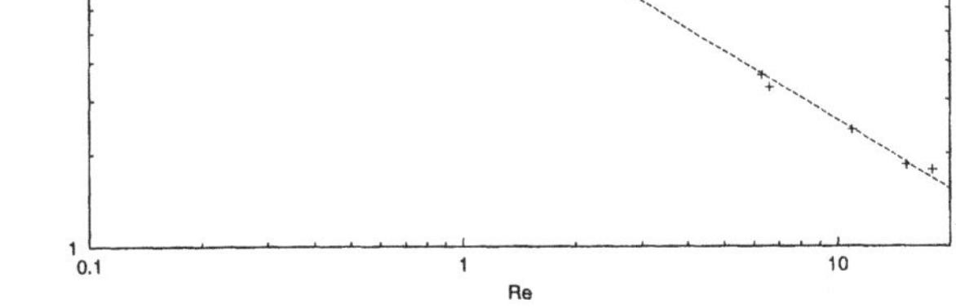

Figure 3. The drag coefficient as a function of the particle Reynolds number for a 2D periodic array of freely moving cylinders. A solid line in included to guide the eye.

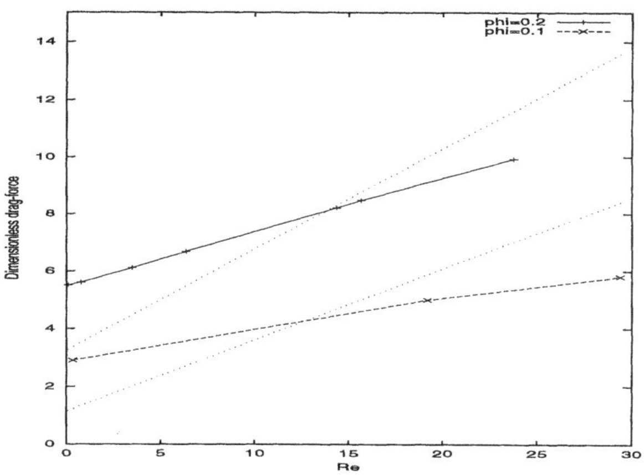

Figure 4. Dimensionless drag-force versus Reynolds-number for a disordered array of spheres. Results presented for solid volume fractions of 0.1 and 0.2, respectively. The dotted lines are corresponding curves for the Ergun equation [3].

5. Conclusions

Our interest is to study drag force closures as a function of the particle Reynolds number and solid fraction in the case of gas-solid flows. For this purpose we use direct numerical simulation of finite-size solid particles suspended in a gas flow by means of lattice-Boltzmann methods on parallel systems. As a validation of the simulation program a few basic benchmark cases in 2D are presented, all showing good agreement with data reported in the literature. Preliminary results on the behavior of the drag force as a function of the Reynolds number are discussed. It is found that for freely moving cylinders the drag-force is on average 30 to 40% higher compared to that of ordered systems and scales according to $Re^{-0.75}$. In three-dimensions we considered random sphere packings. Our results are in good agreement with the recent findings of Hill et al [10]. In the future we will extend our studies to freely moving particles in three-dimensions.

REFERENCES

1. Liang-Shih Fan and Chao Zhu. *Principles of Gas-Solid Flows.* Cambridge University Press, 1998.
2. M.J.V. Goldschmidt, B.P.B. Hoomans and J.A.M. Kuipers, Recent progress towards hydrodynamic modeling of dense gas-particle flows, *Recent Res. Devel. Chemical Eng.* , **4**, 273, 2000.
3. S. Ergun, Fluid flow through packed columns, *Chem. Eng. Prog.*, **48**, 245, 1952.
4. D.L. Koch, Kinetic theory for a monodisperse gas-solid suspension, *Phys. Fluids A*, **2** 1711, 1990
5. J.F. Richardson and W.N. Zaki, Sedimentation and fluidization:part I, *Trans. Instn Chem. Engrs*, **32**, 35, 1954.
6. R. Di Felice, The voidage function for fluid-particle interaction systems, *Int. J. Multiphase Flow*, **20**, 153, 1994.
7. B. Chopard and M. Droz. *Cellular Automata Modeling of Physical Systems.* Cambridge University Press, 1998.
8. O. Behrend, Solid-fluid boundaries in particle suspension simulations via the lattice Boltzmann method, *Phys. Rev. E*, **52**, 1164, 1995.
9. B. Wachmann, S. Schwarzer, and K. Höfler, Local drag law for suspensions from particle-scale simulations, *Int. J. Mod. Phys. C*, **9**, 1361, 1998.
10. R.J. Hill, D.L. Koch, and A.J.C. Ladd, Inertial flows in ordered and random arrays of spheres, *J. Fluid. Mech.*, Submitted, 2000.
11. S. Chen and G.D. Doolen, Lattice Boltzmann method for fluid flows, *Annu. Rev. Fluid Mech.*, **30**, 329, 1998.
12. D. Kandhai, A. Koponen, A. Hoekstra, M. Kataja, J. Timonen and P. Sloot, Lattice-Boltzmann Hydrodynamics on Parallel Systems, *Comput. Phys. Commun.*, **111**, 14 , 2000.
13. P. Raiskinmäki, A. Shakib-Mahesh, A. Koponen, A. Jäsberg, M. Kataja, J. Timonen, Simulations of non-spherical particles suspended in a shear flow, *Comput. Phys. Commun.*, **129**, 185 , 2000.

Parallel CFD Simulations of Multiphase Systems: Jet into a Cylindrical Bath and Rotary Drum on a Rectangular Bath.

M. Khan[a], Clive A. J. Fletcher[a], Geoffrey Evans[b], Qinglin He[b]

[a]CANCES, UNSW, G16 NIC, Australian Technology Park, EVELEIGH, NSW 1430, Australia
mnhka@yahoo.com, Tel: 61-2-9318 0004, Fax: 61-2-9319 2328
[b]Chemical Engineering, University of Newcastle, Australia.

Most of the developed commercial CFD (Computational Fluid Dynamics) packages do not attempt to document (or don't want to publish !!) the detailed algorithm for parallelising the code; even the ordinary solution strategies are tedious to learn sometimes. However, industrial engineers are more concerned about quick and correct solutions of their problems. Key features of this paper are the use of the domain decomposition and encapsulated message passing to enable execution in parallel. A parallel version of a CFD code, FLUENT, has been applied to model some multiphase systems on a number of different platforms. The same models considered for all the platforms to compare the parallel efficiency of CFD in those machines. Two physical models: one is a liquid jet directed into a cylindrical bath to disperse buoyant particles suspended on the top of the bath (3D), and the second one is a rotary drum rotating on a free surface to drag down particles from the free surface. The free surface, high gradient of the velocity, particle-particle, particle-wall collisions make most industrial flow simulations computationally expensive. For many complex systems, like here, the computational resources required limit the detail modelling of CFD. The implementations of computational fluid dynamics codes on distributed memory architectures are discussed and analyzed for scalability. For commercial CFD packages, in many cases the solution algorithms are black boxes, even though parallel computing helps in many cases to overcome the limitations, as shown here. The performance of the code has been compared in terms of CPU, accuracy, speed etc. In short, this research is intended to establish a strategic procedure to optimize a parallel version of a CFD package, FLUENT. The parallelised CFD code shows the excellent efficiency and scalability on a large number of platforms.

Key Words: Parallel computing, multiphase systems, performance.

1. Introduction

Due to the fast growth of chip technology and CFD packages, automatic features of parallelisation and optimization of the codes are likely to be available to achieve a high level of performance with fewer computers and less programming effort. But the availability of these automations seems always to be lagging behind for many industrial applications and so hand-tuning code still plays significant role in achieving an acceptable performance [1].

To predict the solution future of technical problems, fast and correct solutions are being demanded always. Even though 2D axisymmetric is a very easy choice for many multiphase flow simulations, in reality, many practical engineering problems end up with a 3D simulation with different complexities. That's why, parallel computing is widely used in automobile industries, aerodynamics, combustion (spray combustion), multiphase systems [2], nuclear reactor engineering [3]. DNS (Direct Numerical Simulation) requires extensive parallel computing. Lack of suitable parallelising software and lack of standard message inter-process communicators are the main barriers to parallelisation of complex CFD modeling.

A parallel CFD solver is usually more challenging than corresponding sequential solver [4]. The parallel capability of a commercial code, like FLUENT, is very scalable on different platforms. The main aspects in designing parallel algorithms are partitioning of data (domain decomposition), communication across internal boundaries, load balancing and minimizing overhead caused by both computation and communication [5].

This paper has been organized as follow. Section 2 gives the problem description, model used, platforms used. Section 3 describes the methodology applied and section 4 will show the performances. The conclusions have been added at the end of the paper.

2. Sample Test Considered

Figure 1 shows the typical computational domain where the jet inlet is on the top of the liquid cylindrical bath and the outlet is at the bottom attached to a reservoir. There is no accumulation of liquid in the bath. The liquid level in the cylinder is the same as the liquid level in the reservoir in the experimental set up. The nozzle submergence depth is 15 mm, cylinder diameter 180 mm, nozzle diameter 9.6 mm, and jet velocity 5m/s.

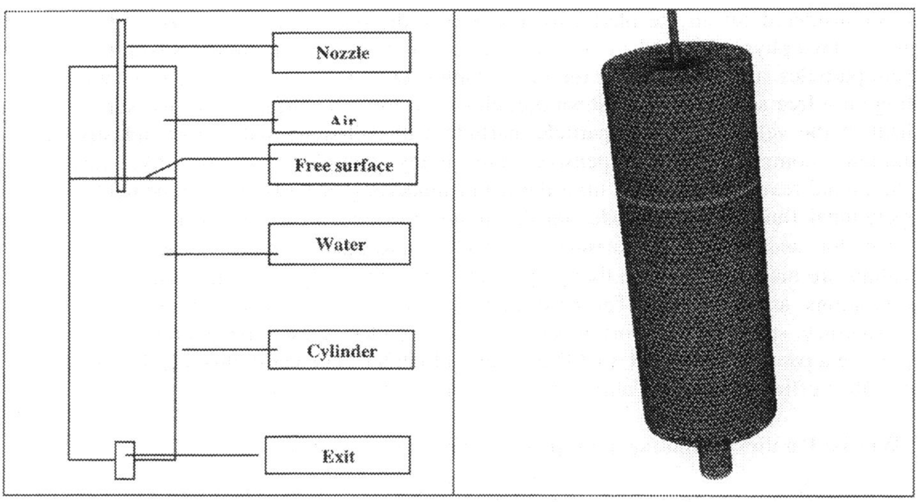

Figure 1: Computational domain: Left Schematic diagram, right 3D computational domain

Figure 2 shows a second sample industrial 3D CFD simulation model. The dimensions are as follows: a water tank of 450mm width, 300 mm thickness (drum length), 280 mm liquid height.

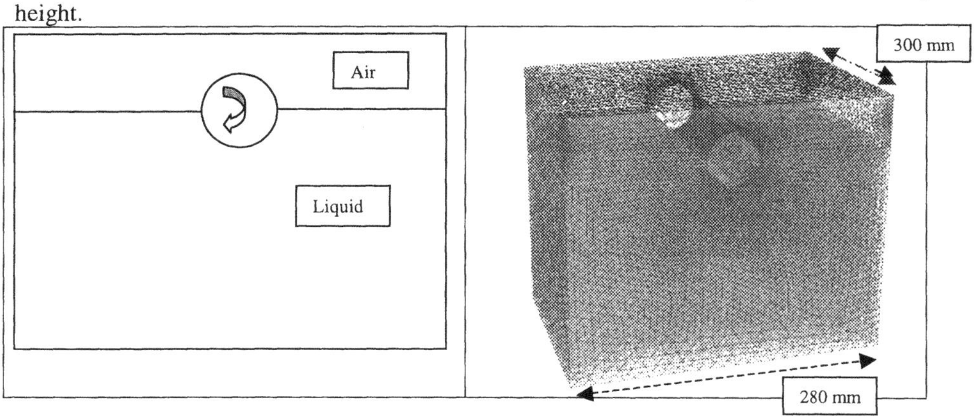

Figure 2: Rotary Drum: Left Schematic diagram, right 3D computational domain

Above the water, there is a free space of 75 mm that is open to the atmosphere. The drum is 75 mm. (0, 0) refers to the center of the cylinder that is located in the center and half submerged in the liquid. In 2D, one side of the domain is 225mmX280mm plus top 75 mm air. For 3D, it is 225mmX280mmX300 mm. The liquid surface is (450-75=) 375 mm for 2D and 2 times 375mmX300mm for 3D. Only 3D simulations are presented here.

Table 1 shows the tests performed and their identifications. For any CFD simulation, it is wise to check the possibility of parallel CFD simulations with the available software and platform, to attack the complexity of the problem in an efficient way.

Table 1: Test cases used

Test Identifications	No of Cells (tetrahedral) and systems	No of Iterations
Test A1	$1.67X10^5$, jet system	1000
Test A2	$1.67X10^5$, jet system	2000
Test A3	$1.67X10^5$, jet system	25
Test A4	$1.67X10^5$, jet system	100
Test B1	$1.59X10^5$, rotary drum system	2000
Test B2	$1.41X10^6$, rotary drum system	25
Test B3	$1.41X10^6$, rotary drum system	100

Test B2 is performed (comparable to Test A2) to confirm that for the same size of mesh and model, the CPU time is nearly the same. Tests B2 and B3 are bigger than all other tests nearly 8 times.

The k and ε transport equations (RNG) have been solved in order to determine the turbulent viscosity [5]. Table 2 shows the modelling parameters.

Table 2: Modelling parameters of primary fluid

Type of parameters	Value/method
Coefficient of the k-ε model (RNG)	C_μ= 0.0845, $C_{1\varepsilon}$=1.42, $C_{2\varepsilon}$=1.68
Discretization	Pressure= Body-force-weighted; momentum, k, ε=1st order upwind; P-V coupling=PISO
Solver	Segregated, implicit, 3D, unsteady (free surface modelling)
Boundary conditions	Jet: inlet, outlet (negative inlet), Rotary: moving wall, pressure outlet (cylindrical and rectangular bath are open to atmosphere)
Time step	Gradually increased from 1.0×10^{-7} sec
VOF parameters	Geometric reconstruction, surface tension (0.0785 N/m)

Table 3 shows the detail specification of the platforms used in this research project.

Table 3: Type of machines used for the test

Identifications	Specifications
M1	IBM RS/6000 SP (www.ac3.edu.au)
M2	SGI Origin 2000 (www.ac3.edu.au)
M3	Compaq Alpha Server SC (nf.apac.edu.au)

3. Partitioning Procedure

The total number of divisions of the computational domain was always an integer multiple of the number of processors. Each processor has approximately the same number of cells. METIS software was used to divide the computational domain (recursive bisection). Attempts are made to minimize the interface ratio variation and global interface ratio [6].

4. Results and Discussions

Table 4 shows the CPU time for the test A1 and A2 on platform M1. This test has been performed to investigate the parallel CFD efficiency on M1. This table shows the natural trend of decreasing the CPU time as the number of processors increased both for 1000 and

2000 number of iterations. As the CPU time is decreasing as the number of processors increased up to 12 processors, the system is limited by the computations, not by communications at least upto 12 processors. The performance of M2 is nearly same as M1.

Table 4: CPU time for on platform M1

No of Processors	CPU time for 1000 iterations (Test A1) (Sec)	CPU time for 2000 iterations (Test A2) (Sec)
4	5994	11005
8	3336	9776
12	3242	8035

The following histogram (Figure 3) shows a better comparison.

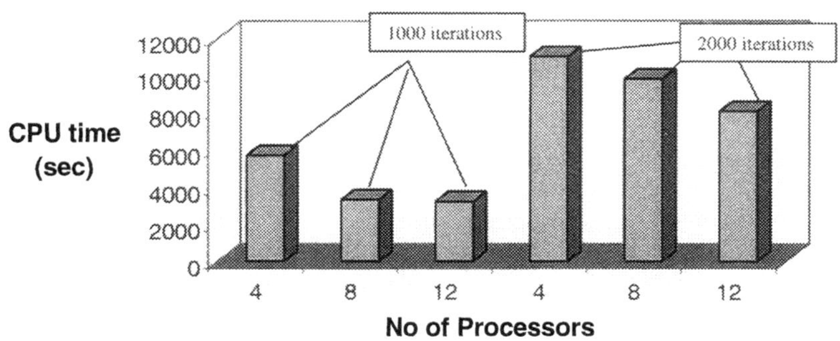

Figure 3: CPU time for test A1 and A2 on platform M1

The following table shows the CPU time for the test A1 and A2 on platform M3

Table 5: CPU time for on platform M1

No of Processors	CPU time for 1000 iterations (Test A1) (Sec)	Speed up	CPU time for 2000 iterations (Test A2) (Sec)	Speed up
1	13876		28177	
2	7128	1.95	14348	1.96
4	3769	3.68	7523	3.75
8	2276	6.10	4582	6.15
10	2008	6.91	4010	7.03

Figure 4 depicts the CPU time and speed up as the number of processors used. The speed up decreases as the number of processors increased, because no further parallelising is possible. For 8 numbers of processors, the efficiency is about 75%, which is quite good for any industrial applications.

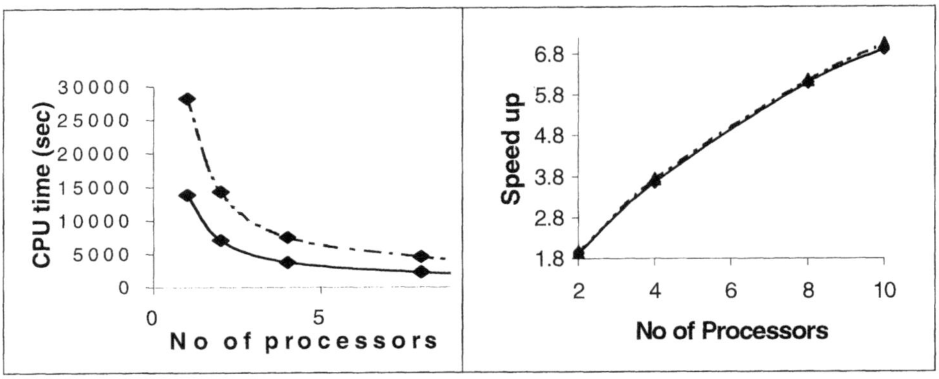

Figure 4: CPU time and speed up for test A1 and A2 on M3 (dotted for higher number of iterations)

Figure 4 also shows that for the same problem (same size of mesh and type and same model), for higher number of iterations, there is no speed up (slightly higher). This is because for the same type of iteration and calculations has been performed all through the domains. However if the number of particles is changed in any particular domain or any refinement of mesh is applied during the larger number of iterations, the speed up may vary.

Figure 5 shows the comparison of test A4 and B3 on platform M3. The mesh of B3 is about 8 times bigger than that of A4. For the same type of model (turbulence and free surface) applied both for the jet and rotary drum system. This test has been performed to investigate, how this platform behave for different size of problem. Figure 5 shows that for larger size of problem (with the same type of model calculations), the speed up is better.

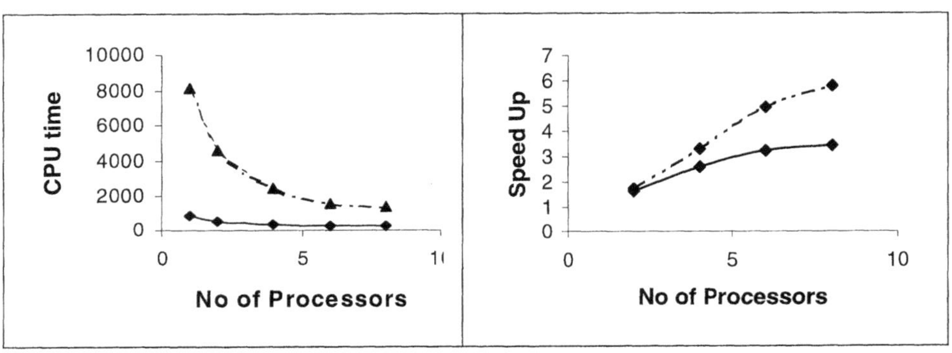

Figure 5: CPU time (sec) comparison and speed up for bigger sized problem (dotted for bigger sized test B3)

The number of divisions used in the CPU time calculations were the same as the number of processors in all the above tests. For a typical case of jet system, where the number of divisions across the principal axis of the jet system was 12 is shown in table 6.

Table 6: Divisions of the 1.67x10^5 cells into 12

Division number	No of cells	Percentage of total cells
1	14045	8.37
2	14268	8.51
3	13572	8.09
4	13590	8.10
5	14337	8.55
6	13907	8.29
7	13991	8.34
8	13584	8.10
9	14395	8.58
10	13839	8.25
11	14287	8.52
12	13914	8.30

The total cells (1.67X10^5) is divided into 12 parts. The deviation from the equal number of cells and number of interface cells between the partitions affect the CPU time and hence the speed up.

The following Figure 6 shows the test case A1 on M1 and M3 machines (1000 number of iterations).

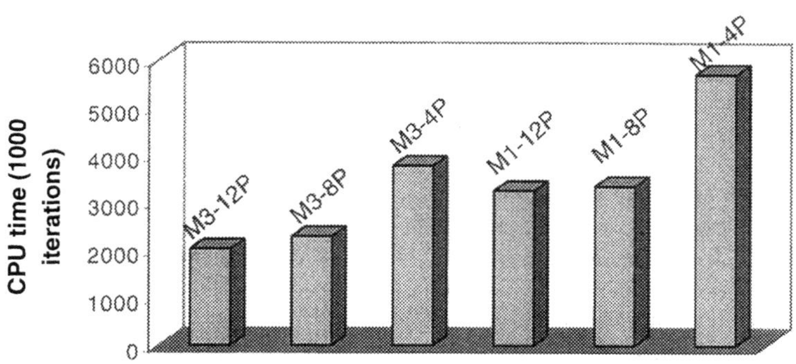

Figure 6: Comparison of CPU time (Sec) between M1 and M3

Test M3-12P refers to 12 processors used on M3 and so on. M3 machine is 1.6 times faster than M1 for this type of CFD simulation.

It should be noted that in M1, the machine operates as a node system and each node contains 4 processors. The maximum speed up from Machine M1 is possible when 1, 2, or 3 nodes have been used.

5. Conclusions

The efficiency of the IBM, SGI and Compaq machines available in some of the biggest computer labs in Australia with respect to the parallelisation of some industrial fluid flow phenomena has been investigated for certain number of iterations. The outcome should help any new practical engineering applications to get a quicker and accurate solution as shown here. The efficiency of parallel CFD for a commercial code, e.g., FLUENT depends significantly on the partitioning method as shown in this paper. For bigger problems, parallel CFD is more efficient. To find a better domain decomposition and number of divisions still require some trial and error; however even the automatic domain decomposition is quite good for instant calculation as shown.

Acknowledgements

This research work has been supported by AC3 (www.ac3.edu.au) and APAC (http://nf.apac.edu.au/). However, the content does not necessarily reflect the overall performance of all software on the parallel machines considered.

References

1. M Behr, D. M. Pressel and Sr. W.B. Sturek, Comput. Methods Appl. Mech. Engrg., Comments on CFD code performance on scalable architectures, 190 (2000), 263-277.
2. G. Tryggvason and B. Bunner, Parallel CFD: Trends and Applications, Direct Numerical Simulations of Multiphase Flows, (2000) 77-84.
3. T. Watanabe and K. Ebihara, Parallel CFD: Trends and Applications, Parallel computation of rising bubbles using lattice Boltzmann method on workstation cluster, (2000), 399-406.
4. H. P. Langtangen and Xing Cai, Parallel CFD: Trends and Applications, A Software Framework for Easy and Parallelization of PDE Solvers, (2000), 43-52.
5. R. Winkelmann, J. Häuser, R. D. Williams, Comput. Methods Appl. Mech. Engrg., Strategies for parallel and numerical scalability of CFD codes, 174 (1999), 433-456.
6. FLUENT 5, User's Guide, Fluent Incorporated, NH, USA, (1998).
7. G. Karypis and V. Kumar, METIS Version 3.0, Manual, University of Minnesota and Army HPC Research Center, (1997).

Parallel Computational Fluid Dynamics – Practice and Theory
P. Wilders, A. Ecer, J. Periaux, N. Satofuka and P. Fox (Editors)

Zooming in on 3D magnetized plasmas with grid-adaptive simulations

R. Keppens[a*], M. Nool[b] and J.P. Goedbloed[a]

[a]FOM-Institute for Plasma Physics 'Rijnhuizen',
P.O. Box 1207, 3430 BE Nieuwegein, The Netherlands
keppens@rijnh.nl – goedbloed@rijnh.nl

[b]CWI,
P.O. Box 94079, 1090 GB Amsterdam, The Netherlands
Margreet.Nool@cwi.nl

We present multidimensional hydro- and magnetohydrodynamic simulations, where the fine-scale dynamics is accurately and efficiently captured through a solution-adaptive meshing procedure. The Adaptive Mesh Refinement (AMR) strategy is implemented for any dimensionality and can be used generally for sets of conservation laws, combined with a suitable conservative high resolution discretization. Specifically, the 2D and 3D MHD scenarios discussed here make use of a level dependent spatial discretization: a fully upwind scheme on the finest grid level is combined with a Total Variation Diminishing Lax-Friedrich method on lower levels. The AMR process ensures that this combination acts as a low-cost hybrid scheme, accurately capturing all sharp flow features. Examples are given of planar evolutions and of 3D Rayleigh-Taylor unstable, magnetized plasma dynamics. Auto-parallelization on multi-processor SGI Origin architectures shows linear to superlinear speedup for non-adaptive 3D MHD calculations, but is as yet unsuccesful in parallelizing multi-level AMR calculations.

1. ALGORITHM AND IMPLEMENTATION ASPECTS

As a promising extension to the Versatile Advection Code initiated by Tóth [9] (VAC, see `http://www.phys.uu.nl/~toth/`), a fully automated Adaptive Mesh Refinement (AMR) [1] scheme is now incorporated and used for multidimensional hydro- and magnetohydrodynamic (MHD) studies. In essence, AMR generates or destructs – both controlled by the ensueing dynamics – hierarchically nested grid levels with subsequently finer mesh spacings. This AMR algorithm and, particularly, all complications associated with ensuring the global conservation property needed for shock-capturing calculations, have been addressed in the work by Berger and Colella [2]. In that pioneering paper, the authors stress that *"... it seems possible to implement a general code where the number of*

*Work done within the association agreement of Euratom and the 'Stichting voor Fundamenteel Onderzoek der Materie' (FOM) with financial support from the 'Nederlandse Organisatie voor Wetenschappelijk Onderzoek' (NWO) and Euratom. It is part of a project on 'Parallel Computational Magneto-Fluid Dynamics', an NWO Priority Program on Massively Parallel Computing. Use of computing facilities is funded by 'Nationale Computer Faciliteiten' (NCF).

dimensions is input". We make use of such a general code – AMRVAC – in simulations presented here, where both the dimensionality and the system of conservation laws is selected in a pre-processing stage.

Noteworthy is the fact that we allow for a level-dependent choice of the discretization scheme employed. We advocate the use of an approximate Riemann solver based method at the highest allowed level(s), in combination with the robust but more diffusive Total Variation Diminishing Lax-Friedrich (TVDLF, see e.g. [10]) method on the coarser levels. This is computationally efficient, and acts as a hybrid scheme since all sharp features are fully captured with the more accurate scheme.

For the MHD studies, we use the eight-wave formulation from Powell, see e.g. [7], to maintain the $\nabla \cdot \mathbf{B} = 0$ constraint to truncation error. The approach modifies the approximate Riemann solver to 'propagate' $\nabla \cdot \mathbf{B}$ errors with the plasma velocity and adds non-conservative source terms to the MHD equations proportional to $\nabla \cdot \mathbf{B}$. In the AMR simulations, adding these corrective source terms in a split fashion inactivates them at regridding operations.

The dimension-independence of the different numerical schemes, equation modules, and of the entire AMR algorithm is realized by the use of the Loop Annotation SYntax or LASY [8]. The code is configured to dimensionality using a single `Perl`-script, prior to compilation. We preferentially run AMRVAC on workstations or shared memory platforms, since the employed data structure still uses statically allocated linear arrays. Improvements to port the code to parallel architectures, as was done succesfully for VAC [6,5] earlier, are under consideration. First attempts to rely on auto-parallelization options, as available on multi-processor SGI Origin architectures, are reported for time-dependent 3D MHD simulations, for both non-adaptive VAC and grid-adaptive AMRVAC simulations.

2. AUTO-PARALLELIZATION RESULTS

The performance and parallel scaling achieved with the Versatile Advection Code [6] on a fully 3D MHD calculation of a Kelvin-Helmholtz unstable magnetized jet, with resolution $50 \times 100 \times 100$ is briefly summarized as follows

- using Fortran 90 on a single processor of a Cray C90, this calculation reached a sustained performance of 388.4 Mflops (40 % peak), executing one time step in 8.4 seconds. With autotasking, a speedup of 3.7 was reached on 4 processors.
- on distributed memory platforms (Cray T3E, IBM SP), the use of High Performance Fortran yielded nearly linear speedup.

In the course of 2001, an SGI Origin 3800 with 1024 processors, **TERAS**, is being installed at the Dutch supercomputing centre SARA in Amsterdam. Ultimately, 512 processors will be accessible in a virtually shared manner using OpenMP. The RISC processors are 500 Mhz, IP35 MIPS R14000, with a peak performance of 1 Gflops. At a minimal effort, automatic generation of parallel code is possible using the compiler option `f90 -apo`, which is subsequently executed on N_p processors by setting the environment variable `OMP_NUM_THREADS`.

Our first attempt to use this auto-parallelization option on the same 3D MHD calculation of a magnetized jet evolution is shown in Fig. 1: the speedup is superlinear up to

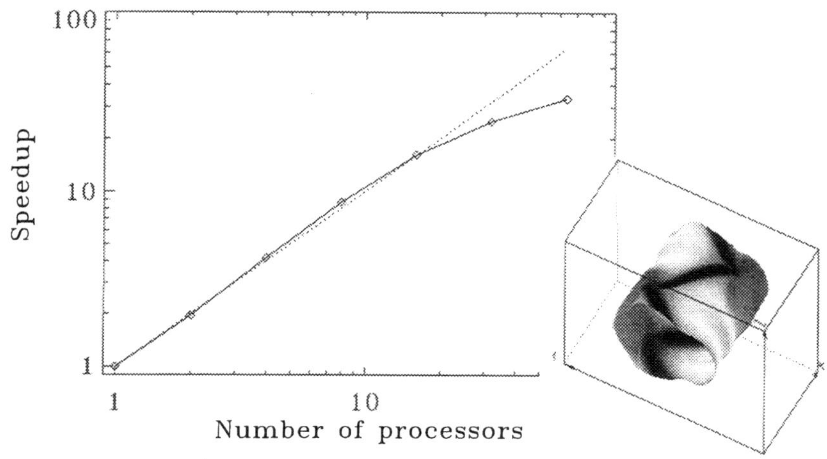

Figure 1. Scaling of VAC up to 64 processors on a SGI Origin 3800 using auto-parallelization (`OpenMP`). A snapshot of the 3D magnetized jet simulation used in the timing experiments is shown as an inset: the jet surface colored by thermal pressure is deformed due to the Kelvin-Helmholtz instability.

16 processors, and then levels off to a value of 33.4 for 64 processors. We note that the memory requirements for this testcase are only 327 Mb. While the speedup results are extremely encouraging, the single processor execution time per time step is 39.8 seconds, a factor of 4.7 longer than on the vector Cray C90 machine (which had the same peak performance per processor). The performance analysis tool `perfex` reports a rather low 53 Mflops on a single CPU of the SGI Origin 3800, which presumably underestimates the actual number of floating-point operations per second[2]. However, judging from the observed differences (Cray versus SGI) in execution time per time step, this particular simulation is currently only achieving up to 10 % of the peak performance on TERAS.

We subsequently used AMRVAC as a Domain Decompositioner (DD) to analyse the obtainable speedup through auto-parallelization on the SGI Origin 3800. To use AMRVAC as a DD, we simply set the maximum number of levels to one, and specify the largest block size at (pre)compile time. The AMR implementation is then only used to decompose the full computational domain in blocks at time $t = 0$, and to ensure the correct filling of internal ghost cell boundaries created by the DD. We then run the same 3D MHD simulation as used for VAC, except that the total resolution is set to $52 \times 104 \times 104$. Running the N-block simulation on N processors yields the speedup curve shown in Fig. 2. Again a superlinear behavior is found up to 8 processors. The speedup reaches 13.8 on 16 processors, but then degrades when using 32 processors. Again the obtained single processor performance is estimated to be at most 10 % peak. While this can certainly be improved, the automated parallelization results are very promising, at least up to several tens of processors.

[2]A two floating-point multiply-add operation counts as one instruction for `perfex`.

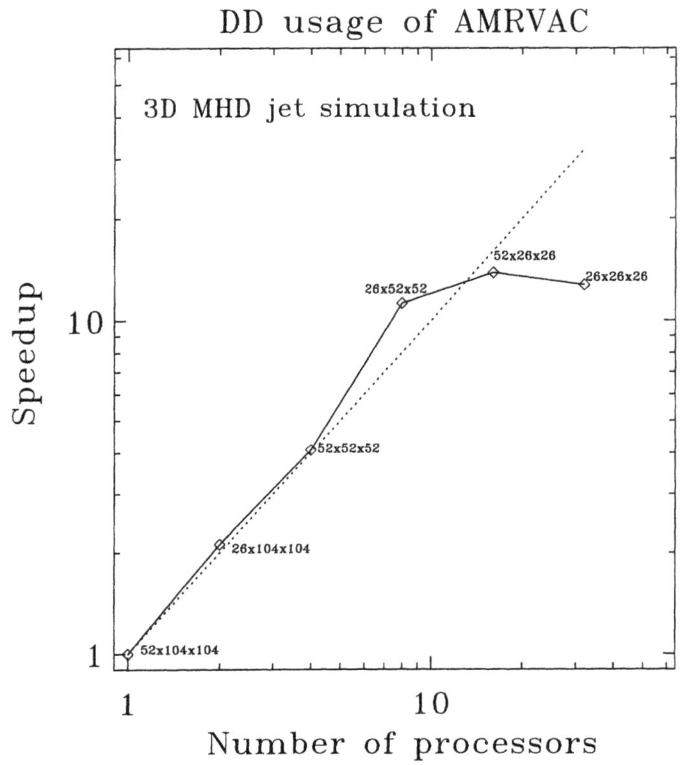

Figure 2. Scaling of AMRVAC up to 32 processors on the SGI Origin 3800 using auto-parallelization. The same 3D MHD jet simulation as for Fig. 1 is domain decomposed with AMRVAC. The block size is indicated: the N block case is run on N processors.

3. PLANAR EVOLUTIONS

To demonstrate the AMRVAC code potentials for grid-adaptive calculations, we start with two examples of planar (2D) evolutions. By selecting the Euler system, we can simulate the interaction of a Mach 2.5 shock with a low density 'bubble'. The pre-shock region has a unit sound speed c_s, except in a circular low density region with $c_s = 5$. This problem is inspired by a similar shock-bubble interaction as found on the website for the AMRCLAW software package at `http://www.amath.washington.edu/~claw/` by LeVeque and coworkers. In the example calculation on that website, the up-down symmetry with respect to the normal from the bubble centre to the shock front was exploited. Here, we let the shock front make an angle of 60° with the x-axis, leaving the symmetry as a check on the calculation. On a 80×40 base grid on the domain $[0,2] \times [0,1]$, we allow for 3 levels. Subsequent refinement ratios are set to 2 and 4, so that the AMR simulation would yield an effective resolution of 640×320. Snapshots of the density (schlieren-plot) and grid structure at time $t = 0$ and at time $t = 0.3$ are shown in

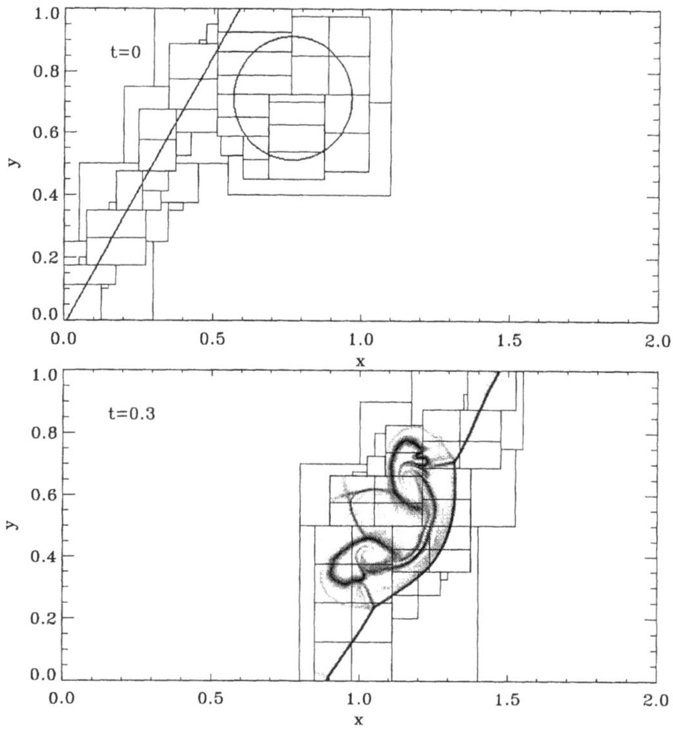

Figure 3. Two-dimensional shock-bubble interaction, allowing for three refinement levels in AMRVAC. The AMR process generates the grids at $t = 0$ at the discontinuities (the shock and the bubble boundary). The density is shown as a schlieren plot.

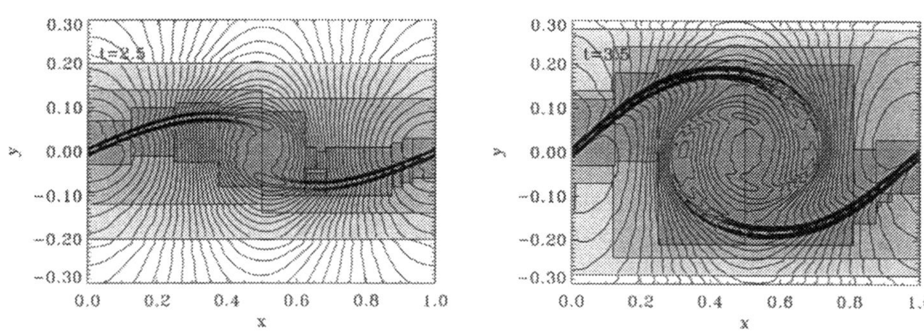

Figure 4. Density evolution in an unstable magnetized shear flow. We show the central region of the computational domain $y \in [-1, 1]$ only. The 4 grid levels are indicated.

Figure 3. We used the TVDLF scheme on levels 1 and 2 and an upwind scheme on the finest level.

A demonstrative MHD problem is a 2D simulation of a magnetized shear flow, in a Kelvin-Helmholtz unstable, subsonic (Mach 0.5) parameter regime. The detailed problem description is borrowed from Keppens *et al.* [4], where high resolution simulations on static grids with VAC are reported. A small perturbation of the system triggers a vortical flow. The magnetic field becomes amplified in a spiral pattern, in turn leading to narrow lanes of low density. Using 4 grid levels on a $[0,1] \times [-1,1]$ domain and a base resolution of 50×100, we recover the high resolution results. Figure 4 shows the density and grid structure at 2.5 and 3.5 sound crossing times. The dynamic regridding process nicely traces the fine-scale density structures. Again, the approximate Riemann-solver is only operative at the highest level.

In summary, the grid-adaptive strategy now allows to efficiently simulate multidimensional HD and MHD scenarios without sacrificing in accuracy. In a recent paper [3], we quantify the obtainable efficiency in computing time for a variety of 1D, 2D, and 3D HD and MHD problems. Indeed, for 2D HD and MHD problems like those given in this section, AMR execution times can be a factor of 10 to 20 times shorter than the corresponding high resolution static grid calculation.

Auto-parallelization experiments on SGI platforms with multi-level grid-adaptive 2D hydrodynamic shock problems were not succesful (no speedup). This is influenced by the involved spatio-temporal intra-level interpolations needed to fill ghost cells for individual grid patches, but is primarily due to the fact that our AMR implementation creates optimally fitted, but very different sized higher level grids. This would suggest algorithmic changes to (1) enforce the same time step on all levels; and (2) change the AMR algorithm to create fixed grid block sizes.

4. 3D SIMULATIONS

We simulated a Rayleigh-Taylor unstable 3D configuration where a heavy plasma rests on top of a lighter plasma in an external gravitational field. With gravity in the $-y$-direction of the unit cube, we run two cases: (1) a pure hydrodynamic scenario where the interface separating the two uniform density regions at time $t = 0$ is given by $y_{\text{int}} = 0.8 + 0.05 \sin 6\pi x \sin 4\pi z$; and (2) a full MHD problem with a uniform horizontal magnetic field $\mathbf{B} = 0.1\hat{e}_x$ and $y_{\text{int}} = 0.8 + 0.05 \sin 2\pi x \sin 2\pi z$. The density contrast is 10. We start with a very coarse base grid of size $20 \times 40 \times 20$, and allow for three refinement levels, achieving a $80 \times 160 \times 80$ resolution locally.

Snapshots of the density structure for the hydro simulation at time $t = 1.2$ in Fig. 5 show that the initial perturbation leads to the formation of 12 'fingers' (spikes) of falling high density plasma. The resulting flow field around the pillars induces roll-up and a clear pairwise interaction along the x-direction causes a distinct asymmetry in the mixing process. Fig. 6 depicts the density structure at $t = 1$ for the MHD case in two vertical cutting planes. In the plane parallel to the initial uniform magnetic field, the formation of fine scale structure is effectively suppressed by the stabilizing magnetic tension. In the plane orthogonal to the field, shorter wavelength features are clearly visible.

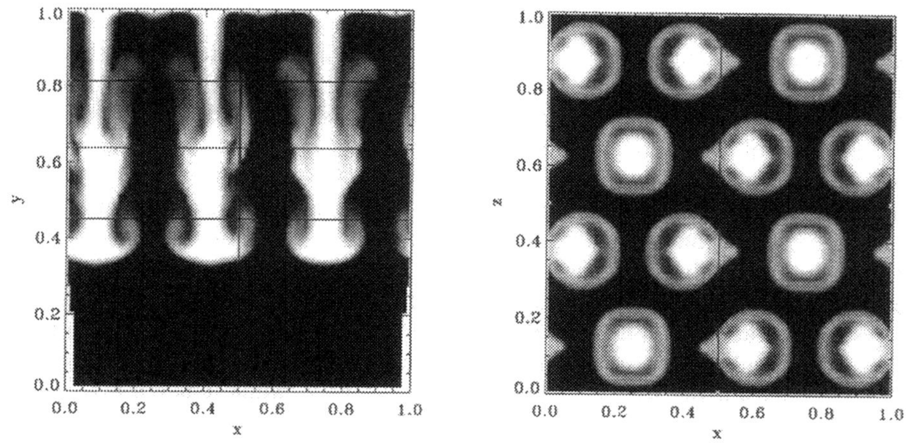

Figure 5. Density structure at $t = 1.2$ in the planes $z = 0.4$ and $y = 0.4$ for a 3D HD simulation of a Rayleigh-Taylor instability. Under the influence of gravity, the heavy plasma mixes into the lighter one underneath. We use AMR with 3 grid levels.

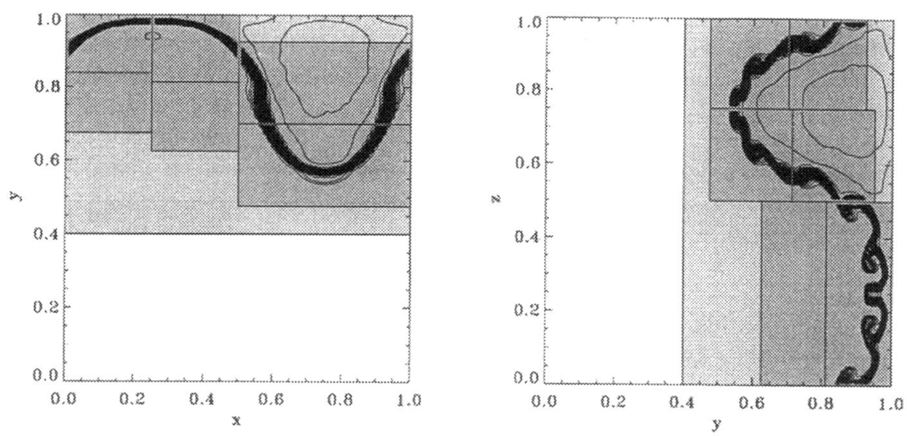

Figure 6. Density structure at $t = 1$ in the planes $z = 0.2$ and $x = 0.3$ for a 3D MHD simulation of a Rayleigh-Taylor instability (gravity along $-y$).

For these 3D Rayleigh-Taylor simulations, the AMR efficiency – defined as the ratio of the time needed to run the corresponding high resolution static grid case to the execution time for the AMR case – remains limited: we only gain a factor of 2.8 for the MHD case. This is because the level one grid resolution is so coarse that level two grids are covering 30 % of the entire computational domain at $t = 0$ (and 16.25 % for level $l = 3$ grids) and this increases as time progresses to reach 60 % (43 % for $l = 3$) at time $t = 1$. Taking that into account, the obtained speedup is indeed optimal, which is also confirmed by the observation that only 1.35 % of the entire CPU time is devoted to AMR-specific calculations (regridding, updating including flux fixings). Much better efficiencies are reachable as soon as the base level grid is not underresolving the physical process.

This latter statement is demonstrated clearly for a 3D advection problem in the table below. We simulated the advection of a sphere of radius 0.2 across the diagonal of a unit cube. Within the sphere, the density was set to 2, while it is 0.5 exterior to the sphere. The advection velocity $\mathbf{v} = (1, 1, 1)$ brings the sphere at $t = 1$ to its original centered position using triple periodic boundary conditions. We compare execution time and memory requirements for a 320^3 static grid simulation on the Origin 3800, with an AMR run exploiting 3 levels, with a 40^3 base grid, and consecutive refinement ratios 2 and 4. Since the coverage of the highest grid level now remains roughly at 10 % throughout the simulation, a significant reduction by a factor of 19.9 in computing time is realized. Note also that the time spent on the regridding process is fully negligible. As this tenfold efficiency is already reached for a 3 level calculation in a pure advection problem, much higher efficiencies will hold in practical 3D MHD simulations. The table also indicates the much reduced memory requirements for the AMR simulations. Our current implementation uses up to 3 solution vectors per level and an order of magnitude in storage is gained. Indeed, the high resolution case could only be run exploiting 4 TERAS processors, while the AMR run fitted in the 1Gb available on a single CPU.

resolution	levels	timing (sec)	eff.	AMR%	solution storage
$320^3 \times [0,1]$	1	172197	-	-	$\mathcal{O}(10^8)$ DP numbers
	3	8653	19.9	4.14	1.68×10^7 DP numbers

At the time of this writing, more efforts are needed on (i) visualizing 3D grid-adaptive magnetized plasma dynamics; and (ii) tuning the software to parallel platforms.

REFERENCES

1. M.J. Berger, *SIAM J. Sci. Stat. Comput.* **7**, 904 (1986).
2. M.J. Berger and P. Colella, *J. Comput. Phys.* **82**, 64 (1989).
3. R. Keppens *et al.*, *J. Comput. Phys.*, submitted (2001).
4. R. Keppens *et al.*, *J. Plasma Phys.* **61**, 1 (1999).
5. R. Keppens, in *Proc. of ParCFD 2000* (Trondheim, Norway), Eds. Ecer *et al.* (2001).
6. R. Keppens and G. Tóth, *Parallel Computing* **26**, 705 (2000).
7. K.G. Powell *et al.*, *J. Comput. Phys.* **154**, 284 (1999).
8. G. Tóth, *J. Comput. Phys.* **138**, 981 (1997).
9. G. Tóth, *Astrophys. Lett. & Comm.* **34**, 245 (1996).
10. G. Tóth and D. Odstrčil, *J. Comput. Phys.* **128**, 82 (1996).

Parallel Computational Fluid Dynamics – Practice and Theory
P. Wilders, A. Ecer, J. Periaux, N. Satofuka and P. Fox (Editors)

Parallel calculations for transport equations in a fast neutron reactor

A.V.Kim*, S.N. Lebedev*, V.N. Pisarev*, E.M.Romanova*,V.V.Rykovanova*, O.V.Stryakhnina*

*Russian Federal Nuclear Center – VNIITF, NTO-2
Snezhinsk, Chelyabinsk region, P.O.Box 245, 456770 RUSSIA

A package for simulation of some severe accidents in reactor is presented. A concept of parallel calculations is considered. The basis of parallel algorithm is geometrical decomposition. The Message Passing Interface (MPI) standard has been used for organizing parallel calculations. The calculations have been performed on a multiprocessor computer system with distributed memory SMP Power Challenge L.

1. INTRODUCTION

The code is intended for mathematical simulation of dynamics of emergency processes in nuclear power installations on fast neutron reactors [1]. The simulated processes are following:
multi component one-speed medium movement, almost non compressible medium movement, elastic-plastic and solid material properties accounting, linear heat conductivity, neutron production, multiplication, absorption and transport in kinetic and diffusion approaches, nuclear burnup, energy release, delayed neutrons and Doppler effect accounting, imposed reactivity insertion (time change of some element isotope concentration in definite part of system).

The medium model takes into account the elasticity, plasticity, compressibility, destruction, fusion, evaporation, multi component structure. The substances can be in various states of aggregation from firm up to gaseous ones or can pass through various states at phenomenon development. The medium can be compositionally heterogeneous or it can become heterogeneous at phenomenon development. The firm fuel rods can contain pores filled by liquid coolant and vapour. The neutron fields and system dynamics interaction is accounted.

For neutron transport equation the diamond model of DSn-method was used. In order to increase calculation accuracy of optically dense materials the dissipation into DSn-method was introduced. The scheme of neutron transport equation in self-conjugate form is considered. The stationary neutron transport equation is solved. The difference analogues of kinetic and diffusion equations for diffusion-synthetic acceleration of iterations convergence are agreed.

The following modules are included into package: the non-stationary hydrodynamics calculation modules, the non-stationary dynamics calculation module accounting elastic-plastic and strength medium properties, the heat conduction equation module, the neutron transport equation module including delayed neutrons accounting and module for stationary transport equation, the modules of neutron-nuclear kinetic equations.

2. CONCEPT OF PARALLEL CODE

The basis of parallel algorithm is geometrical decomposition. It permits to reduce the solution of an initial problem to the serial or parallel calculation of set of more simple ones. The problem initial geometry is splitted into set of subdomains determining the geometry of subproblems, each calculated independently within one time step. The subproblem solves the set of physical processes in giving domain using the suitable for domain difference (space and time) grids. The solution coordination on separation boundaries is executed by interchange of internal boundary conditions describing the system state along the adjacent bounds.

Our code realizes the following calculation scheme [2]. The dynamics, heat conduction, kinetics and neutron transport equations are consistently solved. Each stage is calculated in parallel regime with decomposition of calculation on domains. The calculation coordination on adjacent domain bounds, management and synchronization are proceeded in control process. Messages queues are used for control information transfer and signal synchronizing.

The calculating processes are distributed by scheduler on nodes of computer network according to the amount of calculations. The calculation processes are implemented as modules without the input-output procedures. All interactions with file system are concentrated in one process named Data Manager. In calculation and control processes the input-output is executed through data exchange with Data Manager.

3. ORGANIZATION OF PARALLEL CALCULATIONS

Calculation Manager provides tactics and strategy calculation. It also loads modules, distributes them over network nodes.

For one time step Calculation Manager (CM) executes the following operations:

1. Starting of calculation.

CM receives the message about availability for calculation. Each message contains the information on number of process and domain.

For the pointed domain, the input boundary conditions are calculated and sent to the appropriate process.

2. The process termination in all domains is a synchpoint.

Here the following processes are carried out.

Receiving the message about domain calculation (successful or exception).

Receiving the output data on domain from the appropriate process (output boundary arrays, balance values, etc.).

Processing of data received from domain.

The exchange between Calculation Manager and domain processes, boundary conditions and data describing domain state along bounds is carried out through input and output boundary arrays.

For effective data management we developed the special process named Data Manager (DM).

The localization in DM process of all accesses to problem database excludes deadlock situations caused by computing system external resources use. It allows to consider all distributed nodes as diskless computers and to realize all input-output functions only through Data Manager.

4. TESTS OF THE PARALLEL CODE IN LOCAL COMPUTER NETWORK

Tests of parallel code show that it is possible to receive speed-up and efficiency factors rather close to maximal ones. However on real problems it is impossible to reach the theoretical speed-up factor. The problem fragmentation depends on problem formulation, and the iterative processes are different in each fragment.

Table 1 shows the operating times of separate processes for calculation of 5-fragment problem for different number of solved equations: only heat conduction and total set of equations. The results were obtained for nodes of SMP Power Challenge L.

Table1

	Heat conduction	Calculation of dynamics with account of a heat conduction and neutron processes
1	7.96	43.35
2	10.92	59.66
3	7.96	43.59
4	3.11	13.17
5	5.69	24.85
CM	0.01	0.04
T_{CPU}	35.64	184.66
T_{AST}	38	206

Table 2 presents speed-up factors for this problem on different number of network nodes.

Table2

	Heat conduction			Total set of equations		
	1 node	3 nodes	5 nodes	1 node	3 nodes	5 nodes
T_{ASTR}	38	20	14	206	104	82
$K_{SPEED\ (THEOR)}$	-	2.1	3.27	-	2.3	3.1
$K_{SPEED(REAL)}$	-	1.9	2.71	-	1.98	2.51

$K_{SPEED\ (THEOR)}$ - the maximum speed-up with the given nodes number.
It was received as a ratio of common CPU time to CPU time of the most loaded node.
K_{SPEED} - really speed-up factor.

It is clear that geometrical decomposition is effective only in case when the amount of calculations is distributed uniformly among the processors.

Test for stationary transport equation.
Consider homogeneous sphere (R=2, α=1, β=1, v=1).
In one-speed approache eigenvalue of transport equation was calculated.
This problem was taken from the report of Goldin and Yudintsev [3].
Table 3 presents times (t), speed-up coefficient (k), dependent from number of processes.

Table 3

N proc	T iter s	K_{SPEED} iter	N Iter	T total	K_{SPEED} real
1	4.420	1	14	61.88	1
8 (1*8)	0.585	7.56	14	8.19	7.56
16 (2*8)	0.297	14.89	16	4.75	13.02
32 (4*8)	0.139	31.78	22	3.06	20.22

The first column represents the number of processes. The second column represents calculation time for one iteration. The third column represents the speed-up coefficient for one iteration. The fourth column represents the number of iterations. The fifth and the sixth columns show the real results of calculation. Iteration speed-up coefficients differ from the real ones, because the number of iterations increases depending on number of processors. Note, that for 32 processors the operating system puts data into high-speed memory.

CONCLUSIONS

We developed a new code for simulation of some severe accidents in reactor on liquid metal coolant. The code simulates the phenomena of reactor core destruction, etc. At the same time, the model contains significant number of parameters, connected with physical properties of core materials, to be refined at verification on available experimental data.

We developed the concept of the parallel code on computer network.

We carried out some severe accident simulations using parallel code.

We assume to finish the code for simulation of a wide spectrum of severe accidents, in particular: failures with steam line destruction and failures, connected with uncontrol reactivity in core, which can be accompanied by destruction of core, vessel, the in-core equipment; core melt and its dynamics at accident development.

We investigated the various parallel methods, but not all were brought to practical realization. In particular, the pipeline methods or methods with cluster memory use were not realized. The further researches can give a new theoretical material and base for new parallel algorithms.

REFERENCES

1. Gadzhiev A.D., Gadzhieva V.V., Lebedev S.N.etc. The SINARA software package for mathematical simulation of dynamics of emergency processes in nuclear power installations on fast neutron reactor. J. VANT,. Mathematical simulation of physical processes, 3 (2000) 25.

2. Anikin A.M., Bysjarin A.Ju., Gorbatova I.A., Gribov V.M., Kim A.V.. Experience of creation of the parallel code for the solving of problems of mathematical physics with distributed computing environment. J. VANT,. Mathematical simulation of physical processes, 4 (1996) 8.

3. Goldin V.Y., Yudintsev V.F. Calculation of quasiregular and regular solution for transport equation. J. VANT,. Mathematical simulation of physical processes, 2 (1985) 43.

Parallel Computational Fluid Dynamics – Practice and Theory
P. Wilders, A. Ecer, J. Periaux, N. Satofuka and P. Fox (Editors)

Parallel Large Scale Computations for Aerodynamic Aircraft Design with the German CFD System MEGAFLOW

N. Kroll[a], Th. Gerhold[b], S. Melber[a], R. Heinrich[a], Th. Schwarz[a], B. Schöning[a]

[a]German Aerospace Center, Institute of Aerodynamics and Flow Technology
Lilienthalplatz 7, D-38108 Braunschweig, Germany

[b] German Aerospace Center, Institute of Aerodynamics and Flow Technology
Bunsenstraße 10, D-37073 Göttingen, Germany

Within the framework of the German aerospace program, the national CFD project MEGAFLOW was initiated, which combines many of the CFD development activities from DLR, universities and aircraft industry. Its goal is the development and validation of a dependable and efficient numerical tool for the aerodynamic simulation of complete aircraft. The MEGAFLOW software system includes the block-structured Navier-Stokes code FLOWer and the unstructured Navier-Stokes code TAU. Both codes have reached a high level of maturity and they are being intensively used by the German aerospace industry in the design process of a new aircraft. This paper focuses on the aspects of parallel computing, which is one of the major issues for efficient computations in the industrial framework. The parallelization principle of both the block-structured and unstructured flow solver are outlined. Typical results for industrial applications are given with respect to efficiency and speed-up. Several large scale computations demonstrate the overall efficiency and quality of the MEGAFLOW software.

1 INTRODUCTION

During the last decade, considerable progress has been made in the development and validation of numerical simulation tools for aerodynamic applications. However, even despite recent advances, CFD still suffers from deficiencies in accuracy, robustness and efficiency for complex applications, such as complete aircraft flow predictions. From the aircraft industry's point of view, numerical simulation tools are expected to deliver detailed viscous flow analysis for complete configurations at realistic Reynolds numbers, prediction of aerodynamic data with assured high accuracy and known error bands, fast response time per flow case at acceptable total costs as well as aerodynamic optimization of the main aircraft components [1]. In order to meet these requirements in earlier and earlier stages of the design process, considerable improvements of the CFD methods currently available in industry are necessary.

Within the framework of the German aerospace research program, the national CFD project MEGAFLOW was initiated under the leadership of DLR with the objective of enhancing the capabilities of current CFD methods and supporting the establishment of numerical simulation as an effective tool in the industrial design process [2]. The goal of the project is to produce a dependable, efficient and quality controlled program system for the aerodynamic simulation of complete transport aircraft in cruise as well as take-off and landing configuration.

Viscous flow simulations of high geometric and physical complexity require the discretization of the governing flow equations on very fine grids consisting of several millions of grid points. In order to efficiently solve the corresponding discrete system of equations in an industrial design process at acceptable response time, fast numerical algorithms on cost-efficient high-performance computer hardware are required. Parallel computers are able to provide hardware scalability and enable large scale applications, such as high lift flow around a complete aircraft. Thus, in the MEGAFLOW project special focus is laid upon high efficiency and generality of the parallel flow solvers [3].

2 MEGAFLOW SOFTWARE

The basic components of the MEGAFLOW software are the block-structured flow solver FLOWer and the unstructured hybrid flow solver TAU [2]. Both codes solve the compressible, three-dimensional Reynolds-averaged full Navier-Stokes equations for rigid bodies in arbitrary motion. The motion is taken into account by transformation of the governing equations. For the simulation of aeroelastic phenomena the codes have been extended to allow geometry deformation. Turbulent flow is modeled using different variants either of the k-ω or the Spalart-Allmaras turbulence model. In the following sections the specific features of the Navier-Stokes codes are briefly described.

3.1 Block-structured Navier-Stokes Code FLOWer

The FLOWer code [4] is based on a cell-vertex finite-volume formulation on block-structured meshes. The baseline method employs either central space discretization combined with artificial viscosity or upwind discretization. Integration in time is done using explicit multistage time-stepping schemes. For steady calculations convergence is accelerated by implicit residual smoothing, local time stepping and multigrid. An implicit integration of the turbulence equations ensures efficient calculations on highly stretched cells as they appear in high Reynolds number flows. Preconditioning is used for low-speed flows. For time accurate calculations an implicit time integration according to the dual time stepping approach is employed. A specific feature of the FLOWer code is the Chimera technique, which considerably enhances the flexibility of the block-structured approach [5],[6]. This technique allows the separate generation of component grids that may overlap each other and which are embedded in a Cartesian background grid. This greatly simplifies the generation of structured grids around complex geometries. In combination with flexible meshes, the Chimera technique enables an efficient way to simulate bodies in relative motion. The communication from mesh to mesh is realized through interpolation in the overlapped area. In the case when a mesh overlaps a body which lies inside another mesh, hole cutting procedures have to be used in order to exclude the invalid points from computation. At the hole-boundaries, the flow quantities have to be provided by interpolation.

In close cooperation with the German aeronautical industry and the German National Research Center for Information Technology (GMD), the FLOWer code was extended to be a parallel, fully portable code [7]. The parallelization is based on grid partitioning and the message passing programming model. For messing passing, FLOWer uses the high-level communication library CLIC (Communication Library for Industrial Codes) which was jointly developed by the GMD and the C&C Research Laboratories of NEC [7]. CLIC performs and optimizes all data exchange between the allocated processes and processors guaranteeing a high degree of efficiency and flexibility through load balancing. Since the CLIC library supports the portable communication interface MPI (and since the CLIC library was also developed for sequential computer platforms), FLOWer can be run on any parallel and sequential platform. Recently, CLIC has been extended to support parallel Chimera applications. The library offers parallel search algorithms and automatic hole cutting procedures as well as semi-automatic generation of Cartesian background grids. These features greatly enhance the automation level of Chimera applications.

The code is highly optimized for computer platforms with a moderate number of processors with high performance (vector processor). Using the NEC SX5 computer, typically the FLOWer code achieves 2 GFLOPS on a single processor and its parallel efficiency is about 80%. Fig.1 shows results for a Navier-Stokes calculation for a wing/body configuration with 16 million grid points. Using 8 processors of the NEC SX5 computer, a performance of 11.5 GFLOPS could be achieved. A similar calculation with 6.5 million points demonstrated a performance of more than 7 GFLOPS on the massively parallel computer CRAY-T3E with 128 processors [3].

3.2 Hybrid-Navier-Stokes Code TAU

The Navier-Stokes code TAU [8] makes use of the advantages of unstructured grids. The mesh may consist of a combination of prismatic, pyramidal, tetrahedral and hexahedral cells and therefore combine the advantages of regular grids for the accurate resolution of viscous shear layers in the vicinity of walls and the flexibility of grid generation techniques based on unstructured meshes. The use of a dual mesh makes the solver independent of the type of cells that the initial grid is composed of. Various discretization schemes were implemented, including a central scheme with artificial dissipation and several upwind methods. In order to accelerate convergence, a multigrid procedure was developed based on the agglomeration of the control volumes of the dual grid for coarse grid computations. In order to efficiently resolve detailed flow features, a grid adaptation algorithm for hybrid meshes based on local grid refinement was implemented. With respect to unsteady calculations, the TAU code was extended to simulate a rigid body in arbitrary motion and to allow grid deformation. In order to bypass the severe time-step restriction associated with explicit schemes, the implicit method based on the dual time stepping approach was implemented. For the calculation of low-speed flows, preconditioning of the compressible flow equations similar to the FLOWer code was realized.

The parallelization of the solver is based on a domain decomposition of the computational grid [3]. During the integration process data has to be exchanged between the different domains several time. In order to enable the discretization operator at the domain boundaries, there is one layer of ghost nodes located at interface between two neighboring domains. The data exchange is realized through MPI. For domain decomposition either a simple self-coded algorithm or the more sophisticated public domain software package Metis can be used. In order to reach a high performance on each single subdomain, the code is further optimized either for cache or vector processors through specific edge colouring. Figure 2 shows the performance of the TAU code on the NEC SX5 computer for the viscous calculation around a wing/body/pylon/nacelle configuration with 2 million grid points. The TAU code achieves a parallel efficiency of about 95% with 1 GFLOPS for a single processor.

4 APPLICATIONS

The MEGAFLOW software is intensively used at DLR and German aircraft industry for many aerodynamic problems. Some typical large scale applications are listed below.

4.1 High lift flow

The prediction of a transport aircraft in high-lift configuration is still a challenging problem for CFD. The numerical simulation addresses both complex geometries and complex physical phenomena. The flow around a wing with deployed high-lift devices at high incidence is characterized by the existence of areas with separated flow and strong wake/boundary layer interaction. At DLR effort is spent to explore the capability of the MEGAFLOW system to predict 3D high-lift flows. Calculations for the DLR-ALVAST wing/fuselage combination have been carried out [9]. The deployed high-lift system in take-off configuration consists of nose slat as well as inboard and outboard flaps. Two geometries based on a different level of detail in the CAD-description have been treated (see Figure 3). The first geometry is simplified compared to the wind-tunnel model. In this geometry the gap between the slat and fuselage in the area of the wing junction was not modeled with all details like slat-horn and slat-stump at the leading edge. Moreover, the wing-root fairing on the upper-side of the wing/fuselage junction was likewise not modeled. All these details are captured in the second geometry description. For both geometries unstructured grids were generated using the commercial software package CENTAUR [10]. The grids contain quasi-structured prismatic cell layers around the geometry surface, in order to accurately resolve the viscous effects in the boundary layers. The outer domain of the flow field is covered with tetrahedral cells. For the simplified geometry grid adaptation was used resulting in a grid with about 5.8 million points (grid I). In order to resolve all features of the detailed geometry, a grid with more than 10 million points was used in the second case (grid II). In both grids the grid points in the prismatic layer near the body surface were adapted to ensure a y^+ value of one. In Figure 3 the lift coefficient versus the angle of attack is plotted and compared to experimental data for the simplified geometry. It can be seen that the lift coefficients computed with the hybrid TAU code are consistently overpredicted by about 5 % in the linear range. In the computation the lift breaks down at $\alpha=22^0$. The angle of attack for maximum lift agrees well with measured data. A closer inspection of the flow [9] shows a separation in the area of the wing root at $\alpha=21^0$ which extends streamwise over the whole wing chord. In addition, at $\alpha=22^0$ there is a weak separation on the outer end of the outboard flap which extends to approximately 30% of the local wing chord and which moves with an increasing angle of attack to the inboard-side of the wing. The calculation with the detailed geometry (grid II) at $\alpha=21^0$ reduces the overprediction of the lift coefficient to 1.7% compared to the experiment. In this case the wing/fuselage junction, which has a considerable influence on the stall behavior of the wing, is more accurately described in the numerical simulation. The parallel performance of the TAU code on a Hitachi SR8000 with up to 96 processors is shown in Figure 3.

These studies show that both a detailed description of the geometry and a high grid resolution are required to accurately predict flows around high-lift configurations at maximum lift. The unstructured grid approach and parallel computing are important ingredients for successful numerical simulations with acceptable turn around time.

4.2 Wake vortex encounter

To avoid risks for an aircraft flying in the wake vortex field of a preceding aircraft during take-off and landing, strict rules for minimum required separation distance have been established. These air traffic control separation standards are merely based on the maximum take-off weights of both aircraft. Due to the increasing number of congested airports and the development of very high capacity transport aircraft, in the past few years considerable research effort has been directed towards the wake vortex problem, aiming for a reduction of today's separation standards.

Within the European project WAVENC, investigations for wake vortex encounter have been carried out. A small generic aircraft model has been placed downstream into the vortical flow field of a preceding aircraft. The experimental data base contains forces, moments and surface pressure distributions at several model positions with respect to the vortices generated by the preceding aircraft. Inviscid computations with the Chimera option of the FLOWer code have been performed [11]. Component meshes have been generated separately around the fuselage, the wings and the horizontal and vertical tails of the generic aircraft model (see Figure 4). These component meshes are embedded in a Cartesian background grid, which is well suited for vortex dominated flows. Some details of the mesh are shown in Figure 4. In regions of high gradients additional meshes can be embedded into the background mesh in order to more accurately resolve physical details such as vortices, similar to adaptation techniques. The incoming vortices of the preceding aircraft have been prescribed at the inflow boundary of the computational domain. In order to avoid the diffusion of the vortices, special Cartesian meshes (vortex transport grids) have been incorporated into the Chimera grid structure to transport the incoming vortices to the following aircraft. The grid system consists of 5.5 million points. Computations have been carried out for five positions of the aircraft. For all positions the comparison of the predicted and measured rolling moment coefficient is quite good. Furthermore, the surface pressure distribution at the mid-span position of the port-side wing compares well with experimental data (see Fig. 4). These investigations show that the Chimera technique is well suited for the simulation of a wake vortex encounter situation. The main advantages of this approach are that a mutual interaction of vortices and airplane is captured in the simulation and that a parametric study of the position of the aircraft relative to the vortices is possible without generating new meshes.

The test case of vortex encounter has been used to investigate the newly developed parallel version of the FLOWer Chimera option described above. Figure 5a shows the computing time which is required to search for the donor cells in the overlapped regions of the Chimera grids. These donor cells are used to interpolate the flow variables between the meshes. It can be seen that the parallel performance of the search algorithm on a SGI cluster with four R1000 processors is quite good, whereas on the NEC SX5 computer the gain through parallelization is moderate. It should be noted, that these timings also include the computing time required for cutting the holes in those component grids which overlap the body. The comparison between the SGI and NEC demonstrate a good performance of the implemented search and hole-cutting algorithms on vector processors. Figure 5b shows the total computing time on the

NEC-SX5 of the Chimera calculation after 1000 iterations. The reason for the low parallel performance is that in the current Chimera implementation the flow variables of the whole overlapped regions are exchanged for interpolation between the grids. This results in high communication costs. Since only the boundary data of the overlapped regions are required for interpolation, appropriate data exchange should considerably reduce the communication costs and thus improve the parallel performance of the Chimera option. These modifications are currently implemented in the CLIC library of the FLOWer code.

4.3 Unsteady flows

The prediction of unsteady airloads on wings plays an important role in the aircraft design process. This requires numerical methods which simulate time-accurate flow fields at reasonable costs. In order to bypass the severe time-step restriction associated with explicit time integration schemes, a simple implicit method, known as the dual time stepping approach, was implemented in both MEGAFLOW codes. In combination with multigrid acceleration, this scheme allows efficient calculation of viscous time-accurate flows. With the use of parallel computers, unsteady calculations for three-dimensional flows become feasible.

Figure 7 shows results of an unsteady Navier-Stokes calculation of a delta wing oscillating in pitch at $M_\infty = 0.4$, $Re=3.1\times10^6$. The motion of the wing is described by a sinusoidal harmonic pitching $\alpha(t)=\alpha_0+\Delta\alpha\cdot\sin(\omega t)$ with $\alpha_0=9^0$, $\Delta\alpha=6^0$ and $\omega=0.56$. For the computation shown here the unstructured TAU code was used. Figure 6 shows details of the hybrid grid. After adaptation the grid contains about 1 million grid points. In Figure 7 the hysteresis loops of the lift and pitching moment coefficients are depicted. The results show a fairly good comparison to the experimental data, as well as computational results obtained from the block-structured code FLOWer. The numerical simulation is able to capture the main features of interest in this vortex-dominated flow field. The computation time for one period was 3 hours on a NEC SX5 with 8 processors. Two periods are required to obtain a periodic result in time.

Figure 8 shows results of an inviscid calculation for the pitching oscillation of a complete aircraft using FLOWer. The block-structured mesh consists of 16 blocks and 3.8 million grid points. The simulation for one oscillation cycle costs about 20h on a NEC SX5 (825 MFLOPS). This can be reduced to 7h by using 4 processors.

5 CONCLUSIONS

The main objective of the MEGAFLOW initiative is the development of a dependable, effective and quality controlled program system for the aerodynamic simulation of complete aircraft. Due to its high level of maturity, the MEGAFLOW software system is being used extensively throughout Germany for solving complex aerodynamic problems. Parallelization is one of the important ingredients which make Navier-Stokes simulations standard for industrial applications. However, since industry is still demanding more accurate and faster simulation tools, further development is aimed at improvement of physical modeling, further reduction of problem turn-around time for large scale computations due to advanced algorithms and strategies as well as efficient integration into an interdisciplinary simulation and design system.

6 ACKNOWLEDGEMENT

The authors would like to thank J. Raddatz, M. Widhalm, Th. Schwarz and Hubert Ritzdorf for performing some of the parallel computations.

7 REFERENCES

[1] Raj, P.P., (1998) *Aircraft Design in the 21st Century: Implications for Design Methods*, AIAA 98-2895, 1998.

[2] Kroll, N., Rossow, C. C., Becker, K., Thiele, F., *The MEGAFLOW Project*, Aerosp. Sci. Technol. Vol. 4, 223-237, 2000.

[3] Aumann, P., Barnewitz, H., Schwarten, H., Becker, K., Heinrich, R., Roll, B., Galle, M., Kroll, N., Gerhold, Th., Schwamborn, D., Franke, M., *MEGAFLOW: Parallel Complete Aircraft CFD.* Parallel Computing, Vol 27, 415-440, 2001.

[4] Kroll, N., Radespiel, R., Rossow, C.-C., *Accurate and Efficient Flow Solvers for 3D-Applications on Structured Meshes*, AGARD Report R-807, 4.1-4.59, 1995.

[5] Heinrich, R., Kalitzin, N., *Numerical Simulation of Three-Dimensional Flows Using the Chimera Technique*, Notes on Numerical Fluid Mechanics, Vol 72, Vieweg Braunschweig, 15-23, 1999.

[6] Schwarz, Th., *Development of a Wall Treatment for Navier-Stokes Computations Using the Overset Grid Technique*, 26th European Rotorcraft Forum, Paper 45, 2000.

[7] Schüller, A (Ed.), *Portable Parallelization of Industrial Aerodynamic Applications (POPINDA),* Notes on Numerical Fluid Dynamics, Vol 71, Vieweg Braunschweig, 1999.

[8] Gerhold T., Friedrich, O., Evans J., Galle, M., *Calculation of Complex Three-Dimensional Configurations Employing the DLR-TAU Code,* AIAA 97-0167, 1997.

[9] Melber, S., Rudnik, R., Ronzheimer, A., *3D RANS Structured and Unstructured Numerical Simulation in High-Lift Aerodynamics*, Proceedings of Workshop on EU-Research on Aerodynamic Engine/AirframeIntegration for Transport Aircraft, 13.1 - 13.10, 2000.

[10] Khawaja, A., Kallenderis, Y., *Hybrid Grid Generation for Turbomachinery and Aerospace Applications*, Intern. Journal for Numerical Methods in Engineering, No. 49, 145-166, 2000.

[11] Heinrich, R. *Numerical Simulation of Wake-Vortex Encounters using the Chimera-Technique.* Notes on Numerical Fluid Dynamics, Vieweg Verlag Braunschweig, to be published, 2001.

FIGURES

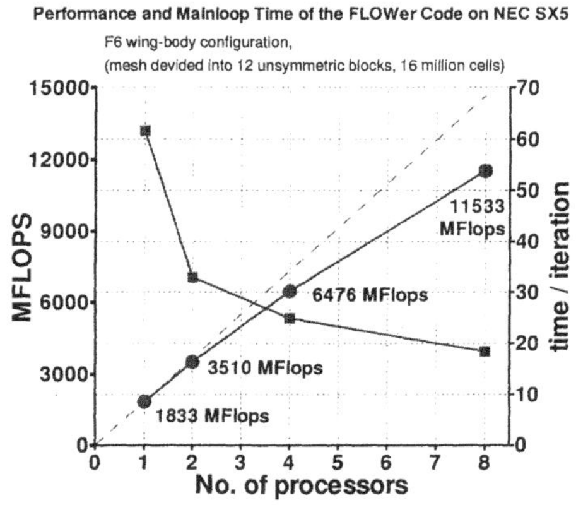

Figure 1: Performance of block-structured FLOWer code, viscous calculation for wing/body configuration.

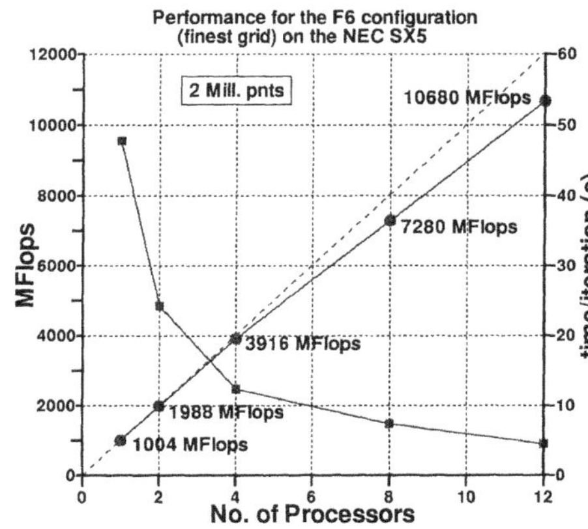

Figure 2: Performance of hybrid TAU code, viscous calculation for wing/body/nacelle configuration.

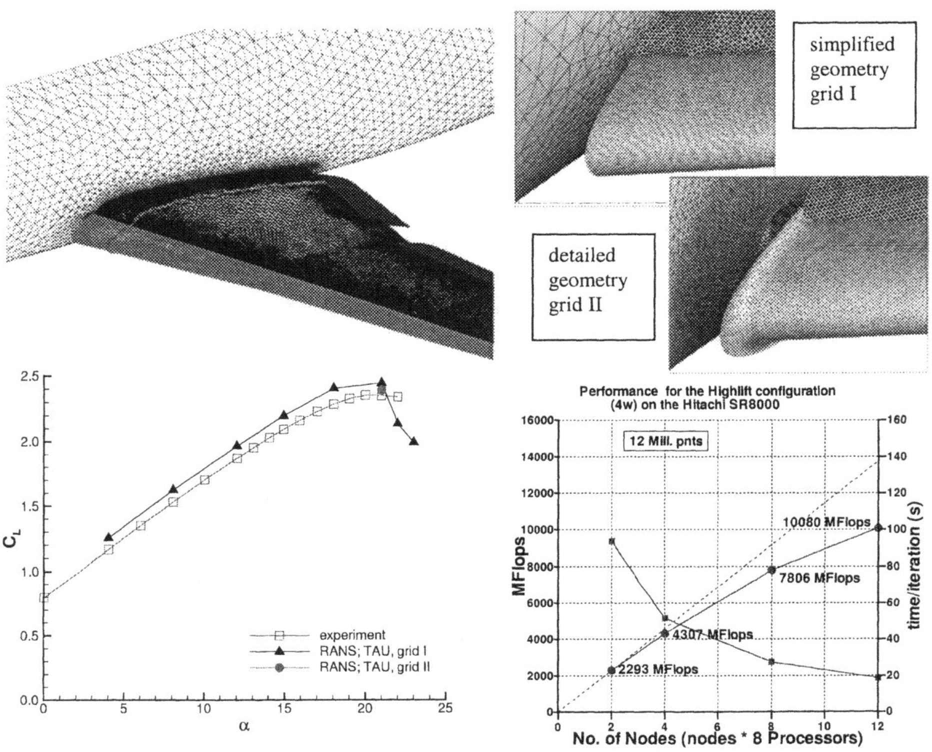

Figure 3: Viscous simulation of DLR-ALVAST high lift configuration with the TAU-Code, $M_\infty=0.22$, $Re=2x10^6$; hybrid surface grid, lift polar, performance of TAU code on Hitachi SR8000.

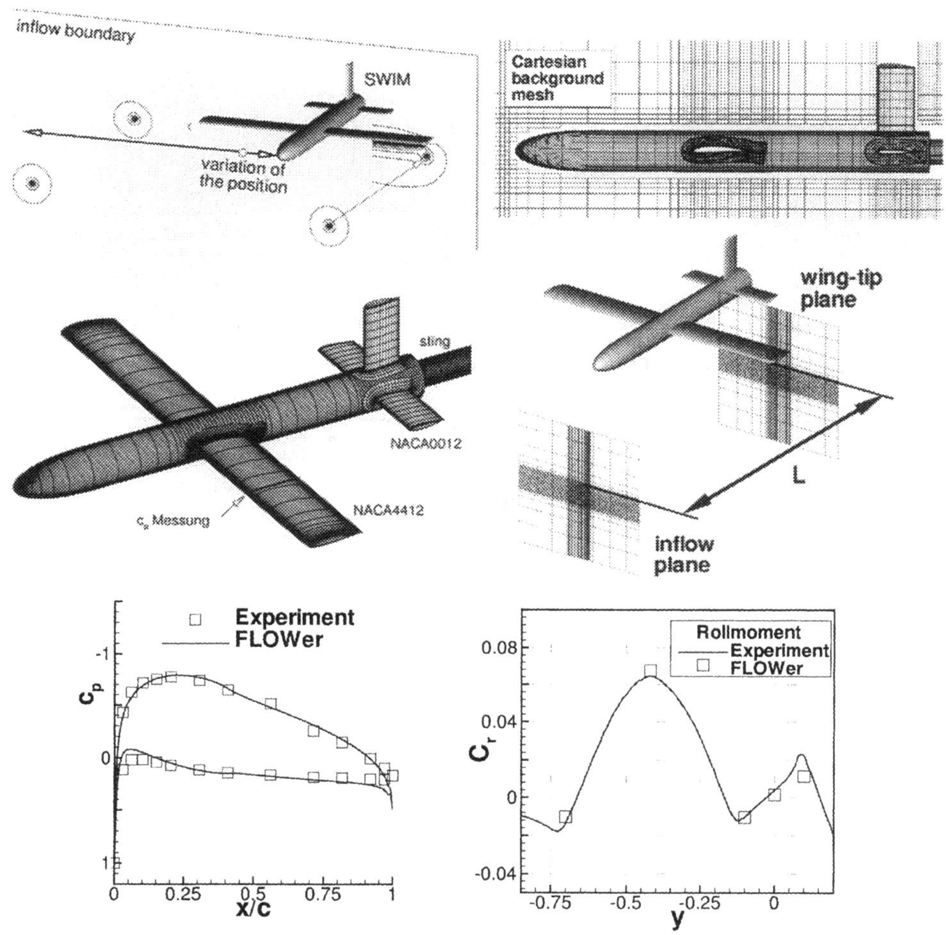

Figure 4: Simulation of vortex encounter using the Chimera option of the FLOWer code.

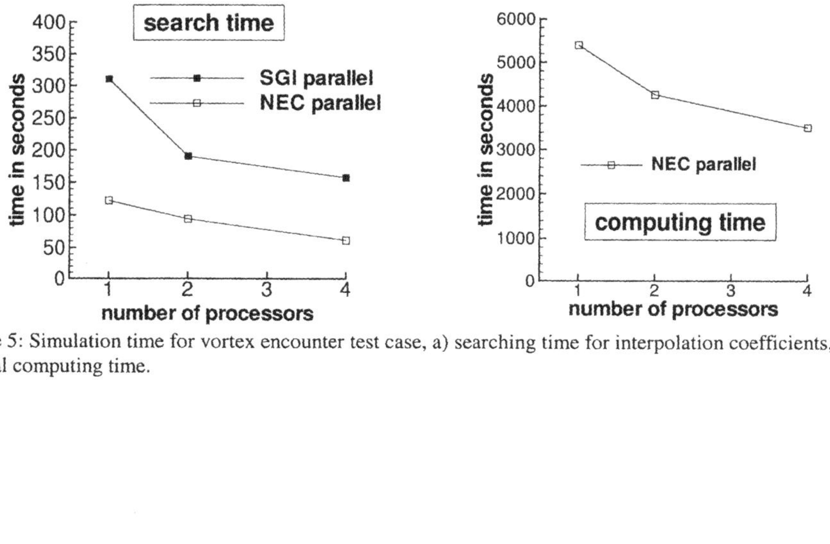

Figure 5: Simulation time for vortex encounter test case, a) searching time for interpolation coefficients, b) total computing time.

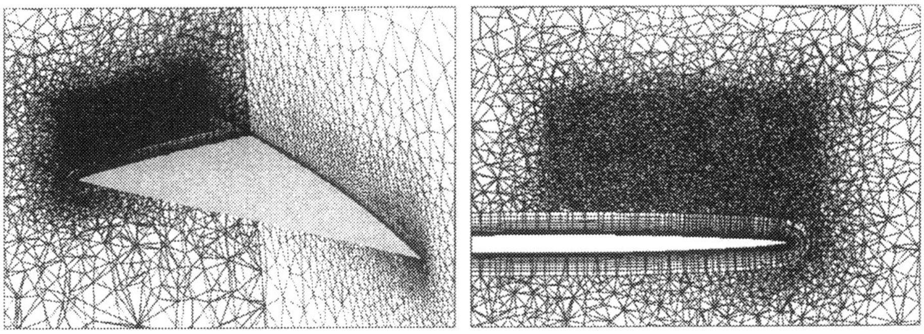

Figure 6: Details of hybrid mesh of oscillating delta wing,

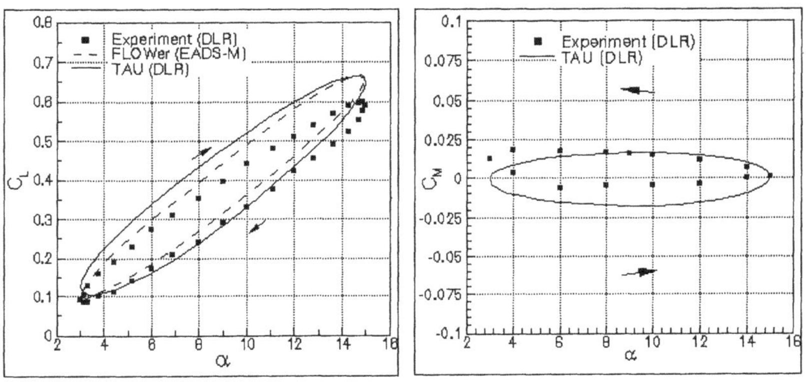

Figure 7: Lift and rolling moment coefficient for oscillating delta wing, M_∞=0.4, Re=3.1x10^6, TAU code.

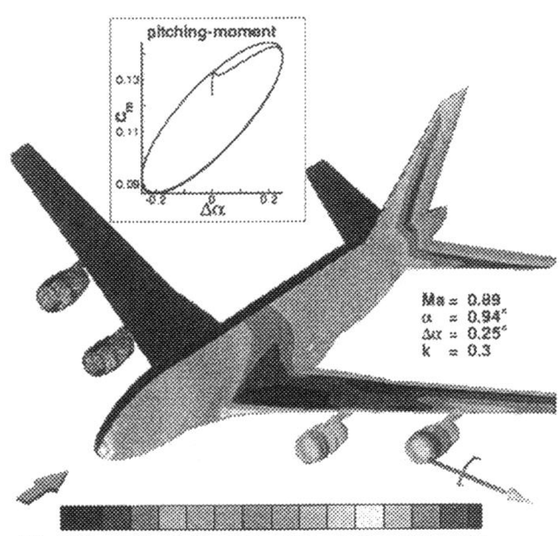

Figure 8: Unsteady inviscid simulation of pitching oscillation of a complete aircraft with the FLOWer code.

Parallel Computational Fluid Dynamics – Practice and Theory
P. Wilders, A. Ecer, J. Periaux, N. Satofuka and P. Fox (Editors)

Towards stability analysis of three-dimensional ocean circulations on the TERAS *

Richard Levine and Fred Wubs [a]

[a]Research Institute for Mathematics and Computing Science,
University of Groningen, P.O.Box 800, 9700 AV Groningen
Email: wubs@math.rug.nl

In this paper we discuss the implementation of a fully implicit numerical model of the three-dimensional thermohaline ocean circulation on the TERAS, i.e., an SGI Origin3800 system with 1024 processors (the new Dutch National Supercomputing facility). Key is the parallelization of the linear system solver MRILU. Therefore we consider important parts of this solver in detail, i.e. the matrix-vector and the matrix-matrix multiplications, where the matrices are all sparse. Data parallel programming using OpenMP is used for the parallelization.

1. Introduction

With ocean circulation models one hopes to contribute to an answer to the central questions in oceanography. Among others there is the question whether there are different circulations possible under the present forcing conditions and if so, whether it is possible that the present circulation turns over into another circulation due to some realistic perturbation. In order to study this in mathematical terms the model is able to follow branches of steady states in parameter space, monitor their linear stability and compute trajectories. In the latter computations much larger time steps can be taken than allowed by explicit schemes. By using the solver MRILU for linear systems of equations and for the generalized eigenvalue problems, results for sufficiently high spatial resolution can be obtained. The emphasis in this paper will be on the parallelization of MRILU, a multilevel ILU preconditioner, for unstructured grids, on the TERAS, i.e., an SGI Origin3800 system with 1024 processors. For scalar equations the parallelization was performed for the CRAY J90, resulting in a speedup of about a factor 3 (see [1]). This was done by data parallel programming using special CRAY directives. For the current parallelization we replaced the CRAY directives by OpenMP directives and generalized the code to handle systems of PDEs. We are aware of the fact that on the TERAS message passing gives better results. In that case domain decomposition techniques are almost unavoidable. This will entail a considerable programming effort, so we decided to find out first what gain is possible by data parallel programming.

*This is part of joint work with Henk Dijkstra and Hakan Oksuzoglu from the Institute for Marine and Atmospheric Research Utrecht of Utrecht University

The TERAS consists of a hypercube structure connecting node boards. Each node board contains 4 processors and a 4GB shared memory. On the TERAS we have exclusive ownership of the requested processors which gives reproducible results. Unfortunately the memory is much slower than the processors which is repaired by using caches. So if a processor wants an element of data it first tries to retrieve it from its cache. If it is not present in cache it fetches a whole chunk of data (cache line), including the element it wants, from memory to cache. The assumption here is that the next element to be processed is close to the current one (locality of data) avoiding a new access of the slow memory. There are two levels of cache, a fast, small L1 cache and a slower, larger L2 cache. In the tests that follow in the next sections round-robin placement of data is used so the data is spread out over all the node boards that are used. The properties above play an important role in the interpretation of the results.

2. The ocean model and continuation code

The ocean model consists of three momentum equations, a continuity equation and equations for heat and salt transport. For the equation for vertical momentum use is made of the shallow layer approximation. Vertical and horizontal mixing of momentum, and of heat and salt, is represented by eddy diffusivities. The discretization is performed on an Arakawa C-grid. For a global model we currently have 120 grid lines in zonal direction, 50 in meriodinal direction and 21 in vertical direction. This leads to a system of about 700,000 unknowns. The flow is forced by annual mean wind stress, heat and fresh water fluxes. In this paper we refrain from further details of this model (see [2] for more on this subject) and turn to the continuation code used in order to explain the type of systems that are to be solved.

By the continuation code we want to trace the solution y from $F(y, \mu) = 0$ as a function of the parameter μ. It is based on the pseudo-arclength method in which the continuation step consists of an Euler predictor and a Newton corrector. The Newton correction equation to be solved is

$$\begin{pmatrix} F_y & F_\mu \\ \dot{y}_0^T & \dot{\mu}_0 \end{pmatrix} \begin{pmatrix} \Delta y \\ \Delta \mu \end{pmatrix} = \begin{pmatrix} b_1 \\ b_2 \end{pmatrix} \tag{1}$$

Here the first equation is split into two systems $F_y z = -F$ and $F_y u = F_\mu$. These two systems can be solved in parallel and then the two solutions are combined in such a way that the last equation in (1) is satisfied.

3. MRILU and Parallelization

In this section the multi-level ILU method MRILU (MR stands for matrix renumbering) will be described in short and we will comment on the parallelization aspects. A more detailed description can be found in [5]. In the algorithm below the basic steps in the factorization process are given. For a sparse matrix the partitioning of step 1 can always be made by extracting a set of unknowns that are not directly connected, the so-called independent set. By also allowing weak connections, which are deleted in step 2, this set can be enlarged. Usually some greedy algorithm is used to find an independent set. After the independent set is found the dropping of steps 2 and 3 and the actual reordering of step 1 has to be done, which is done simultaneously in the program. Since $\tilde{A}_{11}$ is diagonal, also its inverse and consequently the new Schur complement constructed in step 4 will be sparse, which makes it possible to repeat the process. The construction of the Schur complement is the more difficult one, especially the multiplication of the two sparse matrices $\tilde{A}_{21}$ and $\tilde{A}_{11}^{-1}\tilde{A}_{12}$ (the multiplication with the diagonal matrix is easy). So in general we are interested in speeding up the multiplication of two sparse matrices. In the solution phase the L and U factor have diagonal blocks. So solving a system with these matrices amounts to multiplication of a sparse matrix and a full vector. For a typical problem the time spent in the factorization part is 24% for the selection of the nearly independent set, 37% for the reordering of the matrices including dropping, 28% for the computation of the Schur complement, and 11% for other operations. In the solution phase where we use the factorization in a CG-type process, almost all the time is spent we spent in matrix-vector products We will first discuss the data structure before considering the parallelisation of the most time consuming parts on the TERAS.

Set $A^{(0)} = A$.
for i=1..M

1. Reorder and partition $A^{(i-1)}$, to obtain

$$\begin{bmatrix} A_{11} & A_{12} \\ A_{21} & A_{22} \end{bmatrix},$$

such that the matrix A_{11} is sufficiently diagonal dominant.
2. Approximate A_{11} by a diagonal matrix $\tilde{A}_{11}$.
3. Drop small elements in A_{12} and A_{21}.
4. Make an incomplete LU factorization,

$$\begin{bmatrix} I & 0 \\ \tilde{A}_{21}\tilde{A}_{11}^{-1} & I \end{bmatrix} \begin{bmatrix} \tilde{A}_{11} & \tilde{A}_{12} \\ 0 & A^{(i)} \end{bmatrix}$$

where $A^{(i)} = A_{22} - \tilde{A}_{21}\tilde{A}_{11}^{-1}\tilde{A}_{12}$ (Schur complement of $\tilde{A}_{11}$).

endfor

Make an exact (or accurate incomplete) factorization of $A^{(M)}$.

4. Data structure

The sparse matrices in the code are stored in Compressed Sparse Row (CSR) format. For concurrent execution using threads (light-weight processes that share the main memory) we manually split the data representation in several chunks, say T. These chunks consist of consecutive matrix rows that are distributed over the processors for the matrix-vector and matrix-matrix products. In order to represent the chunks, we augment the CSR format with an integer array *par* which has length $T + 1$ (see Figure 1 for an example). This array basically has the same structure as the array *beg*. For chunk c, its starting row is found in $par[c]$, and the last row of its chunk can be found in $par[c + 1] - 1$. The augmented CSR structure will from now on be called the *PCSR* (parallel CSR) structure. The values $par[c]$ are given by

$$par[c] = 1 + ((c - 1) * nr) \textbf{ div } T,$$

where nr is the number of rows in the matrix and a **div** $b = \lfloor a/b \rfloor$. Using this distribution, we have $|par[i] - par[j]| \leq 1$ (for $1 \leq i, j \leq T$), hence the chunks are (almost) equally sized. We assume here that the fill is equally distributed over the rows. In [4] we consider

$$A = \begin{pmatrix} a & 0 & b & 0 & 0 \\ c & d & e & f & 0 \\ g & 0 & h & 0 & 0 \\ i & 0 & 0 & j & k \\ l & 0 & 0 & m & n \end{pmatrix}$$

nr	5													
nc	5													
par	1	2	4	6										
beg	1	3	7	9	12	15								
cf	*a*	*b*	*c*	*d*	*e*	*f*	*g*	*h*	*i*	*j*	*k*	*l*	*m*	*n*
col	1	3	1	2	3	4	1	3	1	4	5	1	4	5

Figure 1. A 5×5 matrix A (left), and its PCSR representation for $T = 3$ (right).

the case that 5-15 elements are distributed randomly over each row. Then for a matrix with 10^4 rows the imbalance of nonzeros using 8 processors and $T = 8$ is less than 6%. So by keeping T small we still expect to have a good load-balance.

5. Matrix-vector multiplication

The matrix-vector multiplication is to be performed on different levels in the solution process, using a sparse matrix stored in CSR-format. One such task is written below.

```
lwb = par(p)
upb = par(p+1)-1
do r=lwb, upb
   y(r) = 0.0
   do nz=beg(r), beg(r+1)-1
      y(r) = y(r) + cfA(nz)*b(colA(nz))
   enddo
enddo
```

The amount of computation for this algorithm is small, so parallel overhead must be reduced as much as possible. This is done by using a small number of tasks T in order to reduce distribution time and avoid any false sharing of cache lines. The indirect addressing in array b is cache unfriendly if the nonzeros in A are positioned irregularly. To study the cache and memory effects we tested two extrema, both with 10 elements per row: (i) all elements are in the first 10 columns and (ii) the elements are randomly distributed over the row. So the first case is cache friendly whereas the second case is not. In Figure 2 plots with the speedup for a range of matrices is depicted for both cases and the run times on one processor are given. In the table we see that for upto $N = 10^5$ the times for MATRIX 1 and MATRIX 2 are almost the same. This means that for these values of N most accesses are from cache. The data requests for values $N \leq 10^4$ are dominated by L1 cache, while the data requests for values $N = 10^5$ are dominated by L2 cache. For $N \geq 10^6$ the data requests are dominated by memory, which causes the decrease in performance for MATRIX 2.

The speedup for MATRIX 1 increases upto $N = 10^5$, drops for $N = 10^6$ and then slightly increases again for $N = 10^7$. In the cases $N = 10^4$ and $N = 10^5$ the speedup is good since on multiple processors a greater percentage of requests is satisfied by the two levels of cache than on 1 processor. The speedup for MATRIX 2 always increases as N gets larger. For the smaller values of N there is less reuse of the cache, so multiple processors will load the same memory segment (cache line), which yields memory conflicts and therefore a degradation of the speedup. As N gets larger there is less chance that the same memory segment has to be loaded into cache simultaneously. So the number of conflicts decreases and hence the speedup increases.

For comparison, on the CRAY J90 the observed speedup [1] is about 6 on 8 processors

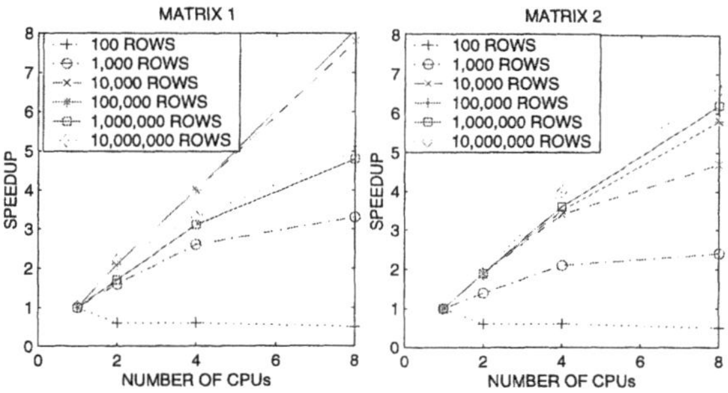

	MATRIX 1	MATRIX 2
N	ms.	ms.
10^2	0.013	0.013
10^3	0.144	0.145
10^4	1.500	1.700
10^5	23.60	28.00
10^6	219.2	1132.2
10^7	2993.5	35348.2

Figure 2. Speedups for CSR matrix-vector multiplication and run times on one processor.

for the problem with $N = 10^4$, independent of the matrix type. On the TERAS we only see a speedup of 5 above for MATRIX 2, so the degradation is less on the CRAY when the memory is accessed randomly.

6. Matrix-matrix multiplication

In these multiplications the amount of computation is larger than in the matrix-vector multiplications, so there is less percentage overhead. This algorithm consists of three phases. First the number of nonzeros in each row of the resulting matrix is computed in parallel, then the starting points of each row in the resulting coefficient array are computed sequentially, and finally the resulting matrix is computed and stored in parallel. In the first part each task p performs the segment left. The array *StorCol* keeps track if there is a nonzero counted in a specific column for each row of C. Therefore each row iteration is started with all elements of *StorCol* set to `false`. The array *nnzRow* keeps track of the number of nonzeros in each row of C. The outer loop in the algorithm is over the rows assigned to the task. Next there is a loop over all the nonzeros in the current row of A. The column number of the nonzero gives the required row number of B. The inner loop runs over all the nonzeros in that row of B. The column numbers of these nonzeros in B give the column number of C where the result will enter. It is checked whether this is a new fill, in which case the number of nonzeros in the row is incremented and the corresponding element of *StorCol* is set to `true`. The array *ColNr* keeps track of the elements of *StorCol* set to `true` and is used to reset those to `false` before the next row iteration is started. In the second (sequential) part the *nnzRow* array is used to compute the *beg* array (see Figure 1) for C.

```
lwb = par(p)
upb = par(p+1)-1
do r=lwb, upb {StorCol = false}
   nnzRow(r) = 0
   do i=begA(r), begA(r+1)-1
      do j=begB(colA(i)), begB(colA(i)+1)-1
         Col = colB(j)
         if not(StorCol(col))
            nnzRow(r) = nnzRow(r) + 1
            StorCol(Col) = true
            ColNr(nnzRow(r)) = Col
         end if
      enddo
   enddo
   do i=1, nnzRow(r)
      StorCol(ColNr(i))=false
   enddo
enddo
```

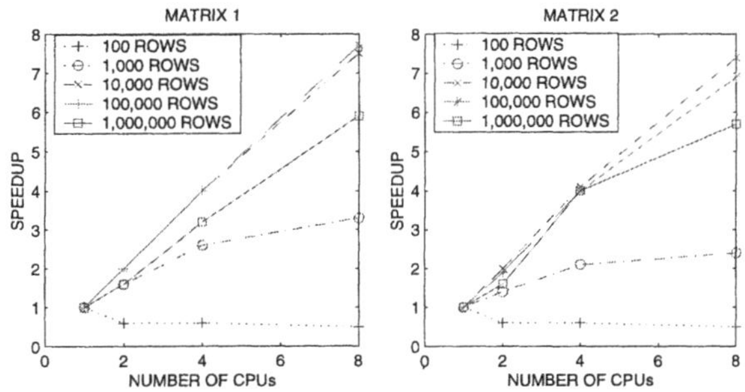

N	MATRIX 1 ms.	MATRIX 2 ms.
10^2	0.41	0.72
10^3	3.90	7.32
10^4	41.0	138
10^5	344	2603
10^6	3516	130251

Figure 3. Speedups for CSR matrix-matrix multiplication and run times on one processor.

The third step consists of computing and storing the results of the multiplication which is done by an extended version of the first part. In Figure 3 we plotted the speedup of the multiplication for a range of matrices and give the results on one processor in a table. We see that on one processor the multiplication of two matrices of type MATRIX 2 is always substantially slower than the multiplication of two matrices of type MATRIX 1 and this aggravates with increasing N. So with MATRIX 2 there must be many cache misses. The resulting memory accesses will become increasingly time consuming if the data cannot be stored on one node board. The speedup in both cases is however almost equal, which indicates that memory conflicts are not a substantial factor. However, it is quite possible that memory conflicts occur but are not seen due to the increased communication time.

For comparison on the CRAY J90 a speedup of about 6 on 8 processors for a problem with $N = 10^3$ is found [1], whereas we only see a speedup of 2 here. Hence, again we see that the TERAS needs much larger matrices than the CRAY J90 in order to benefit from parallelization.

7. The independent set selection and reordering

The nearly independent set selection is done in three steps. Firstly we create a reduced matrix where small elements are dropped according to a sloppy criterion. Secondly, we determine the independent set of this matrix. Finally it is tested whether this set satisfies the more sophisticated dropping criteria for the original matrix. The second part is sequential in nature and therefore hard to parallelize. However, there exist parallelizable variants [3] of the greedy algorithm we use, but often the selected sets are much smaller, which will increase the number of elimination steps in the factorization and make the factorization accordingly more expensive.

The more sophisticated dropping criterion contains a sum of the dropped elements (in absolute values) for respectively rows and columns, which may not exceed a certain magnitude. Since multiple processors may not increase the sum simultaneously the `ATOMIC` directive must be used, which regulates that only one processor at a time accesses a certain array element. This directive restricts the speedup. It turns out that in this way about 30% of the nearly independent set selection is parallelizable.

For the matrix reordering and dropping we have a similar dropping criterion and also there the `ATOMIC` directive is used; timings showed that about 60 % is parallelized.

N	factorization process					solution process					total
	time	1	2	4	8	time	1	2	4	8	8
638,401	27.7	1.	1.3	1.7	1.8	8.27	1.	1.2	1.8	1.7	1.8
2,556,801	113.	1.	1.4	1.5	1.7	51.5	1	1.4	1.8	2.1	1.8
10,233,601	498.	1.	1.4	1.5	1.9	492.	1.	1.4	1.8	2.1	2.0

Table 1
Speedups for solution of Poisson problem using MRILU.

8. Experiments with MRILU on the TERAS

We implemented the ideas above in the MRILU code and performed timings on the TERAS. We did this for the Poisson equation, $\Delta u = f$, which was discretized using the standard 5-point stencil on a non-uniform grid. In the solution process the BiCGSTAB solver was applied. The objective was to gain 6 digits of accuracy. Beforehand, we know that the speedup of the code will be limited due to the limited speedup of the nearly independent set selection and the reordering. Table 1 shows the speedups obtained for different problem sizes; also the timings on one processor are added (column "time"). So we observe a speedup of about a factor 2 for very large problems. On the CRAY J90 we have a similar code running and found a speedup of about a factor 3. Since the processors of the TERAS are much faster, the TERAS' result is still a factor 6 faster.

At this place it also of interest to compare this MRILU version with a variant where the data structure was changed from CSR to Jagged Diagonal form to allow vectorization [6]. Here only the solution process of MRILU for scalar equations was considered for parallelization; a factor 7 speedup on 16 processor for a problem of about 1 million unknowns was found on the CRAY J90. The solution part returned the answer in 1.4 s., which is about 5 times faster than for a comparable problem found from interpolation in Table 1 above. So vectorization is advantageous for MRILU, though also here a considerable effort is necessary to get the desired speed.

9. Results of the ocean circulation model

In Figure 4 (left) a contour plot of a stream function of the global wind-driven ocean model is given, which was obtained from a computation on the TERAS. We measured the time needed (Figure 4(right)) to solve the system corresponding to this flow with MRILU and we split it into three parts: (i) the blockelimination part, which corresponds to the loop of the algorithm in Section 3, (ii) the point-wise incomplete factorization part, which corresponds to the factorization of $A^{(M)}$ in the algorithm, and (iii) the solution part. In the table, these parts are indicated by "block", "point" and "iteration" respectively. Since the construction of the point-wise incomplete factorization and its application were not parallelized, we see only a speedup for the blockelimination and the iteration. The parallelized part of MRILU, i.e. the repeated block elimininination, plays a smaller role in this problem than in the Poisson problem, hence, the benefit of the parallelization is only very modest here. Nevertheless, since the TERAS has a large memory we can run larger problems than before. We conclude that we have to reconsider MRILU for these type of problems. If this is resolved we expect a similar speedup as for the Poisson equation.

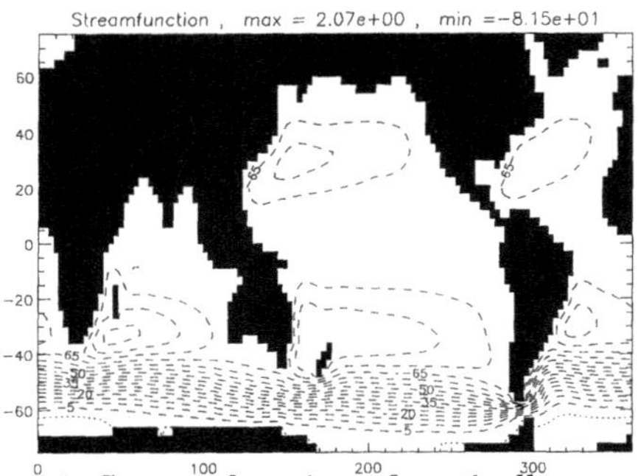

	time (s.)	
	1 CPU	4 CPUs
block	40	20
point	90	90
iteration	60	50

Figure 4. Stream function of vertically averaged flow and run times.

10. Conclusions

The experiments on the TERAS reported in this paper lead us to the following conclusions. Firstly, matrix-vector and matrix-matrix multiplication with CSR-matrices give reasonable speedup. However for large matrices the memory accesses and the communication time between node boards will have a large (negative) impact on the run time. Secondly, the nearly independent set selection and the reordering procedure were only partially parallelizable and therefore gave only limited speedup. Due to Amdahl's law this will also restrict the overall speedup. Hence, though not all parts were parallelized, it is already clear that for MRILU the data parallel programming approach will not give great speedup on distributed memory computers, a factor 2 to 3 at most. This is caused by too fine grained parallelizable parts. Compared to results we had on the CRAY J90 we see that larger chunks are necessary to get reasonable speedup. Moreover we miss the vectorization possibilities of the CRAY which are well usable for our type of problems. Our final conclusion is that for distributed memory computers the code must be changed more drastically in order to get reasonable speedup. We currently think of using domain decomposition ideas in MRILU using message passing for communication between processors.

REFERENCES

1. A. Meijster and F.W. Wubs. Towards an Implementation of a Multilevel ILU Preconditioner on Shared-Memory Computers. *LNCS 1823*, 2000.
2. H.A. Dijkstra, H. Oksuzoglu, F.W. Wubs and E.F.F. Botta, A fully implicit model of the three-dimensional thermohaline ocean circulation, to appear in J. Comput. Phys.
3. M.T. Jones and P.E. Plasman. A parallel coloring heuristic. *SIAM J. Sci. Comput.*, 14(3), 1993.
4. R.C. Levine, Improvements of a numerical continuation code for ocean circulation problems, master's thesis, University of Groningen, 2001.
5. E.F.F. Botta and F.W. Wubs. Matrix Renumbering ILU: An effective algebraic multilevel ILU-preconditioner for sparse matrices. *SIAM J. Matrix Anal. Appl.*, 20(4), 1999.
6. E.F.F. Botta, F.W. Wubs and A. van der Ploeg, A fast linear-system solver for large unstructured problems on a shared-memory parallel computer. Proceedings AMLI'96, eds. O. Axelsson, B. Polman, Nijmegen, 1996.

Parallel Computational Fluid Dynamics – Practice and Theory
P. Wilders, A. Ecer, J. Periaux, N. Satofuka and P. Fox (Editors)

CODE PARALLELIZATION EFFORT OF THE FLUX MODULE OF THE NATIONAL COMBUSTION CODE

Isaac Lopez[*], Nan-Suey Liu[†], Kuo-Huey Chen[±], Erdal Yilmaz[‡], Akin Ecer[‡]

[*]U.S. Army Research Lab., Vehicle Technology Directorate,
21000 Brookpark Rd., Cleveland, OH 44135, USA
email: ilopez@grc.nasa.gov

[†]NASA Glenn Research Center, [*]U.S. Army Research Lab.,
21000 Brookpark Rd., Cleveland, OH 44135, USA
email: nan-suey.liu@grc.nasa.gov

[±]AP Solutions, Inc.,
Solon, OH 44139, USA
email: kuohueychen2@aol.com

[‡]Indiana University Purdue University Indianapolis, Purdue School of Eng. and Techn.,
723 West Michigan Street, Indianapolis, IN 46202-4160, USA
email: {eyilmaz, aecer}@iupui.edu

Abstract. NASA Glenn and Indiana University Purdue University Indianapolis (IUPUI) personnel have teamed together to formulate a collaborative project to parallelize the FLUX module of National Combustion Code (NCC). Previously developed tools at IUPUI were used to parallelize the FLUX module. These tools allow code developers to easily and effectively parallelize Computational Fluid Dynamics (CFD) codes without having to invest significant resources in learning code parallelization programming techniques, which are outside of their area of expertise.

The NCC system of codes can be used to evaluate new combustor design concepts. Once parallelized, NCC can be integrated into a design system to provide a fast turnaround high fidelity analysis of a combustor, early in the design phase. This improvement to the National Combustion Code will contribute to a significant reduction in aircraft engine combustor design time and cost by reducing the hardware builds and tests required, and the accomplishment of the USA national goal to reduce aircraft engine emissions.

1 INTRODUCTION

Within NASA's High Performance Computing and Communication (HPCC) program, the NASA Glenn Research Center is developing an environment for the analysis/design of aircraft engines called the Numerical Propulsion System Simulation (NPSS)[1]. The vision for NPSS is to create a "numerical test cell" enabling full engine simulations overnight on cost-effective computing platforms. To this end, NPSS integrates multiple disciplines such as aerodynamics, structures, and heat transfer and supports "numerical zooming" between 0-dimensional to 1-, 2-, and 3-dimensional component engine codes.

One of such codes is the FLUX module that is being developed as an alternative flow solver for the National Combustion Code (NCC). The National Combustion Code is a system of codes that will enable multidisciplinary simulation of flow and chemistry, in the full combustor from compressor exit to turbine inlet, for modern turbofan engines. Development and validation of the NCC has been done at NASA Glenn in partnership with the U.S. industry. Additional modules have been implemented into the NCC, such as spray, kinetics, and turbulent combustion. It can use both structured as well as unstructured meshes. An unstructured mesh generator enables rapid grid generation of the complex combustor geometries. Reductions in time of grid-generation and the optimization of various code modules resulted in the reduction of total elapsed time for combustor simulation. These features make the NCC a state-of-the-art design tool that enables modeling of the combustor of the jet engine. The U.S. aircraft engine industry is adapting the NCC as a design tool because it enables significant reduction in combustor development time and cost.

To date, the focus of run time reduction has been the flow solver module, CORSAIR-CCD. This paper presents the parallelization efforts of the alternative flow solver, FLUX. Parallelization tools developed at the IUPUI were used for this purpose.

2 THE FLUX CODE

Comprehensive combustion modeling and simulation is an essential, integral part of modern design/optimization of low-emissions, high-performance combustors. An integrated system of computer codes, termed as the National Combustion Code, has been developed by an industry-government team for this purpose [2]. The goal is to perform full combustor simulation on massively parallel computing systems, with the overall turnaround time being overnight.

The current NCC version 1.0 is composed of a set of major modules, among which is a baseline gaseous flow solver known as CORSAIR-CCD. Both PVM and MPI message passing libraries can be used for communication. The targeted computing platforms are networked workstations and PC's. The parallel performance of the current NCC gaseous flow solver, CORSAIR-CCD, has been enhanced over the past several years, and a detailed description of the improvements can be found in [3].

The current gaseous flow solver, CORSAIR-CCD, is based on finite-volume discretization and an explicit four-stage Runge-Kutta scheme. The discretization begins by dividing the spatial computational domain into a large number of contiguous elements composed of triangles and/or quadrilaterals in the 2D case; and tetrahedrons, wedges, and/or hexahedrons in the 3D case. A central-difference finite-volume scheme augmented with second-order and/or fourth-order dissipative operators is used to generate the discretized equations, which are then advanced temporally by the Runge-Kutta scheme. For low Mach number compressible flow, a low Mach number pre-conditioning is applied to the governing equations, and a dual time procedure, along with the Runge-Kutta scheme, is used for temporal advancement. A more detailed description was given in [4,5].

Most recently, an alternative gaseous flow solver for the NCC, termed as "FLUX," is being developed at NASA Glenn Research Center. The FLUX code uses the concept of Conservation Element and Solution Element (CE/SE) to discretize and to solve the conservation equations. The computational domain consists of contiguous triangles in 2D case, and contiguous tetrahedrons in 3D case. By applying the integral form of the conservation equations to the space-time conservation elements, and by evaluating the values of these integrals via the solution elements associated with the solution nodes, a set of generally non-linear algebraic equations become available for temporally advancing the conservative variables and their first-order derivatives at the solution nodes. Numerical damping is usually applied to the evaluation of the first-order derivatives. Pre-conditioning is not needed for low Mach number compressible flows. A more detailed description can be found in [6].

The comparison of results obtained from the FLUX code and the CORSAIR-CCD code suggests that the FLUX code is numerically less dispersive and, by and large, less dissipative. The validation results indicate that FLUX code is capable of calculating compressible flows over a very wide range of Mach numbers. It can calculate low Mach number compressible flows without invoking low Mach number pre-conditioning, and it can crisply capture strong shock waves and other flow discontinuities without resorting to the TVD scheme. When the necessary developments towards turbulent combustion simulation on massively parallel computing platforms are completed, the FLUX code will strengthen the capabilities of the NCC in the areas of high-speed propulsion and unsteady processes.

3 PARALLELIZATION TOOLS

3.1 GPAR (A Grid-based Database System for Parallel Computing)

GPAR is a database management system for CFD codes, developed in CFD laboratory in IUPUI, for the management of interfaces, blocks, and the relations between them. It provides an upper-level simplified parallel computing environment. The blocks and interfaces are basic data groups for parallel computing. In order to achieve an effective parallel computing environment, an appropriate integration of a block and its interfaces is required. Therefore, the blocks and their relations to the interfaces form the computational domain. Information related to these data groups are made available by GPAR throughout the computation. Each block may have several interfaces. Each interface defines the partial boundary of the block connected to one of its neighbor. Each interface has a twin interface belonging to the neighbor block. The block-interface relation is given in Figure 1. Block solver contains no communication and it utilizes the database for data storage and update. Interface solver includes communication between neighbors.

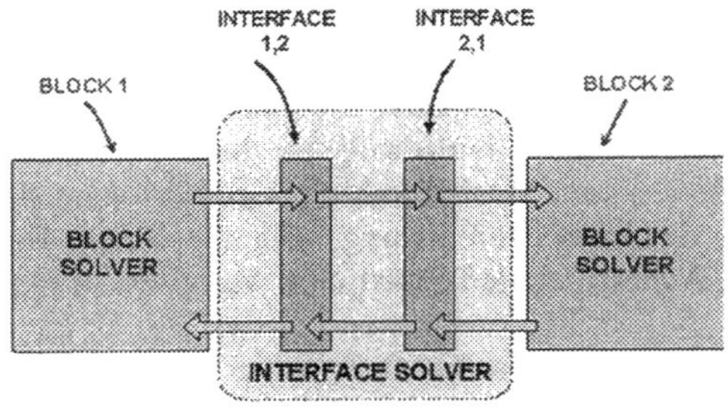

Figure 1. Block-interface relation and communication between blocks.

In very early development of GPAR [7], the users of application programs handle the interface communication. Although it has been designed for several different types of the mesh structures, it was not used for unstructured mesh before. Recently GPAR was brought to a more simplified and compact form [8]. New features were added for different types of CFD solvers. Different data structures were implemented for the unstructured grids. All communication steps between the blocks are handled by GPAR to reduce workload of a user. Therefore, the users are not involved with interface communication issues. Grids for a partition are stored in a very compact form that includes the block, interfaces, and boundary information in a single file. This makes the GPAR files very suitable for moving among machines that are distributed located. A dynamic load balancing application is an example of an application that takes advantage of this feature. Also, the solution file or restart file is stored in a similar manner for each block.

3.2 Parallelization With GPAR

Before starting a parallel solution, a grid for a geometrical domain must be generated and partitioned. For structured grids, a grid division generally does not require any specific tool. However, for unstructured grids there are several different domain decomposition techniques and tools available. The grid information for a partition together with its interfaces is stored in a file, this file is named "gpar_dbfileX", where X stands for partitioned block number.

In parallelization with GPAR, it is assumed that each block is executing a copy of the block solver (i.e., FLUX). The block solver is a sub-program that can be called from a main program. The main program controls whole parallel solution. Parallelization structure used in GPAR is shown in Figure 2. First, parallel GPAR data files are read to the database, which are available at any instant of the solution. Those files contain information related to the grid and the blocks. Then the block solver is called for the solution on the domain. This is followed by communication at the interfaces. After necessary time stepping, the solutions are stored in the GPAR files. For the explicit solvers the interface solver, which performs communication, is activated with just one simple command. Therefore, it is very easy to implement and use GPAR with explicit of solvers.

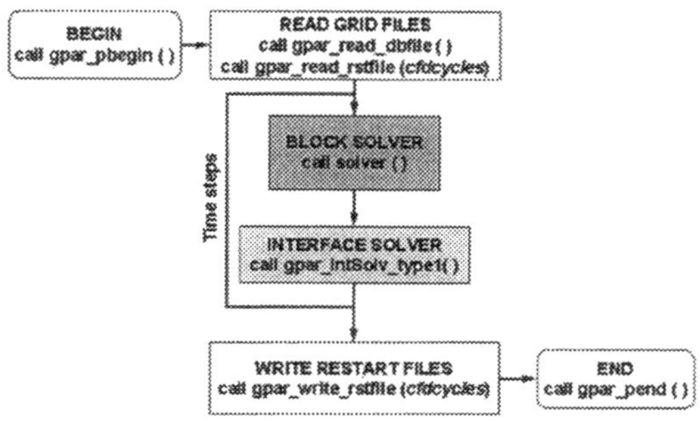

Figure 2. Parallelization structure with GPAR.

In GPAR, there are several retrieve, update, and interface routines to call for parallel execution and data manipulation. These routines have been optimized to bring simplicity and to spend less time in the parallel implementation of a solver. When it is compared with PVM and MPI implementation, GPAR has reduced the whole communication effort to just a single function for CFD based applications. The comparison of such implementation differences between GPAR and PVM/MPI for the present applications is given in Table 1.

Table 1: Comparison of GPAR, MPI, and PVM implementations for communication

	GPAR	MPI	PVM
Actual calls in Fortran	Call IntSolv_Type1(none)	Call mpi_send (..send_array..) Call mpi_recv (..recv_array..)	Call pvmfinitsend (...) Call pvmfpack (...) Call pvmfsend (...) Call pvmfrecv (...) Call pvmfunpack (...)
Overhead		Loops over # of neighbors for send and receive Preparation of send and receive parameter	Loops over # of neighbors for send and receive Preparation of send and receive parameter

4 PARALLEL IMPLEMENTATION OF FLUX CODE

4.1 Objectives of parallelizing FLUX Code

The main objective in parallelizing the FLUX Code with the GPAR is to give more portability options to the serial version of the code. After parallelization, the FLUX code works on computers located at different sites. These computers can have Unix, Linux, and Windows operating systems with distributed and/or shared memory. Extending the current parallel computer resources to heterogeneous environments in a wide area network requires codes to be compatible and portable in all environments. The second objective of the present effort is to provide CFD developers a parallel library that is easy to use without any I/O for further development. The third objective is to maintain the parallel environment independent of the particular code without the knowledge of the code's details. Fourth objective is to provide a capability to develop and test algorithms for multi-physics problems. This multi-physics capability is where different solvers (i.e. flow and combustion) can be tested together and the developer of one code does not need to know the specifics of the other code. Finally, the parallel program should be easily interacted with dynamic load balancing programs. In wide area applications, as well as local area applications, load balancing is an efficient way to improve elapsed time for parallel computing.

4.2 Steps to parallelize the FLUX code

There are several steps followed to parallelize the FLUX code. Before going into details of these steps, two main tasks should be mentioned to give a broad view of parallelizing effort. The first task is to prepare input data for blocks in a suitable form. This data is mainly grid partitioning and any other block related data preparation phase. Figure 3 gives grid structure of the FLUX code with interface elements. The second task is to make parallel implementation of GPAR for the solver.

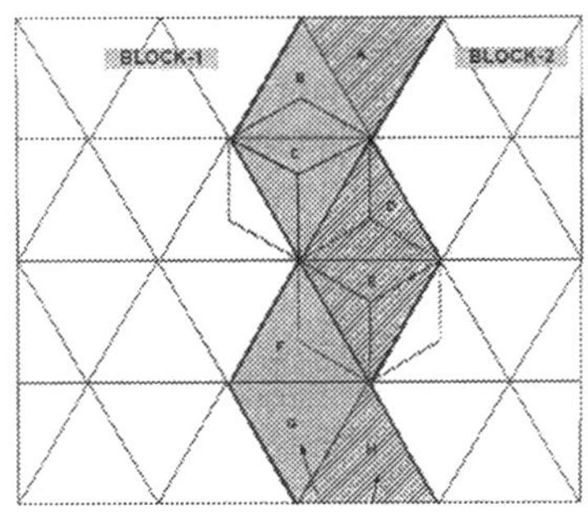

Figure 3. Block and interface structure of the FLUX code

Following are the main steps followed in a parallelization effort:

Step1: Provide a multi-block grid database where each block data and its interfaces to the neighboring blocks are stored individually. GPAR grid data files for the FLUX code have the following format:

For Each Block
- *block number*
- *grid type { structured=0, tetrahedral=1, hexagonal=2, element based FluxC=3 }*
- *number of elements*
- *number of solution variable at each element*
- *maximum node number in an element {4=tetra, 6=hexa, other}*
- *maximum face number in an element*
- For each element
  - *Pid,Sid,Mxnode,x1,x2,x3,x4,y1,y2.y3,y4,z1,z2,z3,z4,Mxface,e1,e2,e3,e4,*
- For each boundary element
  - *element number, element boundary condition index { index= 0,1,2,3 ... }*
- *Number of interface*
- For each interface
  - *interface number*
  - *neighboring block number*
  - *neighboring interface number at the neighboring block*
  - *number of nodes/elements at the interface*
  - For each interface node
    - *element numbers and update index { update=1, no update=0 }*

For each boundary set
    number of boundary elements
    number of boundary values
    element number, value1, value2, value3, ...

Step 2: To eliminate read and write statements from the FLUX code. Read the data directly into database using GPAR outside the CFD code. In the current implementation of the FLUX code all data either for grids or restart are read inside the code with individual read statements. Instead of this, all the data are read with one simple data read command since the format of the FLUX code is known inherently in the grid file and GPAR. Following is actual coding with highlight for GPAR statements:

```
Subroutine read_database
..........
call gpar_read_dbfile()
call gpar_read_rstfile(rstime)
..................
Return
End
```

Step 3: To transfer the data from GPAR to the FLUX code using GPAR calls. Once the reading of the data is accomplished and stored in the database, then it is read to use anytime and at any part in the program. User transfers data from GPAR to local variables, vectors or arrays. A part of the implementation is given below with highlight for GPAR statements:

```
Subroutine read_database
.............
call gpar_getelbasex(1,x1)
call gpar_getelbasey(1,y1)
call gpar_getelbasez(1,z1)
call gpar_putsve(1,u)
call gpar_putsve(2,v)
call gpar_putsve(3,w)
call gpar_putsve(4,p)
..............
Return
End
```

Step 4: To provide a wrapper program which runs the FLUX Code in a parallel environment. It runs CFD solvers for each block, communicates between the solvers, prepares output and restart files. Following is a part of that program with highlight for GPAR statements:

```
Program Main
.............
call gpar_pbegin ()      ! start the parallel environment
call flux
call gpar_pend ()        ! end  the parallel environment
End
Subroutine flux
..........
myid  = gpar_myid ()        ! get proc id for current
nprocs = gpar_nprocs()      ! get number of processors
call casein                 ! Read restart file and make initialization
```

```
      do 300 real_time=1,mtime          ! begin physical time loop
        ..........
        if (nprocs.gt.1) call gpar_intsolv_type2()       ! communicate at interface
        ..........
        if (restarts.gt.0) then
          if (mod(real_time,restarts).eq.0) call gpar_write_rstfile(real_time)  ! write restart file
        endif
300   continue                          ! end physical time loop
      Return
      END
```

5 CONCLUSION

In this paper, an effort to parallelize the FLUX code – which is an alternative flow solver for NCC – is presented. A high-level parallelization tool, GPAR, is used for the implementation of parallel FLUX code. The main objectives of this implementation were to enhance the portability of the FLUX code and to reduce the effort to maintain it on parallel computers of the future. At the same time, it enables the developers of the FLUX code to make changes and additions to the code without getting into the details of parallel programming. Although GPAR uses both MPI and PVM libraries at background, it allows user to write a single processor code and does not require the user to manage the communication between the processors and the data storage. The parallel application program is restructured by identifying and separating a block solver (for computations on a single processor) and interface solver (for data exchange between the processors). Furthermore, all of the parallelization efforts were conducted without accessing the details of the FLUX code. Any testing of future additional developments to the FLUX code can be performed in a parallel computing environment without any further support for parallelization. The next step in this collaboration is to run an application test case to compare the performance of the parallelized code against the original code.

REFERENCES

[1] A. L. Evans, J. Lytle, J., G. Follen, and I. Lopez, An Integrated Computing and Interdisciplinary Systems Approach to Aeropropulsion Simulation, ASME IGTI, June 2, 1997, Orlando, FL.

[2] Liu, N.-S., "On the Comprehensive Modeling and Simulation of Combustion Systems," AIAA Paper 2001-0805, January 2001.

[3] Quealy, A., Ryder, R.C., Norris, A.T., and Liu, N.-S., "National Combustion Code: Parallel Implementation and Performance," AIAA Paper 2000-0336, January 2000.

[4] Ryder, R.C., "The baseline Solver for the National Combustion Code," AIAA Paper 98-3853, July 1998.

[5] Chen, K. -H., Norris, A.T., Quealy, A., and Liu, N. -S., "Benchmark Test Cases for the National Combustion Code," AIAA Paper 98-3855, July 1998.

[6] Liu, N. -S., and Chen, K. -H., "An Alternative Flow Solver for the NCC – The FLUX Code and Its Algorithm," AIAA Paper 2001-0973, January 2001.

[7] Y.P. Chien, A. Ecer, H.U. Akay, and F. Carpenter, "Dynamic Load Balancing on Network of Workstations for Solving Computational Fluid Dynamics Problems," *Computer Methods in Applied Mechanics and Engineering*, Vol. 119, 1994, pp. 17-33.

[8] "*A Data-Parallel Software Environment, GPAR*," CFD Laboratory Indiana University Purdue University Indianapolis, Indianapolis IN, USA (in preparation).

Parallel Computational Fluid Dynamics – Practice and Theory
P. Wilders, A. Ecer, J. Periaux, N. Satofuka and P. Fox (Editors)

Parallelization of a Chaotic Dynamical Systems Analysis Procedure

J. M. McDonough[a] and T. Yang[b]

Department of Mechanical Engineering, University of Kentucky,
Lexington, KY USA 40506-0108
E-mail: [a]jmmcd@uky.edu
[b]tlyang@engr.uky.edu

This paper is focused on the parallelization of a chaotic dynamical system analysis procedure. A curve-fitting method for modeling chaotic time series is parallelized using both a shared-memory programming paradigm and the MPI programming technique. Relatively good parallel performance is obtained from both approaches. An experimental turbulent velocity time series is modeled using the parallelized curve-fitting procedure, and the modeled time series compares favorably with the experimental data.

1. INTRODUCTION

Chaotic dynamical systems arise in a wide variety of physical, biological and social contexts. Turbulent fluid flow, eye movement during reading and fluctuations of the stock market provide respective examples. It is clear from this that modeling such systems is extremely important, and much effort has been devoted to development and study of such models in these, and many other, situations.

The usual approach to modeling chaotic dynamical systems has as its goal accurate short-term prediction (basically, extrapolation) using considerable amounts of previous (in time) data and polynomial approximation (see Casdagli and Eubank [1] for specific examples). The methods used to accomplish this are inexpensive from a computational standpoint, and this approach is appropriate in some circumstances—*e.g.*, stockmarket predictions. On the other hand, in the context of turbulence models, accurate short-term prediction is meaningless simply because the physical system never exactly repeats its previous behaviors, and any model that accurately replicates past behavior is intrinsically flawed. Thus, a different approach is needed.

This was emphasized in McDonough *et al.* [2] where it was demonstrated that chaotic dynamical systems could be curve fit in the sense of reproducing the *structural* "appearance" of the original data and an extensive set of statistical quantities associated with these data, but without attempting a point-by-point exact fit. Such an approach can be applied to any data set represented by a time series, and the outcome of the curve-fitting process is a model that behaves very much like the original data, but with the same intrinsic variation that occurs in, *e.g.*, physical turbulence, as a result of sensitivity to initial conditions. However, because so many statistical quantities must be computed (and for other reasons discussed below), this curve-fitting method is extremely CPU intensive and could benefit significantly from parallelization.

Recently, Hylin & McDonough [3] introduced a new approach to constructing "synthetic velocity" subgrid-scale (SGS) models for large-eddy simulation (LES) techniques based on Kolmogorov scalings and discrete dynamical systems (chaotic maps). Moreover, McDonough & Huang [4] have shown that such discrete dynamical systems (DDSs) can be derived *directly* from the Navier–Stokes (N.–S.) equations, and that these systems exhibit the full range of N.–S. behaviors. This obviously motivates a thorough study of such systems, especially comparison with (curve fits of) experimental data to arrive at truly effective SGS models. Indeed, Mukerji *et al.* [5] presented such a study for soot formation fluctuations in a turbulent diffusion flame and quite successfully fit laboratory data to a linear combination of logistic maps (see May [6] for a detailed discussion of this map), as had previously been done in [2] for numerical data. But we again emphasize that in the past these studies have been so CPU intensive as to severely limit their applicability in practice. In particular, particle image velocimetry (PIV) techniques now make it possible to capture velocity time series for an entire flow field, but it would be essentially impossible to analyze (model) such large amounts of data with the present form of the algorithm despite the obvious utility of doing so. In the current work we are attempting to remove this deficiency by means of parallelization in an effort to obtain an efficient way to construct chaotic-map models of experimental data, especially in the context of turbulent fluid flow.

The remainder of this paper is organized as follows. We begin in Sec. 2 with a brief description of the form of the SGS models that motivate this work. This is followed with a section that presents an overview of the curve-fitting problem, particularly highlighting the arithmetic requirements. Finally, and most importantly in the present context, we describe our parallelization efforts and present results.

2. FORM OF THE SUBGRID-SCALE MODEL

The form of the subgrid-scale model introduced in [3] is

$$q^* = A_q \zeta_q M_q, \tag{1}$$

where A_q and ζ_q are, respectively, amplitude factors and anisotropy corrections derived from Kolmogorov's (mainly K41) theories (see Frisch [7] for detailed discussions), and the M_q are chaotic maps. In early work (including [3]) these were chosen somewhat arbitrarily although the logistic map was widely used, at least in part due to the observation in [7] that quadratic maps of this nature might be viewed as a "poor man's Navier–Stokes equation." (In fact, the coupled DDSs derived and analyzed in [4] are more deserving of this epithet.) Once the SGS result q^* is calculated for each component of the solution vector using (1), it is combined with the corresponding resolved-scale quantity, say $\overline{q}$, to form the *complete* (*i.e.*, large-scale plus small-scale) solution:

$$q(\boldsymbol{x}, t) = \overline{q}(\boldsymbol{x}, t) + q^*(\boldsymbol{x}, t). \tag{2}$$

Because the small-scale, SGS, behavior is now used to directly augment the resolved scale (in contrast to more classical forms of LES where it is used to construct Reynolds stresses—see, *e.g.*, Domaradzki & Saiki [8]), it is imperative that q^* be accurate—at least in a qualitative structural sense, and in turn this implies that the maps M_q must accurately mimic the SGS (high-frequency) temporal behavior. Clearly, the best way to guarantee this is to directly employ experimental data in the model-building process.

3. NATURE OF THE CURVE-FITTING PROBLEM

The form of the chaotic map M_q that will be used in data fitting is given in [2]:

$$M^{(n+1)} = (1-\theta)M^{(n)} + \theta \sum_{l=1}^{k} \alpha_l S_l^{(n+1)}[\omega_l, d_l, m_l(b_l)]. \tag{3}$$

In Eq. (3), α_l is the amplitude, and θ is an "implicitness" factor. The chaotic map $S_l^{(n+1)}[\omega_l, d_l, m_l(b_l)]$ is formulated in the following way [2]:

$$S_l^{(n+1)} = \begin{cases} m_l^{(n)} & \text{if } D_l^{(n+1)} = 0, \\ m_l^{(n+1)} & \text{otherwise,} \end{cases} \tag{4}$$

where $m_l^{(n+1)}$ is the standard logistic map [6]:

$$m_l^{(n+1)} = b_l m_l^{(n)}(1 - m_l^{(n)}), \tag{5}$$

with b_l as the bifurcation parameter. The $D_l^{(n+1)}$ in Eq. (4) are the "duration" functions. They determine how long each realization of (5) remains active once it has been initiated. They are initialized to 1 when the map is activated, incremented by 1 each time the map is evaluated and set to 0 if they exceed d_l — the duration of evaluation of each instance of the modeled map. The mathematical formulation of $D_l^{(n+1)}$ is as follows [2]:

$$D_l^{(n+1)} \equiv \begin{cases} 1 & \text{if } F_l^{(n+1)} = 1, \\ D_l^{(n)} + 1 & \text{if } F_l^{(n+1)} = 0 \text{ and } 0 < D_l^{(n)} < d_l, \\ 0 & \text{otherwise,} \end{cases} \tag{6}$$

with $F_l^{(n+1)}$ as the switching function. The switching function depends on the frequency of evaluation, ω_l, and it sets the map switching frequency. It is defined in [2] as:

$$F_l^{(n+1)} \equiv \begin{cases} 1 & \text{if } (n+1) \bmod \omega_l = 0, \\ 0 & \text{otherwise.} \end{cases} \tag{7}$$

As was emphasized in [2], the goal of modeling an arbitrary chaotic time series via (3) is not to find a time series that exactly coincides with the data but rather one possessing properties to guarantee that the "appearance" of the constructed time series is qualitatively and statistically close to the data. There are many statistical properties for characterizing chaotic data. In the present study, we use 25 different properties of the time series to assess goodness of fit. For a detailed discussion of these, and the methods by which they are calculated, readers are referred to [2].

The method of constructing the map (3) is designed to find the minimum value of the following least-squares functional corresponding to N_p properties (McDonough *et al.* [2]):

$$Q(\alpha_l, b_l, \omega_l, d_l, \theta) = \sum_{i}^{N_p} (\delta p_i)^2, \tag{8}$$

where δp_i is the difference between the modeled map and the data for the ith statistical property. In Eq. (8), α_l, b_l, ω_l, d_l, and θ are unknown parameters of the map (3) to be determined by minimizing Q. In the present study, a gradient method optimization technique was used to search for values of α_l, b_l, and θ, and a direct search was used to find ω_l and d_l due to the fact that these last two quantities can take on only integer values. Our strategy of searching for these unknown parameters is as follows: for each fixed combination of ω_l and d_l, the optimization process is conducted to find the corresponding values of α_l, b_l, and θ. Then values of Q are compared for different sets of ω_l and d_l to obtain the unknown parameters of the modeled time series with smallest value of Q. Based on previous experience reported in [2], the typical value of k in Eq. (3) is three, resulting in a total of 13 unknown parameters. Because of this relatively large number of parameters, the optimization and direct search processes are quite CPU-time consuming. Thus, parallelization is needed to ameliorate this difficulty.

4. APPROACH TO PARALLELIZATION

Parallelization in this study is based on a shared-memory programming paradigm using the HP Fortran 90 HP-UX compiler implementation of Convex-HP compiler directives and, separately, on MPI. The program was parallelized at the "region" level running on the HP N-4000 at the University of Kentucky Computing Center.

The maximum number of threads available on the HP N-4000 is eight, and in the present study each of the eight threads is used to seek values of α_l, b_l and θ using gradient-level optimization for set combinations of ω_l and d_l. Eight optimization processes can be treated simultaneously if eight processors are used. If the number of processors used is less than eight, each processor treats one optimization process first. The rest of the optimization processes will be treated by those processors that finish the first optimization processes earlier. It should be pointed out that, by using this parallel strategy, the speed of convergence of different instances of the gradient optimization processes at different set values of ω_l and d_l is not the same, resulting in unbalanced use of system resources and reduction in parallel efficiency, implying a need for a more sophisticated approach.

To show the speed-up of the parallelization, Figs. 1 and 2 give the speed-up factor plotted against the number of processors. The speed-up factor is calculated using the following definition:

Speed-up factor = (Single-Processor Wallclock)/(Multi-Processor Wallclock)

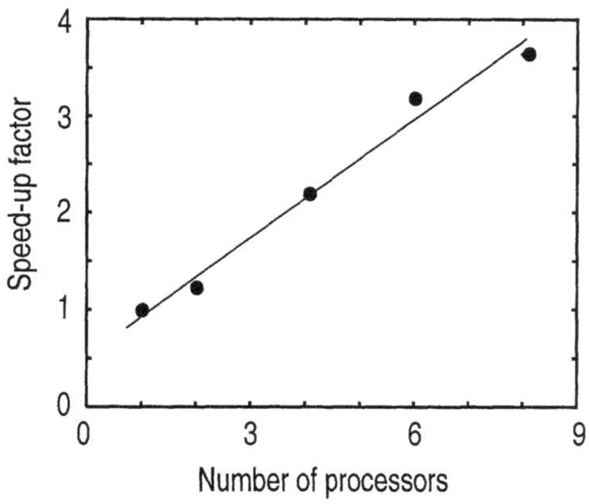

Fig. 1 Speed-up performance by shared-memory programming paradigm on HP N-4000

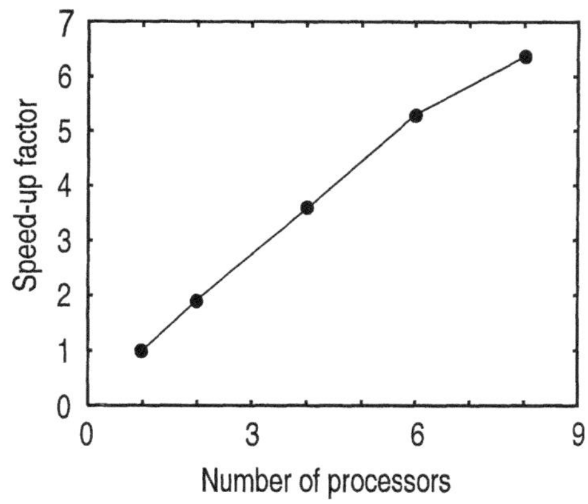

Fig. 2 Speed-up performance of MPI programming technique on HP N-4000

Figure 1 shows the speed-up performance of the shared-memory approach. It can be seen that as the number of processors increases, speed-up increases sub-linearly due to unbalanced use of system resources and the communication between the processors. The maximal speed-up factor for shared-memory parallelization obtained in the present study is 3.86 when *eight* processors are used. Our present work shows that better parallel performance can be obtained using MPI instead of the shared-memory paradigm. Figure 2 shows that use of MPI results in a maximal spead-up factor of 6.37 when eight processors are used. The main contribution to parallel performance of MPI is decreasing the communication time between processors. By using this technique, the comminication between the processors is controlled by the code itself. Therefore, the amount of communication can be easily reduced to its minimum by the code writer. On the contrary, the communication between the processors is controlled by the compiler of the computer in the case of shared-memory, resulting in difficult-to-predict amounts of communication between the processors.

4. RESULTS OF DATA FITTING

In this section, we will present the curve-fitting results obtained from the process described in Section 3. The emphasis is on comparison of the fit results with the experimental time series. We fit a single velocity component obtained from hot-wire anemometry (HWA) measurements [9]. The case of *turbulator* (a turbulence-enhancing bump) flows corresponding to $Re = 1 \times 10^5$ is considered in the present study. For detailed descriptions of the flow configuration and velocity detection method, readers are referred to [9]. The curve-fitting results are compared using three features, *viz.*, the "appearance" of the time series, the power spectral density, and the delay map [2,5,9].

Figure 3 displays three complete velocity time series in parts (a) through (c). The first corresponds to experimental data of the x-component velocity; the second represents evaluation of the model, Eq. 2, with an initial guess of the parameters to be determined in the least-squares fit, and the third shows the data-fitting result. It is observed that the

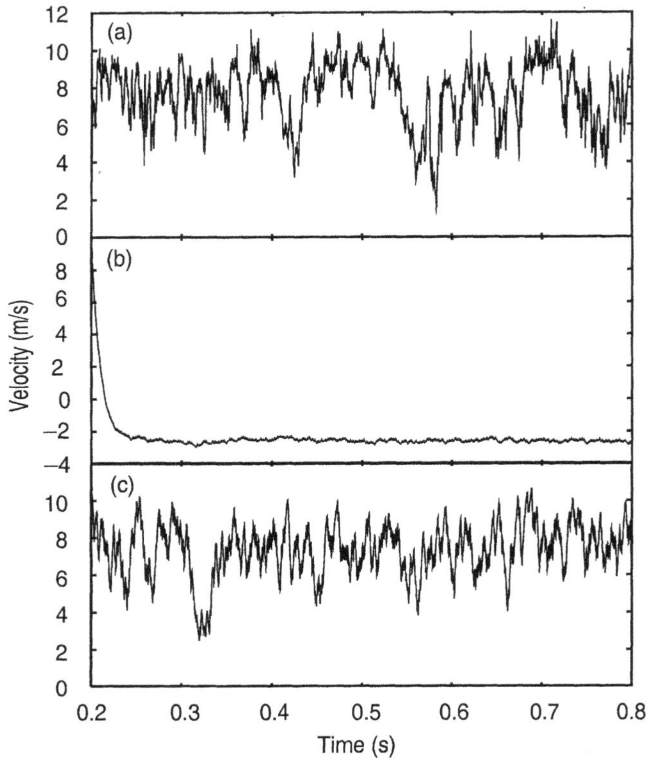

Fig. 3 Velocity time series for $Re=1\times10^5$ (a) Experimental data. (b) Evaluation of model with initial guess of the parameters. (c) The final fit.

Table 1
Initial guess and final fit values of the parameters in the model Eq. (3) for velocity time series

	Initial Guess	Final Fit
ω_1	1	1
ω_2	2	2
ω_3	1	1
d_1	1	1
d_2	1	1
d_3	1	1
α_1	24.0	13.357
α_2	-4.0	-22.907
α_3	-24.0	14.990
β_1	3.80	3.774
β_2	3.80	3.995
β_3	3.80	3.688
θ	0.01	0.0167

"appearance" of the final fit is close to that of the experimental time series. We display part (b) of the figure to emphasize that finding the correct parameters is a nontrival process (and one which requires considerable CPU time). Table 1 lists parameters of both the initial guess and the final fit demonstrating the significant change from the initial guess.

Further comparisons consisting of power spectra and delay maps are presented in Figs. 4 and 5, respectively, with part (a) corresponding to measured data and part (b) to the result of the final fit. The power spectra of the experimental data and the final fit are very similar. But the delay map of the fit does not appear to contain all of the topological features of the measured data. The reason for this may be that the logistic map cannot capture all the behavior of the experimental time series [9]. The poor man's N.–S. equation [4] would possibly lead to a much better fit in this case. A program to study this is currently underway in our research group, and much better fit results have already been obtained. In the present paper, the performance of parallelization of the curve-fitting method is the most important aspect of our presentation.

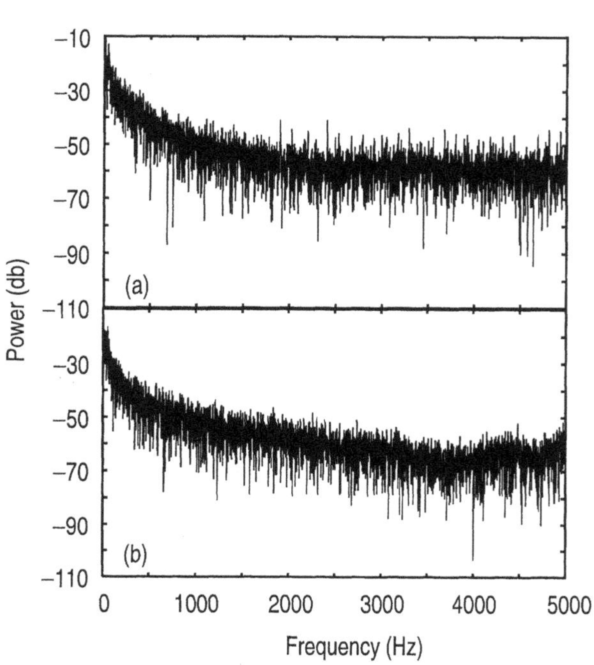

Fig. 4 Power spectra of the velocity time series for *Re=1 ×10*^{5}. (a) Experimental data. (b) The final fit

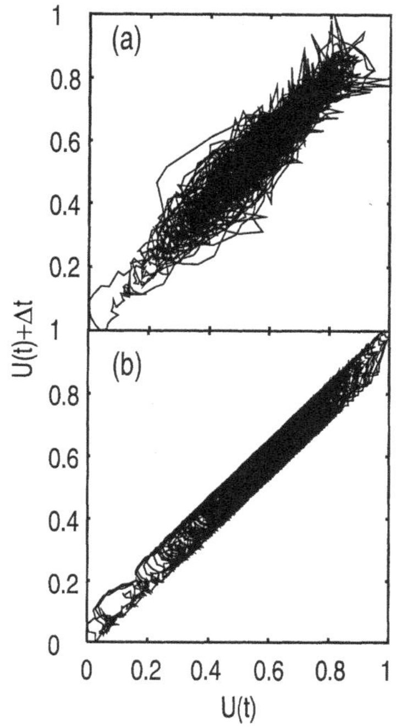

Fig. 5 Delay maps of the velocity time series for *Re=1×10*^{5}. (a) Experimental data. (b) The final fit

5. CONCLUDING REMARKS

The chaotic map curve-fitting method proposed by McDonough *et al.* [2] uses a weighted least-squares functional corresponding to a wide range of statistical quantities as the objective function. A gradient optimization method was used to optimize this complicated objective function, and experimental turbulence data [9] were successfully modeled. The optimization procedure has been parallelized using both the shared-memory programming paradigm and the MPI programming technique. The parallel performances of both techniques were investigated, and it is concluded that MPI provides significantly better parallel speedups than does the shared-memory paradigm in the present case. Fast parallel optimization procedures such as those reported in the present work are expected to enhance the data-fitting of chaotic time series in the near future.

ACKNOWLEDGEMENTS

Financial support from AFOSR under Grant #F49620-00-1-0258, and from NASA/EPSCoR Grant #WKU-522635-00-10 is gratefully acknowledged by both authors. Work of the second author has been partially supported also by the University of Kentucky

Center for Computational Sciences.

REFERENCES

1. M. Casdagli and S. Eubank (Eds.). Nonlinear Modeling and Forecasting, Addison-Wesley Pub. Co. (1992).
2. J. M. McDonough, S. Mukerji and S. Chung. Appl. Math. Comput., Vol. 95 (1998) 219.
3. E. C. Hylin and J. M. McDonough. Int. J. Fluid Mech. Res., Vol. 26 (1999) 539.
4. J. M. McDonough and M. T. Huang. University of Kentucky, Mech. Engr. Report, CFD-03-01 (2001).
5. S. Mukerji, J. M. McDonough, M. P. Mengüç, S. Manickavasagam and S. Chung. Int. J. Heat Mass Transfer, Vol. 41 (1998) 539.
6. R. M. May. Nature, Vol. 261 (1976) 459.
7. U. Frisch. Turbulence: The legacy of A. N. Kolmogorov, Cambridge University Press, Cambridge (1995).
8. J. A. Domaradzki and E. M. Saiki. Phys. Fluids, Vol. 9 (1997) 2148.
9. H. Roclawski, J. D. Jacob, T. Yang, and J. M. McDonough. AIAA Paper #2001-2925, 31st AIAA Fluid Dynamics Conference and Exhibit, Anaheim, CA (2001).

Parallel Computational Fluid Dynamics – Practice and Theory
P. Wilders, A. Ecer, J. Periaux, N. Satofuka and P. Fox (Editors)

Performance Optimization of GeoFEM Fluid Analysis Code on Various Computer Architectures

Kazuo Minami* and Hiroshi Okuda †

*Department of Computational Earth Sciences, Research Organization for Information Science and Technology (RIST),
Tokyo, Japan
e-mail: Minami@tokyo.rist.or.jp

†Department of Quantum Engineering and Systems Science,
The University of Tokyo, Japan
e-mail: okuda@q.t.u-tokyo.ac.jp

We study the fluid analysis module of the finite element package GeoFEM for obtaining a good performance of its parallel linear solver. For this purpose we analyze a basic model loop separately and present some of its execution properties for various computer architectures.

1. Introduction

The Science and Technology Agency (current name: the Ministry of Education, Culture, Sports, Science and Technology), Japan, has begun the Earth Simulator Project in 1997 for predicting various earth phenomena by the 'Earth Simulator' (40 Tflops/peak). As part of this project, a parallel finite element software package, named GeoFEM, has been developed, especially for solving solid earth problems. The current version of GeoFEM includes structural, wave and fluid analysis modules and is being optimized for theEarth Simulator. Considering that various computer architectures exist in the world, GeoFEM, which is mainly targeting the Earth Simulator, must run efficiently on various computers as a multi-purpose parallel finite element code. However, it is not easy for "one code" to run on various machines at high performance, and in general, suitable programming is demanded for each architecture. This paper describes a data structure and coding manner to make the GeoFEM fluid analysis module run efficiently on various computer architectures. For this purpose we have chosen to analyze the basic loop of the fluid analysis module separately.

2. Pre-analysis of computational cost

The fluid analysis module of GeoFEM is roughly divided into two parts. The first part is for calculating the coefficient matrix of the system equation (Matrix Assembling Part). The second part is for solving the linear equation (Solver Part). As is shown in Table 1 over 90% of the computational costs is consumed by the Solver Part.

Table 1: Computational costs of the fluid analysis module

Solver Part	91.2%
Matrix Assembling Part	8.8%

Table 2: Computational costs of the Solver Part

process 1	forward substitution	26.3%	
process 2	backward substitution	26.1%	
process 3	Matrix-Vector Product(lower)	22.2%	
process 4	Matrix-Vector Product(upper)	22.2%	
	others		3.2%
		96.8%	3.2%

Focusing on the Solver Part, most of the computational cost (96.8% of CPU time) is due to four operation processes, listed in Table 2. Thus, these four processes have to be optimized for attaining a high performance. It turns out that these four processes allow for a common loop model.

3. Data Structure, Loop Expansion and Direct Access in Linear Solver

3.1 Performance Factors

In this study, we have mainly used the Hitachi SR8000 system of the University of Tokyo. The system has 128 nodes. SR8000 is a machine of the pseudo vector architecture. A node consists of 8 processors each of which has 1 Gflops peak performance. When using this system, it is important to consider the following five items:

(1) Pseudo vectorization of loops
(2) Software pipelining in loops
(3) Outer loop expansion
(4) Small load/store latency in loops
(5) Direct access to arrays in loops

Item (1) is achieved by switching on the corresponding compiler function on the pseudo vector architecture. Item (2) is achieved similarly both on pseudo vector and scalar architectures. Thus, items (1) and (2) dependent upon compilers only. On the other hand, items (3), (4) and (5) are independent of compilers, but dependent upon coding aspects. In the sequel of this section we will discuss the last three items in relation with a new data structure and a new direct access manner using a reordered array.

3.2 Data Structure/Loop Expansion

The four operation processes depicted in Table 2 have almost the same coding. A model code for these processes is shown in Fig.1.

```
do j=1,2
do i=1,1000
    a(i) = a(i) + b0(i,j)*a(L(i))
enddo
enddo
```

Figure1: Model code for the operation processes in Table 2

In Fig. 1 the array 'b0' is declared as 'dimension b0(1000,2)'. A new data structure: 'dimension b1(2,1000)' is adopted and the model coding is modified as shown in Fig. 2. Small load/store latency of loop is satisfied because 'b1(1,i)' and 'b1(2,i)' are accessed continuously. Both ‘outer loop expansion’ (item (3)) and ‘small load/store latency of loop’ (item (4)) are addressed to simultaneously in this new data structure.

```
do i=1,1000
    a(i) = a(i) + b1(1,i)*a(L(i)) + b1(2,i)*a(L(i))
enddo
```

Figure 2: Modified code using the new data structure and loop expansion

3.3 Direct Access

A good performance of a loop, executing calculations, is often obtained if direct access is used for all arrays in that loop. Therefore, we have tried to achieve this introducing reordening beforehand, taking out the indirect access in our main computational loop. Fig. 3(a) presents the reordering process for array a and Fig. 3(b) is the main calculating process written in direct access manner.

```
do i= 1,1000
    x(i) = a(L(i))
enddo
```

(a) reordering process

```
do i= 1,1000
    a(i)= a(i) + b1(1,i)*x(i) + b1(2,i)*x(i)
enddo
```

(b) main calculating process

Figure 3: Model code using the new direct access manner

4. Performance Evaluation of the Model Loop

Computational performances of the described model loop on various architectures, i.e. SR8000 (pseudo vector and scalar), Alpha (scalar) and VPP5000 (vector), are summarized in Tables 4 and 5. In the Tables the symbol 'o' means that either the feature is switched on during compilation or that the alternative loop structure was adapted. If not, then the symbol 'x' is used. The symbol '-' means that the feature is not adressing the underlying architecture. Note that items (3) and (4) can only be addressed to simultaneously by adapting the loop structure depicted in Fig. 2.

A significant performance improvement (factor 2.5) is obtained by the present loop expansion coding and data structure (Table 4 [A]-->[B]). However, a performance degradation of 25% results from the new direct access manner on pseudo vector architecture (Table 4 [B]-->[C]). This is contrasted by the performance improvement of 70% in scalar mode, using the present direct access manner coding (Table 4 [D-1]-->[D]). A performance improvement was observed by coding (3), (4) and (5) on scalar architectures (Table 4 [D]). Here, it is vital to use the software pipelining option as well (Table 4 [D], [E]).

Table 4: Performance of model loop in fluid analysis code on SR8000 and Alpha

	[A]	**[B]**	**[C]**	**[D]**	**[D-1]**	**[E]**	**[F]**
(1) pseudo vectorization	o	o	o	-	-	-	-
(2) software pipelining	o	o	o	o	o	x	x
(3) outer loop expansion	x	o	o	o	o	o	o
(4) load/store latency	x	o	o	o	o	o	o
(5) direct access	x	x	o	o	x	o	o
performance (MFlops)	68	174	129	145	85	45	83

[A],**[B]**,**[C]**: pseudo on SR8000 / **[D]**,**[E]**: scalar on SR8000 / **[F]**: scalar on Alpha system

Table 5 Performance of model loop in fluid analysis code on VPP5000

	[G]	[H]
(3) outer loop expansion	x	o
(4) load/store latency	x	o
(5) direct access	x	o
performance (MFlops)	1405	1881

[G],[H]: VPP5000 (A single processor has 9.6 Gflops peak performance)

On a vector architecture, a performance improvement of 30% was obtained by coding (3), (4) and (5) (Table 5 [G]-->[H]), reaching 20% of the peak performance of the vector machine (Table 5 [H]).

4. Summary

The basic computational loop in the parallel linear solver underlying the fluid analysis module of GeoFEM has been studied with respect to performance and implementation on scalar, vector and pseudo vector architectures. The proposed data structure and direct access manner has been found to be universally efficient. With this the most important operation parts in the fluid analysis solver of GeoFEM can be optimized for performance over various architectures.

Acknowledgments

This work is a part of the "Solid Earth Platform for Large Scale Computation" funded by the Ministry of Education, Culture, Sports, Science and Technology, Japan through its “Special Promoting Funds of Science and Technology”. The authors would like to thank Professor Yasumasa Kanada, The University of Tokyo, for fruitful discussions on high performance computing.

References

[1] K.Garatani, H.Nakamura, H.Okuda, G.Yagawa, GeoFEM: High Performance Parallel FEM for Solid Earth, Proceedings of 7th High-Performance Computing and Networking (HPCN Europe'99),LNCS-1593,133-140,1999.

Parallel Computational Fluid Dynamics – Practice and Theory
P. Wilders, A. Ecer, J. Periaux, N. Satofuka and P. Fox (Editors)

Large scale CFD computations at CEA

G. Meurant[a], H. Jourdren[a] and B. Meltz[a]

[a]CEA/DIF,
PB 12, 91680 Bruyères le Châtel, France

This paper describes some large scale Computational Fluid Dynamics computations recently done at CEA as well as the parallel computers of the first stage of the Tera project within the Simulation program of CEA.

1. Introduction

The french Commissariat à l'Energie Atomique (CEA) is a governmental agency in charge of all scientific aspects related to the use of nuclear energy in France as well as the study of other sources of energy. It has roughly 16000 employees and 10 locations around the country. To know more about its activities, please give a look at the Web site http://www.cea.fr. The military branch of CEA (CEA/DAM) is in charge of defence applications, that is mainly maintaining the french nuclear stockpile without relying anymore on nuclear testing. This is achieved through a Simulation program which is based on three main components. The first two ones are experimental devices: Airix, an X–ray machine used since 1999 for non nuclear hydro experiments and the Megajoule laser whose first stage is under construction and which will be used for better understanding of the physics of thermonuclear combustion and should lead to ignition through inertial confinement fusion (ICF) experiments. The third component of the Simulation program is made of the parallel computers needed for enhanced numerical simulation. Within CEA this is called the Tera project. Previous numerical computations are going to be improved through better physical models, better numerical schemes, finer meshes and 3D computations.

We have to solve numerically the multimaterial Euler equations for non viscous highly compressible fluids with real equations of state. This system of equations is tightly coupled to some other physics, mainly various transport equations describing particle transport which give sources of energy. Several possibilities exist for handling the CFD part of the equations. One can do Lagrangian computations where the mesh is moving at the velocity of the fluid or Eulerian computations where the mesh is fixed and the fluids flow through the mesh. Another possibility is to have the mesh moving with a velocity different from the one of the fluid; this is denoted as ALE (arbitrary Lagrangian Eulerian). We are also considering AMR (Adaptive Mesh Refinement) techniques. They are going to be described later on in this paper. Notice that for the three last possibilities we have to deal with mixed cells that contain more that one material. This can be handled through interface reconstruction techniques or the solution of concentration equations. All CFD

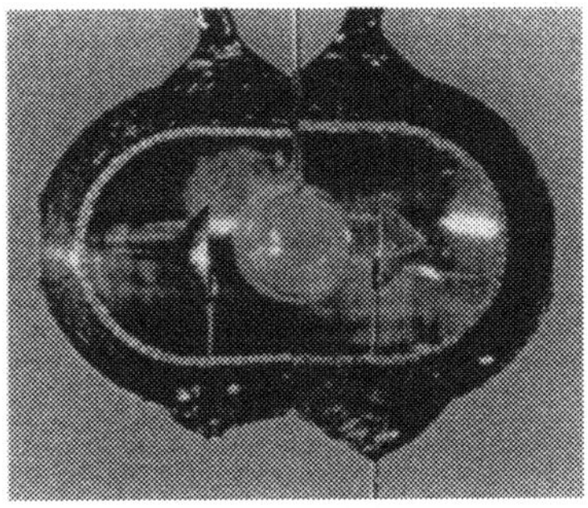

Figure 1. An ICF target

schemes are explicit in time but the particle transport is usually handled implicitly.

An example of unclassified computational problem is provided by the numerical simulation of Inertial Confinement Fusion (ICF) experiments. A spherical target filled with Deuterium and Tritium (hydrogen isotopes) is located in a cylindrical hohlraum that is used to produce an X–radiation "furnace" when illuminated by laser beams. This leads to the implosion of the target putting the material in such conditions of temperature and density that the thermonuclear fusion reaction can start. A cut of one such device that was used in some experiments some years ago is shown in Figure 1.

This is a problem which is extremely difficult for numerical simulation. During the implosion of the spherical target there are some growth of several instabilities that occur and it is extremely important to be able to understand the physics of this phenomena as well as being able to do accurate numerical computations. Examples of instabilities computations related to some other experiments will be given in the next sections.

2. The Tera project

Until 2000, the two main production supercomputers in the CEA/DIF computing center were a Cray T90 and a Cray T3E. They are still in operation. The Cray T90 was the main production computer. It was installed in 96/97 and has 24 vector processors, a peak performance of 1.8 Gflops/processor that is 43 Gflops in total and 512 Mwords of shared memory. The Cray T3E was used for large parallel computations. It was installed in November 96 and has 168 processors, a peak performance of 0.6 Gflops/processor, that is 100 Gflops and 16 Mwords/processor of distributed memory. After several studies, it was decided that for the first step of the Simulation program we need to multiply at least by a factor of 10 this computer capacity. The goal is thus to install a 5 Tflops peak (1 Tflops sustained) computer by the end of 2001. This is the first step of an 8-year project. The winner for the first step request for proposal is Compaq. This computer is going to be installed in three phases described in Table 1. This system is to be complemented by storage capabilities based on HPSS which are scheduled as described in Table 2.

As we said, the computer to be installed at the end of 2001 is the first step in a three

Table 1
The three phases of the Tera–1 installation.

System	Initial	Intermediate	Final
Schedule	April 2000	December 2000	December 2001
Processors	6*4 EV6 (667 Mhz)	75*4 EV6 (833 Mhz)	2500
Peak perf. (sustained)	35 Gflops (6)	500 Gflops (106)	5 Tflops (1.2)
Memory	20 GB	300 GB	2.5 TB
Disks	600 GB	5.5 TB	50 TB

Table 2
The storage capabilities.

System	HPSS server	Level 1	Level 2
Schedule	1999–2001	2001–2002	2002–2003
Type	IBM SP	7 STK silos	RFP in 2002
Capacity	720 GB	1 PB	5 PB

stage project. The next steps of the Tera project are a 10 Tflops sustained machine in 2006 and a 100 Tflops sustained one in 2010. For the 2001 machine, we had to build a new computer room with a floor surface of $2000m^2$.

3. The Tera benchmark

This benchmark is part of the acceptance test of the 2001 final machine. It was devised to be a realistic "demo application" capable of very high parallel performance and to make sure that the whole machine can be dedicated to one single large application that uses the whole memory. Another goal is to learn about hybrid programming (MPI + OpenMP) and explore microprocessor optimization techniques. It was specially written in Fortran 90 by B. Meltz.

It solves the non viscous 3D compressible Euler equations in Eulerian coordinates on a cartesian mesh. The features of this code are a transport plus remapping algorithm, alternating direction splitting X,Y,Z, Z,Y,X, Y,Z,X, X,Z,Y, Z,X,Y, Y,X,Z,..., a second order Godunov scheme, a third order PPM-like remapping and an approximate Riemann solver (à la Dukowicz).

One should ask why using alternating direction which is an old technique? There are some benefits and a few drawbacks. On the benefits' side is memory: work arrays take little space and this allows cache optimization since one line of cells fits in level 2 cache (level 1 for small problems); moreover, work arrays are contiguous (this helps prefetching). Another benefit is simplicity of implementation for optimization and parallelization techniques. The main drawback is that we have to do transposition of the data. An efficient way to do this is to have three copies of 3D global arrays for optimization, one per direction.

Parallelization is done through mesh partitioning. Mesh blocks should be as close as possible to cubes. One block keeps two rows of ghost cells per direction. MPI synchronous

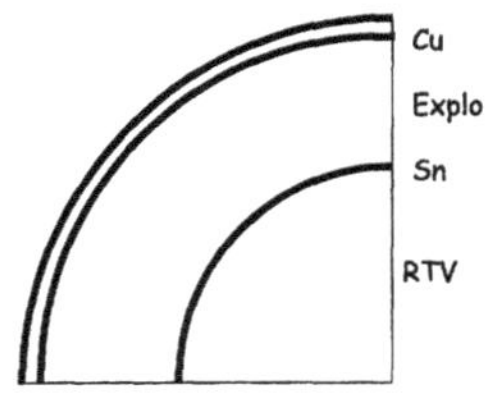

Figure 2. An instability problem

messages are used to send data needed by the other blocks. Notice that some reductions are needed to estimate the next time step. Measured performances on several parallel computers are given in Table 3.

Table 3
Tera benchmark performances.

System	Size (1 dir)	Nb procs	Mflops/p	Gflops	% peak
Cray T90	90	1	112.5	0.112	18.7
Cray T3E	420	168	42.9	7.214	7.2
IBM Pwr3 225 Mhz	270	8	104.1	0.832	11.6
Compaq EV6 667 Mhz	100	8	240.7	1.926	18
Compaq EV6 833 Mhz	1200	300	179.3	106.2	20
Compaq final system	2480	2500	?	1200	25

As seen in Table 3, the intermediate system using MPI gives 106 Gflops. When using a mixed parallel model with OpenMP and MPI, we obtain 87.5 Gflops. The final system will run a problem with $2480^3 = 15.2$ billion cells using 2.4 TB of memory. The largest message is going to be 14 MB and the expected performance 1.2 Tflops.

4. AMR

Work started on parallel Euler computations at CEA in the beginning of the 90s when a code was written to allow out of core large computations on small memory machines. This naturally led to domain decomposition and then parallelism. This code has now parallel capabilities using MPI. In 1996 a parallel computation of an interface instability problem was done on the Cray T3D using 7.8 millions cells and 128 processors. The problem is described on Figure 2. This experiment used a cylinder made of six different shells. The interface between the tin and the RTV is perturbed. During the implosion of the cylinder by a high explosive, a Richtmyer–Meshkov instability grows and this is what we are interested in accurately computing.

For handling these problems we need to have a very fine mesh around the unstable interface. This can be very costly when using a classical Eulerian cartesian mesh because it leads to meshes with a huge number of cells. To try to overcome these problems one can use Adaptive Mesh Refinement (AMR). A code was developed that uses a second order Godunov scheme formulated in total energy. It has multimaterial capabilities with interface reconstruction (à la Youngs). This is a tree-based AMR code written in C++ allowing an arbitrary refinement factor (2x2, 3x3,...) for any cell. It is also relatively easy to incorporate more physics in this code like high explosive reactions rates, material strength, non linear diffusion and MHD. Work on this code has been done by H. Jourdren, P. Ballereau, D. Dureau, M. Khelifi using theoretical results by B. Després. This code uses an acoustic approximate Riemann solver for the Lagrangian hydrodynamic step:

$$p^* - p_R = (\rho c)_R(u^* - u_R), \quad p^* - p_L = -(\rho c)_L(u^* - u_L) \tag{1}$$

$$p^* = \frac{(\rho c)_R p_L + (\rho c)_L p_R + (\rho c)_L(\rho c)_R(u_L - u_R)}{(\rho c)_L + (\rho c)_R} \tag{2}$$

$$u^* = \frac{(\rho c)_L u_L + (\rho c)_R p_R + (p_L - p_R)}{(\rho c)_L + (\rho c)_R} \tag{3}$$

where ρ is the density, u the velocity, p the pressure and c is the sound speed. L and R refer to the left and right states and $*$ to the one we want to compute. A simplified form of this solver is obtained by taking:

$$(\rho c)_L = (\rho c)_R = \rho^* c^* \tag{4}$$

If the internal energy is a jointly convex function of specific volume and entropy (thermodynamic stability), B. Després ([1], [2], [3]) proved the following result: if $c(k,n)(\Delta t/\Delta x) \leq 1$ then

$$T_k^{n+1}(S_k^{n+1} - S_k^n) \geq 0 \tag{5}$$

where S is the entropy and T the temperature, index k refers to the space and index n to the time. With perfect gases this gives $\rho, p, e \geq 0$, the numerical stability condition being

$$\max(\frac{c^2}{c^*}, c^*)\frac{\Delta t}{\Delta x} \leq 1 \tag{6}$$

This type of results have been extended to other hyperbolic systems: elasticity, 3 temperature hydro and MHD. With this AMR code 2D Richtmyer-Meshkov instabilities are 10 times cheaper than a pure Eulerian computation for the same accuracy. An example of shock tube computation is shown on Figure 3. A shock passes through a perturbed interface and an instability develops. The top part of the figure is a pure Eulerian computation and the bottom part is the AMR computation with refinement on the shock and the interface. One can see that the development of the "mushroom" is the same but the AMR computation is much cheaper.

The AMR computation of the cylindrical implosion shown in Figure 2 took one week of one EV6 processor at 667 Mhz. In this computation there are 18 μ cells on the Sn/RTV interface and an average of 700 000 cells, the maximum being 1 700 000. Results are given

Figure 3. A shock tube instability computation

on Figure 4. An Eulerian equivalent with a mesh as fine as this one around the interface would require 64 million cells.

Although the AMR CFD scheme is explicit and "easy" to parallelize, the main problem we have is load balancing since the refinement changes dynamically (that is why the method works!). Domain decomposition of the physical domain and a straightforward MPI implementation do not give very good results since some processors have much more work to do than others. Since there must be only a difference of one level between neighboring cells, propagation of refinement is difficult to parallelize and a dynamic equilibration of work is not easy to do.

On SMP architectures, one can use shared memory within a node. A tentative to use POSIX threads on one node with many more subdomains than processors was done. This is some kind of self scheduling hoping that the operating system would do the job for us. An example was constructed for the advection equation simulating load imbalance by artificially increasing the workload in a part of the domain. Results for the elapsed time and the speed up are shown on Figures 5 and 6 using 4 processors. One can see that when increasing the number of threads to about 32 a good speed up is obtained. This work has been done by H. Jourdren and D. Dureau.

Unfortunately this does not work so well for the AMR code where there are many more communications. We are now looking for a mixed model: MPI and threads and also at modifications of the numerical scheme to increase parallelism. This would hopefully lead us to benefit from memory and computer time savings using the AMR technique and, at the same time, being able to run very large computations with several hundred millions of AMR cells in parallel.

REFERENCES

1. B. Després, Inégalité entropique pour un solveur conservatif du système de la dynamique des gaz en variables de Lagrange, C. R. Acad. Sci. Paris, Série 1, v 324 (1997) 1301-1306.
2. B. Després, Structure des sytèmes de lois de conservation en variables de Lagrange, C. R. Acad. Sci. Paris, Série 1, (1999).

Figure 4. An AMR instability computation

3. B. Després, Invariance properties of lagrangian systems of conservation laws, approximate Riemann solvers and the entropy condition, to appear in Numerische Mathematik.

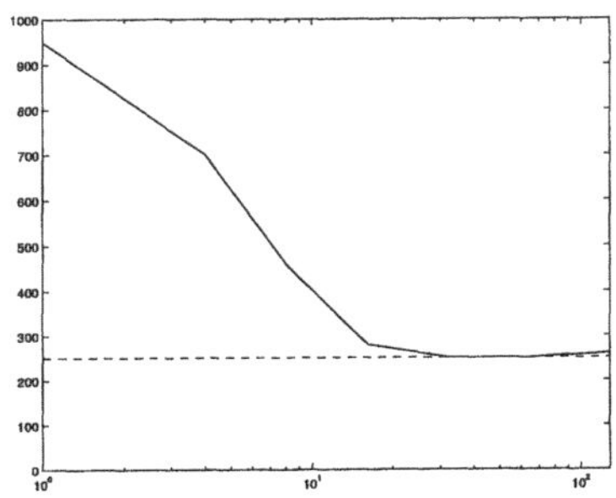

Figure 5. Elapsed time as a function of the number of Posix threads, dashed: 4 balanced MPI processes.

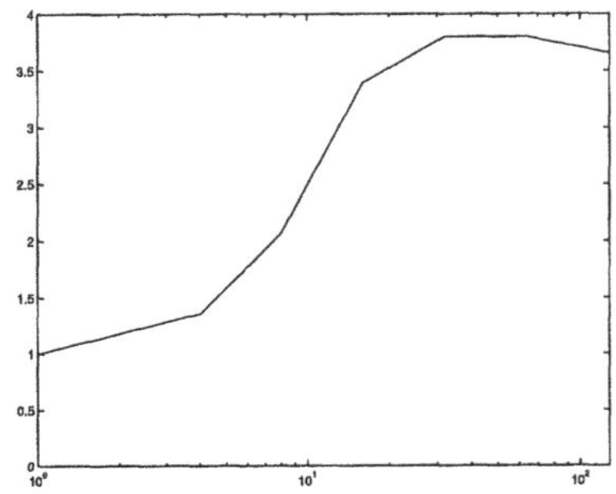

Figure 6. Speed up as a function of the number of Posix threads.

Parallel Computational Fluid Dynamics – Practice and Theory
P. Wilders, A. Ecer, J. Periaux, N. Satofuka and P. Fox (Editors)

Parallel Computation of Gridless Type Solver for Unsteady Flow Problems

K. Morinishi

Department of Mechanical and System Engineering,
Kyoto Institute of Technology,
Matsugasaki, Sakyo-ku, Kyoto 606-8585, Japan

This paper describes a gridless solver for unsteady flow problems and its performance on a parallel computer. Spatial derivative terms of any partial differential equations are gridlessly evaluated using cloud of points. The convective terms can be evaluated with either upwind or central manners. Numerical experiments for Poisson equations show the solver is second order accurate. Reliability and versatility of the gridless type solver have been demonstrated for various computational problems. Parallel computation is carried out using MPI library on the Hitachi SR2201 parallel computer up to 16 processors. The linear speedups are achieved up to 16 PUs.

1. INTRODUCTION

In the gridless type computation, points are first distributed over the computational domain considered. For example, Figure 1 shows the points distributed around a leading edge slat. At each point, spatial derivatives of any quantities are evaluated with linear combinations of certain coefficients and the quantities in the cloud of neighboring points, which is shown in Figure 2. The gridless type solver consists of the gridless evaluation of spatial gradients, upwind evaluation of inviscid flux, and central evaluation of viscous flux [1]. The evaluation is generally second order accurate. The solver may be applied to numerical solutions of any partial differential equations on any grids or points distributed over the computational domain. Reliability and versatility of the gridless type solver have been demonstrated for various computational problems, including Poisson equations, the shallow water equations, the incompressible and compressible Navier-Stokes equations. In this paper, a full description of the solver, some typical numerical results, and parallel computing performance are presented.

2. NUMERICAL PROCEDURE

The gridless evaluation of spatial derivatives, inviscid flux, and temporal discretization are described here.

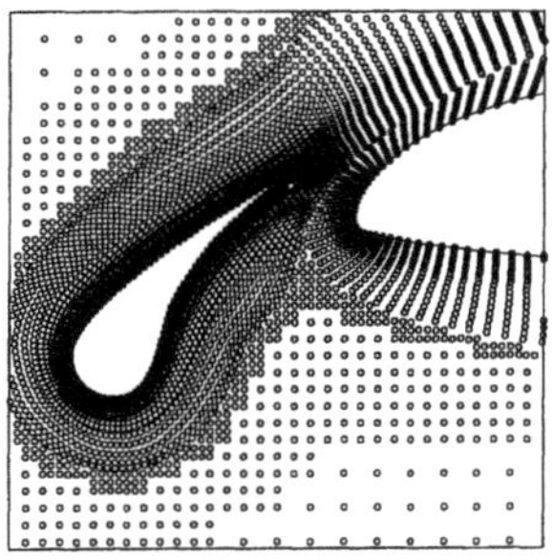

Figure 1. Close-up view of distributed points around a leading edge slat.

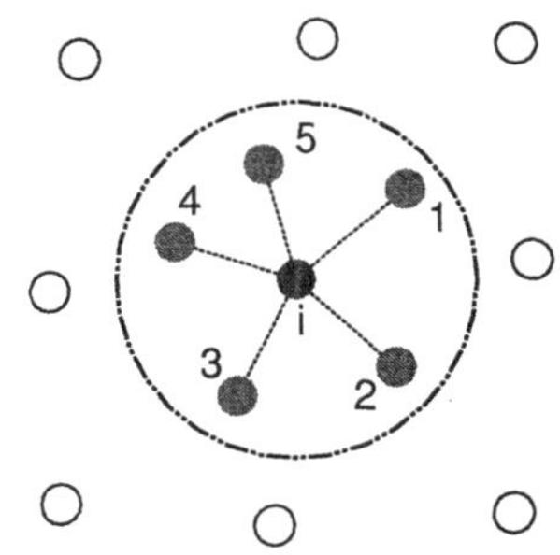

Figure 2. Cloud $C(i)$ for point i.

2.1. Evaluation of the first derivatives

The first derivatives of any function f at a point i can be evaluated with the following linear combination forms in the cloud $C(i)$:

$$\left.\frac{\partial f}{\partial x}\right|_i = \sum_{k\in C(i)} a_{ik} f_k \ , \quad \left.\frac{\partial f}{\partial y}\right|_i = \sum_{k\in C(i)} b_{ik} f_k \tag{1}$$

where the subscript k denotes the index of the point which belongs to the cloud $C(i)$. The sum is obtained over all member points of $C(i)$ including the point i itself. The coefficients a_{ik} and b_{ik} should have the following properties.

$$\sum_{k\in C(i)} a_{ik} = 0 \ , \quad \sum_{k\in C(i)} b_{ik} = 0 \ . \tag{2}$$

Thus the Eqs. (1) can be rewritten in the following forms.

$$\left.\frac{\partial f}{\partial x}\right|_i = \sum_{k\subset C(i)} a_{ik}(f_k - f_i) \ , \quad \left.\frac{\partial f}{\partial y}\right|_i = \sum_{k\subset C(i)} b_{ik}(f_k - f_i) \ . \tag{3}$$

The sum is obtained over all member points of $C(i)$ excluding the point i.

The spatial derivatives of the function f can also be evaluated using the following formulas:

$$\left.\frac{\partial f}{\partial x}\right|_i = \sum_{k\subset C(i)} a_{ik} f_{ik} \ , \quad \left.\frac{\partial f}{\partial y}\right|_i = \sum_{k\subset C(i)} b_{ik} f_{ik} \tag{4}$$

where f_{ik} are evaluated at the midpoint between points i and k. The values may be obtained with a simple arithmetical average. The coefficients a_{ik} and b_{ik} should also have the following properties.

$$\sum_{k\subset C(i)} a_{ik} = 0 \ , \quad \sum_{k\subset C(i)} b_{ik} = 0 \ . \tag{5}$$

The coefficients a_{ik} and b_{ik} can once be obtained using weighted least-squares curve fit methods and stored at the beginning of computation. The weight functions used in this study are given with:

$$w_{ik} = \begin{cases} 1 & r_k \leq \bar{r}_i \\ \left(\frac{\bar{r}_i}{r_k}\right)^2 & r_k > \bar{r}_i \end{cases} \tag{6}$$

where r_k are the relative distances defined with:

$$r_k = \sqrt{(x_k - x_i)^2 + (y_k - y_i)^2} \tag{7}$$

and $\bar{r}_i$ is a reference distance determined for each point i. The distance to the nearest point is a good choice for the reference distance. It should be noted that if the method is applied on a point of any uniform Cartesian grid with usual five point stencil for its cloud, the coefficients are strictly identical to those of the conventional second-order central difference approximations.

2.2. Evaluation of the second derivatives

The second derivatives of the function f can be evaluated with following sequential manner.

$$\left.\frac{\partial^2 f}{\partial x^2}\right|_i = \sum_{k \subset C(i)} a_{ik} \left.\frac{\partial f}{\partial x}\right|_{ik} . \tag{8}$$

The first derivative at the midpoint is evaluated, instead of a simple arithmetical average, using the following equation:

$$\left.\frac{\partial f}{\partial x}\right|_{ik} = \frac{\Delta x}{\Delta s^2}(f_k - f_i) + \frac{1}{2}\frac{\Delta y}{\Delta s^2}\left[\Delta y\left(\left.\frac{\partial f}{\partial x}\right|_i + \left.\frac{\partial f}{\partial x}\right|_k\right) - \Delta x\left(\left.\frac{\partial f}{\partial y}\right|_i + \left.\frac{\partial f}{\partial y}\right|_k\right)\right] \tag{9}$$

where

$$\Delta x = x_k - x_i \ , \quad \Delta y = y_k - y_i \ , \quad \Delta s^2 = \Delta x^2 + \Delta y^2 \ .$$

A Laplace operator as well as the second derivatives can also be evaluated directly as follows:

$$\left(\frac{\partial^2 f}{\partial x^2} + \frac{\partial^2 f}{\partial y^2}\right)_i = \sum_{k \in C(i)} c_{ik} f_k \ . \tag{10}$$

The coefficients c_{ik} can be obtained and stored at the beginning of computation with solving the following system of equations using QR or singular value decompositions.

$$\sum_{k \in C(i)} c_{ik} f_k^{(m)} = d^{(m)} \ . \tag{11}$$

The components of $f^{(m)}$ and $d^{(m)}$ are given with:

$$f^{(m)} \in \left(1, x, y, x^2, xy, y^2, \cdots\right) \tag{12}$$

and

$$d^{(m)} \in (0, 0, 0, 2, 0, 2, \cdots) \ . \tag{13}$$

2.3. Evaluation of convective flux

A scalar convective equation may be written as:

$$\frac{\partial f}{\partial t} + \frac{\partial uf}{\partial x} + \frac{\partial vf}{\partial y} = 0 \tag{14}$$

where u and v are velocity field given. The convective term can be evaluated using the gridless evaluation of the first derivatives Eqs.(4) as:

$$\begin{aligned} \frac{\partial uf}{\partial x} + \frac{\partial vf}{\partial y} &= \sum a_{ik}(uf)_{ik} + \sum b_{ik}(vf)_{ik} \\ &= \sum g_{ik} , \end{aligned} \tag{15}$$

where the flux term g at the midpoint is expressed as:

$$g = Uf \qquad (U = au + bv) . \tag{16}$$

Similar expression can be obtained for vector equations. For example, the two-dimensional compressible Euler equations may be written as:

$$\frac{\partial \mathbf{q}}{\partial t} + \frac{\partial \mathbf{E}}{\partial x} + \frac{\partial \mathbf{F}}{\partial y} = 0 \tag{17}$$

The inviscid terms can be evaluated as:

$$\begin{aligned} \frac{\partial \mathbf{E}}{\partial x} + \frac{\partial \mathbf{F}}{\partial y} &= \sum a_{ik}\mathbf{E}_{ik} + \sum b_{ik}\mathbf{F}_{ik} \\ &= \sum \mathbf{G}_{ik} \end{aligned} \tag{18}$$

The flux term $\mathbf{G}$ at the midpoint is expressed as:

$$\mathbf{G} = \begin{pmatrix} \rho U \\ \rho u U + ap \\ \rho v U + bp \\ U(e+p) \end{pmatrix} \tag{19}$$

An upwind method may be obtained if the numerical flux on the midpoint is obtained using Roe's approximate Riemann solveras:

$$\mathbf{G}_{ik} = \frac{1}{2}\left(\mathbf{G}(\tilde{\mathbf{q}}_{ik}^{+}) + \mathbf{G}(\tilde{\mathbf{q}}_{ik}^{-}) - |\tilde{\mathbf{A}}|(\tilde{\mathbf{q}}_{ik}^{+} - \tilde{\mathbf{q}}_{ik}^{-})\right) \tag{20}$$

where $\tilde{\mathbf{q}}$ are the primitive variables and $\tilde{\mathbf{A}}$ are the flux Jacobian matrices. The second order accurate method may be obtained, if the primitive variables at the midpoint are reconstructed with:

$$\tilde{\mathbf{q}}_{ik}^{-} = \tilde{\mathbf{q}}_i + \frac{1}{2}\phi^{-}\delta\tilde{\mathbf{q}}_{ik}^{-} , \quad \tilde{\mathbf{q}}_{ik}^{+} = \tilde{\mathbf{q}}_k - \frac{1}{2}\phi^{+}\delta\tilde{\mathbf{q}}_{ik}^{+} \tag{21}$$

where $\tilde{\mathbf{q}}_{ik}^{-}$ and $\tilde{\mathbf{q}}_{ik}^{+}$ are defined with:

$$\delta\tilde{\mathbf{q}}_{ik}^{-} = \nabla\tilde{\mathbf{q}}_i \cdot \mathbf{r}_{ik} , \quad \delta\tilde{\mathbf{q}}_{ik}^{+} = \nabla\tilde{\mathbf{q}}_k \cdot \mathbf{r}_{ik} . \tag{22}$$

The flux limiters ϕ^- and ϕ^+ are defined as:

$$\phi^- = \frac{\delta\tilde{\mathbf{q}}_{ik}\delta\tilde{\mathbf{q}}_{ik}^- + |\delta\tilde{\mathbf{q}}_{ik}\delta\tilde{\mathbf{q}}_{ik}^-|}{\delta\tilde{\mathbf{q}}_{ik}^2 + \delta\tilde{\mathbf{q}}_{ik}^{-2} + \epsilon}, \quad \phi^+ = \frac{\delta\tilde{\mathbf{q}}_{ik}\delta\tilde{\mathbf{q}}_{ik}^+ + |\delta\tilde{\mathbf{q}}_{ik}\delta\tilde{\mathbf{q}}_{ik}^+|}{\delta\tilde{\mathbf{q}}_{ik}^2 + \delta\tilde{\mathbf{q}}_{ik}^{+2} + \epsilon} \tag{23}$$

where ϵ is very small number which prevents null division in smooth flow regions and $\delta\tilde{\mathbf{q}}_{ik}$ are defined as:

$$\delta\tilde{\mathbf{q}}_{ik} = \tilde{\mathbf{q}}_k - \tilde{\mathbf{q}}_i \,. \tag{24}$$

The monotonous quality of the solver may further improved if $\delta\tilde{\mathbf{q}}_{ik}^-$ and $\delta\tilde{\mathbf{q}}_{ik}^+$ are replaced with

$$\delta\tilde{\mathbf{q}}_{ik}^- = 2\nabla\tilde{\mathbf{q}}_i \cdot \mathbf{r}_{ik} - \delta\tilde{\mathbf{q}}_{ik}\,, \quad \delta\tilde{\mathbf{q}}_{ik}^+ = 2\nabla\tilde{\mathbf{q}}_k \cdot \mathbf{r}_{ik} - \delta\tilde{\mathbf{q}}_{ik}\,. \tag{25}$$

The third order accurate method may be obtained, for example, if $\tilde{\mathbf{q}}_{ik}^-$ are evaluated with the following reconstruction after weighted ENO schemes [2].

$$\tilde{\mathbf{q}}_{ik}^- = \tilde{\mathbf{q}}_i + \frac{1}{2}\left(\omega_0^-\delta\tilde{\mathbf{q}}_{ik}^- + \omega_1^-\delta\tilde{\mathbf{q}}_{ik}\right) \tag{26}$$

The weight ω_0^- and ω_1^- are defined with

$$\omega_0^- = \frac{\alpha_0^-}{\alpha_0^- + \alpha_1^-}, \quad \omega_1^- = \frac{\alpha_1^-}{\alpha_0^- + \alpha_1^-} \tag{27}$$

where

$$\alpha_0^- = \frac{1}{3}\frac{1}{\left[(\delta\tilde{\mathbf{q}}_{ik}^-)^2 + \epsilon\right]^2}, \quad \alpha_1^- = \frac{2}{3}\frac{1}{\left[(\delta\tilde{\mathbf{q}}_{ik})^2 + \epsilon\right]^2}\,. \tag{28}$$

The gradients of primitive variables $\nabla\tilde{\mathbf{q}}$ are obtained using Eqs.(4) at each point. These gradients of primitive variables are also used for the evaluation of the viscous stress in the Navier-Stokes equations.

2.4. Temporal discretization

Explicit Runge-Kutta methods or implicit sub-iteration methods can be used for the temporal discretization of the gridless type solver. For example, an implicit sub-iteration method may be written for the Euler equations (17) as:

$$\left(\frac{1}{\Delta\tau}\mathbf{I} + \sum_{k\subset C(i)} \mathbf{A}_{ik}^+(\mathbf{q}_i^n)\right)\Delta\mathbf{q}_i + \sum_{k\subset C(i)} \mathbf{A}_{ik}^-(\mathbf{q}_k^n)\Delta\mathbf{q}_k = -\frac{\partial\mathbf{q}}{\partial t} - \mathbf{W}(\mathbf{q}_i^{n+1,m}) \tag{29}$$

where τ is a pseudotime and $\mathbf{W}$ is the gridless evaluations of the flux terms. The correction $\Delta\mathbf{q}$ is defined as:

$$\Delta\mathbf{q} = \mathbf{q}^{n+1,m+1} - \mathbf{q}^{n+1,m} \tag{30}$$

where n and m are physical and pseudo time indexes. The second order solution may be obtained if the time derivative is evaluated as:

$$\frac{\partial\mathbf{q}}{\partial t} = \frac{3\mathbf{q}^{n+1,m} - 4\mathbf{q}^n + \mathbf{q}^{n-1}}{2\Delta t}\,. \tag{31}$$

The solution of this linear system of equation can be obtained with LU-SGS method [3].

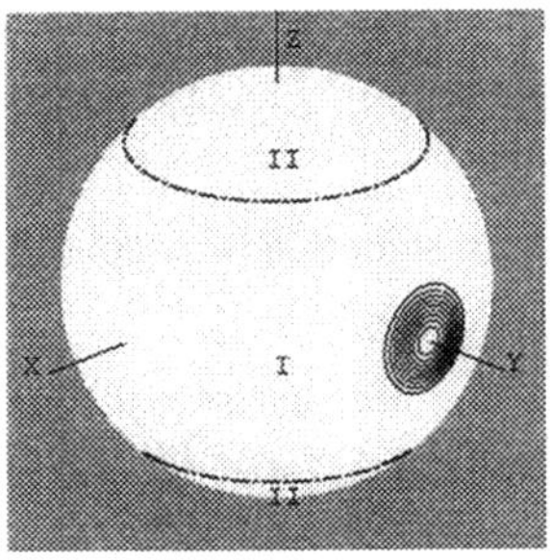

Figure 3. Initial cosine bell and coordinate zones.

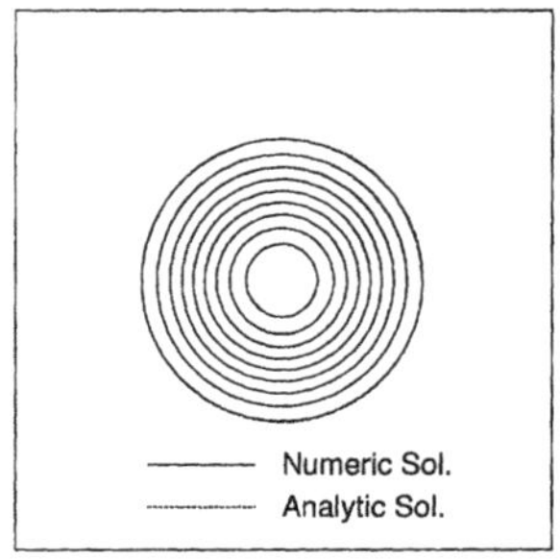

Figure 4. Comparison of numeric solution with analytic one after a full rotation.

3. NUMERICAL RESULTS FOR FUNDAMENTAL TESTS

In this section, reliability of the gridless type solver is examined in numerical results for fundamental test problems.

3.1. Advection of cosine bell on a spherical surface

A scalar convective equation on a spherical surface may be written as [4]:

$$\frac{\partial f}{\partial t} + \frac{1}{a\cos\theta}\left[\frac{\partial uf}{\partial \lambda} + \frac{\partial vf\cos\theta}{\partial \theta}\right] = 0 \tag{32}$$

where, if the sphere is the earth, λ is the longitude, θ the latitude, and a is the radius of the earth ($6.37122 \times 10^6 m$), respectively. The initial cosine bell is at λ of $\pi/2$ on the equator as shown in Fig. 3. The velocity field is given so that the cosine bell is advected around the earth through the poles as:

$$u = u_0 \cos\lambda \sin\theta\ , \quad v = -u_0 \sin\lambda \tag{33}$$

where advecting velocity u_0 is given by:

$$u_0 = 2\pi a/(12 days)\ . \tag{34}$$

The convective equation (32) is singular at θ of $\pm\pi/2$. In order to avoid the singularity, the following two coordinate zones are introduced for gridless computing of Eq. (32).

Zone I	$\lvert z\rvert \le \frac{1}{\sqrt{2}}$	$\theta_I = \sin^{-1} z$	$\lambda_I = \tan^{-1}\frac{y}{x}$
Zone II	$\lvert z\rvert > \frac{1}{\sqrt{2}}$	$\theta_{II} = \sin^{-1} x$	$\lambda_{II} = \tan^{-1}\frac{z}{y}$

Computational points are unstructurally distributed on the sphere. The total number of points used for this test case is 49154. The numerical solution after a full rotation (12 days) is compared with the analytic one in Fig. 4. The solution is obtained using the third order reconstruction. The comparison is very good so that it is hard to distinguish the numerical solution from analytic one.

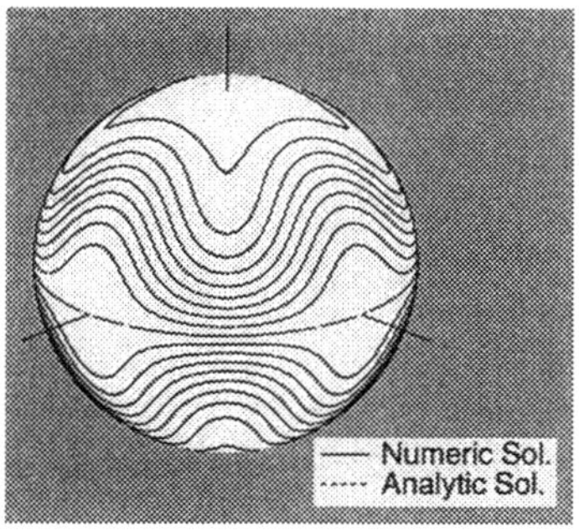

Figure 5. Comparison of stream function obtained for Rossby-Haurwitz waves with analytic one.

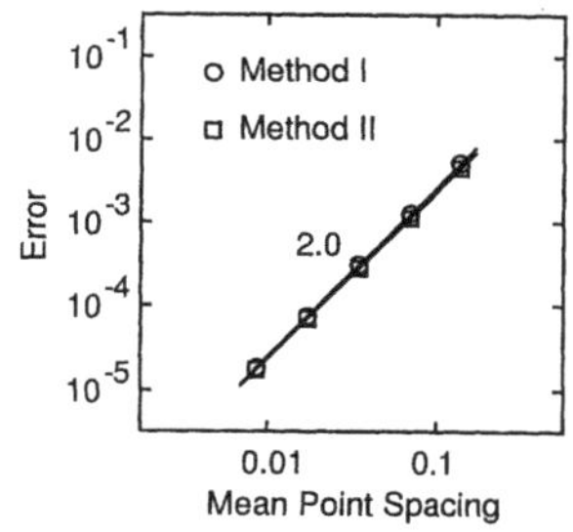

Figure 6. L_2 errors as a function of mean point spacing.

3.2. Application to a Poisson equation on a spherical surface

The Poisson equation for the stream function of Rossby-Haurwitz waves on the spherical surface may be written as [4]:

$$\frac{1}{a^2 \cos^2\theta} \frac{\partial^2 f}{\partial \lambda^2} + \frac{1}{a^2 \cos\theta} \frac{\partial}{\partial \theta}\left(\cos\theta \frac{\partial f}{\partial \theta}\right) = \zeta \tag{35}$$

Here ζ is the following vorticity

$$\zeta = 2\omega \sin\theta - K(R^2 + 3R + 2) \sin\theta \cos^R\theta \cos R\lambda \tag{36}$$

and f is the stream function of which analytic solution is given with:

$$f = -a^2\omega \sin\theta + a^2 K \sin\theta \cos^R\theta \cos R\lambda \tag{37}$$

where ω, K, and R are the following constants.

$$\omega = K = 7.848 \times 10^{-6} s^{-1}, \quad R = 4\,. \tag{38}$$

Numerical solutions of the Poisson equation are obtained with GMRES method on five different point density. Figure 5 shows the comparison of numerical solution with analytic one. The numerical solution is obtained with directly evaluating the Laplace operator on the spherical surface as:

$$\frac{1}{a^2 \cos^2\theta} \frac{\partial^2 f}{\partial \lambda^2} + \frac{1}{a^2 \cos\theta} \frac{\partial}{\partial \theta}\left(\cos\theta \frac{\partial f}{\partial \theta}\right) = \sum_{k \in C(i)} c_{ik} f_k \tag{39}$$

The total number of points used for the solution is 3074. Again the comparison is so good that no difference can be found between numerical and analytic solutions. The L_2 errors obtained on different point density are plotted as a function of normalized mean point spacing in Fig. 6. Two series of numerical data are plotted in the figure. One is obtained with sequential evaluation of the second derivatives (Method I) and the other is obtained with the direct evaluation of the Laplace operator (Method II). From the figure, both gridless evaluating methods are effectively second order accurate because both the slopes of error curves are about 2.0.

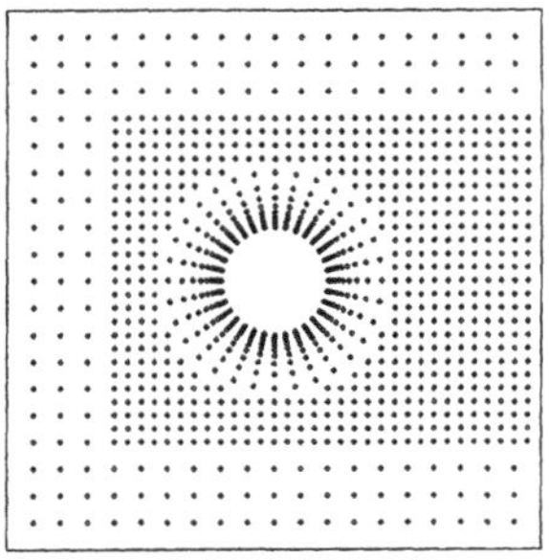

Figure 7. Sample point distributions for a circular cylinder.

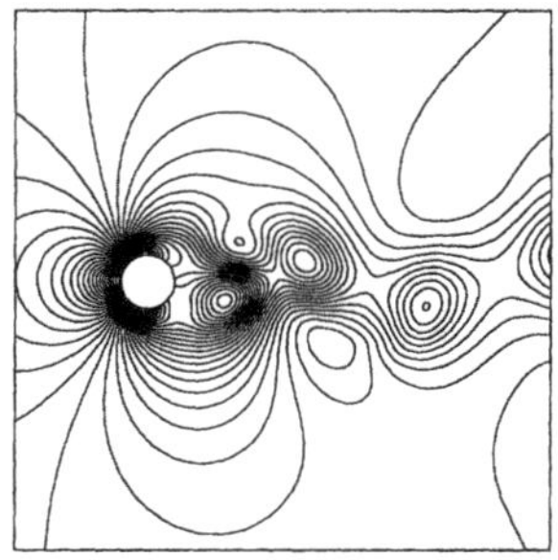

Figure 8. Instantaneous C_p contours around a circular cylinder.

3.3. Unsteady flow over a circular cylinder

Unsteady flows over a circular cylinder are typical test problems of numerical methods for the incompressible Navier-Stokes equations. Figure 7 shows a schematic sample of point distributions for the gridless type computation. The actual computation is made for the total point number of 41204 and the point spacing normal to the circular cylinder surface of 0.005. Figure 8 shows instantaneous pressure contours at a free stream Reynolds number of 100. The solution is obtained for the incompressible Navier-Stoke equations with artificial compressibility. The amplitude of lift coefficient, mean drag coefficient, and Strouhal number computed are 0.326, 1.337 and 0.166, respectively. These computed values agree well with many other results reported in literature.

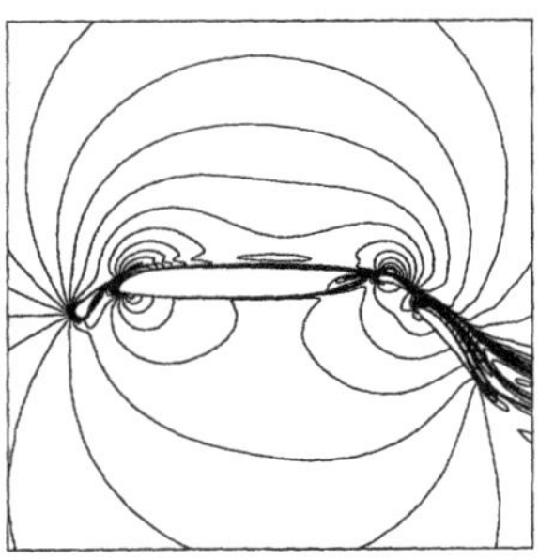

Figure 9. Mach number contours around a four-element airfoil.

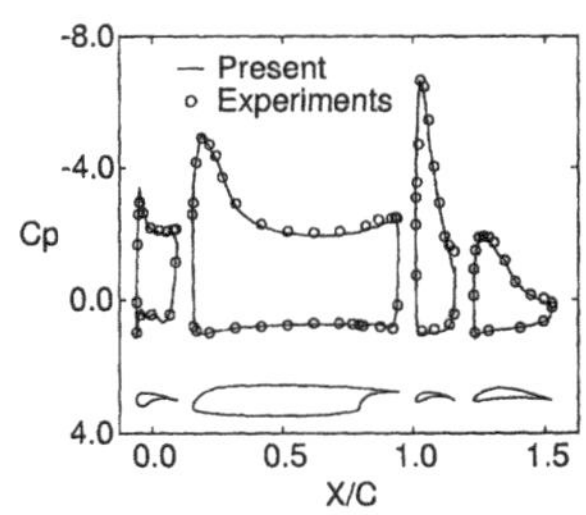

Figure 10. Comparison of surface pressure distributions.

3.4. Turbulent flow over a NASA four-element airfoil

Versatility of the gridless type solver can be demonstrated for numerical simulation of turbulent flows over a NASA four-element airfoil. Figure 1 shows a close-up view of points distributed around the leading edge slat. The whole computational domain is a square with a length of 32 chord lengths. The total number of points distributed in the domain is 38273, including 209 points on the main element surface, 129 points on the leading edge slat, and 97 and 129 points on the two trailing-edge flaps, respectively. The point spacing normal to the airfoil surface is about 10^{-5}. The computation is carried out for the compressible Navier-Stokes equations. Figure 9 shows the Mach number contours obtained for a free stream Mach number of 0.201, an attack angle of 0.01°, and a Reynolds number of 2.83×10^6. The Baldwin and Lomax turbulent model is used only on the points near the airfoil. The pressure distributions obtained along the airfoil surfaces are compared with experimental data in Fig. 10. The agreement of the prediction with experiments is quite satisfactory.

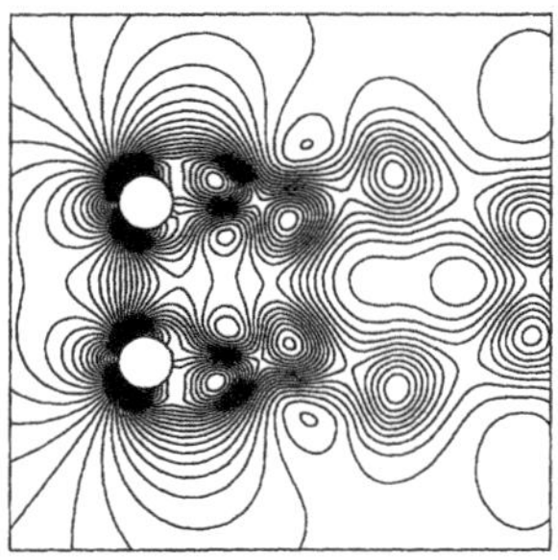

Figure 11. Instantaneous C_p contours around two circular cylinders.

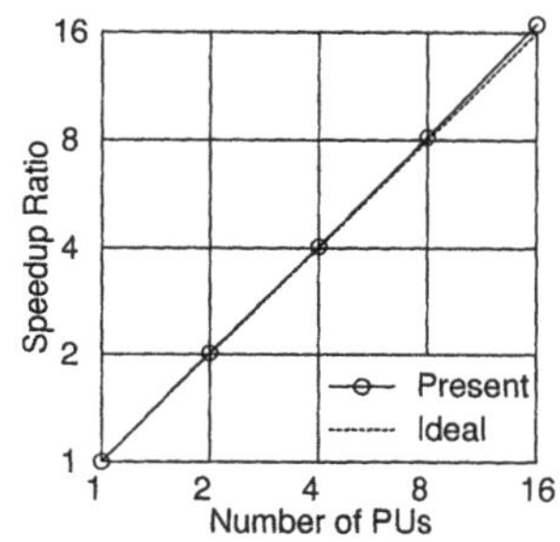

Figure 12. Speedup ratios as a function of the number of processors.

4. PARALLEL COMPUTING PERFORMANCE

In this section, parallel computing performance of the gridless solver based on domain decomposition is presented for two unsteady flow problems. Numerical experiments are carried out using MPI library on the Hitachi SR2201 parallel computer at Kyoto Institute of Technology. The computer has 16 PUs which are connected by a crossbar network. Each PU consists of a 150MHz PA-RISC chip, a 256MB memory, and two cascade caches.

Figure 11 shows instantaneous pressure contours for a flow over two circular cylinders at a free stream Reynolds number of 100. A typical flow pattern of antiphase vortex shedding is clearly captured in the figure. The total number of points used for the solution is 60735. Figure 12 shows speedup ratios as a function of the number of processors. The linear speedup is achieved up to 16 processors.

Figure 13 shows a typical numerical result of the shallow water equations on a rotating sphere. Williamson et al. [4] proposed a standard test set for validating numerical methods to the shallow water equations in spherical geometry. The numerical result is obtained for the case 6 of the standard test set. The total number of points used for this computation is 196610. The height solution at the day of 14 is plotted in the figure. The standard spectral transform solution of T213 spectral truncation [5], which corresponds to 640×320

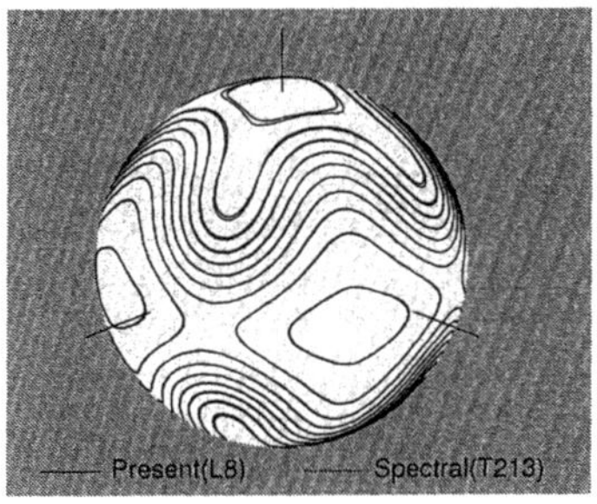

Figure 13. Comparison of height solution at day 14.

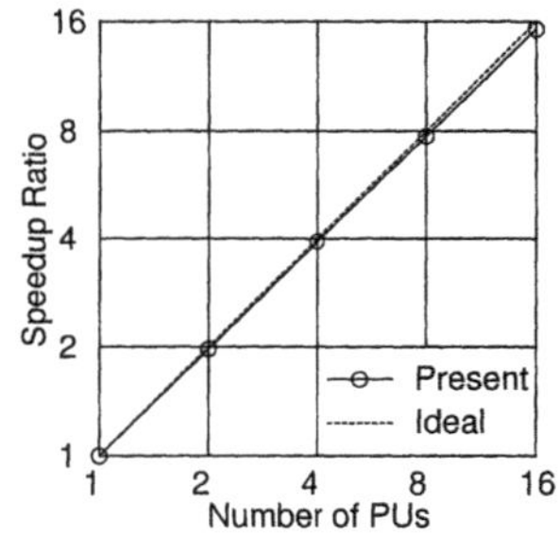

Figure 14. Speedup ratios as a function of the number of processors.

(= 204800) grid, is also plotted for comparison. The comparison between both solutions is generally good. Figure 14 shows speedup ratios as a function of the number of processors. The speedup ratio of 15.3 is achieved with 16 processors.

5. CONCLUSIONS

Parallel computing performance of the gridless type solver for unsteady flow problems are presented. Numerical experiments of various computational problems validate the reliability and versatility of the gridless type solver. Parallel computation is carried out using MPI library on the Hitachi SR2201 parallel computer. The linear speedups are achieved up to 16 processors.

This study was supported in part by the Research for the Future Program (97P01101) from Japan Society for the Promotion of Science and a Grant-in-Aid for Scientific Research (12650167) from the Ministry of Education, Science, Sports and Culture of the Japanese Government.

REFERENCES

1. Morinishi, K., A Gridless Type Solver - Generalized Finite Difference Method -, Computational Fluid Dynamics for the 21st Century, Notes on Numerical Fluid Mechanics, **78**, Springer, 43-58 (2001).
2. Jiang, G-H. and Shu, C-W., Efficient Implementation of Weighted ENO Schemes, *Journal of Computational Physics*, **126**, 202-229 (1996).
3. Morinishi, K., Parallel Computing Performance of an Implicit Gridless Type Solver, Parallel Computational Fluid Dynamics, Trends and Applications, Elsevier, 315-322 (2001).
4. Williamson, D.L., Drake, J.B., Hack, J.J., Jakob, R., and Swarztrauber, P.N., A Standard Test Set for Numerical Approximations to the Shallow Water Equations in Spherical Geometry, *Journal of Computational Physics*, **102**, 211-224 (1992).
5. Jakob-Chienm R., Hack, J.J, Williamson, D.L., Spectral Transform Solutions to the Shallow Water Test Set, *Journal of Computational Physics*, **119**, 164-187 (1995).

Parallel Computational Fluid Dynamics – Practice and Theory
P. Wilders, A. Ecer, J. Periaux, N. Satofuka and P. Fox (Editors)

Clusters in the GRID: Power Plants for CFD

Michael M. Resch [a]

[a]High Performance Computing Center Stuttgart, University of Stuttgart
Allmandring 30, 70550 Stuttgart, Germany, e-mail: resch@hlrs.de

Two trends have emerged in hardware architecture for computational science and engineering in the last years. First there is the idea of a computational grid [1]. The approach of the GRID is often described as: to provide computational power in the same way as the power grid provides electrical power. Harnessing the cycles of idle computers located somewhere in the internet computational power is at the disposal of scientists and engineers like electrical power. It has therefore become common to compare the computational GRID to the power grid. Computers serve as power plants in this environment. They should be transparent to the user and should provide a constant level of power over time.

The trend in computer hardware on the other hand is towards clusters [2,3,8,13]. Built from commodity parts they are at least theoretically able to provide supercomputing power at much lower costs. While traditional supercomputers are expensive and limited resources, clusters are said to be cheap and flexible. An increase in compute power can easily be achieved by increasing the number of PCs. A number of projects have proven the feasibility of such clusters [9,12,15,17].

This paper discusses the role of such clusters in a GRID environment from the point of view of computational fluid dynamics (CFD). It gives emphasis to the three main stages of simulation: pre-processing, processing and post-processing.

1. Introduction

Clusters of workstations and PCs have gradually replaced supercomputers in recent years. While in Japan at least NEC with its SX-series was able to keep up with the price/performance ratio of clusters, in the US vector computers were for some time almost extinct due to the prohibitively high import taxes imposed by the US government. Only recently and following pressure from HPC users [4] an agreement between Cray, NEC and the US government was achieved that opens the US market for Japanese vector supercomputers again [16]. It was in this environment that clusters were able to develop a dominance which from a US point of view seemed to be complete. Especially the key projects like ASCI [5] and NPACI [6] were focusing exclusively on clusters. As a consequence this architecture can be found today on each level of computing from desktop systems to supercomputers.

The concept of a cluster was stretched somewhat further by projects that aimed to couple geographically dispersed systems to form a single resource [7]. Although usage of such distributed systems for a single application showed some problems the concept itself was further pursued to become what is nowadays called GRID-computing [1]. The basic idea is to create an infrastructure that provides dependable, consistent, pervasive, and inexpensive access to high-end computational capabilities. The key elements are: Usability as easy as a PC, coordination of resources, pervasiveness, consistency, dependability, flexibility, security, low cost, access to high speed resources. Summarizing that concept one might say that the GRID attempts to provide CPU power as simple as electrical power with the GRID resembling the power grid.

Given the enormous needs in compute power of CFD the theoretically low price/performance ratio of clusters as well as the ubiquitous and cheap access to CPU power as heralded by the grid community have become of interest for the CFD community. The vision is to have a GRID where clusters serve as computational power plants and users access whatever resource they need to solve their problems.

In the following we work out in more detail the features of clusters and the GRID relevant to CFD (chapters 2 and 3). In chapter 4 we categorize the requirements of CFD simulations. Chapter 5 describes the role of clusters as power plants in the GRID for CFD applications. From these findings we extract an ideal cluster which is described in chapter 6.

2. The Cluster Concept

The term cluster was quite clearly defined for some time as being a couple of stand-alone systems that are coupled very loosely by a slow network. This notion has changed a little bit with the cluster concept becoming a more main stream approach and with high speed networks becoming cheaper [13]. Today a cluster can be characterized by the following components:

- Some kind of node that may be either a PC or workstation with 1 or more processors. A single node can also work as a stand-alone system.
- Some kind of network that takes care of the communication requirements.
- Some kind of software environment that helps to manage and use the cluster as a single resource although probably not providing a full single system image.

The typical cluster, as can be found in many research departments, looks as follows:

- A PC node with one or two Intel-Pentium processors
- Fast Ethernet, Gigabit Ethernet or Myrinet for communication
- Beowulf [9] or SCore [17] software as a 'glue' for all components

The goals of such systems are:

- Low price resulting in a potentially high performance/price ratio: With the performance level of Pentium systems coming close to that of workstation processors and prices staying rather low, due to the high sales volume, this goal is achieved at least for the theoretical performance level.

- Flexibility of the overall system: Clusters can easily be used for multiple purposes. Being available for teaching during day time they can be converted to computing resources during night time and weekends [12]. Furthermore, the concept of a cluster allows to easily add both nodes and network connections without having to change the system. This allows to adapt to changing needs and changing budgets.

- Scalability: In principal a cluster can scale to thousands of processors as long as only raw processor power is considered. For a number of applications this is the key issue.

These systems – as long as they are small (with node numbers in the range of 32-128) – are quite cheap and can easily be installed and maintained by small groups of researchers. They are designed to work for one or a few applications. Sometimes they are also used for teaching. Users benefit from the availability of the resource but suffer from the loose integration of the components. The main drawbacks are:

- Low memory bandwidth: This has recently been improved by the Pentium IV.

- Low network bandwidth: Even if Myrinet is used the speed of the PCI-bus limits the bandwidth.

- High latency: Only with special implementations of MPI a competitive latency can be achieved for Myrinet.

Furthermore, there are a number of other approaches to cluster computing. They all try to make use of standard components and are distinguished from each other in the level of integration they attempt to achieve. Usually, however, higher integration increases the price. Such big systems therefore are found in large projects and differ from small clusters mainly in their substantial price level [5,6].

3. The GRID Concept

The notion of the GRID was first coined by Foster and Kesselman [1] who tried to summarize the state of the art in the field. The basic idea is to make distributed computational resources available to users without geographical limitations. Initially the driving force was metacomputing which was aiming at coupling systems to increase the level of performance [7,18]. Experience, however, shows that the most likely positive impact of coupling systems is less to be found in increased performance and more in better usage of compute cycles and other resources [14]. The basic idea is to create an infrastructure that provides:

- dependable

- consistent

- pervasive

- inexpensive

access to high-end computational systems. The key elements are:

- Usability as easy as a PC: The final goal is to fully hide away the complexity and heterogeneity of distributed resources from the user.
- Coordination of resources: Intentionally the GRID includes thousands of computers and other resources. These will have to be coordinated in a reasonable way.
- Flexibility: The more resources the GRID includes the more complex the structure may become. It will have to cope dynamically with new resources and with removed ones.
- Security: Authorization, Authentication and Access control are critical issues once the GRID leaves the academic non-profit community.
- Low cost: A key to success for the GRID is the reduction of costs. Only if it's cheaper to have work done in the GRID rather than do it locally the GRID does have any meaning.

A number of projects try to meet these requirements. The most relevant ones today are probably GLOBUS [19], Legion [20], UNICORE [21] and TME [22]. None of these projects is currently able to solve all problems but they help at least to lay the foundation for a future GRID infrastructure.

4. Requirements of CFD

The requirements for pre-processing are typically low with respect to performance of the system. But they are higher with respect to main memory. A lot of software for preprocessing has been ported to PCs during the last years. This has also been a result of the spreading of Linux. A typical PC configuration with an IA32 processor can provide memory only up to 4 GByte. Although of competitive speed IA32 processors will face a memory problem there.

Processing requirements for a standard CFD application are much harder to fulfill. Typically a CFD-code requires high memory bandwidth to allow for a substantial speed of the computation. Traditionally this high memory speed has been provided by vector-computers like the Cray T90 or the NEC SX-series while microprocessors lagged behind substantially. The next generation of microprocessors claims to address the problem of memory bandwidth. First announcements and preliminary results actually show a substantial increase. However, this relative increase in bandwidth does by far not meet the requirements of CFD applications. While these usually need a bandwidth of about 1.5 words per flop typical microprocessors still deliver only about 0.5 or less words per flop. Technically spoken this means that the sustained performance even of optimized codes will be less than one third of the peak performance. First experience shows that even with new microprocessors we can not expect to see more than about 10-20 percent of the peak performance in a typical application. Results in the range of 5 percent are more likely.

What is worse is the high latency for memory access. Due to complaints of users vendors have increased memory bandwidth. At the same time, however, latencies have not been improved in the same way. What we will face in the near future is a further increase in speed of processors, accompanied with only a moderate increase in bandwidth. But latencies are expected to remain pretty much the same, such that more and more it is the high latencies that further reduce sustained performance.

Post-processing for some years was done mainly on sgi systems. With new graphics hardware integrated into standard PCs this may change dramatically in the future. Currently a number of projects are aiming at the development of visualization software for PCs at a level of quality that is comparable to traditional sgi systems. Again the bottleneck for IA32 processors may be the small main memory. An additional problem is the synchronization of images if PCs are used to drive VR environments like caves.

5. Clusters in the GRID

Summarizing these first results for the three production stages in CFD one could claim that a cluster is a potential power plant for a computational GRID to serve the needs of CFD. However, there are still some open issues to be resolved.

The key issue is performance. With the US open again for advanced Japanese technology and Cray working intensively on a new generation of systems vector computers have come back into the game. Clusters with their low memory bandwidth and high memory access latencies are hardly able to compete with such high-end systems. CFD is not typically an embarrassingly parallel application which can easily be distributed across thousands of loosely coupled nodes. The key to success mostly is shared memory with high bandwidth. Clusters will have a problem to even compete with the performance level of vector supercomputers.

The price/performance aspect is another issue which does not allow to give a clear answer. Clusters definitely allow small groups to get substantial levels of performance without investing too much money. Such systems can be run locally as long as they are kept small. With increased number of nodes in the cluster, however, the 'total cost of ownership' is more dominated by costs for maintenance and infrastructure rather than the price of the system itself. The big cluster projects of the ASCI program [5] show exactly this problem.

Although a lot of effort has been put into stabilizing software environments for clusters [9–11,17] they still are not as easy to handle as a traditional supercomputer. Although Linux has become rather mature it is not as stable as expected by users. So far hardly any efforts have been made to integrate several different systems into a single cluster that can serve as a pool of resources for different purposes. Furthermore the stability of the hardware has become a real issue. With a rate of failure of 1 system out of 100 per month bigger clusters may loose one component as often as every 8 hours. It will take software measures to make sure that such problems are hidden away from the user and do not affect performance.

6. A Cluster for CFD

Given the requirements of CFD it is obvious that the speed of access to memory (both with respect to latency and bandwidth) is one of the key factors. For parallel applications the speed of network communication between processes is another important factor (again with respect to latency and bandwidth). These requirements are best fulfilled by a shared memory system based on vector processors.

However, these systems are limited in scalability. While the number of processors to be found in clusters is always increasing, the number of CPUs in shared vector systems is decreasing. So when it comes to building bigger systems even traditional vector computers are clustered as is the case in the Japanese Earth Simulator Project [23]. We end up with all supercomputers being a bundle of boxes, connected by some high-speed communication network.

Knowing this, the key issue in the design of a system is in the data management. On a single system, access to data is already a problem. With hundreds or even thousands of boxes handling of data becomes tricky. The core of a cluster that we propose is therefore a global parallel file system. The cluster itself takes care of the requirements of CFD and integrates a supercomputer, pre-processing servers and visualization servers.

The role of pre-processing servers is to offload the standard work from the supercomputer. The role of the visualization server is to offload post-processing work. Data should be kept in the global file system during the production cycle of a CFD simulation. Although high speed wide area networks are in place today the sustained bandwidth is hardly high enough to transfer terabytes of data to the end user. It is therefore much better to keep data in place and to only transfer images created by the visualization server.

The supercomputer has to consider the requirements of CFD simulations. Given the still tremendous needs for vector supercomputers [4] it should integrate both vector systems and microprocessor systems. There are two reasons why microprocessors should be included. First, it is impossible to achieve a competitive level of performance with a pure vector system without compromising budget limitations. Even tough price/performance may be better for small and medium sized installations microprocessors are the only technology to scale at reasonable prices to the thousands which is the only way today to reach tens of TFLOPS. Second, – due to the lack of vector computers in the US – there are a number of new algorithms that have been specifically tuned for clustered systems. Such algorithms may in the future well be able to exploit thousands of processors even though they may not perform very well on vector processors.

7. Conclusion

We have briefly gone through the requirements of CFD. We have shown the potential of clusters and the expected benefits of the GRID. Our comparison of requirements and available functionality shows that loosely coupled clusters of PCs are by no means fit to serve as power plants for CFD in the GRID. Still vector processors are the only technology to provide the necessary memory bandwidth and low latency. Still the highest possible integration of processors and memory is the way to achieve scalability at the application performance level. However, with the economical pressure on hardware vendors rising we have to face the challenge of price/performance ratios of big clusters. We therefore propose as power plants in the GRID a mixture of vector processors and microprocessors. This allows to serve the needs of traditional CFD users and at the same time helps to further develop new scalable methods.

REFERENCES

1. Ian Foster and Carl Kesselman, "The Grid: Blueprint for a New Computing Infrastructure", Morgan Kaufmann San Francisco/California, 1999.
2. Rajkumar Buyya, "High Performance Cluster Computing, Volume 1: Architectures and Systems", Prentice Hall, 1999.
3. Rajkumar Buyya, "High Performance Cluster Computing, Volume 2: Programming and Applications", Prentice Hall, 1999.
4. U.S. Global Change Research Program, Subcommittee on Global Change Research, "High-End Climate Science: Development of Modeling and Related Computing Capabilities", A Report to the USGCRP from the ad hoc Working Group on Climate Models, December 2000.
5. Accelerated Strategic Computing Initiative (ASCI), http://www.llnl.gov/asci/ (6.9.2001)
6. National Partnership for Advanced Computational Infrastructure (NPACI), http://www.npaci.edu/ (6.9.2001)
7. T. DeFanti, I. Foster, M. E. Papka, R. Stevens and T. Kuhfuss, "Overview of the I-WAY: Wide Area Visual Supercomputing", International Journal of Supercomputing Applications, 10, 123-131, 1996.
8. Uwe Harms, "Clusters triumph over custom systems", Scientific Computing World, 49, October/November 1999.
9. Daniel Ridge, Donald Becker, Phillip Merkey, Thomas Sterling, "Beowulf: Harnessing the Power of Parallelism in a Pile-of-PCs", Proceedings IEEE Aerospace, 1997.
10. Lorna Smith, "Comparison of Code Development Tools on Clusters", Technology Watch Report, Edinburgh Parallel Computing Centre, The University of Edinburgh, Version 1.0, 1999.
11. Kenneth Cameron, "Harnessing the Power of PC Clusters", Technology Watch Report, Edinburgh Parallel Computing Centre, The University of Edinburgh, Version 1.0, 1998.
12. F. Rauch, C. Kurmann, B. M. Müller-Lagunez, T. M. Stricker, "Patagonia - A Dual Use Cluster of PCs for Computation and Education", Proc. of the second workshop on Cluster-Computing, 25./26. March 1999, Karlsruhe, Germany, 1999.
13. SMABY Group, "Complex Scalable Computing - Trends and Analysis in the Technical Market", Executive Summary, Market Audit & Forecast, May 1999.
14. B. Engquist, L. Johnsson, F. Short (Eds.), "Simulation and Visualization on the Grid", Lecture Notes in Computational Science and Engineering, Springer, 2000.
15. Panagiotis A. Adamidis and Michael M. Resch, "Low cost computing in casting industries", European Congress on Computational Methods in Applied Sciences and Engineering, ECCOMAS 2000, Barcelona, 11-14 September 2000.
16. Earl Joseph II, Christopher Willard, Debra Goldfarb and Nicholas Kaufmann, "Capability Market Dynamics Part 1: Changes in Vector Supercomputers - Will the Cray/NEC Partnership Change the Trend?", IDC, March 2001.
17. Yutaka Ishikawa, Hiroshi Tezuka, Atsuhi Hori, Shinji Sumimoto, Toshiyuki Takahashi, Francis O'Carroll, and Hiroshi Harada, "RWC PC Cluster II and SCore Cluster System Software – High Performance Linux Cluster." In Proceedings of the 5th Annual

Linux Expo, pages 55-62, 1999.
18. Michael Resch, Dirk Rantzau and Robert Stoy, "Metacomputing Experience in a Transatlantic Wide Area Application Testbed", Future Generation Computer Systems (15)5-6 (1999) pp. 807-816, 1999.
19. Ian Foster, Carl Kesselman, "GLOBUS: A Metacomputing Infrastructure Toolkit", International Journal of Supercomputer Applications, 11, 115-128, 1997.
20. Andrew Grimshaw, Adam Ferrari, Fritz Knabe, Marty Humphrey, "Legion: An Operating System for Wide-Area Computing", Technical Report, University of Virginia, CS-99-12, 1999.
21. D. Erwin, "UNICORE and the Project UNICORE Plus", Presentation at ZKI working group Supercomputing meeting on May 25, 2000.
22. Hiroshi Takemiya, Toshiyuki Imamura and Hiroshi Koide, "TME a Visual Programming and Execution Environment for a Meta-Application", JAERI internal report, 2000.
23. Keiji Tani, "Status of the Earth Simulator System", Proceedings of Supercomputer 2001, Heidelberg/Germany, June 2001.

Parallel Computational Fluid Dynamics – Practice and Theory
P. Wilders, A. Ecer, J. Periaux, N. Satofuka and P. Fox (Editors)

An efficient parallel algorithm for solving unsteady Euler equations

Wilson Rivera[a], Jianping Zhu[b] and David Huddleston[c]

[a]Electrical and Computer Engineering Department, University of Puerto Rico at Mayaguez, P.O. Box 9042, Mayaguez, PR 00680, U.S.A.

[b]Department of Mathematics & Statistics and Engineering Research Center, Mississippi State University, Mississippi State, MS 39762, U.S.A.

[c]Department of Civil Engineering, Mississippi State University, Mississippi State, MS 39762, U.S.A.

When solving time dependent partial differential equations on parallel computers using the non-overlapping domain decomposition method, one often needs numerical boundary conditions on the boundaries between subdomains. These numerical boundary conditions can significantly affect the stability and accuracy of the final algorithm.

In this paper, a new approach for generating boundary conditions based on explicit predictors and implicit correctors is used to solve unsteady Euler equations for airfoil flow field calculations. Numerical results demonstrate that the new algorithm is highly scalable and more accurate than the existing algorithms.

1. Introduction

Domain decomposition method has been widely used for solving time dependent PDEs. It dates back to the classical Schwartz alternating algorithm with overlapping subdomains for solving elliptic boundary value problems. Advantages of using domain decomposition approach include high level of parallelism, efficient treatment of complex geometries, and reduction of computational complexity and storage.

When solving time dependent PDEs using non-overlapping subdomains, the domain decomposition method could either be used as a preconditioner for Krylov type algorithms, or as a means to decompose the original domain into subdomains and solve the PDEs defined in different subdomains concurrently. When it is used as a preconditioner, the relevant PDE is discretized over the entire original domain to form a large system of algebraic equations, which is then solved by Krylov type iterative algorithms. The preconditioning step and the inner products involved in the solution process often incur a significant amount of communication overhead that could significantly affect the scalability of the solution algorithms.

If the original domain Ω is decomposed into a set of non-overlapping subdomains Ω_k, $k = 1, \cdots, K$, it would be ideal that the PDEs defined in different subdomains could be solved independently. This often requires numerical boundary conditions at the boundaries between subdomains as shown in Fig. 1, where a one-dimensional domain Ω is decomposed into M subdomains Ω_i, $i = 1, \cdots, M$, and numerical boundary conditions at the points $r_i, i = 1, \cdots, M-1$, are needed if the PDEs defined in $\Omega_i's$ are to be solved concurrently. These numerical boundary conditions are not part of the original mathematical model and the physical problem. One way to generate those numerical boundary conditions is to use the solution values from the previous time step t_n [1,4,5]. This is often referred to as time lagging (TL). The other way to generate numerical boundary conditions is to use an explicit algorithm to calculate the solutions at the boundaries between subdomains based on the solutions from the previous time step, and then solve the PDEs defined on different subdomains concurrently using an implicit method [2,3]. This is referred to as the explicit predictor (EP) method.

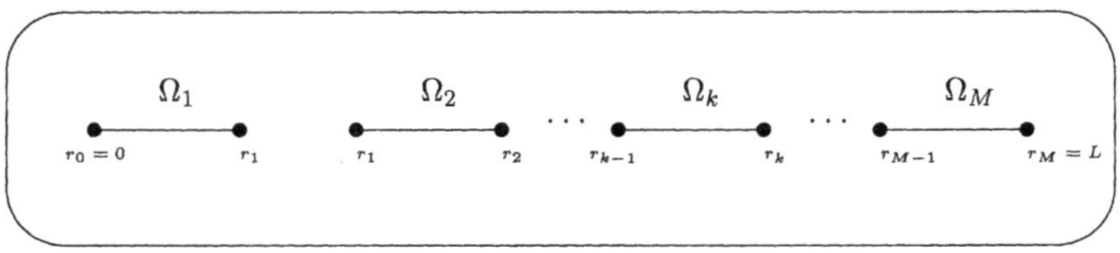

Figure 1. The original domain Ω is decomposed into M subdomains $\Omega_k, k = 1, \ldots, M$. Numerical boundary conditions are needed at the points r_i between subdomains, $i = 1, \cdots, M-1$.

Table 1
Maximum errors : $\frac{\Delta t}{\Delta x^2} = 1$, $T = 1.2$, $M = 2$.

Δx	BTCS	TL	EP
0.1000	0.2506E-02	0.2796E-02	0.2504E-02
0.0500	0.6263E-03	0.6796E-03	0.6261E-03
0.0250	0.1566E-03	0.1717E-03	0.1565E-03
0.0100	0.2505E-04	0.2936E-04	0.2505E-04
0.0050	0.6262E-05	0.8195E-05	0.6262E-05
0.0025	0.1566E-05	0.2482E-05	0.1566E-05

In [6,7], we have demonstrated, using a linear one-dimensional heat equation, that the stability and accuracy of the solution algorithms could be significantly affected by the TL and EP methods. Table 1 shows the maximum errors between the exact solution and the numerical solutions obtained using three different methods for the following equation

$$\frac{\partial u}{\partial t} = \frac{\partial^2 u}{\partial x^2} - 12x^2, \quad x \in \Omega = (0,1), \quad t \in [0,1], \tag{1}$$

Table 2
Maximum errors : $\frac{\Delta t}{\Delta x^2} = 1$, $T = 0.3$, $M = 2$.

Δx	BTCS	TL	EP
0.1000	0.1128E-01	0.8482E-01	0.8032E-02
0.0500	0.2804E-02	0.3918E-01	0.2395E-02
0.0250	0.6999E-03	0.1827E-01	0.6487E-03
0.0100	0.1119E-03	0.6928E-02	0.1086E-03
0.0050	0.2789E-04	0.3396E-02	0.2757E-04
0.0025	0.6995E-05	0.1680E-02	0.6944E-05

Table 3
Maximum errors: $\frac{\Delta t}{\Delta x} = 0.1$, $T = 0.3$, $M = 2$.

Δx	BTCS	TL	EP
0.1000	0.1128E-01	0.8482E-01	0.8032E-02
0.0500	0.4707E-02	0.8235E-01	0.1448E+06
0.0250	0.2122E-02	0.8132E-01	0.3704E+45
0.0100	0.7933E-03	0.8078E-01	0.5644E+201
0.0050	0.3874E-03	0.8060E-01	∞
0.0025	0.1914E-03	0.8052E-01	∞

with the initial and boundary conditions

$$u^*(x,0) = \sin \pi x + x^4, \quad u(0,t) = 0, \quad u(1,t) = 1,$$

and the exact solution

$$u^*(x,t) = e^{-\pi^2 t} \sin \pi x + x^4.$$

Three different algorithms are used to calculate numerical solutions at $T = 1.2$. The BTCS refers to the use of the backward time central space algorithm without domain decomposition. There is no need for numerical boundary conditions in this case. TL and EP refer to the use of time lagging and explicit predictor methods, respectively, to generate a numerical boundary condition at the middle point of the domain Ω, which is decomposed into two subdomains. Note that, in this particular case, the errors from all methods are similar, with TL method being slightly inaccurate, which may lead one to conclude that all three methods have comparable accuracy.

Table 2 has similar contents as those in Table 1, except that the solution is calculated to the time level of $T = 0.3$. It now appears that the EP method, with similar errors as those from the BTCS algorithm, is more accurate than the TL method. However, the results from Table 3, also calculated to $T = 0.3$ using $\frac{\Delta t}{\Delta x} = 0.1$, shows that both EP and TL fail to deliver solutions with similar accuracy as those from the BTCS algorithm.

Note that although both EP and TL methods become quite inaccurate in some cases, the behaviors of the errors are quite different for the two methods: The errors from the EP method demonstrate exponential growth in some cases, indicating lose of stability. However, as long as the computation stays in the stable region, the accuracy is similar to the BTCS algorithm without domain decomposition. On the other hand, the errors from the TL method do not grow exponentially. They just decrease at a much slower rate than the BTCS algorithm in some cases. In particular, when the grid is being refined with $\Delta t = \Delta x$, the errors from the TL method remains roughly a constant.

The conclusion is that the TL algorithm is more stable than the EP method, but in general reduces accuracy for calculating unsteady (transient) solutions. On the other hand, the EP method is more accurate than the TL method, but only conditionally stable. A new method based on explicit predictor and implicit corrector (EPIC) for generating numerical boundary conditions was discussed in [6,7]. The EPIC method combines the advantages of both the TL (stability) and EP (accuracy) methods (see Figure 2(a)). Table 4 contains similar results as those in Table 3, except for the additional column for the results from the EPIC algorithm. While both the TL and EP methods fail to deliver reasonably accurate solutions for many cases, the EPIC method produces results with comparable accuracy in all cases as those from the BTCS algorithm without using domain decomposition.

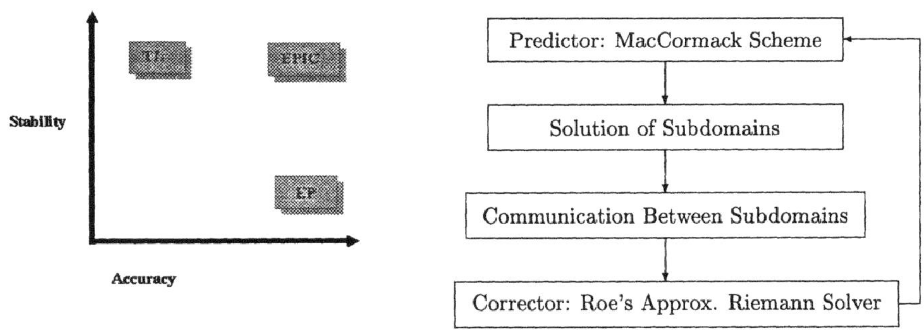

(a) (b)

Figure 2. (a). The EPIC method combines the advantages of both the TL and EP methods. (b). Flow Chart of the EPIC method for solving Euler equations.

In this paper, we will discuss the application of the EPIC method to the solution of two-dimensional nonlinear Euler equations in flow simulation for an airfoil on parallel computers. Numerical results demonstrate that the EPIC method is scalable and more accurate for solving nonlinear problems in CFD applications than the TL and EP methods.

Table 4
Maximum errors: $\frac{\Delta t}{\Delta x}=0.1$, $T=0.3$, $M=2$.

Δx	BTCS	TL	EP	EPIC
0.1000	0.1128E-01	0.8482E-01	0.8032E-02	0.6620E-02
0.0500	0.4707E-02	0.8235E-01	0.1448E+06	0.2712E-02
0.0250	0.2122E-02	0.8132E-01	0.3704E+45	0.1213E-02
0.0100	0.7933E-03	0.8078E-01	0.5644E+201	0.4546E-03
0.0050	0.3874E-03	0.8060E-01	∞	0.2222E-03
0.0025	0.1914E-03	0.8052E-01	∞	0.1099E-03

2. Solution of Euler Equations Using Domain Decomposition

The two-dimensional Euler equations in body-fitted curvilinear coordinates can be written as

$$\frac{\partial Q}{\partial \tau}+\frac{\partial E}{\partial \xi}+\frac{\partial F}{\partial \eta}=0, \tag{2}$$

with equation of state given by

$$p=(\gamma-1)\{e-\frac{\rho}{2}(u^2+v^2), \tag{3}$$

where ρ is the mass density, u and v are the velocity components in the x and y directions, respectively, e is the total specific energy of the fluid, p is the pressure, and γ is the ratio of specific heats.

Eq. (2) is discretized by an implicit finite volume algorithm, which leads to

$$\mathcal{F}(Q^{n+1})=\frac{Q_{i,j}^{n+1}-Q_{i,j}^{n}}{\Delta\tau}+\frac{E_{i+\frac{1}{2},j}^{n+1}-E_{i-\frac{1}{2},j}^{n+1}}{\Delta\xi}+\frac{F_{i,j+\frac{1}{2}}^{n+1}-F_{i,j-\frac{1}{2}}^{n+1}}{\Delta\eta}=0, \tag{4}$$

where the flux vectors are evaluated at cell faces. The Roe's approximate Riemann solver was used to calculate fluxes at cell surfaces [8]. The algorithm is first order accurate in time and up to third order accurate in space [9]. The original spatial domain Ω is decomposed into non-overlapping subdomains $\Omega_k, k=1,\cdots,K$. The numerical boundary conditions between subdomains are generated by using a predictor scheme based on the two-step MacCormack scheme:

$$\hat{Q}_{i,j}^{n+1}=Q_{i,j}^{n}-\Delta\tau\delta_i E(Q^n)-\Delta\tau\delta_j F(Q^n), \tag{5}$$

$$Q_{i,j}^{n+1}=\frac{1}{2}\left[\hat{Q}_{i,j}^{n+1}+Q_{i,j}^{n}-\Delta\tau\delta_{i-1}E(\hat{Q}^{n+1})-\Delta\tau\delta_{j-1}F(\hat{Q}^{n+1})\right]. \tag{6}$$

Once the numerical boundary conditions have been generated, the equations in different subdomains can be solved concurrently by different processors. The subdomain solutions are obtained using the Newton's method [9]:

$$\mathcal{F}'(Q^{n+1,m})(Q^{n+1,m+1}-Q^{n+1,m})=-\mathcal{F}(Q^{n+1,m}),\quad m=1,2,\ldots. \tag{7}$$

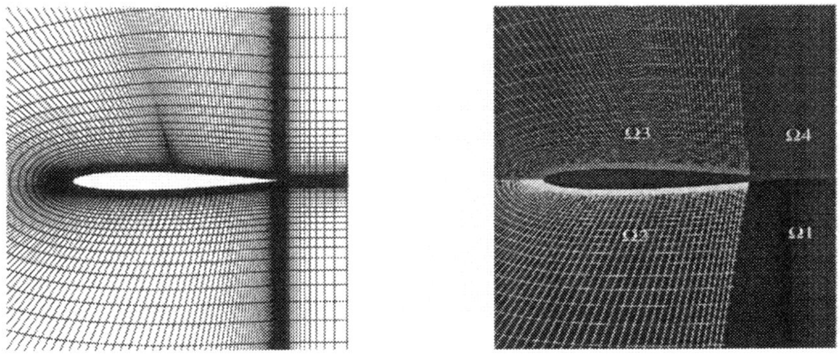

(a) (b)

Figure 3. (a). A 290 × 81 grid for the NACA0012 airfoil. (b). A 4-subdomain decomposition (Ω_1, Ω_2, Ω_3, and Ω_4) for the NACA0012 airfoil.

After the subdomain solutions have been calculated, the corrector based on the Roe's approximate Riemann solver

$$Q_{i,j}^{n+1} = Q_{i,j}^{n} - \Delta\tau\delta_\xi E(Q^{n+1}) - \Delta\tau\delta_\eta F(Q^{n+1}) \tag{8}$$

is used to update the numerical boundary conditions. Figure 2(b) summarizes the computation process of the EPIC method. The communications between different processors are carried out by using the MPI to ensure maximum portability.

3. Results

In order to demonstrate the performance and accuracy of the EPIC method for solving nonlinear Euler equations, a series of computations for transonic flow around a NACA0012 airfoil were carried out. Figure 3(a) shows the grid used for these calculations. It is a 290 × 81 C-grid with 200 points along the airfoil surface. Figure 3(b) illustrates the decomposition of the original grid into four subdomains.

The unsteady calculations correspond to the flow over the NACA0012 airfoil pitching about the quarter chord point. The movement of the airfoil is prescribed such that the angle of attack varies sinusoidally according to the relation $\alpha(t) = \alpha_m + \alpha_0 \sin(M_\infty kt)$, where α_m is the mean angle of attack, α_0 is the amplitude of the unsteady angle of attack, and k is the reduced frequency defined as $k = \frac{wc}{V_\infty}$, where w is the frequency, c is the chord length, and V_∞ is the freestream velocity.

For our calculations, these parameters are $M_\infty = 0.755$, $k = 0.1628$, $\alpha_m = 0.016$, and $\alpha_0 = 2.51$. The numerical results are compared with the experimental data by Landon [10]. Figure 4 shows the pressure distributions of the upper and lower surfaces for $\alpha = -2.41$ and $\alpha = 2.34$, respectively. Eight subdomains are used in this test case. It

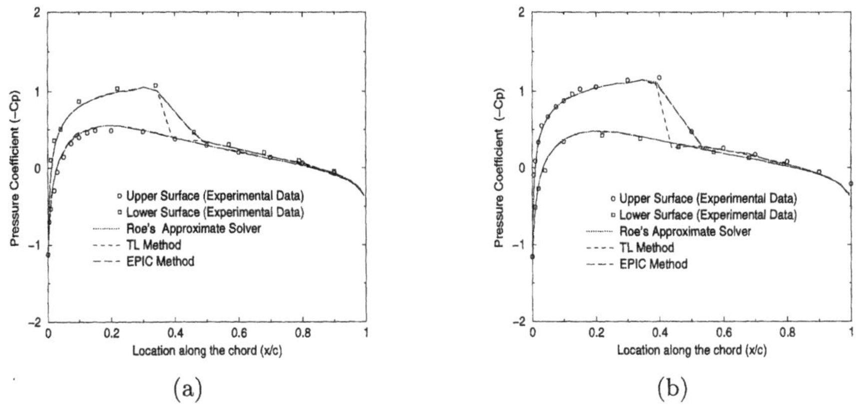

Figure 4. (a). NACA 0012 unsteady pressure distribution: $\alpha = -2.41$. (b). NACA 0012 unsteady pressure distribution: $\alpha = 2.34$.

is clear that the pressure distributions obtained using the EPIC method match well with the experiment data and that obtained without using domain decomposition. The TL approach, on the other hand, produces a considerable error in resolving the shock. The results from the EP approach is omitted here since it was very unstable in this test case.

In order to evaluate the influence of the CFL condition on the behavior of different methods discussed here, the calculations were also carried out using different CFL numbers. Numerical results indicate that the TL method is more sensitive to variations of CFL number than the EPIC method. The EPIC method demonstrated the same high quality results for a wide range of CFL numbers, while the TL method demonstrated a reduction in quality as the CFL number was increased.

Figure 5 shows the speedup for the calculation of solutions using the TL and EPIC methods, respectively. The computations were carried out on an SGI Power Challenge XL parallel computer with 16 processors. It is clear from the figure that the EPIC method is highly scalable with almost ideal speedup, which can be maintained on more processors by increasing the problem size.

4. Conclusions

A series of unsteady flow computations were carried out to study the performance and accuracy of the EPIC method for solving nonlinear Euler equations. While both the TL and EPIC methods are highly scalable, numerical results demonstrate that for transient problems the EPIC method is much more accurate than the traditional TL and EP methods.

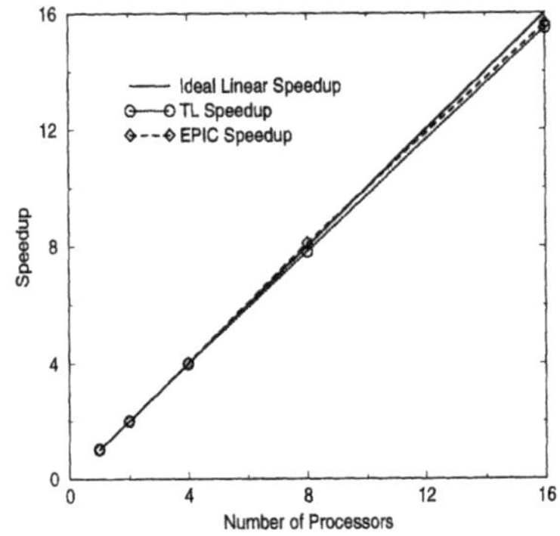

Figure 5. Speedup: 290 × 81 grid for the NACA0012 airfoil.

REFERENCES

1. S. Barnard, S. Saini, R. Van der Wijngaart, M. Yarrow, L. Zechtzer, I. Foster, and O. Larsson, Proceedings of the 7th Symposium on the Frontiers of Massively Parallel Computation, 1999.
2. C. N. Dawson, Q. Du, and T. F. Dupont, Math. Computation, 57(1995) 63.
3. Y. A. Kuznetsov, Sovietic J. of Num. Analy. and Math. Modeling, 3 (1988) 99.
4. Y. Lu and C. Y. Shen, IEEE Tran. Antennas and Propagation, 45 (1997) 1261.
5. R. Pankajakshan and W. R. Briley, Parallel Comp. Fluid Dynamics: Implementation and Results Using Parallel Computers, Elseiver Science, Amsterdam, 1996.
6. H. Qian and J. Zhu, Proc. of the 1998 International Conference on Parallel and Distributed Processing Technology and Applications, CSREA Press, Athens, GA, 1998.
7. W. Rivera and J. Zhu, Proc. of the 1999 International Conference on Parallel and Distributed Processing Technology and Applications, CSREA Press, Athens, GA 1999.
8. P. L. Roe, Journal of Computational Physics, 43 (1981) 357.
9. D. L. Whitfield, J. M. Janus, and L. B. Simpson, Engineering and Industrial Research Report MSSU-EIRS-ASE-88-2, Mississippi State University, 1988.
10. R. H. Landon, Compendium of Unsteady Aerodynamic Measurements, Advisory Report 702, AGARD, 1982.

Parallel Computational Fluid Dynamics – Practice and Theory
P. Wilders, A. Ecer, J. Periaux, N. Satofuka and P. Fox (Editors)

Parallel Kalman Filtering for a Shallow Water Flow Model

Mark Roest & Edwin Vollebregt

Faculty of Information Technology and Systems
Large Scale Systems Group (ITS/WAGM)
Delft University of Technology
M.R.T.Roest@its.tudelft.nl, E.A.H.Vollebregt@its.tudelft.nl *

This paper discusses the parallelization of one of the tools of Rijkswaterstaat/RIKZ, the Dutch governmental organization that is responsible for the maintenance of the coast and waterways. This tool, the RRSQRT Kalman filter, is used to improve simulations of water flow in a coastal area through corrections based on observations of the actual water flow. Computing these corrections is computationally extremely expensive, and thus parallelization is needed to keep the simulation time within reasonable limits.

The parallelization is particularly interesting because there are different forms of parallelism in the RRSQRT Kalman filtering algorithm. At some points in the algorithm it is best to do computations for different parts of the water flow area in parallel, whereas at other points it is best to do different computations for the whole water flow area in parallel. To accommodate for this, the software that implements the algorithm has been split into components, where each component has only one form of parallelism. The different components were then parallelized.

The system is now operational and a major experiment is underway to validate the parallelization and to explore the potential of RRSQRT Kalman filtering for large-scale 3D water flow models.

1. Overview

This paper discusses the parallelization of one of the tools of Rijkswaterstaat/RIKZ, the Dutch governmental organization that is responsible for the maintenance of the coast and waterways. This tool, the RRSQRT Kalman filter, is used to assimilate observations into the numerical models that simulate the surface water flow in coastal areas, estuaries and rivers. The assimilation makes run-time corrections on the simulation results to compensate for modeling errors.

The parallelization of the RRSQRT Kalman filter is the most recent in a series of efforts to introduce parallelism and domain decomposition into the software that is used by Rijkswaterstaat/RIKZ. The preceding efforts have given a lot of experience and have motivated many of the design choices that have been made in the parallelization of the RRSQRT Kalman filter.

*This research that is presented in this paper has been ordered by Rijkswaterstaat/RIKZ.

This paper will first briefly discuss the way in which parallelism is introduced in RIKZ's simulation software. This discussion will place the current parallelization effort in its proper context. Next, a short introduction will be given to the way in which the Kalman filter operates, followed by a description of the way in which it has been parallelized and some preliminary performance results. The paper will be concluded with an overview of the current situation and the work that will be done in the next few months.

2. Parallelization at Rijkswaterstaat/RIKZ

To perform its tasks, RIKZ relies on its WAQUA/TRIWAQ package which simulates shallow water flow in coastal areas, estuaries and rivers using finite differences. Models simulated by WAQUA/TRIWAQ can be either 2D (for WAQUA) or 3D (for TRIWAQ). The size of models is typically in the order of 10^5 grid points in the horizontal plane. Up to 10 layers are used for 3D modeling. A simulation run usually involves in the order of 10^4 time steps.

A parallel version of this package has long been available and is operationally used for the simulation of large scale models and in situations where the results are time-critical. Recently, the parallel version has been extended to also allow domain-decomposition: the parts of the model that are simulated in parallel can now have non-matching meshes.

The relative ease with which domain decomposition could be built based on the parallel version is the result of a fortunate design choice that was made at the very beginning. Instead of developing parallel constructs within the WAQUA/TRIWAQ code, the code has been extended to allow multiple instances of the program to cooperate on solving a problem. The parallelization thus comes down to (automatically) splitting the simulation model into submodels and simulate these submodels by cooperating instances of WAQUA/TRIWAQ. This choice made it relatively easy to accommodate also submodels that were not created by splitting a simulation model, but instead were created as distinct submodels in their own right.

A second advantage of this design choice has been that it became possible to quickly develop a version of WAQUA/TRIWAQ that could be coupled to the morphological model MORSYS of WL|Delft Hydraulics. This has demonstrated that the two institutes could be using each others software, each benefiting from the others expertise.

An extensive discussion of parallelization efforts at Rijkswaterstaat/RIKZ is given by Vollebregt, Roest and Lander [1].

3. Data assimilation: the RRSQRT Kalman filter

The basic idea of data assimilation is to use observations from real life to improve (the results of) numerical simulation. It is a well established method in meteorology, where uncertainty in the results of models is strongly related to poorly known initial and boundary conditions. Using data assimilation, this uncertainty can be reduced significantly. But also in the simulation of surface water flow, data assimilation has proved to be quite useful.

The kind of data assimilation that is the subject of this paper is the so called on-line data assimilation: observations are fed into a running simulation which adapts its state (flow fields, water levels) at time-instances for which observations are available so as to better match these observations.

Adapting the state to better match the observations can be done in a large number of ways. One of these is by using a Kalman filter. This filter maintains a matrix that models the uncertainties in the model in terms of the covariance of the uncertainties in the mesh points. This covariance matrix is used to compute a gain matrix, which in turn is used to determine the way in which the model state is updated. At places with high uncertainty the update will give preference to the observed value and modify the state at that place to better match the observation. At places with low uncertainty, the filter will tend to ignore the observation.

Obviously, for large models the covariance matrix becomes huge: for a 2D model of 10^5 grid points with 4 uncertain state variables per grid point it would be in the order of 10^{11} elements. Given the fact that it must be updated at every time step to account for changes in the state, it will be clear that in its basic form, Kalman filtering is prohibitively expensive in terms of computation time. Therefore, an approximate but much smaller covariance matrix is used. In the RRSQRT variant of the Kalman filter, the approximation is a reduced rank approximation of the square root of the full covariance matrix. The reduced rank approximation has the full number of rows but only a very limited number of columns. The RRSQRT Kalman filter algorithm is discussed in detail by Verlaan [2] and by Roest and Vollebregt [3].

Rijkswaterstaat/RIKZ has developed an RRSQRT Kalman filter for its WAQUA/TRIWAQ software. But even with the RRSQRT approximation, the filter is still computationally too expensive to be used for operational purposes. It typically takes in the order of 50-100 times as much computating time than normal WAQUA/TRIWAQ runs, which are themselves usually time-consuming in their own right. Given the favorable experiences with the parallel version of WAQUA/TRIWAQ, the parallelization of the RRSQRT Kalman filter has been an obvious step.

4. Parallelization of the RRSQRT Kalman filter

The original RRSQRT Kalman filter program is a version of WAQUA/TRIWAQ that is extended with filtering functionality. The initial idea for the parallelization was to create multiple instances of the program, each operating on a submodel, and make these instances cooperate. This would be in line with the approach taken for the parallelization of WAQUA/TRIWAQ. But it has turned out to be a difficult approach, because the parallelism in the Kalman filtering stage requires a distribution of data over the processes that is very different from that during the parallel simulation. Hence, there is not a clearly defined 'submodel' on which an instance of the program could work.

The solution has been to split the program into multiple components first (see second stage of development in Figure 1). There is a component for the flow model (which is basically the same as normal WAQUA/TRIWAQ), a filter-component (which adapts the state of the model to the available observations) and two components that together perform the time-propagation of the covariance matrix. These last two components are firstly the PROPWAQ component, which performs the time-stepping for columns of the covariance matrix (which is essentially a time step for the flow model but with a modified initial state) and secondly the COVMAT component that performs the computations

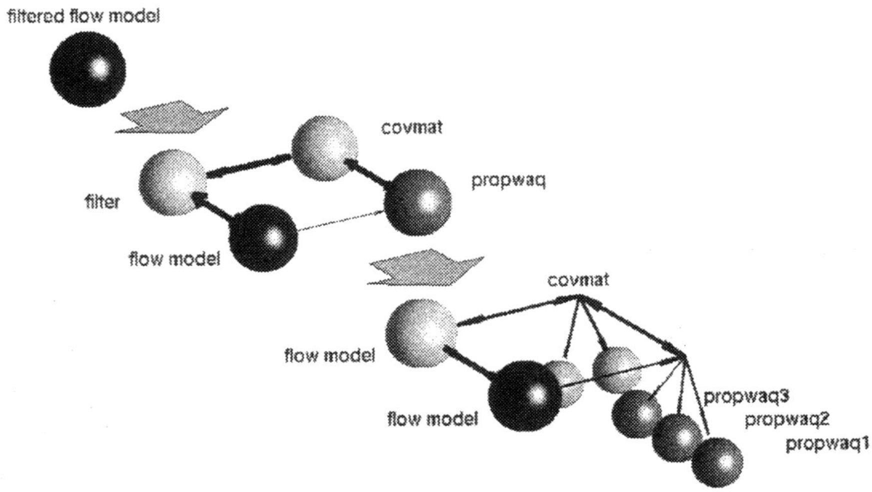

Figure 1. Overview of the parallelization of the Kalman filtering software. Top-left is the original situation, then comes the componentization into four components, each with a different form of parallelism. The parallelization of the two major components is shown at the bottom-right.

related to the square root approximation. In all, this gives four different components.

These components are coupled through communications: the flow simulation component sends its state to the filtering component, which obtains the gain matrix from the COVMAT component that holds the latest square root approximation of the covariance matrix. The filtering component modifies the state and sends it back to the flow simulation component. The COVMAT component sends the columns of the covariance matrix to the PROPWAQ component, which performs a timestep propagation for the columns and sends them back. The PROPWAQ component obtains some of its data directly from the flow model, like boundary conditions for the model to be simulated. These are not part of the columns in the covariance matrix, but are needed to perform a timestep propagation for the columns.

Each of these components has been parallelized in the most appropriate way. The approach taken to parallelize each component is the same approach as has been used for WAQUA/TRIWAQ: create multiple cooperating instances of the component, each operating on a distinct part of the computational problem. The parallel instances of the COVMAT component (which manages the covariance matrix) each operate on a number of rows of the matrix. The parallel instances of the PROPWAQ component each operate on a number of columns of the covariance matrix. Thus, the way in which the computational problem is divided into subproblems depends on the kind of computational problem that is handled by each of the components. At present not all computations of the COVMAT component are parallelized: COVMAT performs an eigenvalue determination on a relatively small matrix which has not been parallelized. The performance results that will be shown below illustrate that is one of issues that will have to be addressed in the near future, as it limits the performance in some cases.

In the communications from one (parallelized) component to another, a redistribution of data must take place. The parallel instances of the component that manages the RRSQRT matrix each handle a number of rows, but they communicate with parallel instances of the component that performs the timestep for columns of the covariance matrix, which each handle a number of columns. The communication library that is used to couple the components takes care of the proper shuffling of data so that each component needs to send only what it has and receives only what it needs.

In fact, the communication library is centered around two main operations: the `AVAIL` operation, which is called by a component to specify that a particular set of data is ready to be used by other components (i.e. is available for use by the other components) and the `OBTAIN` operation, which is called by a component to specify that subsequent operations can not proceed without a particular set of data.

The `AVAIL` and `OBTAIN` operations do not need a specification of where the data must be sent to or must be received from. The operations just need the data itself and a specification of its so called *index set*. To understand the concept of an index set, the reader may think of it as for example the range of indices of the array in which the data is stored. So the index set of an array `VELOCITY(MMAX,NMAX)` would be the index set $[1, \text{MMAX}] \times [1, \text{NMAX}]$. In a parallel context, components may specify that they hold only part of an array (e.g. ranges $[2,5] \times [8,10]$ in the example). Now one component might call the `AVAIL` operation specifying that it has `VELOCITY` available for index set $[2,5] \times [8,10]$ and another component could call the OBTAIN operation specifying that it needs `VELOCITY` for index set $[1,10] \times [8,9]$. In this case, the communication library would send `VELOCITY` from indices $[2,5] \times [8,9]$ from the availing component to the obtaining component. The obtaining component will have to receive the data for the rest of its indices from other components and will remain in the `OBTAIN` operation until all data has arrived or until it has been established that the data will never arrive.

5. Results and Ongoing work

The work on the parallelization is ongoing. A first parallel version was completed last year. In this version, the original Kalman filter software has been split into components and the two computationally most intensive components have been parallelized (see Figure 1).

The performance of this first parallel version has been evaluated for two different simulation models: the CSM8 model, which is a model of the North-West European Continental Shelf, and the Coast model, which is a model of the coastal waters for the entire Dutch sea-coast. Both models are relatively small. The CSM8 model is a 2D model with around 20.000 grid points. The Coast model has only 1.560 grid-points in the horizontal plane, but has 5 layers for vertical resolution, amounting to a total of some 7.500 grid-points.

The performance has been evaluated on a HP K460 four-processor shared memory

Table 1
Speedup with respect to componentized sequential run (i.e. a single PROPWAQ and COVMAT, second row of table). Speedup is given both in terms of measured CPU-time and in terms of measured wall-clock time.

Model	CSM8 model		Coast model	
	CPU-time	WC-time	CPU-time	WC-time
original program	1.06	1.06	0.75	0.74
1 PROPWAQ, 1 COVMAT	1.00	1.00	1.00	1.00
2 PROPWAQ, 2 COVMAT	1.69	1.70	1.94	1.94
4 PROPWAQ, 4 COVMAT	2.55	2.52	3.49	3.42

system. Table 1 lists the speedup that is attained, both in terms of CPU-time and in terms of wall-clock time. Speedup is given with respect to a run with the four-component system consisting of a single instance of each type of component (second stage of development as shown in Figure 1). This is essentially a sequential run because the dependencies between the components makes it impossible for them to be doing computations at the same time.

The performance results clearly show that the scalability is not yet satisfactory. The poor scalability is caused by the fact that the flow model and the eigenvalue computation in COVMAT were not yet parallelized. The eigenvalue computation is far more significant for the CSM8 model than for the Coast model, thus explaining the better scalability for the Coast model. Even so, several simple optimizations have been found after the experiments were finished that would bring the speedup for the CSM8 model to an estimated 3.4 when using four PROPWAQs and four COVMATs.

Recently, a parallel version of the flow simulation component has been introduced by replacing the original component by the already available parallel version of WAQUA/TRIWAQ. This parallel version of WAQUA/TRIWAQ is also used to further parallelize the PROPWAQ component, which performs the time step for the columns of the covariance matrix. These improvements have alleviated one of the remaining sequential bottlenecks. The only sequential bottleneck that remains now is the eigenvalue computation in the component that manages the covariance matrix. The amount of time taken for this computation depends on the model input (see the discussion of the performance results above) and will not pose much of a problem for the models that will be used in the near future.

One last improvement that is foreseen is that the component that manages the covariance matrix will also be parallelized along columns rather than just along rows. This will allow a comparison of the performance using a row-wise distribution of the matrix versus that using a column-wise distribution.

Once it is fully functional, the parallelized RRSQRT Kalman filter will be applied to a large 3D model. This experiment will firstly validate the parallel version of the software. But more important is that it will show the feasibility to apply Kalman filtering to the large scale three dimensional models that are typically used at Rijkswaterstaat/RIKZ. This is expected to lead to a significant further improvement in the quality of the results

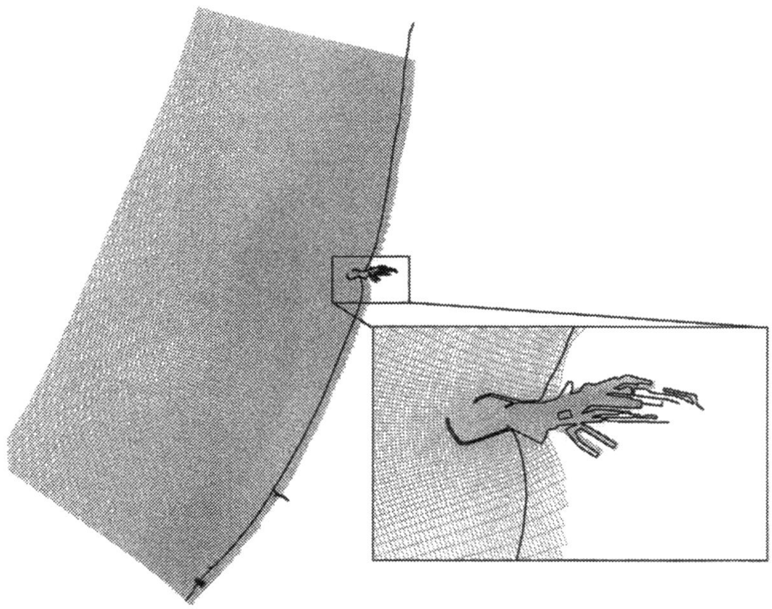

Figure 2. An overview of the grid of the Ymond model with a detail of the grid near the harbor.

that are obtained from these models.

The model that has been selected for this experiment is the so called Ymond model, which covers an area of the sea of about 70×40 kilometer near the city of IJmuiden in the Netherlands (see Figure 2). The grid consist of 28.000 grid points in each of 4 layers, leading to a reduced rank approximation of the square root of the covariance matrix of about 700.000×100 elements. The time step is 0.5 minutes and the simulation period is a month (December 1998), so the total number of time steps is around 89.000. Without parallelization, the filtered run would take more than a year on a modern LINUX PC, which is clearly unpractical. Using parallelization and a fast, large-scale parallel computer, we expect to show that the runtime can be reduced to a few tens of days.

REFERENCES

1. E.A.H. Vollebregt, M.R.T. Roest and J.W.M. Lander, Large Scale Computing at Rijkswaterstaat, submitted to Parallel Computing.
2. M. Verlaan, Efficient Kalman Filtering Algorithms for Hydrodynamic Models, Ph.D. thesis, Delft University of Technology, 1998.
3. M.R.T. Roest and E.A.H. Vollebregt, Decomposition of Complex Numerical Software into Cooperating Components, to be published in Proc. HPCN99 conference.

Parallel Computational Fluid Dynamics – Practice and Theory
P. Wilders, A. Ecer, J. Periaux, N. Satofuka and P. Fox (Editors)

A Parallel Solenoidal Basis Method for Incompressible Fluid Flow Problems*

Sreekanth R. Sambavaram and Vivek Sarin
Department of Computer Science
Texas A&M University
College Station, TX 77843

The convergence of iterative methods used to solve the linear systems arising in incompressible flow problems is sensitive to flow parameters such as the Reynolds number, time step and the mesh width. Incompressibility of the fluid makes the systems indefinite, and poses difficulty for the iterative solvers. This paper outlines a class of solenoidal basis methods that use local solenoidal functions to restrict fluid velocity to divergence-free subspace. An optimal preconditioner based on the Laplace operator is used to solve the resulting ill-conditioned reduced system. Experimental results for two and three dimensional problems show that the convergence of the proposed algorithm is optimal across the range of flow parameter variation. Scalability of the algorithm is suggested by the experiments on the SGI Origin 2000.

1. INTRODUCTION

Large-scale simulation of incompressible flow is one of the most challenging application. Realistic simulations are possible only with the use of sophisticated modeling techniques, preconditioned iterative methods and advanced parallel architectures. The motivation for this work is to develop an effective approach for solving the linear systems arising in incompressible flows with high efficiency on a multi-processor platform.

The principles of classical mechanics, thermodynamics, and laws of conservation of mass, momentum, and energy govern the motion of the fluid. Law of conservation of momentum for incompressible, viscous flow in a region Ω with boundary $\partial\Omega$ is captured by Navier-Stokes equation given by

$$\frac{\partial \mathbf{u}}{\partial t} + \mathbf{u} \cdot \nabla \mathbf{u} = -\nabla p + \frac{1}{R}\Delta \mathbf{u}, \tag{1}$$

where $p = p(\mathbf{x}, t)$ is the pressure, R is the Reynolds number, and $\mathbf{u} = \mathbf{u}(\mathbf{x}, t)$ is the velocity vector at $\mathbf{x}$. The law of conservation of mass for incompressible fluids gives rise to the continuity equation

$$\nabla \cdot \mathbf{u} = 0 \quad \text{in } \Omega. \tag{2}$$

*This work has been supported in part by NSF under the grants NSF-CCR 9984400, NSF-CCR 9972533, and NSF-CTS 9873236. Computational resources were provided by NCSA, University of Illinois at Urbana-Champaign.

Appropriate boundary conditions may be specified for fluid velocity. Suitable discretization and linearization of the equations (1)–(2) result in the following linear system

$$\begin{bmatrix} A & B \\ B^T & 0 \end{bmatrix} \begin{bmatrix} u \\ p \end{bmatrix} = \begin{bmatrix} f \\ 0 \end{bmatrix}, \tag{3}$$

where B^T is the discrete divergence operator and A is given by

$$A = \frac{1}{\Delta t} M + C + \frac{1}{R} L, \tag{4}$$

in which M is the mass matrix, L is the Laplace matrix, and C is the matrix arising from the convection term. When operator splitting is used to separate the linear and non-linear terms, we obtain the generalized Stokes problem (GSP) with a symmetric positive definite A given as

$$A = \frac{1}{\Delta t} M + \frac{1}{R} L. \tag{5}$$

The linear system (3) is large and sparse. Although direct methods can be used to solve this system, they require prohibitively large amount of memory and computational power. The inherent sequential nature of these techniques limits the efficiency on parallel architectures. In contrast, iterative methods require significantly less memory and are well suited for parallel processing. These methods can be made more reliable by using preconditioning techniques which accelerate convergence to the solution. In order to make iterative methods more competitive, one must devise robust preconditioning techniques that are not only effective but parallelizable as well.

This paper presents a preconditioned solenoidal basis method to solve the linear system (3) arising in the generalized Stokes problem. Section 2 describes the solenoidal basis method and section 3 outlines the preconditioning scheme. Experiments for the driven cavity problem in 2D and 3D are presented in section 4.

2. A SOLENOIDAL BASIS METHOD

Incompressible fluid flow can be viewed as compressible flow with additional constraint that fluid velocity should be divergence free. This incompressibility constraint in (3) $B^T u = 0$, makes the linear system indefinite. This indefinite nature is the main cause of difficulty in solving the system via preconditioned iterative methods. The degree of difficulty also depends upon the nature of matrix A which is affected by the Reynolds number R and the choice of time step Δt and mesh width.

Solenoidal basis methods are a class of techniques that use a divergence-free or solenoidal basis to represent velocity. A discrete solenoidal basis can be obtained by computing the null space of the divergence operator B^T. A matrix $P \in R^{n \times (n-m)}$ that satisfies the condition $B^T P = 0$ is used to compute divergence-free velocity via the matrix-vector product $u = Px$, for an arbitrary $x \in R^{(n-m)}$. Clearly, such a velocity satisfies the continuity constraint $B^T u = 0$. By restricting u to the column space of P and pre-multiplying the first block of (3) by P^T, we get the following *reduced system*

$$P^T A P x = P^T f, \tag{6}$$

which may be solved by a suitable iterative method such as the conjugate gradients (CG) method, GMRES, etc. (see, e.g., [6]). Once x has been calculated, velocity is computed as

$$u = Px, \tag{7}$$

and pressure is recovered by solving the least squares problem iteratively

$$Bp \approx f - APx. \tag{8}$$

The success of the solenoidal basis method depends on a number of factors. First, the matrix-vector product with the reduced system must be computed efficiently. Second, one must develop a robust and effective preconditioner for the reduced system. Finally, these computations must be implemented efficiently on a parallel processor.

At each iteration, the matrix-vector product with P^TAP is computed as a series of three matrix-vector products with P, A, and P^T, respectively, in that order. Each column of P represents a solenoidal function with a localized region of influence on the mesh. As an example, consider a uniform 3D mesh to discretize a driven cavity problem via the Marker-and-Cell (MAC) scheme. The MAC scheme assigns pressure unknowns to each node and velocity unknowns to each edge. One can construct a local circulating flow by assigning appropriate velocity to the edges forming a face of a given cell in this mesh. Such a flow is represented as a vector, and the set of these vectors form the columns of P. The localized nature of these flows leads to a sparse P with a nonzero structure resulting from the underlying mesh. This may be used to compute Px and P^Ty efficiently in parallel. Furthermore, one can apply P and P^T to a vector without actually constructing P itself. This feature has been exploited to develop a matrix-free implementation.

We proposed the use of local solenoidal functions for 2D flows in [7], where we presented a scheme to construct a solenoidal basis derived from circulating flows or vortices on uniform meshes. We also outlined an optimal preconditioning technique for the generalized Stokes problem. In [8], we introduced a linear algebraic technique to construct a hierarchical basis of solenoidal functions which is applicable to the generalized Stokes problem on arbitrary meshes. This approach was successfully applied to 2D particulate flow problems using structured meshes [3,5,9] and was extended to unstructured meshes [3]. Details of a distributed memory parallel implementation were presented in [3]. Several schemes have been proposed for computing discrete solenoidal functions [2,1]. Unlike other schemes, our approach can be formulated as a linear-algebraic method which is applicable to arbitrary discretization schemes including finite element and finite volume methods.

In this paper, we extend the solenoidal basis method to 3D problems defined on uniform meshes. In 3D, the solenoidal basis P constructed from local solenoidal functions turn out to be rank-deficient due to linear dependence between the local circulating flows. However, it can be shown that the space of discrete solenoidal functions is contained within the column space of P. Since the reduced system is consistent despite the rank-deficiency of the system matrix, it can be solved by preconditioned CG or GMRES. Iterative solution of the driven cavity problem via the MAC scheme exhibits optimal convergence rate. The parallel implementation demonstrates good speed improvement on a medium-sized multiprocessor.

3. ACCELERATING CONVERGENCE BY PRECONDITIONING

Effective preconditioning of the reduced system is critical to the overall success of the solenoidal basis method. The design of the preconditioner becomes challenging because the reduced system matrix P^TAP is ever explicitly formed. One can take advantage of the analogy between matrix vector products involving P in the solenoidal basis method with vortex methods to construct the preconditioner.

Vortex methods are a class of techniques that solve the vorticity transport equation instead of the Navier-Stokes equation. Vorticity field ξ is expressed in terms of velocity $\mathbf{u}$ and velocity is in turn obtained by applying curl operator on scalar stream function ψ. In particular, vorticity is given as $\xi = \nabla \times \mathbf{u}$ and velocity is expressed as $\mathbf{u} = \nabla \times \psi$. The relation between ξ and ψ is given by the Poisson equation $\Delta\xi = -\psi$ The matrix vector product $u = Py$ computes the velocity vector function $u = \nabla \times \mathbf{y}$ and the matrix vector product $w = P^T u$ computes the vorticity vector function $\mathbf{w} = \nabla \times \mathbf{u}$. Further more, y and w are analogous to the velocity potential ψ and ξ. The relation $\mathbf{w} = \nabla \times \nabla \times \mathbf{y}$ can be implemented via matrix vector product $w = P^T P y$. Since, w and y are assumed divergence-free, it is easy to show that $-w = \Delta y$, which is identical to the relation between ψ and ξ.

Observing that the product Px and $P^T y$ compute the discrete curl of the functions represented by x and y, respectively, it can be inferred that the product $y = P^T P w$ represents $\nabla \times \nabla \times \mathbf{w}$ in a discrete setting. Thus, the matrix $P^T P$ can be shown to be equivalent to the Laplace operator on the solenoidal function space. This suggests the following preconditioner for the generalized Stokes problem:

$$G = \left[\frac{1}{\Delta t} M + \frac{1}{R} L_s\right] L_s, \tag{9}$$

where L_s is the Laplace operator for the local solenoidal functions. Since the preconditioned system is spectrally equivalent to a symmetric positive definite matrix, one can use preconditioned CG to solve the reduced system (6).

4. EXPERIMENTS

In this section, we present results of numerical experiments for the driven cavity problem. The preconditioned solenoidal basis method was used to solve the linear system arising in the generalized Stokes problem. The experiments were conducted for 2D unit square and 3D unit cube domains. In each case, the MAC scheme was used to discretize the domain. The linear system was solved under various physical conditions by changing the ratio $h^2 R/\Delta t$ which determines the condition number of A. For the 3D driven cavity problem, the condition number of A is approximated by

$$\kappa(A) = \frac{h^2 R/\Delta t + 12}{h^2 R/\Delta t + h^2}. \tag{10}$$

Hence, $\kappa(A) < 2$ when $h^2 R/\Delta t > 12$, and $\kappa(A) \approx h^{-2}$ when $h^2 R/\Delta t \ll 12$. This ratio also captures the difficulty associated with solving the linear system when parameters such as mesh width (h), Reynolds number (R) and time step (Δt) are changed.

The first set of experiments highlights the effectiveness of the preconditioner in accelerating convergence of the CG method. The linear system in the preconditioning step was solved by CG as well, resulting in an inner-outer scheme. The iterations were terminated when the relative residual was reduced below 10^{-4}. A much larger tolerance (10^{-2}) was used for the inner iterations. Table 1 presents the iterations required by the preconditioned CG method for several instances of $h^2R/\Delta t$. The preconditioner ensures a stable convergence rate independent of the values of various parameters h, R, and Δt, suggesting optimality of the preconditioner. The overall computation time was reduced significantly by using a large threshhold for the inner CG iterations. This choice did not adversely effect convergence of the outer iterations.

Table 1
Convergence rate independent of mesh width, R and Δt.

Mesh	Reduced system size	$h^2R/\Delta t$		
		10^{-2}	10^{0}	10^{+2}
2D				
128 × 128	16,128	12	7	4
256 × 256	65,024	12	7	4
512 × 512	261,120	12	7	4
3D				
8 × 8 × 8	1,176	8	6	5
16 × 16 × 16	10,800	12	8	6
32 × 32 × 32	92,256	16	9	7

4.1. Parallel Performance

The solution methodology can be effectively parallelized. On a multi-processor machine with q processors, the domain is partitioned into q partitions, and the underlying mesh is distributed across processors. Parallelization of the computation of P and matrix vector products with P and P^T is fairly straight forward (see, e.g., [7]). Other operations such as vector additions, inner-products and matrix vector products with A are also easy to parallelize. The reader may refer to the texts [6,4]. The linear system in the preconditioning step can be solved via parallel versions of fast poisson solvers, domain decomposition, multi-grid, and multi-level methods.

A second set of experiments focused on the parallelization of the proposed algorithm. All the elementary matrix-vector products were parallelized by distributing the grid equally among the processors. The algorithm was parallelized using OpenMP. Table 2 indicates that the algorithm can be parallelized with high efficiency on a multi-processor platform such as the SGI Origin 2000.

5. CONCLUSIONS

This paper presents a high performance algorithm for solving the linear systems arising from incompressible flow problems. The proposed solenoidal basis method uses discrete

Table 2
A parallel implementation using OpenMP demonstrates good speed improvement on 16 processors of SGI Origin 2000.

	Mesh size = 256 × 256 × 256	
Processors	Time	Speedup
1	2684.10	1.0
2	1552.95	1.7
4	961.71	2.8
8	473.67	5.7
16	278.88	9.5

local solenoidal functions to represent divergence-free velocity. A reduced system is solved in the divergence-free subspace via a preconditioned iterative scheme. An optimal preconditioner has been suggested which assures stable convergence regardless of parameters such as the mesh width, Reynolds number, and the time step. An inexpensive low accuracy iterative solve for the preconditioner appears to be sufficient for optimal convergence. The method is parallelizable with high efficiency.

REFERENCES

1. K. Gustafson and R. Hartman. Divergence-free bases for finite element schemes in hydrodyna mics. *SIAM J. Numer. Anal.*, 20(4):697–721, 1983.
2. C. A. Hall, J. S. Peterson, T. A. Porsching, and F. R. S. ledge. The dual variable method for finite element discretizations o f Navier-Stokes equations. *Intl. J. Numer. Meth. Engg.*, 21:883–898, 1985.
3. M. G. Knepley, A. H. Sameh, and V. Sarin. Design of large scale parallel simulations. In D. Keyes, A. Ecer, J. Périaux, N. Satofuka, and P. Fox, editors, *Parallel Computational Fluid Dynamics.* Elsevier, 2000.
4. V. Kumar, A. Grama, A. Gupta, and G. Karypis. Introduction to parallel computing: Design and analysis of algorithms. Benjamin/Cummings, 1994.
5. T.-W. Pan, V. Sarin, R. Glowinski, A. H. Sameh, and J. Périaux. A fictitious domain method with distributed Lagrange multipilers for the numerical simulation of particulate flow and its parallel implementation. In C. A. Lin, A. Ecer, N. Satofuka, P. Fox, and J. Périaux, editors, *Parallel Computational Fluid Dynamics, Development and Applications of Parallel Technology*, pages 467–474. North-Holland, Amsterdam, 1999.
6. Y. Saad. Iterative methods for sparse linear systems. PWS publishing company, 1996.
7. V. Sarin. Parallel linear solvers for incompressible fluid problems. In *Proceedings of the SIAM Conference on Parallel Processing for Scientific Computing.* Portsmouth, VA, Mar. 2001.
8. V. Sarin and A. H. Sameh. An efficient iterative method for the generalized Stokes problem. *SIAM Journal of Scientific Computing*, 19(1):206–226, 1998.
9. V. Sarin and A. H. Sameh. Large scale simulation of particulate flows. In *Proceedings of the Second Merged Symposium IPPS/SPDP*, pages 660–666, Apr. 1999.

Parallel Computational Fluid Dynamics – Practice and Theory
P. Wilders, A. Ecer, J. Periaux, N. Satofuka and P. Fox (Editors)

A Multilevel, Parallel, Domain Decomposition, Finite-Difference Poisson Solver

Albert W. Schueller[a] and J. M. McDonough[b]

[a]Department of Mathematics, Whitman College,
Walla Walla, WA USA 99362
E-mail: schuelaw@whitman.edu

[b]Department of Mechanical Engineering, University of Kentucky,
Lexington, KY USA 40506-0108
E-mail: jmmcd@uky.edu

We present a new implementation of multilevel domain decomposition algorithms based on use of a simple successive line overrelaxation technique for the basic solver, and an approximate Schur complement procedure that naturally provides subdomain overlaps in a formally non-overlapping construction. The procedure is parallelized using OpenMP on HP SPP-2200 and N-4000 SMPs, leading to essentially linear speedups through eight processors, and the ability to obtain solutions to elliptic Dirichlet problems in run times independent of problem size on the latter of these two machines.

1. INTRODUCTION

Domain decomposition methods (DDMs) have been under development since at least the mid 1980s, and early intuitive forms of the basic approach, originally due to Schwarz [1], had been used in computational fluid dynamics (CFD) in the guise of multi-block structured methods considerably earlier than 1985 (see Thompson *et al.* [2] and references therein). Evolution of multi-processor parallel supercomputers began by the late 1980s to early 1990s and has served as one of the prime motivators for studying DDM algorithms. By the early 1990s DDM theory had become well developed, and it was recognized that DDMs share many features with multigrid methods. Moreover, it is now known that in the context of elliptic partial differential equations (PDEs) optimal DDM convergence rates can be achieved only by versions that incorporate some multigrid technology, namely coarse-grid correction (see, *e.g.*, Smith *et al.* [3]).

It is well known that solving the pressure Poisson equation (PPE)

$$\Delta p = \nabla \cdot \boldsymbol{U}, \qquad (1)$$

with p denoting *pseudo pressure—a velocity potential* and $\boldsymbol{U}$ the velocity vector, can consume as much as 90% of the total time expenditure in computing solutions to the incompressible Navier–Stokes (N.–S.) equations. Thus, being able to solve the PPE efficiently is crucial. This paper presents further work on an algorithm for this purpose, first described by McDonough *et al.* [4] and consisting of a multilevel DDM. The ideas embodied in the approach are similar to those presented in [3], but the implementation employed is rather different. In particular, although we view the overall DDM as preconditioning of a simple algorithm, as is done in [3], we here utilize finite-difference discretization rather than the finite elements used in [3], and we employ an easily parallelized red-black ordered successive line overrelaxation (SLOR) as our basic algorithm in place of the Krylov subspace-based methods of [3]. In addition, we introduce overlap of subdomains through an efficient approximate Schur complement procedure.

The principal goal of this study is to produce a general elliptic solver having the following properties: *i)* total arithmetic scales "nearly" linearly with number of unknowns; *ii)* run times are essentially independent of number of equations when parallelization is used; *iii)* properties *i)* and *ii)* hold even without symmetry or positive definiteness of the system matrix, and hence for problems posed in generalized coordinates with other than Dirichlet boundary conditions; *iv)* the method retains its favorable properties for three-dimensional problems. In the present work efforts will be focused on properties *i)* and *ii)*.

The remainder of this paper is organized as follows. In Sec. 2 we provide a fairly detailed discussion of our multilevel DDM algorithm, and in Sec. 3 we present a model problem with an exact solution. Parallelization results are presented in Sec. 4, followed by summary and conclusions in a final section.

2. ALGORITHM DESCRIPTION

The goal of producing this algorithm is to minimize the time necessary to solve the pressure Poisson equation common in the numerical solution of the incompressible N.-S. equations of fluid dynamics and other elliptic boundary-value problems. Figures 1a,b provide a schematic to aid in understanding the features of this three-level algorithm consisting of the following parts.

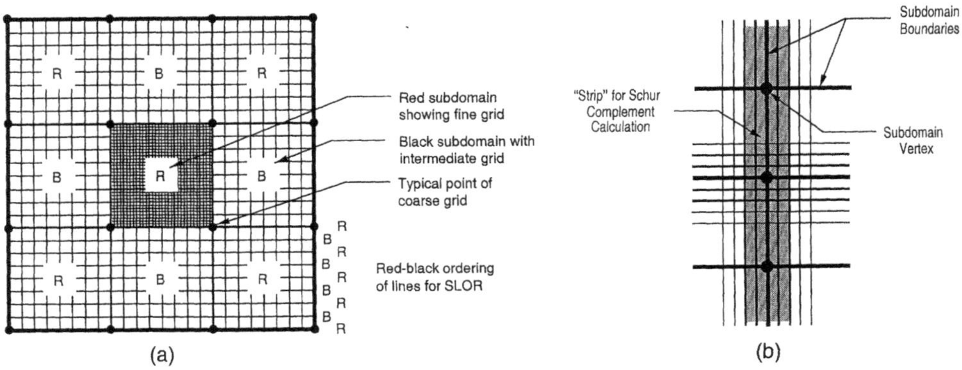

Figure 1. Three-level red-black ordered grid arrangement: (a) basic grid decomposed in red and black regions; (b) approximate Schur complement construction near interior domain boundaries.

A simple multilevel interpolation scheme equivalent to the nested iteration of a full multigrid (FMG) method is used to start the fine grid iteration with a better approximation of the solution and thus save iterations. Domain decomposition is then employed to speed the convergence of the full fine grid SLOR. It is well known that the convergence rate of SLOR slows as the problem size increases; indeed, for 2-D Dirichlet problems the required number of iterations is proportional to $N^{1/2}$, where N is the total number of equations, *i.e.*, grid points, in the system (see, *e.g.*, Young [5]). Thus after a few iterations on the entire fine grid, the algorithm shifts to artificial Dirichlet problems on subdomains using the current iterate from the fine grid plus an approximate Schur complement to determine boundary conditions on the interior (artificial) boundaries of the subdomains. Because the subdomains are generally much smaller than the original problem and because of stability of the finite-difference equations with respect to perturbations in boundary data, the SLOR on the subdomain shows relatively rapid convergence and overall accuracy is quickly improved on the subdomain.

But because the interior boundary data are in error, there is a limit to how much improvement the subdomain relaxation provides. Thus, after the subdomain relaxations, we return to the full grid for a few more iterations to improve the approximation near the interior boundaries. Indeed, before we return to the full grid, we relax on strips around the vertical and horizontal artificial boundaries (analogous to constructing a Schur complement), again taking advantage of the smaller problem size and stability of the finite-difference equations to perturbations in boundary data.

Details of this algorithm are outlined for the three-grid (coarse, intermediate, fine) multilevel structure displayed in Fig. 1.

Algorithm *(Multilevel Approximate Schur Complement DDM)*

1. *Exactly solve problem on coarsest grid.*

2. *Interpolate this solution to the intermediate grid, and apply (parallelized) RB-ordered SLOR until the asymptotic convergence rate is observed.*

3. *Interpolate this result to the finest grid, and again apply parallelized RB-ordered SLOR iterated until asymptotic convergence rate is observed.*

4. *Apply domain decomposition of the finest grid decomposed into red and black subdomains (see Fig. 1). Process all red domains simultaneously on separate processor complexes employing parallelized RB-ordered SLOR within each subdomain. Then repeat for the black subdomains.*

5. *Compute approximate Schur complements to update subdomain boundary values using RB-ordered SLOR iterated to the overall problem final required convergence tolerance.*

6. *Perform fixed number of iterations on the global fine grid using parallelized RB-ordered SLOR; check residual at last iteration.*

7. *If residual satisfies convergence tolerance, stop; otherwise compute approximate Schur complements and return to step 4.*

The rationale and specific details associated with this algorithm are contained in the following remarks.

***Remark* 1.** The first three steps of the algorithm correspond to "nested iteration," and hence comprise the right-hand side of a multigrid **V** cycle.

***Remark* 2.** The convergence tolerance employed in steps 2 and 3 may need to be relaxed in practice. In any case it is easily checked via a theorem proved in Hageman and Young [6] that

$$\lim_{n\to\infty} \frac{\|d^{(n+1)}\|}{\|d^{(n)}\|} = \lambda_1, \tag{2}$$

where λ_1 is the spectral radius of the iteration matrix of the method, and $d^{(n)}$ is the iteration error defined for the nth iteration by

$$d^{(n)} = u^{(n)} - u^{(n-1)}. \tag{3}$$

Hence, the solution process actually yields the spectral radius of the iteration matrix via the power method. Values of λ_1 begin to stabilize (converge) as the asymptotic convergence rate is approached, and we terminate iterations in steps 2, 3 and 4 based on this.

***Remark* 3.** The red-black ordering of the DDM subdomains is not actually necessary, but it does reduce the amount of shared storage required by separate processors. Furthermore, on systems with only a small number of processors this provides a well-defined way to assign them to subdomains. Clearly, load balancing considerations will dictate that these subdomains be of similar sizes.

***Remark* 4.** We also comment here that subdomain solves produce an effect on convergence rate similar to that of coarse-grid corrections in the multigrid context. In particular, recall that the the spectral radius of, *e.g.*, the Jacobi iteration matrix, $\rho(\boldsymbol{B})$, corresponding to a second-order centered discretization of the Dirichlet problem for Laplace's equation on $\Omega \equiv (0,a) \times (0,b)$ is

$$\begin{aligned} \rho(\boldsymbol{B}) &= \frac{1}{2}\left(\cos\frac{\pi h}{a} + \cos\frac{\pi h}{b}\right) \\ &\simeq 1 - \frac{1}{4}\left[\left(\frac{\pi h}{a}\right)^2 + \left(\frac{\pi h}{b}\right)^2\right], \end{aligned} \tag{4}$$

where $h = h_x = h_y \sim \mathcal{O}(N^{-1/2})$ with $N = N_x \times N_y$, the total number of grid points. It is clear that reducing a and/or b (*i.e.*, restricting calculations to subdomains) has a similar effect to increasing h, the grid spacing (*i.e.*, coarsening the grid).

***Remark* 5.** The method employed to compute the approximate Schur complement not only updates subdomain boundary values in our nonoverlapping decomposition, but also provides *effective* overlap. Thus, the overall method achieves multiplicative Schwarz convergence rates even though it is formally an additive Schwarz method.

***Remark* 6.** We have employed RB-ordered SLOR as our basic solver throughout this algorithm. This can be replaced with one's favorite elliptic solver, but we have in the past found this method to be very robust (even for systems that are neither symmetric nor positive definite) and easily parallelizable.

We also note that of necessity non-parallelized versions of SLOR were employed within the subdomain solves, which themselves are done in parallel, because current hardware seems unable to handle multilevel parallelization in practice, especially in a multi-user environment, despite its theoretical attractiveness.

Two additional items associated with this algorithm deserve comment. First, it should be noted that one of the main shortcomings of FMG in the context of N.–S. solutions is that the coarse-grid solutions tend to be highly underresolved, and the solutions (even on the fine-grid) are not necessarily smooth. It is then unclear whether anything will actually be gained by prolongating nonsmooth, extremely inaccurate solutions to the fine grid. Except during the initiation phase, our algorithm avoids this difficulty completely because once the domain decomposition loop (steps 4 through 7 of the algorithm) is begun all calculations are performed at the resolution of the fine grid. Second, it is also important to note that our overall goal is to obtain a method for solving elliptic PDEs for which run times (*i.e.*, wall-clock times) are nearly independent of problem size. For this to be possible on current machine architectures (with considerable communication overhead), it is essential that the underlying method be able to compute solutions in $\mathcal{O}(N)$ arithmetic operations, where N is the total number of grid points.

Then if the problem size increases by, say a factor of four, only a factor of four increase in number of processors should be required—which is to some extent possible on current SMP-type machines. In the context of our implementation, this leads to the requirement that subdomain sizes and number of subdomains per processor remain constant as the problem size increases. If the algorithm is performing as we would hope, then in this scenario, run times should be nearly independent of problem size.

3. TEST PROBLEM

To test the performance of the algorithm described in the preceding section we employ a basic 2-D elliptic Dirichlet problem

$$-\Delta u = f, \quad (x, y) \in \Omega \equiv (0,1) \times (0,1), \tag{5a}$$

with

$$u(x, y) = 0, \quad (x, y) \in \partial\Omega. \tag{5b}$$

We have chosen the function f to be

$$f(x, y) = 13\pi^2 \sin 3\pi x \sin 2\pi y. \tag{6}$$

Then the exact solution to Prob. (1) is

$$u(x, y) = \sin 3\pi x \sin 2\pi y, \qquad (x, y) \in \overline{\Omega}, \tag{7}$$

shown in Fig. 2. This solution is C^{∞} but sufficiently complicated to provide a nontrivial model.

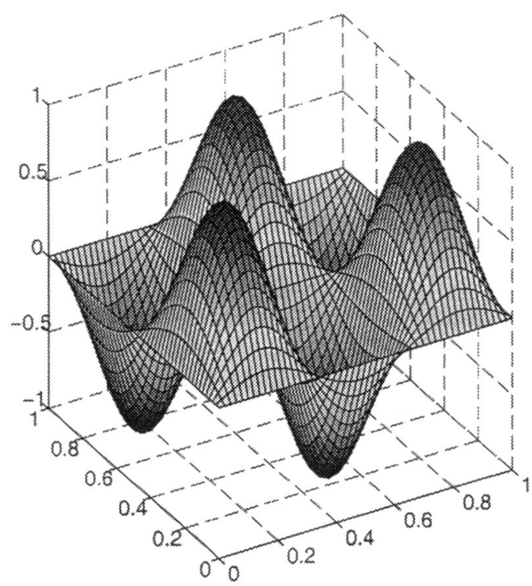

Figure 2. Exact solution to model problem.

4. RESULTS

Figures 3 and 4 display results obtained for the above problem. Figure 3 shows that the required "nearly" linear increase in run time with problem size is not being exceeded, even for very large problems. The data points shown were obtained in single-processor runs in non-dedicated (hence, realistic practical conditions) mode on a HP SPP-2200 at the University of Kentucky Computing Center. However, it should be noted that these results do not represent fully-converged solutions, but instead are for a fixed decrease in residual. It is also worth noting that despite the increase in overall algorithmic complexity, compared with the base SLOR solver alone (applied to the fine-grid problem), total run times are approximately 33% faster for the complete algorithm due to improved convergence rate of the multilevel DDM combination.

Parallelization was performed with OpenMP compiler directives (shared-memory mode) on the HP SPP-2200 and HP N-4000 at the University of Kentucky Computing Center. The former was a 64-processor machine that has now been replaced by the latter, which is a 96-processor SMP. For this part of the study Problem (1), discretized using a standard second-order, centered-difference approximation, was solved on fine grids ranging in size from 101×101 to 401×401 points. The domain decomposition aspect of the algorithm employed 101×101 non-overlapping subdomains of the fine grid. The multilevel portion of the scheme utilized a 9×9 **coarse** grid, an 81×81 **intermediate** grid and the aforementioned 401×401 **fine** grid, with analogous grid-size ratios used for the 101×101 and 201×201 grids. Actual speedups for fixed multilevel grid arrangements are displayed in Fig. 4, for both SPP-2200 and N-4000 parallelizations. Results from the N-4000 show very good scaling through four processors, and continue to exhibit fairly good speedups even through eight processors. In fact, one sees essentially the same *slope* ($\simeq$ 1.0) in speedup even between five and eight processors, but with actual speedup shifted (lower). The precise reason for this shift in these data is not currently known.

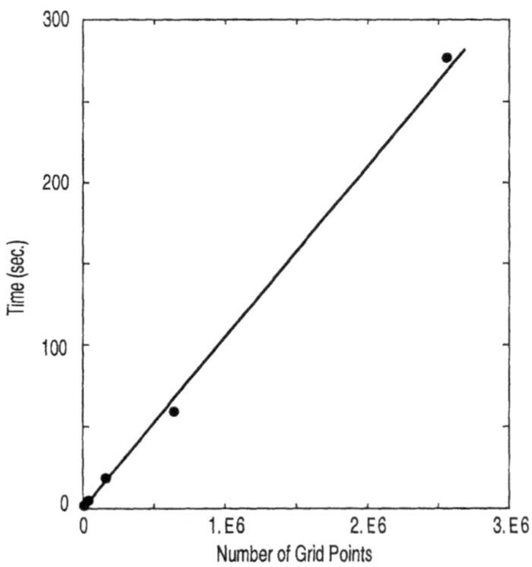

Figure 3. Run time vs. number of grid points.

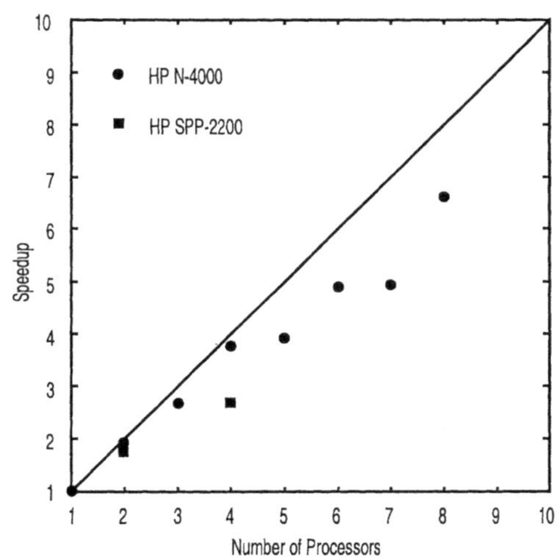

Figure 4. Speedup vs. number of processors.

With the ratio $\rho_{s/p}$ of number of subdomains n_s to number of processors n_p fixed, we obtained the following results in dedicated runs on the SPP-2200. For $\rho_{s/p} = 1$ the run time increased by only a factor 1.556 in going from a 101×101 to a 201×201 grid (factor of four increase in total number of grid points). When $\rho_{s/p} = 2$, using 201×201 and 401×401 grids, the factor was 0.885, and for $\rho_{s/p} = 4$ using these same grids led to a factor of 0.685. On the HP N-4000 with $\rho_{s/p} = 2$, the factor is 1.187, again employing dedicated runs. A value of 1.0 for this factor implies run times independent of problem size. Hence, we conclude that rather generally, the goal of run times independent of problem size is being achieved, at least in an average sense. Moreover, this is being accomplished with a very simple widely-used (in the context of commercial CFD codes) algorithm, SLOR.

5. SUMMARY AND CONCLUSIONS

In this paper we have presented a new implementation of the well-known multilevel DDM concept. The main advantage of the present method is its overall simplicity, being based on successive line overrelaxation, and thus ease of parallelization. We have tested the algorithm with a model Dirichlet problem having known exact C^∞ (but nontrivial) solution and demonstrated good parallel speedups in addition to nearly $\mathcal{O}(N)$ total arithmetic scaling with problem size. This combination leads to run times essentially independent of problem size when parallelization is applied.

Continuing efforts with this algorithm will address problems in generalized coordinates, non-Dirichlet boundary conditions and further parallelization to move the method toward applicability in actual N.–S. equation solution procedures.

ACKNOWLEDGEMENTS

Support of the second author (JMM) by AFOSR Grant #F49620-00-1-0258 and NASA/EPSCoR Grant #WKU-522635-00-10 is gratefully acknowledged.

REFERENCES

1. H. A. Schwarz, Vierteljahrsschrift Naturforsch. Ges. Zürich, 15, (1870) 272.
2. J. F. Thompson, Z. U. A. Warsi and C. W. Mastin, J. Comput. Phys. 47, (1982) 1.
3. B. Smith, P. Bjørstad and W. Gropp, *Domain Decomposition,* PARALLEL MULTILEVEL METHODS for ELLIPTIC PARTIAL DIFFERENTIAL EQUATIONS, Cambridge University Press, Cambridge, (1996).
4. J. M. McDonough, V. E. Garzon and A. W. Schueller, presented at PARALLEL CFD '99, Williamsburg, VA, (1999).
5. D. M. Young, Iterative Solution of Large Linear Systems, Academic Press, New York, NY, (1971).
6. L. A. Hageman and D. M. Young, *Applied Iterative Methods*, Academic Press, New York, 1981.

Parallel Computational Fluid Dynamics – Practice and Theory
P. Wilders, A. Ecer, J. Periaux, N. Satofuka and P. Fox (Editors)

Parallelization of a large scale Kalman filter: comparison between mode and domain decomposition

A.J. Segers[a], A.W. Heemink[b]

[a]KNMI (Royal Netherlands Meteorological Institute), P.O. Box 201, 3730 AE De Bilt, The Netherlands; email: Arjo.Segers@knmi.nl

[b]Faculty of Information Technology and Systems, Delft University of Technology, P.O. Box 5031, 2600 GA Delft, The Netherlands

A large scale Kalman filter for assimilation of measurements in an air pollution model has been implemented in parallel on a massive parallel platform. The technique of Kalman filtering requires a large number of independent model propagations, and therefore owns a large amount of natural parallelism.
Two different methods for parallelization have been implemented and compared. First, each processors has been equipped with a complete copy of the model, leading to efficient implementation of multiple model propagations. Second, the filter has been implemented in combination with a parallel, domain decomposed version of the model, which leads to less efficient propagations, but more efficient matrix algebra in other filter stages. The performance of the parallelization of different filter components has been analyzed in terms of speedup and total execution time. Both methods are shown to perform almost equally, with the domain decomposition slightly favored for flexibility and simple implementation.

1. Introduction

Application of a data-assimilation tool to large scale model puts a large demand on computing power. In practice, the demand is always ahead of the available computing devices, because the need for assimilation has grown, and the underlying models have become more extensive. The capacities of available computing machine's provide a limit to the assimilation problems which can be solved, and the chosen implementation should explore the capacities of the platform as much as possible.

The growing availability of fast multiprocessor machines encouraged the application of the Kalman filter technique to large models. In fact, without the use of these machine, some filter problems can not be solved at all. Online application of filter techniques for periodic forecasts are bounded by the time period in which the problem has to be solved, and this task is often not feasible for a single processor. Besides, some applications do simply not fit in the memory which can be addressed by a single processor.

An early example of parallel implementation of a Kalman filter is found in [6]. Measurements from the UARS satellite were assimilated in a global 2D model, with either the model parallelized over its domain, or through processing the model on different parts of the covariance matrix in parallel. In both methods, the model had to be evaluated n^2 times, with n denotes the number of elements in the state (here $\mathbf{O}(10^4)$). The analysis of measurements turned out to be a bottleneck for the speedup of the parallel filter.

A fundamental change in application of filter techniques to large models was the introduction of low rank filters. Instead of operating and storing a full rank covariance matrix, a limited set of state vectors (modes) is used to describe the correlations. In the context of the Ensemble Kalman filter, Evensen (1994) proposed to integrate ensemble members independently by multiple processors. This strategy leaves the model intact, and could therefore be applied to any available model immediately. The approach was used in [5] for a large 2 layer shallow water model, assigning one mode (ensemble member) to each processor. The analysis equations are however not easily solved in this configurations, since analysis of each single measurement requires data from all ensemble members.

The rather inefficient implementation of the analysis step is now recognized as a major disadvantage of an ensemble-decomposed parallel filter, especially when large amounts of measurements are to be analyzed. Houtekamer and Mitchell (2000) proposed a method for online analysis of $\mathbf{O}(10^4)$ measurements. Batches of observations are analyzed in disjunct area, where spatial correlations between grid cells and observations are ignored after some distance. These batches are efficiently processed in parallel, if the ensemble is decomposed over the domain rather than over the ensemble members. Application of this approach is therefore limited to models for which a domain decomposition is available.

The two different approaches for a parallel filter, decomposition over the modes or over the model domain, have both been implemented for a Kalman filter around the air pollution model LOTOS [7,3]. The performance of the parallelization has been examined in detail when running on a massive parallel platform.

2. Filter equations and data decomposition

The LOTOS model [7] simulates hourly concentrations of 26 gas-phase components in the lowest 2 km of the troposphere. The domain is covered by three layers of grid cells, each with a size of approximately 60 × 60 km. The location and extend of the domain is flexible; in a typical configuration, the model covers western and central Europe with a grid of 40 × 40 cells, leading to a state vector with $n = 1.7 \cdot 10^5$ elements. Ozone measurements from about 50 sites are available on a regular base, and will be assimilated with the model simulations. The stochastic model for LOTOS and the observations is given by:

$$\mathbf{x}[k+1] = \mathbf{M}(t[k], \mathbf{x}[k], \mathbf{w}[k]) \tag{1}$$

$$\mathbf{y}^o[k] = \mathbf{H}'\mathbf{x}[k] + \mathbf{v}[k] \tag{2}$$

where $\mathbf{x}[k]$ contains the concentrations at time $t[k]$, $\mathbf{M}$ is the LOTOS model, $\mathbf{y}^o$ contains available measurement, $\mathbf{H}'$ is an interpolation operator from state vector to measurement locations, and $\mathbf{v}$ is the measurement or representation error (zero mean, covariance $\mathbf{R}$). The noise input $\mathbf{w}$ is used to quantify uncertainties in model parameters such as emissions, deposition, and photolysis rates [3]. The Kalman filter computes the optimal mean $\hat{\mathbf{x}}$ and covariance $\mathbf{P}$ of the concentrations given all measurements that have become available:

$$\hat{\mathbf{x}}_{[k]} = \mathrm{E}\left[\mathbf{x}_{[k]} | \mathbf{y}^o{}_{[k]}, \ldots\right] \quad , \quad \mathbf{P}_{[k]} = \mathrm{E}\left[(\mathbf{x}_{[k]} - \hat{\mathbf{x}}_{[k]})(\mathbf{x}_{[k]} - \hat{\mathbf{x}}_{[k]})' | \mathbf{y}^o{}_{[k]}, \ldots\right] \tag{3}$$

The covariance matrix $\mathbf{P}$ has size $n \times n = \mathbf{O}(10^{10})$, and is therefore far to large to handle. Instead, the low rank factorization $\mathbf{P} = \mathbf{S}\mathbf{S}'$ is used, where the matrix $\mathbf{S}$ has shape $n \times m$ with m the number of modes in the covariance matrix. Each mode represents a possible direction for the difference between true state and mean. In practical applications, the covariance is dominated by a limited number of modes; a value of about 60 is sufficient for the filter around LOTOS [7,3].

The particular filter implementation used in this research is a Reduced Rank SQuare RooT (RRSQRT) filter, see for example [3] for the details of the algorithm. Each filter step consists of a sequence of four operations. During the *forecast* stage, an ensemble of state vectors ξ_j is formed from the best available mean and covariance square root. The ensemble members are formed from linear interpolations of the columns of $\mathbf{S}$, and represent a number of possible values of the true state. The ensemble is propagated by the model, forced by special samples of the noise input:

$$\xi_{j[k]} = \mathbf{x}^a{}_{[k]} + \mathbf{S}_{[k]}\,\omega_j \quad , \quad \xi^f_{j\,[k+1]} = \mathbf{M}(\xi_{j[k]}, \mathbf{w}_{j[k]}) \qquad j = 1, 2, \ldots \tag{4}$$

The propagated ensemble provides a forecast of the mean and covariance at $t_{[k+1]}$; the forecasts $\hat{\mathbf{x}}^f{}_{[k+1]}$ and $\mathbf{S}^f{}_{[k+1]}$ are computed with an operation inverse to (4). The model is called once for each ensemble member, and the forecast is therefore the most expensive operations in the filter. Whenever measurements become available, the *analysis* computes a new mean and covariance given the new information, in a sequence of linear algebra operations:

$$\Theta' = (\Psi'\Psi + \mathbf{R})^{-1}\Psi' \quad , \quad \Psi' = \mathbf{H}'\,\mathbf{S}^f \tag{5}$$

$$\mathbf{x}^a = \mathbf{x}^f + \mathbf{S}^f\,\Theta\,(\mathbf{y}^o - \mathbf{H}'\,\mathbf{x}^f) \tag{6}$$

$$\mathbf{S}^a = \mathbf{S}^f\,\mathbf{B} \quad , \quad \mathbf{B}\mathbf{B}' = \mathbf{I} - \Theta\Psi' \tag{7}$$

The matrix multiplication $\mathbf{S}^f\mathbf{B}$ in (7) is applied in a separate *transformation* stage, where all multiplications are applied in a single operation $\mathbf{S}^a = \mathbf{S}^f(\mathbf{B}\tilde{\mathbf{V}}\Omega)$. The matrix $\tilde{\mathbf{V}}$ is part of a *rank reduction* stage, which approximates the covariance square root by its largest singular vectors, necessary if the number of columns has grown during the forecast. The columns of $\tilde{\mathbf{V}}$ contain the eigenvectors of $(\mathbf{SB})'(\mathbf{SB})$ corresponding to the largest eigenvectors. The interpolation coefficients used in (4) to form the forecast ensemble are stored in Ω.

The workload in the filter is strongly related to the square root matrix $\mathbf{S}$. A parallel version of the filter should therefore distribute the tasks involved with $\mathbf{S}$, which is thereto decomposed and distributed over the processors. Two options are considered: a decomposition over the columns, equivalent to each processor owning a number of modes, and a decomposition over the rows, where each processor is responsible for a certain part of the model domain. Both options are illustrated in figures 1–2.

The filter around LOTOS has been parallelized with both methods, described in detail in the following sections. The performance has been measured for an implementation on a massive parallel machine (Cray T3E) running on 1, 2, 4, 8, 16, or 32 processors. The total filter problem turned out to be to large to fit in the memory of a single processor; to be able to measure the performance for 1 pe to, the size of the domain has been limited to 32×32 grid cells for these experiments.

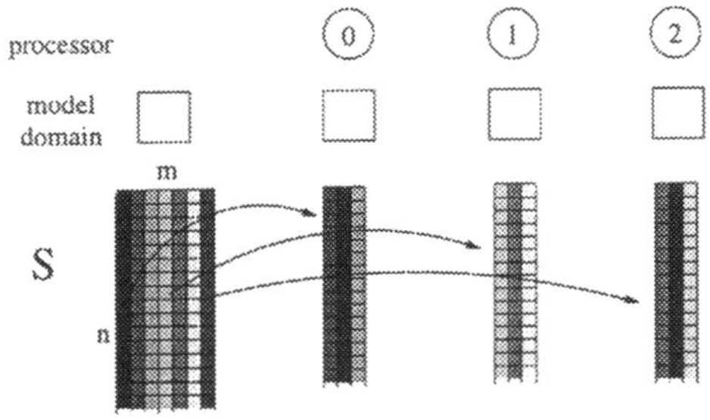

Figure 1. *Decomposition of the Kalman filter over the modes. Each processor owns a complete copy of the model, and a number of columns of the covariance square root.*

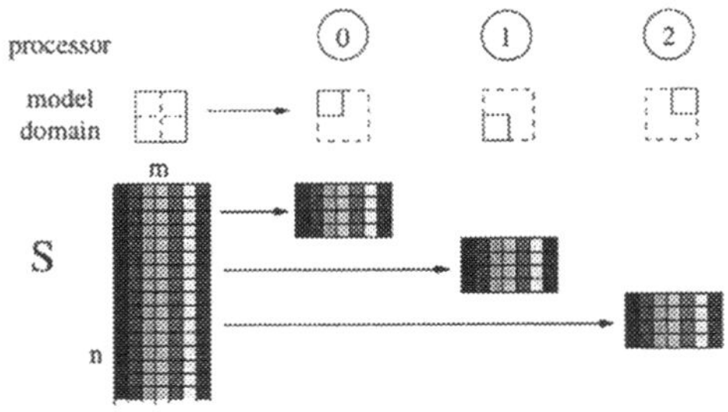

Figure 2. *Decomposition of the Kalman filter over the domain. Each processor owns a part of the model domain, and the rows of the covariance square root related to this sub domain.*

3. Decomposition over the modes

The mode decomposition is based on on exploring the natural parallelism of the forecast. Eq. (4) requires a large number of more or less similar model propagations; these could be processed in parallel without interaction. An almost optimal speedup is expected for the forecast stage, and since this is the major time consumer, the speedup of the filter is expected to be large. Each processor should manage the propagation and analysis of a certain number of ensemble members; therefore, each processors owns a complete copy of the model dynamics and model data (emissions, wind fields, etc). If the ensemble is distributed equally, the time required for a single filter step is the same for each processor.

Operations between different columns of **S** are however complicated by the distribution over the processors. Especially the transformation requires a lot of communication: each single column of **S** is to be replaced by a linear combination of all columns, from which a large number is stored in remote memory.

Figure 3 shows the speedup of the different stages and the total filter; the computation of the diagonal of the covariance matrix is added as an extra stage (required for output). As expected for a mode decomposed filter, the best speedup is achieved for the forecast stage, when each processor performs an equal number of model integrations. The speedup is not perfect, since each processor has to spent some time on the initialization of the model. The speedups measured for the rank reduction and the transformation are comparable with that of the forecast. Communicational overhead for growing number of processors seems to be compensated for by the spread of the workload. The costs of the rank reduction are completely determined by computation of $\mathbf{S}'\mathbf{S}$, which requires a large amount of communication since the columns of $\mathbf{S}$ are distributed. The actual eigenvalue decomposition of this rather small matrix takes less than 5% of the costs.

The worst speedup is measured for the analysis stage. In the current setup, hardly any parallelism is present; each processor computes a copy of the vector $\mathbf{a}$ and matrix $\mathbf{B}$ from eq. (6-7), and this part of the analysis has therefore no speedup at all. Some speedup is achieved from the interpolation of the modes to the measurement locations (5). The total costs of computing this entities is however small, since the number of measurements is limited (less than 50 each hour). For larger numbers of measurements, eq. (5) and (7) might be solved in parallel, or a domain decomposition strategy could be considered.

The total speedup is the sum of the speedup of the different filter stages, weighted with their execution time. For 8 processors, the total speedup is almost perfect (7.5), while for 16 and 32 processors, the speedup is still very good (14 and 24 respectively). The speedup curve is not flattening, thus running the filter on more than 32 pe's will still increase the total speed. The total speedup is strongly related to the speedup of the forecast. Running on 32 pe's, about 80% of the computation time is spent on the forecast, while the other stages consume only a fraction. The less efficient parallelization of the analysis, and less important, of the diagonal computation, do therefore hardly hamper the total speedup, but make it only slightly smaller than the speedup of the forecast.

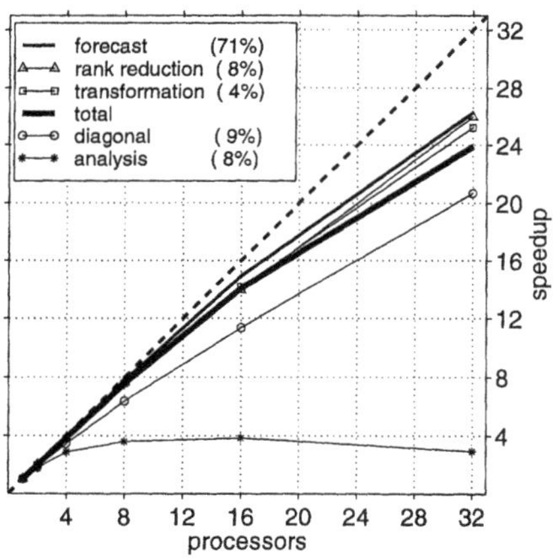

Figure 3. *Speedup of different stages and total filter for the mode decomposed filter. The percentages in the legend denote the fraction of the total execution time spent on certain stage, for evaluation on 32 pe's.*

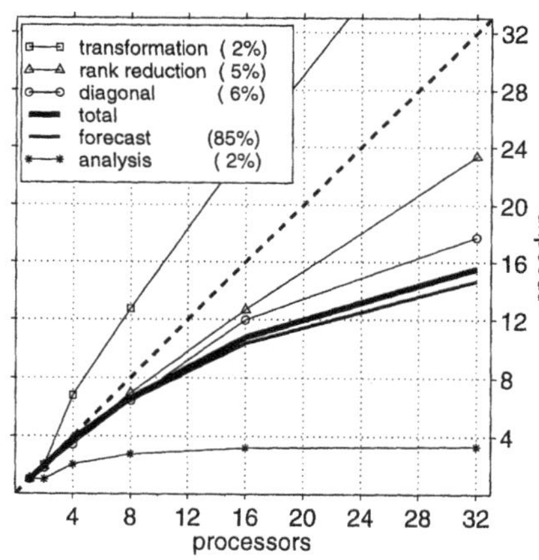

Figure 4. *Speedup of total filter and different stages for the domain decomposed filter.*

4. Decomposition over the domain

A domain decomposition of the LOTOS model could be made in a straight forward manner, since the grid is very regular (rectangular, and almost equidistant). The domain selected for a certain application is divided into a number of sub domains, each of them assigned to a different processor. If each sub domain covers the same number of cells, operations such as chemistry, vertical exchange and deposition will require the same amount of computation time. The advection scheme (Runge-Kutta method) requires concentrations of two shells of boundary cells, to be copied from sub domains stored on other pe's.

The straight forward domain decomposition has been implemented in LOTOS and tested on the Cray T3E. Thanks to the very fast communication on this machine, the speedup of the domain-decomposed advection is very good in spite of the simple approach. With decomposition in 8 domains, the speedup of the advection is about 6.3, which is reasonable in comparison with other examples of parallel Runge-Kutta schemes [2]. The total speedup is even better, since the other operations in the model (chemistry, deposition, vertical exchange) are not influenced by the domain decomposition, and thus show a perfect speedup. A total speedup of 7.4 for 8 processors and 13.4 for 16 processors was measured for a model configuration with 40×40 grid cells. For larger numbers of processors, the speedup decreases strongly, since the ratio between the number of local grid cells and remote boundary cells becomes inefficient.

The data structures of the parallel filter should follow the domain decomposition of the model. Each processor is assigned to a single domain, and should own a copy of all entities of the filter that have any relation with this domain. The covariance square root is therefore distributed in blocks of rows (figure 2). A number of smaller entities that are of interest for all sub-domains are not decomposed, but each processor owns a copy; their contents have to be synchronized. The implementation of the filter equations was found to be very simple for the domain decomposition, since almost all operations act row-wise on $\mathbf{S}$. The parallelized filter code hardly differs from a single processor code, and is therefore easily ported to other platforms.

Figure 4 shows the speedups measured for the domain decomposed filter. The most apparent result is the observed super linear speedup of the transformation stage. Thanks to smaller arrays in the decomposed $\mathbf{S}$, the computation of $\mathbf{S}^f(\mathbf{B}\tilde{\mathbf{V}}\Omega)$ is 44 times faster if evaluated on 32 processors. The net effect of this super linear speedup is however limited, since the total computation time spent on the transformation is only a few percent of the total. Another apparent result is the not very optimal speedup of the forecast stage. Evaluated on 16 processors, the forecast is only 10 times faster, while speedups above 13 were measured for the LOTOS model without filter. The parallel LOTOS model suffered in this experiment from the smaller grid size of 32×32, necessary for running the filter on a single processor too.

The analysis shows a similar lack of speedup as observed for the mode decomposition. The performance of the domain decomposed rank reduction and computation of the covariance diagonal are comparable with that of the mode decomposed filter. This is remarkable, since the operations on the covariance square root used in these stages require a lower amount of communication in case of domain decomposition. The mode decomposition benefits from the fast communication network in the Cray computer, where transfers of large blocks of memory are rather cheap.

The reduction of the model domain required for running the filter on a single processor decreased the performance of the domain decomposed filter significantly. Therefore, the performance of the parallel filters have been compared again for the original configuration with 40×40 grid cells. In addition, the source codes have been compiled with full optimization. The optimization increase the memory consumption, and had been turned off to produce an executable able to run on a single processor.

Figure 5 shows the relative differences in computation time for the two methods, for a filter running on 8 processors. With this number of pe's, the filter problem is solved within a reasonable time. The domain decomposition is now about 10% faster than the mode decomposition, completely due to a faster computation of the covariance diagonal. The other linear algebra operations on the covariance square root are slightly cheaper too, while the forecast is slightly more expensive. Both observations are in agreement with what was expected for the domain decomposition.

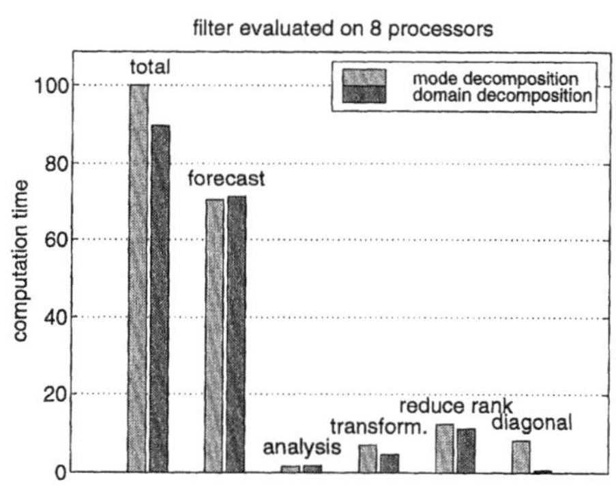

Figure 5. *Total computation time spent on different filter stages for mode or domain decomposed filter, for a LOTOS configuration with* 40×40 *grid cells and evaluation on 8 processors. Both filters are configured and compiled in their most optimal form.*

5. Discussion and conclusions

Evaluation of the experiments with the mode- and domain-decomposed shows that both methods deliver an efficient parallel filter. The mode decomposition guarantees a good speedup even for large numbers of processors, independent of the model. The performance of the domain decomposed filter depends completely on the performance of the parallel model, which decreases strongly for growing number of processors. For a number of 8 processors, which is a reasonable number for the assimilation experiments with LOTOS, the performance of both methods is almost indifferent. The amount of communication required for the domain decomposition is substantial less than for the mode decomposition, where complete state vectors need to be transfered. On a platform with relative slow communication such as a Beowulf cluster, the domain decomposition is therefor probably more efficient.

The domain decomposition is favored when large numbers of measurements are to be assimilated. Implementation of the algorithm proposed in [4] explores the fact that the spatial impact of measurements is limited, and this is efficient implemented in a domain decomposed model. In the LOTOS application, the number of measurements was to limited to consider these options.

Implementation of a mode decomposed filter turned out to be more complicated than a domain decomposed version. Almost all stages in the filter are influenced by the parallelization, while in the domain decomposition the parallelization is hidden in the model. Building a parallel LOTOS model was rather simple since only the advection part needed to be parallelized, and this was facilitated by the regular grid.

Concluding, for the parallelization of the filter around LOTOS the domain decomposition is slightly favored. Since the performance in our applications is almost indifferent, this choice is based on more or less subjective criteria such as easy implementation, and potential positive effects when running on other platforms or when large numbers of measurements are to be analyzed. For application of the filter to another model, these advantages are not relevant if an efficient domain decomposition is not available. Spending some time on improvement of the parallel model might be useful here; having an efficient parallel model is useful anyway. If this is not possible or the result is not satisfactory, the mode-decomposition is the only alternative, and the parallelism inherent to the filter algorithm still ensures a proper efficiency.

Acknowledgment

The Delft University *Center for High Performance Applied Computing* is acknowledged for providing access to the Cray T3E.

REFERENCES

1. Geir Evensen. Sequential data assimilation with a nonlinear quasi-geostrophic model using monte carlo methods to forecast error statistics. *Journal of Geophysical Research*, 99(C5):10143–10162, 1994.
2. G. Fritsch and G. Möhres. Multistage simulations for turbomachinery design on parallel architectures. In D.R. Emerson, A. Ecer, J. Periaux, N. Satofuka, and P. Fox, editors, *Proceedings of the Parallel CFD'99 Conference, Manchester, 1997*, pages 225–238. Elsevier Science B.V., 1998.
3. A.W. Heemink and A.J. Segers. Modeling and prediction of environmental data in space and time using kalman filtering. *submitted to Stochastic Environmental Research and Risc Assesment*, 2000.
4. P.L. Houtekamer and Herschel L. Mitchell. A sequential ensemble kalman filter for atmospheric data assimilation. *Monthly Weather Review*, 129(1):123–137, january 2001.
5. Christian L. Keppenne. Data assimilation into a primitive-equation model with a parallel ensemble kalman filter. *Monthly Weather Review*, 128(6):1971–1981, 2000.
6. P.M. Lyster, S.E. Cohn, R. Ménard, L.-P. Chang, S.-J. Lin, and R. G. Olsen. Parallel implementation of a kalman filter for constituent data assimilation. *Monthly Weather Review*, 125(7):1674–1686, 1997.
7. M. van Loon, P.J.H. Builtjes, and A.J. Segers. Data assimilation applied to lotos: First experiences. *Environmental Modelling and Software*, 15(6–7):603–609, 2000.

Parallel Computational Fluid Dynamics – Practice and Theory
P. Wilders, A. Ecer, J. Periaux, N. Satofuka and P. Fox (Editors)

A direct algorithm for the efficient solution of the Poisson equations arising in incompressible flow problems

M.Soria, C.D.Pérez-Segarra, K.Claramunt, C.Lifante [a] *

[a]Laboratori de Termotecnia i Energetica
Dept. de Maquines i Motors Termics
Universitat Politecnica de Catalunya
Colom 9, E-08222 Terrassa, Barcelona (Spain)
manel@labtie.mmt.upc.es

In parallel simulations of incompressible flows, the solution of the Poisson equations is typically the main bottleneck, specially on PC clusters with conventional networks. The communications should be reduced as much as possible in order to obtain a high parallel efficiency. With this purpose, a specific version of Schur Complement algorithm has been developed. It is based on the use of direct algorithms for each subdomain and for the interface equation. The inverse of the interface matrix is evaluated and stored in parallel. Only one all-to-all communication episode is needed to solve each Poisson equation almost to machine-accuracy. Turbulent buoyancy-driven natural convection has been used as a benchmark problem. The speed-up obtained on a 24-nodes Beowulf cluster is presented, as well as a breakdown of the computing costs.

1. Introduction

A new class of parallel computers, the so-called *Beowulf clusters* of personal computers (PCs) running Linux has recently become available. Compared with conventional parallel computers, their cost is significantly lower and their computing power per-processor is comparable. However, they are *loosely coupled* computers: their communication performance is poor compared with their computation performance, specially if conventional (100 Mbits/s) networks are used. This is a serious obstacle for the efficiency of the parallel CFD algorithms on PC clusters. Parallel algorithms tolerant to loosely coupled computers must be developed in order to use efficiently their CPU potential.

Here, our attention is restricted to incompressible flows and segregated algorithms. The discrete momentum and energy equations can be treated explicitly or easily solved with a conventional parallel iterative algorithm [1]. This is not the case of the (*Pressure-correction equation* or *Poisson equation*), that must be treated implicitly:

$$\nabla^2 p = \nabla \cdot \mathbf{u}^* \tag{1}$$

*This work has been financially supported by the "Comisión Interministerial de Ciencia y Tecnología", Spain (Project TIC99-0770) where p is the scalar field to be determinate and $\mathbf{u}^*$ is a known vector field. The details and the terminology depend on the segregated algorithm used, but the linear equations to be solved are equivalent.

In this paper, our aim is to design an algorithm to allow the efficient solution of Poisson equations arising in incompressible CFD, using loosely-coupled parallel computers. The approach proposed is a specific version of the so called *Schur Complement methods* (SC). In SC algorithms the subdomains are non-overlapping and separated by an interface that is solved implicitly. For the case of incompressible flows discretized with staggered meshes, as it is frequently done for DNS and LES, an important property of the discrete Poisson equation is that *its matrix A remains constant during all the fluid flow simulation.* This is, our problem is not to solve a single equation but actually to solve for x_i in each of the equations:

$$Ax_i = b_i \tag{2}$$

At least one these equations must be solved per time step/iteration. Let M be the total number of equations to be solved. For time-accurate problems, M is a large number[2]. Hence, the hypothesis that has been used to develop the present implementation of SC is that any *reasonable* amount of computing time can be spent in a *pre-processing* stage, where only A is used, if then it allows us to reduce substantially the cost of solving each of the M equations. The data generated in the pre-processing stage can even be saved and reused for other simulations, as matrix A only depends on the mesh.

2. Schur Complement algorithm

2.1. Overview

Schur complement (or Dual Schur Decomposition) [2–7] is a direct parallel method, based on the use of non-overlapping subdomains with implicit treatment of interface conditions. It can be used to solve any sparse linear equation system: no special property of the matrix or the underlying mesh is required (except non-singularity). In order to obtain the exact solution of the linear equation system, each processor has to solve twice its own subdomain. All the processors cooperate to solve an interface equation. In our implementation, *only one communication operation is needed to solve accurately each equation system.* The algorithm will be described using two-dimensional problems for clarity, but its extension to three-dimensions is straightforward (except for RAM memory limitations, this aspect is discussed in [7]). The unknowns in vector x are partitioned into a family of P subsets, called *inner domains*, that are labeled from 0 to $P-1$ (to follow the MPI convention), plus one *interface*, labeled s. The interface s is defined so that for any pair of subsets p_1 and p_2, no unknown of p_1 is directly coupled with any unknown of p_2, but only to its own unknowns and to s.

[2]As an example, to solve our benchmark problem from initial conditions to statistically steady flow, $M \approx 10^6$.

With the new ordering, equation (2) becomes:

$$\begin{bmatrix} A_{0,0} & 0 & \cdots & A_{0,s} \\ 0 & A_{1,1} & \cdots & A_{1,s} \\ \vdots & & & \vdots \\ 0 & \cdots & A_{P-1,P-1} & A_{P-1,s} \\ A_{s,0} & A_{s,1} & \cdots & A_{s,s} \end{bmatrix} \begin{bmatrix} x_0 \\ x_1 \\ \vdots \\ x_{P-1} \\ x_s \end{bmatrix} = \begin{bmatrix} b_0 \\ b_1 \\ \vdots \\ b_{P-1} \\ b_s \end{bmatrix} \tag{3}$$

For the solution of the reordered system, the interface unknowns are isolated using Gaussian elimination to obtain:

$$\begin{bmatrix} A_{0,0} & 0 & \cdots & A_{0,s} \\ 0 & A_{1,1} & \cdots & A_{1,s} \\ \vdots & & & \vdots \\ 0 & \cdots & A_{P-1,P-1} & A_{P-1,s} \\ 0 & 0 & \cdots & \tilde{A}_{s,s} \end{bmatrix} \begin{bmatrix} x_0 \\ x_1 \\ \vdots \\ x_{P-1} \\ x_s \end{bmatrix} = \begin{bmatrix} b_0 \\ b_1 \\ \vdots \\ b_{P-1} \\ \tilde{b}_s \end{bmatrix} \tag{4}$$

The last block equation, that involves only unknowns in x_s is the *interface equation*:

$$\tilde{A}_{s,s} x_s = \tilde{b}_s \tag{5}$$

where

$$\tilde{A}_{s,s} = A_{s,s} - \sum_{p=0}^{P-1} A_{s,p} A_{p,p}^{-1} A_{p,s} \tag{6}$$

and

$$\tilde{b}_s = b_s - \sum_{p=0}^{P-1} A_{s,p} A_{p,p}^{-1} b_p \tag{7}$$

The key point of SC is that the interface equation can be solved to obtain the interface nodes x_s *before* the inner nodes x_p. After the determination of x_s, each processor can solve its inner nodes x_p from its row of Eq. 4.

2.2. Pre-processing

The preprocessing algorithm is used to evaluate data that depends only on matrix A. The first step is the evaluation of $\tilde{A}_{s,s}$. Eq. (6) is rewritten as:

$$\tilde{A}_{s,s} = A_{s,s} - \sum_{p=0}^{P-1} \tilde{A}_{s,s}^{p} \tag{8}$$

Here each of the terms $\tilde{A}_{s,s}^{p} = A_{s,p} A_{p,p}^{-1} A_{p,s}$ is the contribution from processor p to $\tilde{A}_{s,s}$. To evaluate $\tilde{A}_{s,s}^{p}$ we proceed column by column, solving t from $A_{p,p} t = [A_{p,s}]_c$. Then, column c is evaluated as $\left[\tilde{A}_{s,s}^{p}\right]_c = A_{s,p} t$. When the contribution of each processor is available, they are added and the result subtracted from $A_{s,s}$, to obtain $\tilde{A}_{s,s}$. To solve the equations that involve matrix $A_{p,p}$, a band-LU decomposition algorithm is used [8].

The solution of the interface equation is critical for the global efficiency of the SC algorithm. In our implementation, the solution time is minimized by using the inverse of the interface matrix $\tilde{A}_{s,s}$. It is evaluated and stored in parallel in the preprocessing stage. A row-based distribution is used. Each of the processors stores the rows of $\tilde{A}_{s,s}^{-1}$ that it needs to evaluate its part of the product $x_s = \tilde{A}_{s,s}^{-1}\tilde{b}_s$. In order to determinate row r, the equation system to be solved is $\tilde{A}_{s,s}^{t}\left[\tilde{A}_{s,s}^{-1}\right]^{r} = e_r$. A total of N_s (the number of interface nodes) must be solved. To solve them, the LU decomposition of $\tilde{A}_{s,s}^{t}$ is evaluated.

As the algorithm has been designed for loosely coupled computers, a block LU decomposition is employed, instead of a conventional LU decomposition. The factorization is evaluated by means of a Block Gaussian elimination with pivoting. The criteria used for pivot selection is the condition number of the sub-matrices, that is explicitly evaluated. Other methods can be used for the evaluation of the LU decomposition.

2.3. Solution

The solution of each pressure-correction is done in three steps: *(i)* Evaluation of right-hand-side of interface equation (that can not be evaluated during preprocessing as it depends on b) *(ii)* Solution of interface nodes and *(iii)* Solution of the inner domains.

Solution of $Ax = b$
- S.1-Evaluate $\tilde{b}_s = b_s - \sum_{p=0}^{P-1} \tilde{b}_s^p$:
  - S.1.1-Solve $A_{p,p}t = b_p$
    using the pre-evaluated LU dec. of $A_{p,p}$
  - S.1.2-Evaluate $\tilde{b}_s^p \leftarrow A_{s,p}t$
  - S.1.3-Evaluate $\tilde{b}_s \leftarrow \sum_{p=0}^{P-1} \tilde{b}_s^p$ (MPI_Allreduce)
  - S.1.4-Evaluate $\tilde{b}_s \leftarrow b_s - \tilde{b}_s$
- S.2-Solve the interface nodes x_s from $\tilde{A}_{s,s}x_s = \tilde{b}_s$:
  - S.2.1-Evaluate $x_s = \tilde{A}_{s,s}^{-1}\tilde{b}_s$ where needed
    using the required subset of $\tilde{A}_{s,s}^{-1}$ rows (evaluated in P.7)
- S.3-Solve the inner nodes x_p from $A_{p,p}x_p = b_p - A_{p,s}x_s$:
  - S.3.1-Evaluate $t = b_p - A_{p,s}x_s$
  - S.3.2-Solve $A_{p,p}x_p = t$
    using the pre-evaluated LU dec. of $A_{p,p}$

In order to save RAM memory and CPU time, it is important to store matrices $A_{s,p}$, $A_{p,s}$ and $\tilde{A}_{s,s}^{-1}$ as sparse. In our implementation, a data structure oriented to do fast matrix-vector products is used (as it is needed in step S.2.1). For each row the non-null columns and its indices are stored.

3. Benchmark

3.1. Benchmark problem

Incompressible, two-dimensional, time-accurate Navier-Stokes equations plus energy transport equation have been used as a problem model to benchmark the parallel solver developed. The flow is assumed to be single-phase, single-component and incompressible.

Viscous dissipation and thermal radiation have been neglected. The fluid is assumed Newtonian and the physical properties constant. Under these hypothesis, the governing equations are:

$$\nabla \cdot \mathbf{u} = 0 \tag{9}$$

$$\frac{\partial u_x}{\partial t} + \mathbf{u} \cdot \nabla u_x = -\frac{1}{\rho}\frac{\partial p}{\partial x} + \nu \nabla^2 u_x + f_x \tag{10}$$

$$\frac{\partial u_y}{\partial t} + \mathbf{u} \cdot \nabla u_y = -\frac{1}{\rho}\frac{\partial p}{\partial y} + \nu \nabla^2 u_y + f_y \tag{11}$$

$$\frac{\partial T}{\partial t} + \mathbf{u} \cdot \nabla T = \frac{k}{\rho c_p} \nabla^2 T \tag{12}$$

To account for the density variations the Boussinesq approximation has been used. The components of the body force are $f_x = 0$, $f_y = g\beta(T - T_0)$. $\beta = -\frac{1}{\rho}\frac{\partial \rho}{\partial T}$ is the thermal volumetric expansion coefficient. The governing equations are discretized using a staggered mesh [9]. SMART [10] numerical scheme is employed for the discretization of the convective terms. The deferred correction procedure is used [11].

The simulation of two-dimensional buoyancy-driven turbulence in a tall ($A = L_y/L_x = 4$) rectangular cavity, as described in [12], has been used carried out. The vertical walls are isothermal and the horizontal walls adiabatic, with a temperature gradient of ΔT. The initial conditions are cold, steady fluid. The Prandtl number is $Pr = 0.71$ and the Rayleigh number based on the cavity width is $Ra = 10^9$. In these conditions, the flow is known to be turbulent. Therefore, instantaneous fields are *not* two-dimensional. However, the two-dimensional simplification, studied by many authors (e.g. [13]), is still useful for us as a benchmark problem. A mesh of $N = 121 \times 241$ control volumes and a dimensionless time step $\Delta\tau = 1.25 \times 10^{-7}$ were used in [12]. For the benchmarks, the same time step has been used, while meshes from $N = 50 \times 100 = 5000$ to $N = 600 \times 1200 = 720000$ have been used. The meshes must be strongly concentrated near the boundaries to resolve the boundary layers.

3.2. Computing times and speed-up

The *speed-up*, S_p defined as the ratio between the computing time needed by one processor and by P processors[3], has been measured using a cluster with 24 standard PCs (AMD K7 CPU at 900 Mhz and 512 Mbytes of RAM) and a conventional network (100 Mbit/s 3COM network cards and a 3COM switch). They run Debian Linux 2.1, kernel version 2.2.17. The MPI implementation used is LAM 6.1, and the C compiler is GCC version 2.7.2.3.

For the measure, the full Navier-Stokes algorithm was executed and the wall times spent in the solution of pressure correction equation measured during 20 time steps and then averaged [4]. As an indication, for $N = 400 \times 10^3$ unknowns and $P = 24$ processors, the computing time is ≈ 0.35 seconds. The speed-ups, for different mesh sizes are presented in Fig.1.

[3]For the case of a single processor, a band-LU algorithm is used.

[4]As the algorithm proposed is totally direct, the computing times should theoretically be identical for each execution. The averaging is done to minimize the errors in the measure.

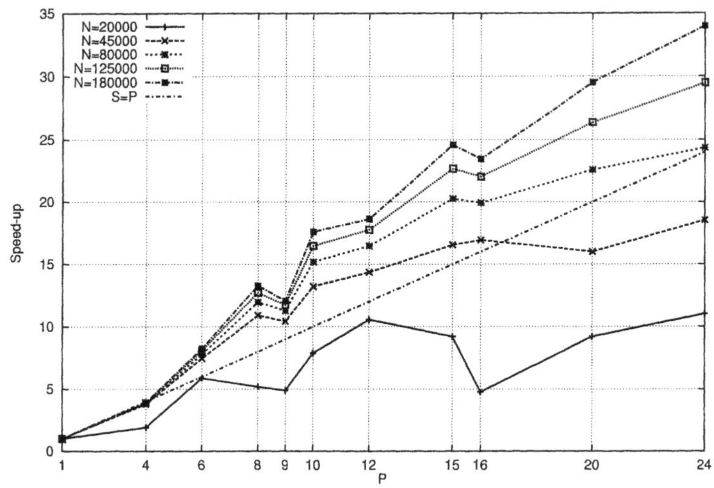

Figure 1. Speed-up of the SC algorithm in a PC cluster.

Considering the computing system used, the speed-up obtained is high. For large meshes, it is even super-linear ($S > P$). As each processor has to solve twice its own domain, the maximum speed-up that should be expected is only $\frac{P}{2}$. To clarify the reason of this behavior, a breakdown of the SC computing costs has been carried out.

3.3. Breakdown of computing costs

More than 95% of the computing time of SC algorithm is due to three operations: the solution of both local problems (steps S.1.1 and S.3.2), the matrix-vector product needed to solve the interface nodes (step S.2.1), and the communications needed to evaluate the right-hand-side of the interface equation, $\tilde{b}_s$ (step S.1.3). The fraction of the total computing time spent in each of these operations has been represented versus the number of unknowns N for $P = 24$ in Fig.2.

In spite of the poor performance of the network, communication time is not the dominant factor, except for the smaller meshes and higher number of processors considered. For the case of $N = 405 \times 10^3$ and $P = 24$, an ideal parallel computer with infinite bandwidth and null latency would only be about 10% faster. The cost of the solution of the interface equation, that is between 30% and 40% for $P = 24$, is lower if less processors are used.

Typically, the majority of the time is spent in the solution of the band-LU systems. The cost of this operation grows more than linearly with the number of unknowns. For a large P, the time to solve twice a system with $\frac{N}{P}$ unknowns is significantly smaller than the time to solve once a system with N unknowns.

In our case, the super-linear speed-ups can actually be an indication that a sequential

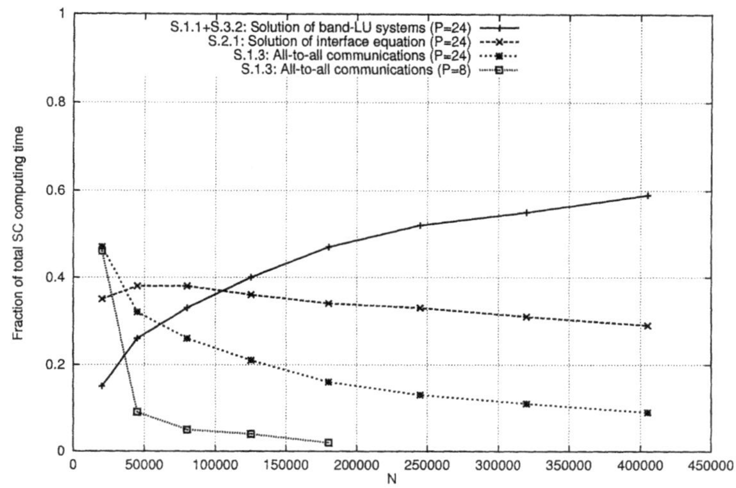

Figure 2. Breakdown of the processing time of the SC algorithm in a PC cluster, for $P = 24$. The network time has also been represented for $P = 8$.

Schur decomposition rather than a band-LU would be a faster alternative for the solution with $P = 1$. As a sequential computer can always be used to emulate the behavior of a parallel computer, a sequential algorithm with a higher operation count than a parallel algorithm is not the optimal.

To allow a better discussion of the computing times and precision obtained with the algorithm proposed in this work, it is worth comparing it with the well-known sequential ACM method. For our problem model, the band-LU solver has been compared [6] with ACM for a mesh of $N = 110450$ control volumes and a precision to stop the iterative solver $\epsilon \approx 10^{-3}$ (relative to the initial residual). Their computing times are comparable: band-LU can be slightly better or worse depending on the parameters used for ACM. On the other hand the accuracy of the algorithm here proposed is $\epsilon = 10^{-13}$ and, for this mesh size and using 24 processors, it is about 30 times faster than band-LU.

4. Conclusions

A direct parallel algorithm to solve the pressure-correction equation of incompressible fluid flow problems, based on the Schur Complement algorithm, has been described. As the matrix of the equation remains constant during all the fluid flow simulation, the algorithm proposed relies on a relatively long pre-processing step that is then reused to solve the pressure-correction equation for each iteration, using only one all-to-all communication episode.

The algorithm proposed has been benchmarked on a PC cluster, with a conventional

100 Mbit/s network.In spite of the low communications performance of the cluster, the algorithm described provides high speed-ups and, more important, low absolute computing times.

For the problem model, using 24 processors, with a mesh of $N = 125 \times 10^3$ control volumes, the algorithm proposed is about 30 times faster than sequential band-LU, which for this application and mesh size is comparable to Additive Correction Multigrid.

REFERENCES

1. J.M.Ortega, Introduction to Parallel and Vector Solution of Linear Systems, Plenum Press, 1988
2. D.Vanderstraten and R.Keunings, International Journal of Numerical Methods in Fluids, vol. 28, pp. 23–46, 1998
3. S.Kocak et al., Parallel Computational Fluid Dynamics: Proc. '98 PCFD Conference, pp. 353–360, North Holland, 1999
4. S.Kocak and H.U.Akay, Parallel Computational Fluid Dynamics: Proc. '99 PCFD Conference, pp. 281-288, North Holland, 2000
5. E.Simons, An Efficient Multi-domain Approach to Large Eddy Simulation of Incompressible Turbulent Flows in Complex Geometries, Ph.D. thesis, Von Karman Institute for Fluid Dynamics, Brussels, Belgium, 2000
6. M.Soria, Parallel Multigrid Algorithms for Computational Fluid Dynamics and Heat Transfer, Ph.D. thesis, Universitat Politecnica de Catalunya, Barcelona, Spain, 2000 (available from the author).
7. M.Soria, C.D.Pérez-Segarra, A.Oliva, Numerical Heat Transfer - Part B, vol. 41, 2002
8. W.H.Press et al., Numerical Recipies in C. The art of Scientific Computing, Cambridge University Press, 1994
9. S.V.Patankar. Numerical Heat Transfer and Fluid Flow, McGraw-Hill, 1980
10. P.H.Gaskell et al., International Journal of Numerical Methods in Fluids, vol. 8, pp. 1203–1215, 1988
11. M.S.Darwish and F.Moukalled, Numerical Heat Transfer - Part B, vol. 26, pp. 79–96, 1994
12. M.Farhangnia et al., International Journal of Numerical Methods in Fluids, vol. 23, pp. 1311–1326, 1996
13. S.Paolucci, Journal of Fluid Mechanics, vol. 215, pp. 229–262, 1990

Parallel Computational Fluid Dynamics – Practice and Theory
P. Wilders, A. Ecer, J. Periaux, N. Satofuka and P. Fox (Editors)

Current Status of CFD Platform - UPACS -

Ryoji Takaki[a], Mitsumasa Makida[a], Kazuomi Yamamoto[a], Takashi Yamane[a], Shunji Enomoto[a], Hiroyuki Yamazaki[a], Toshiyuki Iwamiya[a] and Takashi Nakamura[a]

[a]CFD Technology Center, National Aerospace Laboratory,
7-44-1 Jindaiji-higashi, Chofu, Tokyo 182-8522, Japan
E-mail:ryo@nal.go.jp

UPACS, U*nified* **P***latform for* **A***erospace* **C***omputational* **S***imulation*, is a project to develop a common CFD platform since 1998 at National Aerospace Laboratory. This project aims to overcome the increasing difficulties in recent CFD code programming on parallel computers for aerospace applications which includes complex geometry problems and coupling with different types of simulations such as heat conduction in materials and structure analysis. Moreover, it aims to accelerate the development of CFD technology by sharing a common base code among research scientists and engineers. The development of UPACS for compressible flows with multi-block structured grids is complete and further improvements are planned and being carried out. The current status of UPACS is described in this paper.

1. INTRODUCTION

Recent rapid development of computer hardware and software technology has made it possible to conduct massive numerical simulations of three dimensional flows around complicated aircraft with realistic configurations, complicated unsteady flows in jet engines and physically complicated flows such as chemically reacting flows around re-entry vehicles. It has also made it possible to conduct direct optimization of aerodynamic design including structure or heat conduction analysis. However, various problems arise from the increased complexity in computer programs due to the adaptation to complex configurations, the parallel computation and the multi-physics couplings. The parallel programming is one of the biggest problems making the program more complicated. Moreover, the portability is sometimes lost due to the parallelization because many computer vendors supply their own language or compiler directives for parallelization. Although the programming has been accomplished by great efforts of a few researchers and engineers for each specific application area, it is actually inefficient not only for writing programs but also for the code validation and the accumulation of know-how among CFD researchers.

In order to overcome the difficulty in such complicated programming and to accelerate the code development, some researches have been trying to separate flow solver parts from parts which handle parallel procedures. DLR has developed a simulation program *TRACE* [1] for turbomachinary flows since early 1990s. An object oriented framework for parallel CFD using C++ has been proposed by Ohta[2,3]. NAL started a pilot project

UPACS (Unified Platform for Aerospace Computer Simulation)[4] in 1998. This project aims at development of common CFD codes that can be shared among CFD researchers and code developers. Moreover, UPACS is expected to make CFD researchers free from the parallel programming if they do not want to know details about parallelization.

2. CONCEPT OF UPACS CODE

After several conceptual studies and development of several pilot codes, the UPACS code (version 1.0) was released on Oct. 2nd, 2000. The UPACS code is based on the following design concepts and approaches;

- Multi-block structured grid method
- Separation of CFD solvers from multi-block/overset and parallel procedures
- Portability
- Hierarchical and encapsulated structure

2.1. Multi-block structured grid method

Considering the applicability to complex configurations and higher resolution at a boundary layer, a multi-block structured grid method is chosen as the first step. The multi-block approach can be easily applied to parallel computing. However difficulty exists in how to make connection information for every block. A pre-process program *createConnect3D*, which searches connected block faces automatically, has been developed to construct the connection information table.

The multi-block approach is still difficult to construct for very complicated configurations and sometimes it requires much time and effort to make it. An overset grid method combined with a multi-block grid has also been used. There is a similar problem how to find indices and calculate coefficients for interpolation of variables. A pre-process program *createOversetIndex*, which finds donor cells and calculates interpolation coefficients for each cell, has been developed to construct overset information. From the viewpoint of parallelization, the overset grid method can be considered to be a special case of the multi-block grid method. Therefore, both (multi-block and overset) can be treated similarly.

2.2. Separation of CFD solvers from multi-block/overset and parallel procedures

Based upon the connection and overset information obtained from the pre-process programs *createConnect3D* and *createOversetIndex*, necessary data which are required for calculation at each block are transferred. Two kinds of data transfer for inter-block communication can be considered. The first is process-to-process using MPI library. The second is inside-one-process using memory copy, exactly speaking, pointer switching. The parallel procedures and multi-block/overset data control procedures are clearly separated from the CFD solver modules so that the solver can be modified without considering the parallel processes and the multi-block/overset data handling as if the solver modules are only for single block problems. With this framework, solvers inside each block can be changed by block(Figure 1). In Figure 1, coupling analysis between flow and heat conduction in objects are realized by changing a flow solver to a heat conduction solver in some blocks which are assigned to the object material. The number of blocks in each process is defined by users freely, but each block can not be assigned over several processes.

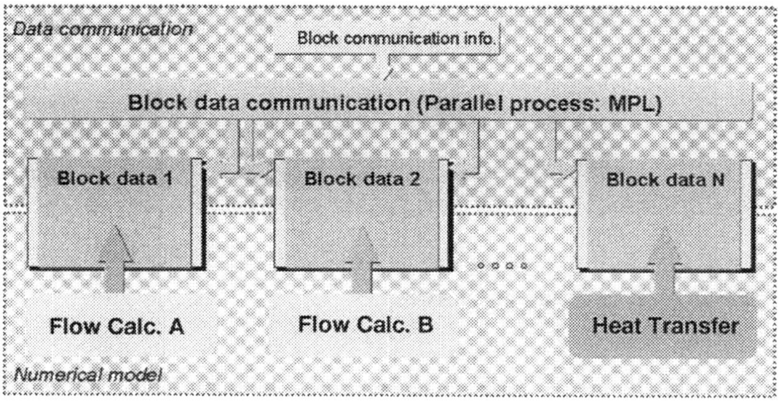

Figure 1. Concept of the separation of parallel and multi-block process from solvers inside each block

2.3. Portability

A parallelization method based on a domain decomposition concept using the message passing interface (MPI) is used to minimize the dependency on hardware architectures. MPI might be more difficult than other parallel languages like High Performance Fortran. However, most CFD researchers and code developers who want to modify the solver (at single block level) of UPACS are not required to know about MPI. This portability is shown by the fact that the UPACS code can be run upon various parallel computers from super computers, workstations to PC-UNIX clusters.

2.4. Hierarchical and encapsulated structure

One of the key features of the UPACS code is the hierarchical structure of program and data being clearly defined. Every module is encapsulated so that the code can be shared and modified more easily among CFD researchers and developers. Figure 2 shows the hierarchical structure of data and procedures. The UPACS code consists of three layers as follows.

- **Top layer**: Main loop level
  The main loop level corresponding to the main program of general CFD codes, is prepared for code extension, determines the framework of iteration algorithm that would be dependent on the solution methods or numerical models. Some of the parallel procedures such as initialization and control of data transfer between blocks are written at this top level. Actual data transfer is treated by the multi-

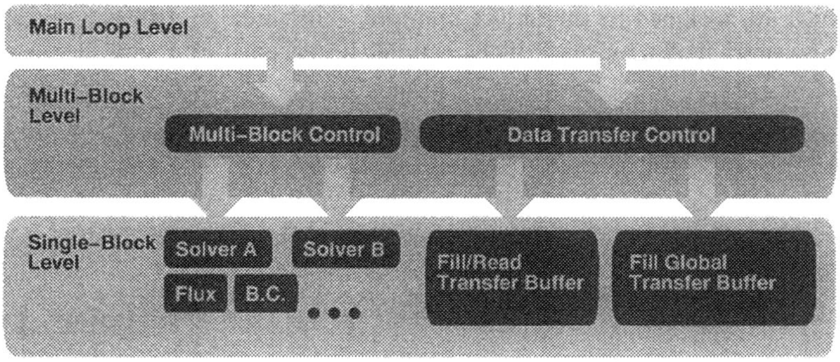

Figure 2. Hierarchical structure of UPACS

block level in the middle layer which separates the main loop level from the single block level. Users may want to modify this level according to their own purpose.

- **Intermediate layer**: Multi-block level
  The multi-block level has to handle the complicated multi-block/overset data controls for the multi-processor (assignment of blocks to each processor and so on) and the data transfer between the blocks. These procedures are independent of calculations inside each block. Therefore this layer can be generalized and prepared as a library so that one can achieve complicated calculations without getting into the details of data handling algorithms for parallel computation or multi-block/overset approach.

- **Bottom layer**: Single-block level
  The lowest level, the single-block level, consists of the CFD solver modules for a single block, which can be easily prepared for several numerical models, like numerical flux schemes, time integration methods, turbulent models, physical boundary conditions and so on. Thus calculations inside each block can be conducted independently because data of neighboring blocks are already prepared as values at imaginary cells by the data transfer conducted in the intermediate layer. Therefore the solver modules can be written as if they are for a single block grid. Variables which are necessary in single block calculations are assembled into a structure which is only referred by name at the multi-block level. Therefore, modification of variables does not affect the multi-block level.

The concept of *Hierarchical structure of programs and encapsulation of data and calculation procedure* is partly based on the object oriented programming. C++ is one of the most popular languages for object oriented programs, however, Fortran 90 has been chosen for UPACS. One of the reasons is that the most CFD researchers at NAL have used Fortran 77 and are not familiar with C or C++ because considering the run-time performance, Fortran 77 is the most excellent choice for scientific computations on vector computers. Moreover, Fortran has more functions especially to handle arrays, which are necessary for scientific computations, than C or C++ . Comparing several points of view, performance, functions, familiarity, new features (object oriented approach, encapsulation and structure), Fortran 90 has been chosen as a programming language to develop UPACS.

3. APPLICATIONS

The released version has been applied to several problems as well as used in NAL's projects. Extensions for the multi-disciplinary and multi-physics problems have been conducting.

3.1. 2D airfoil

The UPACS code has been successfully applied to viscous calculations with the Spalart-Allmaras turbulence model to predict stall phenomena for 2D airfoils (NACA $63_1 - 012$, NACA $63_3 - 018$ and NACA 64A $-$ 006) for the first step of code validation. For two kinds of stall type, trailing edge stall (NACA $63_3 - 018$) and leading edge stall (NACA $63_1 - 012$) comparison between experimental results and calculated results, which were conducted by the UPACS code, shows good agreement. Figure 3 shows comparisons of $C_{L\alpha}$ for NACA $63_1 - 012$ and NACA $63_3 - 018$.

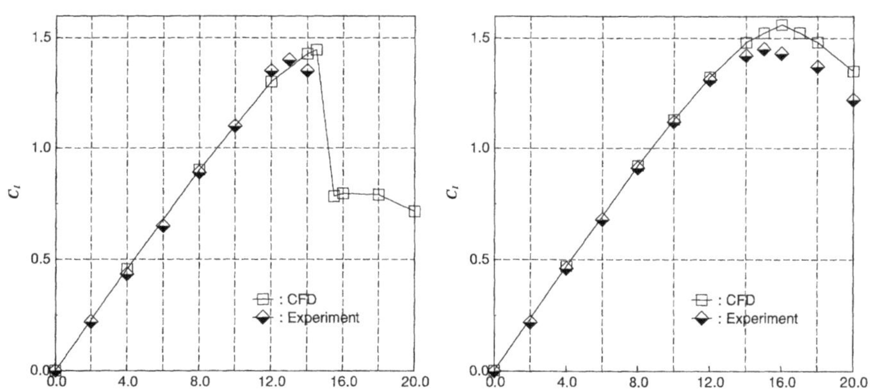

a) NACA$63_1 - 012$ b) NACA$63_3 - 018$

Figure 3. Comparison of $C_{L\alpha}$ for stall prediction

3.2. Flow around turbomachine

Figure 4(a) shows a computational grid around NASA Rotor 35 transonic compressor rotor including 1.1 million grid points and 17 blocks. Figure 4(b) and Figure 4(c) show relative Mach number distributions at 99% span and stream lines in almost peak efficiency operating condition at design speed. The complicated flow structures are clearly captured.

3.3. Flow around SST

NAL has been conducting a project of supersonic research airplane to establish design technology based on the CFD. The UPACS code is expected to play an important role in this project. Figure 5 shows surface pressure contours of the experimental vehicle (without tail wing configuration).

3.4. Flow around turbine blades

The code is being extended for multi-disciplinary problems such as conjugate simulations of flow with combustion, heat conduction, or structure analysis. Figure 6 shows a multi-block grid around turbine blades for the coupling analysis between flow and heat conduction in blades. In this grid, the structured grid is made not only in the flow regions but also inside the turbine blades. The UPACS code can handle these conjugate simulations with heat conduction in objects by replacing a flow solver with a heat conduction solver in applicable blocks.

4. CONCLUSION

A common CFD platform UPACS which separates flow solvers from multi-block and overset treatments and parallel processes has been developed in order to accelerate the CFD research and simulation program development. It has succeeded to get better portability on various computers and easily share simulation programs and knowledges among CFD researchers and users. The UPACS solver can effectively simulate flows around complicated geometries with multi-block or overset structured grids using some assisting tools for UPACS. It has been successfully applied to some of the aerospace simulations. These results show good performance of the UPACS solver. UPACS continues its evolution to become a better CFD based tool.

REFERENCES

1. K. Engel and et. al. Numerical investigation of the rotor-stator interaction in a transonic compressor stage. AIAA Paper 94-2834, 1994.
2. T. Ohta. An object-oriented framework for parallel computational fluid dynamics. In *Proceedings of Aerospace Numerical Simulation Symposium '98*, pages 367–372. NAL SP-46, February 1998. (in Japanese).
3. T. Ohta. An object-oriented programing paradigm for parallel computational fluid dynamics on memory distributed parallel computers. In *Parallel Computational Fluid Dynamics*, pages 561–568. Elsevier Science B.V., 1998.
4. T. Yamane and et. al. Development of a common cfd platform - upacs -. In *Parallel Computational Fluid Dynamics -Proceedings of the Parallel CFD 2000 Conference Trondheim, Norway (May 22-25, 2000)*, pages 257–264. Elsevier Science B.V., February 2001.

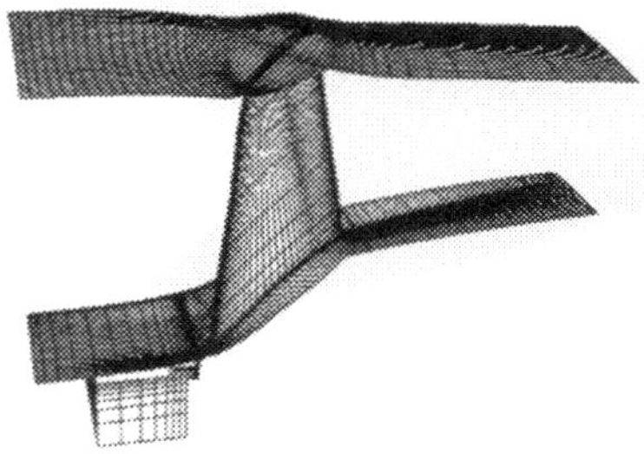

(a) Grid

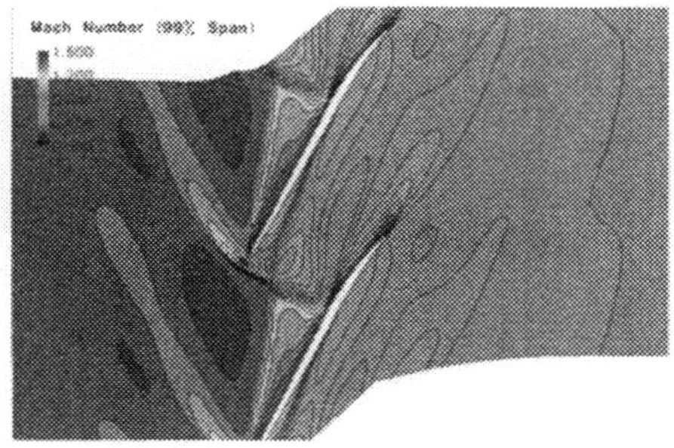

(b) Relative Mach number distributions

(c) Stream lines

Figure 4. Flow simulation around NASA Rotor 35

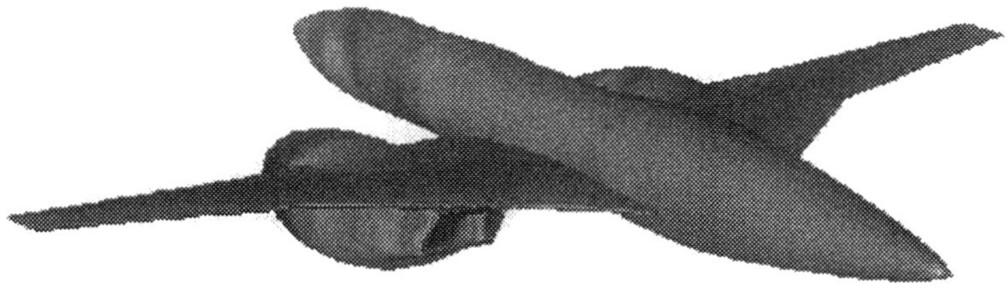

Figure 5. Supersonic flow around experimental SST

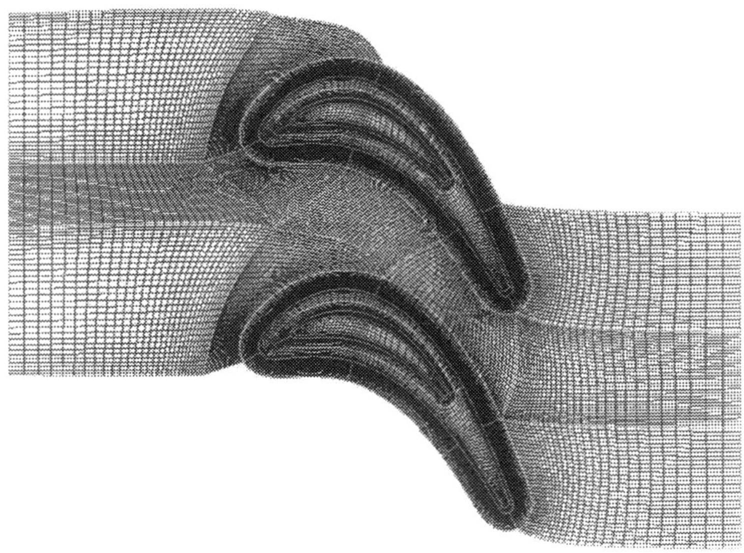

Figure 6. Multi-block grid around turbine blades for coupling analysis between flow an heat conduction

Parallel Computational Fluid Dynamics – Practice and Theory
P. Wilders, A. Ecer, J. Periaux, N. Satofuka and P. Fox (Editors)

A symmetry preserving discretization method, allowing coarser grids

A. Twerda[a] A.E.P. Veldman[b] G.P. Boerstoel[a*]

[a]TNO-Department of Applied Physics
Systems and Processes Division
P.O. Box 155, 2600 AD Delft, The Netherlands

[b]Department of Mathematics, University of Groningen,
P.O. Box 800, 9700 AV Groningen, The Netherlands

This paper presents a fourth order discretization method based on Richardson extrapolation, which is stable on relatively coarse grids. The method we describe is inspired by the method introduced by [9] which is they called 'symmetry preserving'. Our method turns out to be an extension of the Jameson cell-centred method [3]. The resulting fourth order method enables us to use coarser grids and therefore reduces computing time.

1. INTRODUCTION

TNO-Department of Applied Physics has developed a software package X-stream. X-stream is a Computational Fluid Dynamics (CFD) code specifically aimed towards the glass industry. Using this code it is possible to model complete *glass melting furnaces* and *forehearths* including both the glass tank and combustion chamber. The program contains sub-models for:

- Batch melting and conversion
- Electrical boosting
- Turbulence
- Radiation models glass and combustion chamber
- Combustion of natural gas/oil with air/oxygen

These complex simulations require locally fine meshes resulting in a slow convergence rate or instabilities during the computation. Figure 1 shows a grid of

*G.P. Boerstoel died March 7 2001.

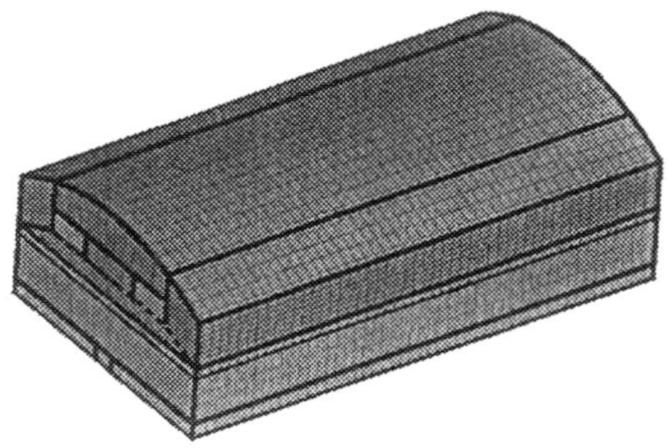

Figure 1. Typical grid of a simulation using various sizes of control volumes

one of the test-cases of the X-stream program consisting of a large difference between the smallest and largest control volumes. To enable these simulations to take an active part in a design phase of a glass furnace the computation time of these calculations should be in the order of hours in stead of days.

One way of decreasing computing time is to use parallel computing in combination with efficient solvers such as multi-grid [7]. Another approach to decrease computing time is the reduction of grid points, in other words, reducing the number of Mflops. The optimal method would be a combination of the both approaches.

In this paper a fourth order discretization method is described, which is stable on coarse grids. The method is an analog of the method which is described by Verstappen and Veldman [9]. The resulting method is called symmetry preserving because the (skew)-symmetry properties of the original equations are still conserved after discretization.

Other succesfull attempts to construct higher order schemes were undertaken by [4] and [2]. The first one describes the fluxes with a higher order interpolation method. The latter also aimed at conserving conservative properties of the original equations.

2. MATHEMATICAL DESCRIPTION

The incompressible Navier-Stokes are considered within the Boussinesq approximation. The following conservation laws can be specified. Conservation of mass,

$$\frac{\partial(\rho u_i)}{\partial x_i} = 0 \tag{1}$$

where ρ denotes the density, t time and u_i the velocity component in the x_i direction. Conservation of momentum implies

$$\frac{\partial(\rho u_i)}{\partial t} + \frac{\partial(\rho u_i u_j)}{\partial x_j} = -\frac{\partial p_i}{\partial x_j} + \frac{\partial}{\partial x_j}(\mu \frac{\partial u_i}{\partial x_j}) + \rho b_i \tag{2}$$

Here, p is the pressure, μ the viscosity and b_i the buoyancy term. Finally, conservation of energy imposes

$$\frac{\partial(\rho C_p T)}{\partial t} + \frac{\partial}{\partial x_j}(u_j \rho C_p T) = \frac{\partial}{\partial x_j}\left(\alpha \frac{\partial T}{\partial x_j}\right) \tag{3}$$

with T being the temperature and α the thermal diffusivity. The buoyancy term can be expressed as:

$$b_i = g_i \beta (T - T_0) \tag{4}$$

with g_i the gravitational acceleration and β the thermal expansion coefficient.

3. SYMMETRY PRESERVING DISCRETIZATION

3.1. Second order method

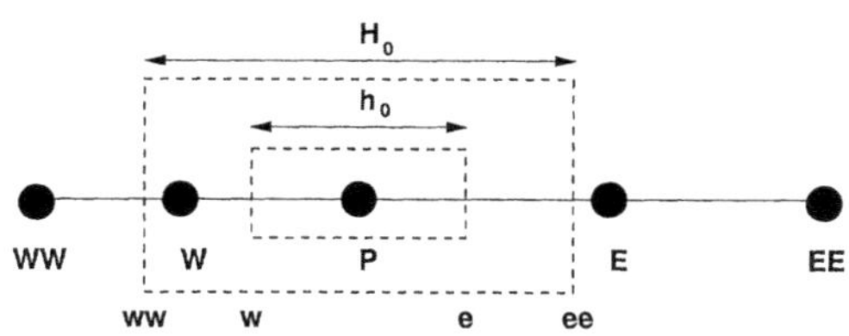

Figure 2. Collocated cell centered grid arrangement. Fine (h_0) and coarse (H_0) control volumes. Volume faces are located at positions ww, w, e, and ee.

To explain our choice of discretization, we will start with a one dimensional time dependent transport equation

$$\frac{\partial \phi}{\partial t} + \frac{\partial}{\partial x} F(\phi) = 0. \tag{5}$$

Finite volume discretization of the second term on the left hand side of equation (5) around point P (Figure 2) can be written as,

$$h_0 \frac{\partial F}{\partial x} = F_e - F_w + O(h_0^3). \tag{6}$$

with h_0 the cell size of the control volume (CV). The flux through face e (Figure 2) is given by

$$F_e = [\rho u_e \phi_e - \Gamma_e (\frac{d\phi}{dx})_e] , \tag{7}$$

with Γ being the diffusion coefficient.

Equation (7) implies that the value of ϕ and its derivative must be approximated at position e using the nodal values in P and E.

There are several methods to discretize these inter-nodal values. We will use the *Jameson cell-centered* method [3]. Here, ϕ_e is computed with

$$\phi_e \approx \frac{1}{2}(\phi_E + \phi_P) . \tag{8}$$

The derivative is discretized using

$$(\frac{d\phi}{dx})_e \approx \frac{\phi_E - \phi_P}{x_E - x_P}. \tag{9}$$

To be more concrete let us consider the semi-discretized version of equation (5)

$$\mathbf{M}_h \frac{d\boldsymbol{\phi}_h}{dt} + \mathbf{L}_h \boldsymbol{\phi}_h = 0 , \tag{10}$$

with $\mathbf{M}_h$ representing a diagonal matrix with the cell-sizes (h_0 in Figure 2) on the main diagonal. The matrix $\mathbf{L}_h$ results from the combined diffusive and convective fluxes through the cell-faces. The vector $\boldsymbol{\phi}_h$ contains the nodal values of the solution. The time evolution of the 'energy' of the solution, $||\boldsymbol{\phi}_h||_h^2 = \boldsymbol{\phi}_h^* \mathbf{M}_h \boldsymbol{\phi}_h$, behaves as

$$\begin{aligned} \frac{d}{dt}||\boldsymbol{\phi}_h||_h^2 &= \frac{d\boldsymbol{\phi}_h^*}{dt}\mathbf{M}_h \boldsymbol{\phi}_h + \boldsymbol{\phi}_h^* \mathbf{M}_h \frac{d\boldsymbol{\phi}_h}{dt} \\ &= -(\mathbf{L}_h \boldsymbol{\phi}_h)^* \boldsymbol{\phi}_h - \boldsymbol{\phi}_h^* \mathbf{L}_h \boldsymbol{\phi}_h \\ &= -\boldsymbol{\phi}_h^* (\mathbf{L}_h + \mathbf{L}_h^*) \boldsymbol{\phi}_h . \end{aligned} \tag{11}$$

A closer examination of equation (11) shows that the energy is conserved if $\mathbf{L}_h$ is skew-symmetric. Furthermore, the energy decreases if $\mathbf{L}_h$ is positive real. Looking at the discretization scheme corresponding with equation (8) we see that for the Jameson cell-centred method in equation (8) the convection term results in a skew-symmetric operator and therefore preserves the energy on any grid. In particular, the convective term has no contribution to the main-diagonal. This feature is called *symmetry-preservation.* This is not always the case for other discretization methods, which can produce negative convective contributions to the main-diagonal of $\mathbf{L}_h$, possible making the method unstable. The discretization of the diffusive term poses no problem and is found to be symmetric. Veldman and Verstappen showed in several papers [8,9] that if skew-symmetry of the convective term is preserved, the resulting method is stable.

3.2. Fourth order method

To extend the second order method to a fourth order method, a coarse CV which is twice as large as the fine control volume is considered, see Figure 2. For this coarse CV we apply the same discretization as for equation (6)

$$H_0 \frac{\partial F}{\partial x} = F_{ee} - F_{ww} + O(H_0^3) \,, \tag{12}$$

with $H_0 \equiv 2h_0$ is the size of the coarse CV. The flux through the east face of the coarse CV is given by

$$F_{ee} = [\rho u_{ee} \phi_{ee} - \Gamma_{ee} (\frac{\partial \phi}{\partial x})_{ee}] \,. \tag{13}$$

To ensure a symmetry-preserving property of our discretization the following approximations for the value of ϕ and its derivative at location ee are used.

$$\phi_{ee} = \frac{1}{2}(\phi_{EE} + \phi_P) \tag{14}$$

$$(\frac{\partial \phi}{\partial x})_{ee} = (\frac{\phi_{EE} - \phi_P}{x_{EE} - x_P}) \tag{15}$$

The leading truncation error of equation (6) can now be removed using Richardson extrapolation. Since the errors in equation (6) and equation (13) are of the third order on a uniform grid, taking the linear combination 8∗(6) - 1∗(13) results in a fourth order method.

$$(8h_0 - H_0) \frac{\partial F}{\partial x} \approx 8(Fe - Fw) - (F_{ee} - F_{ww}) \,. \tag{16}$$

Similarly, on non-equidistant grids the constant coefficients 8 and -1 are used. This is to ensure that the resulting matrix is and remains symmetry preserving.

In multi-dimensional computations the truncation error is not of order three but of order 2^{d+2} with d being the number of spatial dimension.

4. NUMERICAL EXAMPLES

4.1. Lid driven cavity

The symmetry-preserving method is tested on standard problem, the 2-D lid driven cavity with $Re = (U_{lid}L)/\nu = 1000$ with $\nu = \mu/\rho$ [4,6]. Because of the steep gradients near the walls a non-uniform grid is used with refinement close to the walls.

Figure 3 shows the profiles along the centre line of the cavity for the horizontal velocity component U. The grid has been refined from (10×10) to (160×160) CVs. For the second-order method a grid independent solution is achieved on the (80×80) grid while for the fourth-order method this is achieved already on the (40×40) grid. This means that a factor of 4 less grid points are needed for the fourth-order scheme to achieve the same accuracy of the solution.

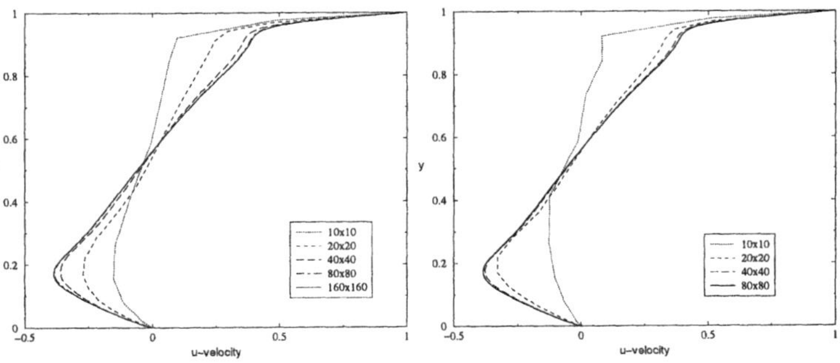

Figure 3. Grid dependence of vertical velocity profile at the centre line for the lid driven cavity with $Re = 1000$. **Left**: Second-order scheme. **Right**: Fourth-order scheme.

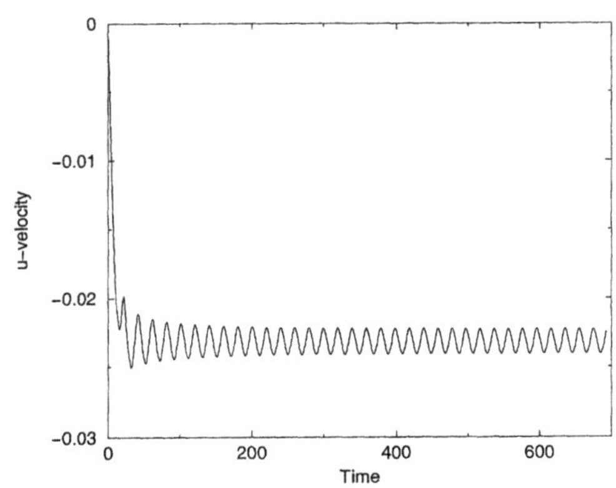

Figure 4. Time history of the U-velocity in a monitor point for the thermally driven cavity

4.2. Thermally driven cavity

Furthermore, the symmetry preserving method is applied on a 2-dimensional time dependent test-case. This is a thermally driven cavity. For time integration a second order backward difference formula is used. Here the Navier-Stokes equation are solved together with the energy equation. For a special value of the Grasshoff number ($Gr = (\beta \Delta T g H^3)/\nu^2 = 1.5 \cdot 10^{-5}$) and Prandtl number ($Pr = \nu/\alpha = 0$) the solution has a periodic solution which can been seen in Figure 4. More details on the particular test case can be found in [1].

The frequency of the solution for our method is shown in Table 1 together with some other calculations taken from literature. Our method performs well, especially taking into account the small number of grid points used in the simulation.

Table 1
Frequency f for different methods of the thermally driven cavity

Method	grid-size	f
2nd order method	(80×20)	0.05058
2nd order method	(120×30)	0.05112
4th order method	(80×20)	0.05208
4th order method	(120×30)	0.05198
Nobile (1996)	(128×32)	0.04981
Nobile (1996)	(256×64)	0.05078
Behnia *et. al.* (1990)	(160×40)	0.05124
Behnia *et. al.* (1990)	(320×80)	0.05159

5. DISCUSSION & CONCLUSION

The symmetry preserving discretization method for cell-centred collocated grids, described in this paper, results in a robust method for solving the Navier-Stokes equations.

Using this method for laminar flow simulations grid-independent solutions are obtained on coarser grids and therefore reducing the computational costs. Typically the grid-spacing can be doubled. For a 3D simulation this results in a reduction in grid-points of a factor of eight. For time-dependent solutions, not only the grid-spacing can be increased but, as a result, the time step can be doubled as well, when the same CFL number is applied.

The number of iterations required to calculate a steady state solution or per time-step in a time-dependent simulation is approximately the same when using either the second-order method or the fourth-order method. Using the fourth order method instead of the second order method, the computational cost per iteration is increased by approximately 35%. The memory storage requirements for the fourth-order method is approximately 30% greater than the memory required for the second-order method.

Until now only orthogonal grids have been applied. Future research will include extending the method to non-orthogonal grids and using parallel computers to decrease the simulation time even further.

REFERENCES

1. M. Behnia and G. de Vahl Davis. Fine mesh solutions using stream function-vorticity formulation. In B. Roux, editor, *Notes on Numerical Fluid Mechanics*, 27:11–18, 1990.
2. F. Ducros, F. Laporte, T. Soulères, V. Guinot, P. Moinat and B. Caruelle. High-Order Fluxes for Conservative Skew-Symmetric-like Schemes in Structured Meshes: Application to Compressible Flows. *Journ. Comp. Phys*, 161:114–139, 2000.
3. A. Jameson, W. Schmidt, and E. Turkel. Numerical solutions of the Euler equations by finite volume methods using Rung–Kutta time–stepping schemes. *AIAA*, 81-1259, 1981.
4. Z. Lilek, M. Perić. A fourth-order finite volume method with colocated variable arrangement *Computers and Fluids*, 3:239–252, 1995.
5. E. Nobile. Simulation of time-dependent flow in cavities with the additive-correction multi-grid method, part II: Application. *Num. Heat Transfer, Part B*, 30:351–370, 1996.
6. M. Thompson, J. Ferziger. An adaptive multi-grid technique for the incompressible Navier-Stokes equations *Journ. Comp. Phys*, 82:94–121, 1989.
7. A. Twerda., R.L. Verweij, T.W.J. Peeters, and A.F. Bakker *The need for multi-grid for large computations* In D. Keyes, A. Ecer, J. Periaux, N. Satofuka and P. Fox Eds. *Proc. of the parallel CFD'99 Conference*, Elsevier, 2000.
8. A.E.P. Veldman and E.F.F. Botta. Grid quality: an interplay of grid and solver. In *Numerical Methods for Fluid Dynamics*, 4:329–335. Oxford University Press, 1993.
9. R.W.C.P. Verstappen and A.E.P. Veldman. Direct numerical simulation of turbulence at lower costs. *J. Engineering Mathematics*, 32:143–159, 1997.

Parallel Computational Fluid Dynamics – Practice and Theory
P. Wilders, A. Ecer, J. Periaux, N. Satofuka and P. Fox (Editors)

Multitime multigrid convergence acceleration for periodic problems with future applications to rotor simulations

H. van der Ven, [a*] O.J. Boelens,[a] and B. Oskam[a]

[a]National Aerospace Laboratory NLR, P.O. Box 90502, 1006 BM Amsterdam, The Netherlands

The simulation of certain time-dependent flow phenomena can pose a grand challenge to Computational Fluid Dynamics. In this paper, a multitime multigrid convergence acceleration algorithm is introduced which significantly reduces the turnaround time to reach time-periodic solutions. The application of this convergence acceleration algorithm will lead to time-efficient simulations of helicopter rotors in forward flight. A comparison between multitime multigrid acceleration and the classical serial time multigrid acceleration shows that an order of magnitude reduction in turnaround time can be achieved at the expense of an order of magnitude increase in memory use.

1. Introduction

The simulation of certain time-dependent flow phenomena can pose a grand challenge to Computational Fluid Dynamics (CFD). Turnaround times of time-accurate simulations can be large if the characteristic bandwidth of the time dependent periodic flow solution is large. Bandwidth is defined as the ratio of the maximum and minimum frequency of the Fourier spectrum of the time periodic solution. In this paper, a new convergence acceleration algorithm is introduced which significantly reduces the turnaround time for time-periodic solutions with large bandwidth. The new algorithm will be applied to a space-time discontinuous Galerkin (DG) method [10].

1.1. Applications

Main application area for the DG method is the simulation of the flow around helicopter rotors, in both hover and forward flight conditions. Rotor flows are characterised by complicated aerodynamic phenomena, such as blade-vortex interaction and compressibility effects at the advancing blade, leading to high noise levels. Present day CFD technology is sufficiently mature to accurately resolve these aerodynamic phenomena (for details, see [2]), but at a prohibitively high computational cost. For a typical helicopter rotor in steady state forward flight the mimimum frequency is the blade passing frequency, and the maximum frequency is dominated by the short duration Blade-Vortex Interaction events, given rise to a bandwidth of the order of one hundred.

*venvd@nlr.nl

1.2. Parallel algorithm development

For the development of parallel algorithms the ultimate goal is reduction of turnaround time.In a first approximation, turnaround time is computed as work divided by speed, where work is the required number of floating point operations per simulation, and speed is the number of floating point operations per second achieved by the algorithm on a certain architecture. So one can either increase the speed, or decrease the required flop count (increase the algorithm efficiency), in order to decrease the turnaround time per simulation.

It is important to recall that optimising only speed, or work, may lead to suboptimal algorithms. Maximising speed may lead to foolish algorithms, which for instance keep recomputing storable data to increase the flop rate, whereas the turnaround time stays the same. On the other hand, algorithms minimising work may not benefit from the peak speed of existing hardware architectures. If a 'minimum work' algorithm is neither vectorisable nor parallelisable, nobody will even consider it. More realistically, dynamic grid algorithms usually are more efficient in terms of flop count since the same solution can be obtained with less grid cells, but the dynamically evolving grids imply load balancing on parallel machines, which is poorly scalable. Hence, parallel algorithm development should be motivated by the miminisation of the *quotient* work over speed, resulting in a minimum turnaround time.

1.3. Outline

The outline of the paper is as follows. In Section 2 rotor flow simulations are briefly presented. In Section 3 the new convergence acceleration algorithm is presented and results are given in Section 4. Conclusions with respect to forward flight simulations with a minimum turnaround time and parallel algorithm development are drawn in Section 5.

2. Helicopter rotor in forward flight

In recent years at NLR the flow solver based on a discontinuous Galerkin finite element discretisation of the unsteady compressible Euler equations, which was originally designed for fixed wing applications [9,10], has been modified to enable the simulation of helicopter rotor flows [1].

The vortex capturing capability through grid adaptation has successfully been used in the simulation of the Caradonna-Tung helicopter rotor in hover, and the simulation of the Operational Loads Survey (OLS) helicopter rotor in forward flight.

Before turning to the forward flight simulation, which is the relevant measure used to assess the minimum turnaround time of a CFD code, the steady simulation of the Caradonna-Tung rotor in hover is discussed, since the efficiency of the grid adaptation algorithm for this simulation is exemplary of what we want to achieve for time-accurate forward flight simulations.

A rotor in hover is considered as a steady state problem in the rotating reference frame. A simulation of the Caradonna-Tung rotor [4] using a collective pitch of twelve degrees and a tip Mach number of 0.61 has been performed (for details see [1]). The simulation was performed on both a fine grid and a locally refined coarse grid. Both grids demonstrated the same vortex persistence, with the locally refined mesh containing only 15% of the number of grid cells of the fine mesh. Hence, local grid refinement displays an *algorithmic* speedup of six.

A similar simulation has been performed for the OLS rotor in forward flight, with the difference that this simulation is time-dependent in the inertial frame, and that the vortex position in the grid changes over time. The OLS rotor has been simulated using a tip Mach number of 0.664, an advance ratio of 0.164, and a thrust coefficient of 0.054. The blade motion schedule was obtained by trimming the rotor as a whole using the CFD code, and not a lifting line vortex code. Good agreement with experiments was obtained [2].

On the downside, the complete simulation using 288 time steps per period for only three periods on a final refined mesh of 1.2 million grid points took 60 hours (20 hours per period) on the eight processor NEC SX-5/8B, at a congregate speed of 24 Gflop/s (about 40% peak) and using 13 GB of memory. For an analysis tool such a computing time is too large, and inhibits its use in helicopter performance analysis tool sets.

3. Multigrid acceleration

3.1. Background

Multigrid acceleration [3] combines classical iterative techniques, such as point relaxation or local time stepping, with coarse level corrections to yield a method superior to the iterative techniques alone. In the present paper two types of multigrid acceleration are considered:

- STMG, serial time multigrid, where the coarse level corrections only pertain to the spatial operator,

- MTMG, multitime multigrid, where the coarse level corrections pertain to both space and time discretisations.

For periodic time-dependent problems STMG acceleration yields an asymptotic convergence rate of $1 - O(\Delta t/T)$, with Δt the time step and T the period, which is the same rate as classical iterative techniques for steady-state problems. MTMG acceleration restores the superior convergence acceleration, where the asymptotic convergence rate is bounded away from one, independent of space and time discretisation. The above will be illustrated in Section 4.

3.2. A multitime multigrid algorithm

Following the ideas of Horton, Vandewalle, and Worley for parabolic equations [5,6,11] we propose to solve the problem for all time steps simultaneously. Periodic problems, such as the rotor in forward flight, can be considered as steady state problems in the space-time domain, so this would seem like a feasible idea.

Horton and VandeWalle [6] considered parabolic PDE's, and introduced a space-time multigrid method in which all time-levels are treated simultaneously. They showed that for parabolic equations it is not possible to treat the time as just another space dimension, and they had to revert to semi-coarsening techniques in order to maintain multigrid convergence. Their restriction and prolongation operators in time take the direction of time into account. As smoother they applied a coloured pointwise Gauss-Seidel relaxation.

These ideas are extended to hyperbolic PDE's in a straightforward way. The equations to be solved remain the same, but all time steps are solved simultaneously. The system can be solved in parallel. The solution strategy of the system is basically the same as for a system of equations derived using a spatial discretisation. A pseudo-time is introduced and the solution is marched to a steady state using standard acceleration techniques such as local time stepping, grid sequencing and multigrid. One important difference is that the multi-level techniques will be applied to the space-time grid, and not be restricted to the space grid only. Hence the method is called a *multitime multigrid* (MTMG) acceleration algorithm.

In contrast with Horton et al. we treat the time as just another space dimension. The restriction and prolongation operators in time are identical to the space operators. A five stage Runge-Kutta scheme is used as a smoother. No semi-coarsening techniques have been applied in the multigrid algorithm.

Let $\mathcal{L}_h$ be the DG discretisation operator for a given time slab, U_h^n the solution in the time slab $[t^n, t^{n+1}]$, which satisfies the equations

$$\mathcal{L}_h(U_h^n, U_h^{n-1}) = 0,$$

where U_h^{n-1} is the solution in the previous time slab. For a periodic problem the equations are

$$\begin{cases} \mathcal{L}_h(U_h^i, U_h^{i-1}) = 0, & 1 \leq i \leq N, \\ U_h^0 = U_h^N, & \end{cases}$$

if the period is divided into N time slabs.

In the STMG algorithm the equations are solved in pseudo-time τ as

$$\frac{\partial \tilde{U}_h^n}{\partial \tau} + \mathcal{L}_h(\tilde{U}_h^n, U_h^{n-1}) = 0, \qquad n = 1, 2, \ldots, \infty$$

where U_h^0 is the inital solution, upon convergence we set $U_h^n = \tilde{U}_h^n$, and proceed to the next time step. In the MTMG approach the equations are solved as

$$\frac{\partial \tilde{U}_h^i}{\partial \tau} + \mathcal{L}_h(\tilde{U}_h^i, \tilde{U}_h^{i-1}) = 0, \qquad 1 \leq i \leq N,$$

simultaneously, with the periodic boundary condition $\tilde{U}_h^0 = \tilde{U}_h^N$.

The L_2-residual ϵ_{per} of a solution obtained with the MTMG algorithm is defined as

$$\epsilon_{\text{per}}^2 = \frac{1}{N} \sum_{i=1}^{N} ||\mathcal{L}_h(\tilde{U}_h^i, \tilde{U}_h^{i-1})||_2^2, \tag{1}$$

where $||\cdot||_2$ is the L_2-norm in the time slab. In the STMG approach the single time step residual $\epsilon_s^{(n)}$ is measured:

$$\epsilon_s^{(n)} = ||\mathcal{L}_h(\tilde{U}_h^n, U_h^{n-1})||_2, \qquad (n \geq 1) \tag{2}$$

but this residual does not measure the convergence to a periodic solution. Given a series $\tilde{U}_h^{kN+1}, \dots, \tilde{U}_h^{(k+1)N}$ of solutions in the k-th period, define the solution vectors $\tilde{V}_h^i = \tilde{U}_h^{kN+i}$, $1 \le i \le N$. Considering $\tilde{V}_h$ as a periodic solution, we compute ϵ_{per} as follows:

$$\begin{aligned}
\epsilon_{\text{per}}^2 =& \frac{1}{N}\left(\|\mathcal{L}_h(\tilde{V}_h^1, \tilde{V}_h^N)\|_2^2 + \sum_{i=2}^{N} \|\mathcal{L}_h(\tilde{V}_h^i, \tilde{V}_h^{i-1})\|_2^2\right) \\
=& \frac{1}{N}\left(\|\mathcal{L}_h(\tilde{U}_h^{kN+1}, \tilde{U}_h^{(k+1)N})\|_2^2 + \sum_{i=2}^{N} \|\mathcal{L}_h(\tilde{U}_h^{kN+i}, \tilde{U}_h^{kN+i-1})\|_2^2\right) \\
=& \frac{1}{N}\left(\|\mathcal{L}_h(\tilde{U}_h^{kN+1}, \tilde{U}_h^{(k+1)N})\|_2^2 + \sum_{i=2}^{N} (\epsilon_s^{(kN+i)})^2\right). \qquad (3)
\end{aligned}$$

4. Results

The MTMG method has been applied to two-dimensional transonic flow over an harmonically oscillating NACA0012 foil. The freestream Mach number is 0.8, the angle of attack oscillates between -0.5^o and 4.5^o with a non-dimensional frequency of 0.314. The space-time mesh consists of $256 \times 128 \times 20$ elements, where the last dimension is the time dimension. In Figure 1 the convergence of the MTMG method is compared with that of the conventional STMG method. For the STMG method at each time step the implicit system is solved in pseudo-time, and six time steps are shown. At each time step n the residual $\epsilon_s^{(n)}$ is converged to $5 \cdot 10^{-7}$. For the MTMG method full multigrid has been applied, with 150 iterations on the coarsest mesh with $64 \times 32 \times 5$ grid cells, 200 multigrid cycles on the next finer level, and 500 multigrid cycles on the fine mesh. For both methods V cycles with one prerelaxation and one postrelaxation have been used. The convergence rate of the MTMG method is comparable to the convergence rate of the multigrid algorithm for the space DG discretisation operator for steady state problems.

The time-dependent pressure distribution on the upper side of the airfoil is shown in Figure 2(a). The motion of the shock is clearly visible. Moreover, the comparison of MTMG with STMG is good. In Figure 2(b) the polar plots of the lift coefficient C_L are shown for both methods: agreement is good, but not excellent, considering the fact that both methods solve the same set of discrete equations. Upon convergence both STMG and MTMG schemes should result in solutions that are equal up to machine accuracy.

In order to compare the flow solutions of the different methods, we compute the periodic residual for the solution obtained with the STMG method over the first five periods. Using (3) we have

$$\epsilon_{\text{per}}^2 = \frac{1}{N}\left(\|\mathcal{L}_h(\tilde{U}_h^{kN+1}, \tilde{U}_h^{(k+1)N})\|_2^2 + (N-1)(5 \cdot 10^{-7})^2\right). \qquad (4)$$

In Table 1 this residual is shown for five consecutive periods of the STMG iterates. Clearly, convergence in L_2-norm to a time periodic solution is slow, and the residual $\|\mathcal{L}_h(\tilde{U}_h^{kN+1}, \tilde{U}_h^{(k+1)N})\|_2$ dominates the time step residual $\epsilon_s^{(n)}$. This may cause the differences in the flow results. Note that based on the aerodynamic coefficients in Figure 2 one would conclude convergence in about three periods.

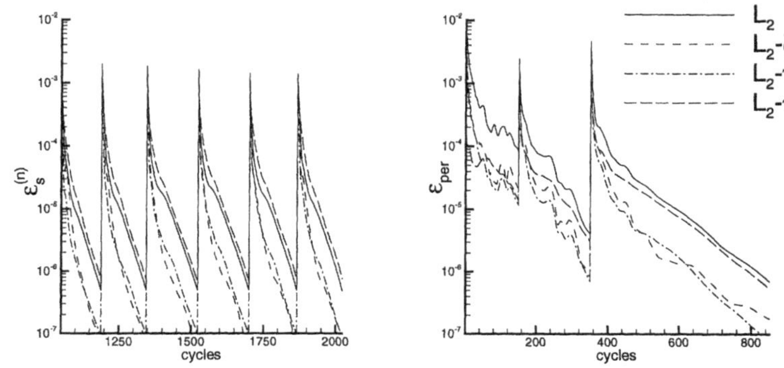

Figure 1. Convergence history of the residual $\epsilon_s^{(n)}$ of six time steps of the STMG method (left) and the complete convergence history of the residual ϵ_{per} of the MTMG method with full multigrid (right).

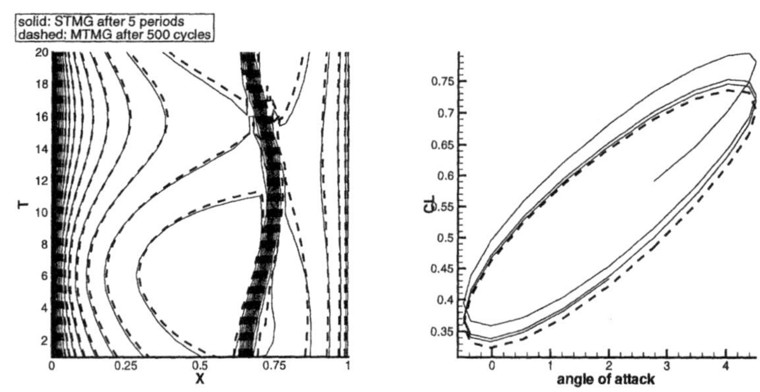

(a) Pressure distribution

(b) Polar plot of C_L

Figure 2. Comparison of itwo final iterates, one reached after five periods with STMG, and one after 500 cycles of MTMG

period	ϵ_{per}
1	$0.89 \cdot 10^{-3}$
2	$0.61 \cdot 10^{-3}$
3	$0.35 \cdot 10^{-3}$
4	$0.24 \cdot 10^{-3}$
5	$0.20 \cdot 10^{-3}$

Table 1
Residual ϵ_{per} defined in (4) of the STMG simulation.

Since the discretised equations are the same for both the STMG and the MTMG acceleration algorithms, the number of floating point operations per grid cell per time step per iteration are equal. There is only negligible difference in the computational cost of the multigrid algorithm, since the coarse grids in the MTMG algorithm are smaller than in the STMG algorithm, since the MTMG grid levels are also coarsened in the time direction. Hence, 150 multigrid cycles for MTMG require the same amount of floating point operations as 150 multigrid cycles per time step for a complete period in the STMG algorithm.

Based on the L_2-norm, the MTMG algorithm would require only 25 fine grid cycles to reach the same residual level as five periods of the STMG algorithm. Since the average number of fine grid cycles per time step of the STMG algorithm is 150, this would imply that MTMG is $(150 \times 5)/25 = 30$ times faster than the STMG algorithm. The speedup is this large since the STMG acceleration performs poorly in terms of reaching the periodic steady state. Since an extensive study of the convergence of time-periodic problems is beyond the scope of the present paper, we will not go into further details. By engineering standards one would require a decrease in the residual of three to four orders in magnitude, depending on the spatial and temporal resolution. To satisfy this standard, the MTMG method requires 250 cycles, while the STMG method would require at least 50 periods to reach the same level of periodicity, again resulting in a speedup of thirty.

The qualitatively greater efficiency of the MTMG method can partly be explained by the fact that it presupposes the existence of a periodic solution, and partly by the fact that the multigrid algorithm is applied to the space-time system, and not only to the space system. Moreover, the full multigrid algorithm provides better initial solutions for the implicit system, whereas the time serial algorithm uses the solution of the previous time step as the initial solution.

5. Conclusions

5.1. Time-periodic simulations

The standard way of obtaining a periodic solution by time integration is a slow process. For an oscillating transonic airfoil, there is hardly any convergence in L_2-norm over five periods, indicating an asymptotic convergence rate of $1 - O(\Delta t/T)$.

Considering the poor convergence of the STMG acceleration, it is difficult to make a definitive comparison between the convergence rate of the MTMG acceleration with the performance of the STMG acceleration.

Considering the computational complexity, MTMG has the following properties:

- the number of periods required to resolve the transient is reduced to one, reducing the work to be done per simulation with a factor in the order of the number of time steps per period,
- the algorithm has increased scalability since the grid size is increased by a factor equal to the number of time steps,
- the memory use increases with a factor proportional to the number of time steps.

The increased scalability and the increase in memory use make the method ideally suited for MPP machines. Especially for time periodic applications with large bandwidth, and for which a large number of time steps is required. The simulation of the rotor in forward flight is such an example, requiring 288 time steps per period.

5.2. Helicopter rotor in forward flight

If we would apply the MTMG algorithm to the simulation of the flow field of a rotor in forward flight, we estimate the following performance increase:

- MTMG versus STMG for k periods yields a speedup of at least k,
- since the time-periodic flow is now treated as a steady state problem in space-time, we can apply local grid refinement to the space-time grid, where the grid is only refined *where* and *when* a vortex is present. A similar reduction in grid size as for the rotor in hover can be expected, yielding a speedup of 6,
- an MPP machine with 1000 processors of 1 Gflop/s each (at a sustained performance of 10% peak speed), would be four times faster than the NEC SX-5/8B (at a sustained performance of 40% peak speed). Since the MTMG algorithm is a static algorithm, it is easily scalable even beyond a 1000 processors, so a speedup of at least four is feasible.

Combining these three improvements, the turnaround time of the simulation of a rotor in forward flight is decreased by a factor $24k$: $20k$ hours for k periods are reduced to less than an hour to obtain a periodic solution using MTMG. Considering the slow convergence to a periodic solution of the STMG method, one should even doubt that seven periods are sufficient to obtain a periodic solution for the rotor in steady state forward flight, further increasing the speedup of the MTMG method. Based on the memory requirements of the discontinous Galerkin method, it is expected that the memory requirements for the simulation using MTMG and local grid refinement is about 100 GB.

5.3. Parallel algorithm development

The MTMG algorithm has shown an algorithmic speedup by a factor of the order of the number of time steps per period with respect to STMG for a two dimensional, time periodic simulation. In the context of parallel computing, however, it is more important that a dynamic algorithm is turned into a static algorithm. *All* grid manipulations are performed in a preprocessing phase, and not at each time step during the simulation. Grid deformation to accomodate the body motion is performed during the grid generation. Local grid refinement has to be performed only two or three times during the simulation, which is the standard procedure for grid refinement for steady state problems.

As an explicit, static method, the MTMG method is easily scalable beyond 1000 processor MPP machines, as has been demonstrated in the American ASCI project [7,8].

Hence, a combination of an increase in algorithm efficiency and algorithm speed is projected to lead to forward flight simulations with a turnaround time of less than an hour on an MPP machine with 1000 processors of 1 Gflop/s each.

REFERENCES

1. O.J. Boelens, H. van der Ven, B. Oskam and A.A. Hassan, *Accurate and efficient vortex-capturing for a helicopter rotor in hover*, in the proceedings of the 26th European Rotorcraft Forum, The Hague, 2000.
2. O.J. Boelens, H. van der Ven, B. Oskam and A.A. Hassan, *The boundary conforming discontinuous Galerkin finite element approach for rotorcraft simulations*, submitted to Journal of Aircraft, 2001.
3. A. Brandt, *Multi-Level adaptive solutions to boundary value problems*, Math. of Comp. **31**, 333-390, 1977.
4. F.X. Caradonna and C. Tung, *Experimental and analytical studies of a model helicopter rotor in hover*, NASA Technical Memorandum 81232, 1981.
5. G. Horton, S. Vandewalle and P. Worley, *An algorithm with polylog parallel complexity for solving parabolic partial differential equations*, SIAM J. Sci. Comput., **16(3)**, 531-541, 1995.
6. G. Horton and S. Vandewalle, *A space-time multigrid method for parabolic PDEs*, SIAM J. Sci. Comput., **16 (4)**, 848-864, 1995.
7. D.E. Keyes, D.K. Kaushik, and B.F. Smith, *Prospects for CFD on Petaflops Systems*, NASA/CR-97-206279, 1997.
8. D.J. Mavripilis, *Large-scale parallel viscous flow computations using an unstructured multigrid algorithm*, NASA/CR-1999-209724, 1999.
9. J.J.W. van der Vegt and H. van der Ven, *Discontinuous Galerkin finite element method with anisotropic local grid refinement for inviscid compressible flows*, J. Comp. Physics, **141**, 46-77, 1998.
10. J.J.W. van der Vegt and H. van der Ven, *Space-Time discontinuous Galerkin finite element method with dynamic grid motion for inviscid compressible flows. Part I: General formulation.* Submitted to J. Comp. Physics, 2001.
11. P.H. Worley, *Parallelizing across time when solving time-dependent partial differential equations*, in Proc. 5th SIAM Conference on Parallel Processing for Scientific Computing, Eds. J. Dongarra, K. Kennedy, P.Messina, D. Sorensen, and R. Voigt, SIAM, 1992.

Parallel Computational Fluid Dynamics – Practice and Theory
P. Wilders, A. Ecer, J. Periaux, N. Satofuka and P. Fox (Editors)

DIRECT NUMERICAL SIMULATION OF TURBULENCE ON A SGI ORIGIN 3800

R.W.C.P. Verstappen[a] and R.A. Trompert[b]

[a]Research Institute for Mathematics and Computing Science
University of Groningen, P.O.Box 800, 9700 AV Groningen, The Netherlands

[b]SARA Computing and Networking Services
Kruislaan 415, 1098 SJ Amsterdam, The Netherlands

This contribution concerns the parallel solution of the incompressible Navier–Stokes equations for direct numerical simulation (DNS) of turbulent flow. The parallelization is based on the idea of splitting the flow domain into smaller subdomains which can be treated independent of each other. To compute the convective and diffusive term in parallel, the flow domain is partitioned in the streamwise direction. The computational time of this part of the algorithm scales superlinear with the number of processors, as the cache is more efficiently used as the problemsize per processor decreases. To solve the pressure in parallel, we make explicitly use of the fact that the turbulent flow under consideration is statistically homogeneous in the spanwise direction. The Poisson equation for the pressure can then be solved using a combination of a Fast Fourier Transform method in the spanwise direction and an Incomplete Choleski Conjugate Gradient method in the spectral space. The FFT is computed in parallel by treating the unknowns simultaneously in the streamwise direction, while the ICCG is computed in parallel by treating the unknowns simultaneously in the spanwise direction. The MPI implementation of the FFT/ICCG solver shows a very good scalability that is very close to ideal speed-up.

1. INTRODUCTION

Unraveling the complicated non-linear dynamics of turbulence is a major scientific and technological challenge. The vast majority of human energy consumption, for example, is related to the turbulent transport of mass, heat and momentum [1]. Turbulent flow typically involves a large range of dynamically significant scales of motion. Direct numerical simulation (DNS) is not yet a realistic possibility in most cases, since the cost of computing all scales of motion from the Navier–Stokes equations is beyond our means.

To obtain an acceptable computational effort in engineering and environmental applications, the computation is restricted to the large scale(s) of motion, where a turbulence model takes care of the integral effect of the non-resolved scales on the resolved scales, see *e.g.* [2]. In many applications, however, the required simulation accuracy cannot be reached with existing turbulence models. DNS results play a key role in improving our understanding of turbulence and in obtaining better turbulence models. The continuous increase of computer power, as well as the sustained improvements of numerical methods strengthen the role of DNS as a path-finding tool in turbulence research.

In the late '80-ies and the '90-ies direct numerical simulations were performed on vector computers, in those days the fastest available computers. The algorithms were chosen and tuned to perform well on these machines and achieved about 40% of peak performance. Nowadays the scenery has changed. Top 500 leading positions are hold by massively parallel computers with one thousand or more processors [3].

This paper concerns the performance of our DNS code on a 1024-processor SGI Origin 3800. Currently, the machine is subdivided into several partitions consisting of 64, 128, 256 or 512 400MHz MIPS R12K processors. Each partition has Cache-Coherent Non Uniform Memory Access (CC-NUMA). From the user's point of view, CC-NUMA machines behave like shared-memory computers: the system shows only a single memory image to the user even though the memory is physically distributed over the processors. This eases the porting of programs, like our DNS code, that are originally developed and tuned for shared-memory parallel vector machines.

In our approach the parallelization is based on the idea of splitting the flow domain into smaller subdomains which can be treated independent of each other. We consider two ways of enforcing the parallelization. To start, we use the autoparallelization option of the Fortran compiler. This method has the great advantage that the existing code basically remains unchanged and that it is portable. Its disadvantage is that the user has no other means than compiler directives to optimize the efficiency. Secondly, we have put more effort in the parallelization by applying message passing. The advantage of this communication model is that it also applies to distributed-memory computers, and that the programmer can steer the parallelization such as to achieve the highest efficiency. In this paper, both approaches are compared for a DNS of a three-dimensional turbulent flow that is statistically homogeneous in one spatial direction.

The paper is organized as follows. The numerical algorithm is briefly outlined in Section 2. After that, the parallelization is discussed (Section 3), results are presented (Section 4), and conclusions are drawn (Section 5).

2. ALGORITHM

The smallest scales of motion in a turbulent flow result from a subtle balance between convective transport and diffusive dissipation. Therefore, in a numerical simulation method it is important that numerical diffusion (from the convective discretization) does not interfere with physical dissipation. With this in mind, we have developed a spatial discretization method which, in mathematical terms, takes care that (skew)symmetry of the differential operators that are approximated is preserved: convection is approximated by a skew-symmetric discrete operator, whereas diffusion is discretized by a symmetric, positive-definite operator. The temporal evolution of the discrete velocity vector $\boldsymbol{u}_h$ is governed by a finite-volume discretization of the incompressible Navier–Stokes equations:

$$\boldsymbol{\Omega}\frac{\mathrm{d}\boldsymbol{u}_h}{\mathrm{d}t} + \boldsymbol{C}(\boldsymbol{u}_h)\,\boldsymbol{u}_h + \boldsymbol{D}\boldsymbol{u}_h - \boldsymbol{M}^*\boldsymbol{p}_h = \boldsymbol{0}, \qquad \boldsymbol{M}\boldsymbol{u}_h = \boldsymbol{0}, \tag{1}$$

where the vector $\boldsymbol{p}_h$ denotes the discrete pressure, $\boldsymbol{\Omega}$ is a (positive-definite) diagonal matrix representing the sizes of the control volumes for the discrete velocities, $\boldsymbol{C}(\boldsymbol{u}_h)$ is built from the convective flux contributions through the control faces, $\boldsymbol{D}$ contains the diffusive fluxes, and $\boldsymbol{M}$ is the coefficient matrix of the discretization of the integral form of the law of conservation of mass. The essence of symmetry-preserving discretization is that the coefficient matrix $\boldsymbol{C}(\boldsymbol{u}_h)$ is skew-symmetric whereas $\boldsymbol{D}$ is positive-definite. Under these two conditions, the evolution of the discrete energy $\boldsymbol{u}_h^*\boldsymbol{\Omega}\boldsymbol{u}_h$ of any solution $\boldsymbol{u}_h$ of (1) is governed by

$$\frac{\mathrm{d}}{\mathrm{d}t}(\boldsymbol{u}_h^*\boldsymbol{\Omega}\boldsymbol{u}_h) \stackrel{(1)}{=} -\boldsymbol{u}_h^*(\boldsymbol{C}+\boldsymbol{C}^*)\boldsymbol{u}_h - \boldsymbol{u}_h^*(\boldsymbol{D}+\boldsymbol{D}^*)\boldsymbol{u}_h = -\boldsymbol{u}_h^*(\boldsymbol{D}+\boldsymbol{D}^*)\boldsymbol{u}_h \leq 0. \tag{2}$$

So, the energy is conserved if the diffusion is turned off. With diffusion (that is for $\boldsymbol{D} \neq \boldsymbol{0}$) the right-hand side of (2) is negative for all $\boldsymbol{u}_h \neq \boldsymbol{0}$, since $\boldsymbol{D} + \boldsymbol{D}^*$ is positive definite. Consequently, the energy of the semi-discrete system (1) decreases unconditionally in time. In other words, (1) is stable, and there is no need to add an artificial damping mechanism to stabilize the spatial discretization. For more details the reader is referred to [4]-[6].

The pressure gradient and the incompressibility constraint are integrated implicitly in time; the convective and diffusive fluxes are treated explicitly. The computation of one time step is divided into two substeps. First, an auxiliary velocity $\hat{\boldsymbol{u}}_h$ is computed by integrating the convective and diffusive transport of momentum over one step in time. For this, the following one-leg method is applied:

$$\boldsymbol{\Omega}\left((\beta+\tfrac{1}{2})\hat{\boldsymbol{u}}_h - 2\beta\boldsymbol{u}_h^n + (\beta-\tfrac{1}{2})\boldsymbol{u}_h^{n-1}\right) = -\delta t\left(\boldsymbol{C}(\boldsymbol{u}_h^{n+\beta})\boldsymbol{u}_h^{n+\beta} + \boldsymbol{D}\boldsymbol{u}_h^{n+\beta}\right) \tag{3}$$

where $\boldsymbol{u}_h^{n+\beta} = (1+\beta)\boldsymbol{u}_h^n - \beta\boldsymbol{u}^{n-1}$. The parameter β is taken equal to 0.05 in order to optimize the convective stability of the one-leg method [4].

Next, the pressure gradient is added to the auxiliary velocity $\hat{\boldsymbol{u}}_h$ such that the resulting velocity field satisfies the law of conservation of mass. Therefore, the pressure need be computed from the Poisson equation

$$\boldsymbol{M}\boldsymbol{\Omega}^{-1}\boldsymbol{M}^*\boldsymbol{p}_h = -\frac{\beta+\frac{1}{2}}{\delta t}\boldsymbol{M}\hat{\boldsymbol{u}}_h \stackrel{def}{=} \boldsymbol{r}_h. \tag{4}$$

3. PARALLELIZATION

This section concerns the parallelization of the two main ingredients of the computational procedure: the evaluation of the discrete convection-diffusion equation (3) and the solution of the Poisson equation (4).

3.1. Convection-diffusion equation

The auxiliary velocity $\hat{\boldsymbol{u}}_h$ can be computed from Eq. (3) by means of a sparse matrix-vector multiplication. Grid partitioning leads straightforward to parallelization while keeping the algorithm unchanged. We choose to partition the flow domain in the streamwise direction. The number of streamwise grid points N_x is taken to be an integer multiple of the number of processors p. Then, the streamwise subdomains are equally large and can be divided equally over the processing nodes. The partitions can be treated independent of each other. The only aspect to consider is that each streamwise subdomain has to share data with its right- and left-hand streamwise neighbours to perform its part of the sparse matrix-vector multiplication (3).

3.2. Poisson equation

Turbulent flows that are statistically homogeneous in one spatial direction can be handled very well using periodic boundary conditions in that direction. Here, we make explicitly use of periodicity in the spanwise direction. The Poisson equation (4) for the pressure can then be solved by means of a Fast Fourier Transform method in the spanwise direction and an Incomplete Choleski Conjugate Gradient method in the resulting spectral space. After the Fourier transformation, the discrete Poisson equation (4) falls apart into a set of mutually independent equations of the form

$$\left(\boldsymbol{M}\boldsymbol{\Omega}^{-1}\boldsymbol{M}_{\mathrm{xy}} + \boldsymbol{\Lambda}_{\mathrm{z}}\right)\hat{\boldsymbol{p}}_h = \hat{\boldsymbol{r}}_h, \qquad (5)$$

where the non-zero entries of the diagonal matrix $\boldsymbol{\Lambda}_{\mathrm{z}}$ are given by the spanwise eigenvalues $2\cos(2\pi k/N_{\mathrm{z}})$ of the Poisson matrix, and $\boldsymbol{M}\boldsymbol{\Omega}^{-1}\boldsymbol{M}_{\mathrm{xy}}$ denotes the restriction of the Poisson matrix $\boldsymbol{M}\boldsymbol{\Omega}^{-1}\boldsymbol{M}$ to the (x, y)-plane. The complex vectors $\hat{\boldsymbol{p}}_h$ and $\hat{\boldsymbol{r}}_h$ are the spanwise Fourier transforms of the solution $\boldsymbol{p}_h$ and the right-hand side $\boldsymbol{r}_h$ of Eq. (4) respectively. The spanwise dimension N_{z} is taken to be a power of 2, and the transforms are computed with the standard Fast Fourier Transform method. Their calculation is divided into equal chunks. The chunks correspond to the streamwise subdomains that are used to compute the discrete convection-diffusion equation in parallel. Obviously, the parallelization is perfect: the spanwise FFT's are fully independent of each other, *i.e.* can be computed without any mutual communication.

The set of equations (5) is solved in parallel by treating the unknowns $\hat{\boldsymbol{p}}_h$ simultaneously in the spanwise direction. This implies that the parallel direction changes from the streamwise to the spanwise direction. This change requires `alltoall` communication. The set (5) consists of mutually independent 2D Poisson equations, where depending on the Fourier mode the diagonal of the coefficient matrix is strengthened. Each equation is solved iteratively by means of an Incomplete Choleski Conjugate Gradient (ICCG) method. As the diagonal increases with the frequency, the ICCG-iterations for the high frequencies converge faster than those for the low frequencies. The resulting potential unbalance in the work load is to a large extent counterbalanced by the accuracy of the initial guess: the pressure at the previous time level forms a much better initial guess for the low frequencies than for the high ones, since the low frequencies change less per step in time.

Once the pressure $\hat{\boldsymbol{p}}_h$ in the Fourier space has been computed, it has to be transformed back into the physical space, so that the pressure gradient can be added to the auxiliary velocity. For this, the parallel direction has to be changed back from the spanwise to the streamwise direction. Again this calls for `alltoall` communication.

4. RESULTS

The flow problem solved was a turbulent flow past a long cylinder with a square cross-section at a Reynolds number of Re $= 22,000$. The computations were carried out on two grids, with $240\times192\times128$ and $480\times284\times128$ gridpoints. Velocity fields as obtained from the numerical simulations can be found in [5]. Here we report on the parallel performance of the code on a 1024-processor SGI Origin 3800 system.

A scalability test has been conducted for 1, 2, 4, 8, 16 and 32 processors on the coarse grid and for 4, 8, 16, 32, 64 and 128 processors on the fine grid. The results for the coarse

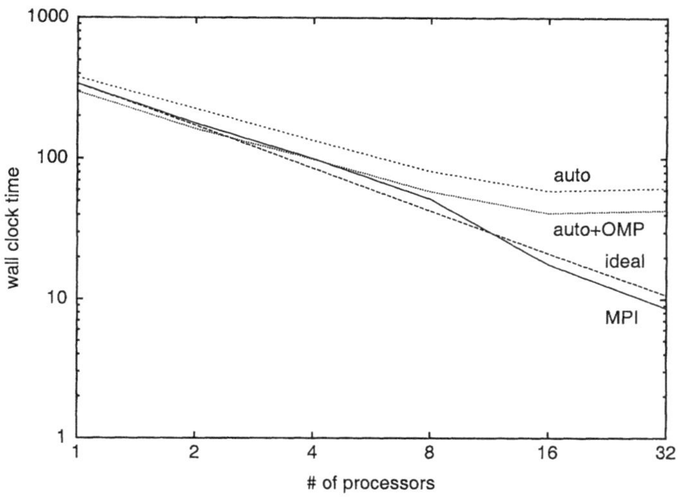

Figure 1. Wall-clock time versus the number of processors; $240 \times 192 \times 128$ grid.

grid are displayed in Figure 1. The scalability of the code with automatic parallelization is poor due to excessive synchronization. The compiler failed to recognize the trivially parallelizable outer loops in the Poisson solver. Instead of computing the set of mutually independent equations (5) in parallel, some inner loops in the ICCG-iterations were parallelized. We have corrected this by forcing the parallelization of the outer loops with OpenMP compiler directives. This improved the performance significantly. For a small number of processors the autoparallelization corrected with OpenMP directives performs as good as the MPI implementation. For a larger number of processors, however, the scalability of the MPI version is superior. Likely, the weak performance of the autoparallelization version may be boosted further by identifying the remaining trouble spots and inserting appropriate compiler directives. Yet, this requires an effort larger than that needed for the MPI implementation.

As explained in Section 3, the computation of one step in time consists of two parts. Table 1 shows that the ratio of the convection-diffusion part over Poisson solver is increasing for an increasing number of processors. This is accompagnied by an increasing ratio of the main communications (`sendrecv` for the convection-diffusion part and `alltoall` for the Poisson solver). This may be explained by the following simple analysis.

The time for one `sendrecv` is $\alpha + n\beta$, where α denotes the latency, β is 1/bandwidth and n is the size of the message. The message size n for a `sendrecv` is of the order of $N_y \times N_z$ bytes. In an `alltoall` operation a processor has to send a block of data to all the other $p-1$ processors. As this can be done in parallel by cyclically shifting the messages on the p processors, the time needed to complete an `alltoall` is $(p-1)(\alpha + n\beta)$. In our application, one processor has to treat $N_x \times N_y \times N_z/p$ grid points. In an `alltoall` operation data blocks containing an $1/p$-th part of it are sent to the other processors. So, the message size n for an `alltoall` operation is of the order of $N_x \times N_y \times N_z/p^2$. Hence,

Table 1
Relative time needed to solve the Poisson equation for the pressure, and to perform the main communications (`sendrecvs` and `alltoalls`) for the $240 \times 192 \times 32$ grid. The wall-clock time for the convection-diffusion part is scaled to 1.

# of processors	Poisson	`sendrecv`	`alltoall`
1	3.25	0.00	0.02
8	2.33	0.20	0.27
32	1.65	0.49	0.28

for a constant number of grid points and an increasing number of processors the time needed to do a `sendrecv` operation remains constant, whilst the time taken by `alltoall` decreases with $1/p$ (assuming that α is sufficiently small compared to $n\beta$). This explains the relative decrease of the communication time for the Poisson solver. Figure 2 shows that the measured wall clock time needed to perform one complete `sendrecv` and `alltoall` confirms our simple scalability analysis.

Figure 3 displays the wall clock time versus the number of processors for the fine grid computations. The test for the automatically parallelized versions starts at 8 processors because, unlike the MPI code, the autoparallelized code does not fit into the memory of 4 processors. Also for the fine grid, the MPI version is superior to the automatic parallelization version. OpenMP directives in the ICCG subroutine improve the performance, but as for the coarse grid the option autoparallelization with OpenMP directives performs less than MPI (for more than 16 processors). The speedup of the MPI version is close to ideal apart from a peculiar kink at 16 processors. At 16 processors the bandwidth of the communication between level-1 and level-2 cache and between the cache and memory was substantially less than the bandwidth at 8 and 32 processors.

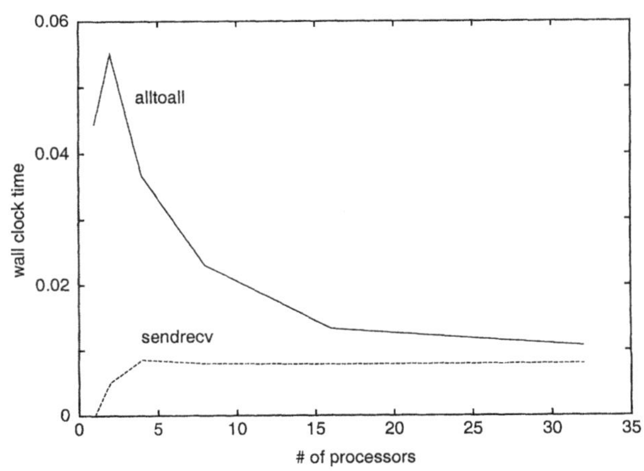

Figure 2. The wall clock time (in seconds) needed to perform an `alltoall` and a `sendrecv` operation on the $240 \times 192 \times 32$ grid.

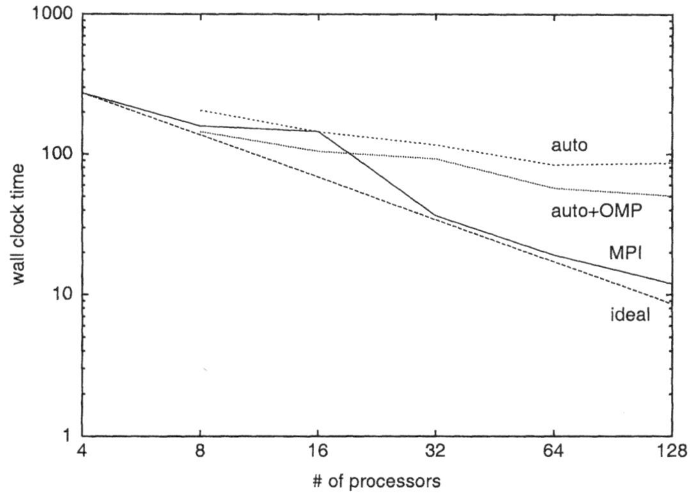

Figure 3. The wall clock time versus the number of processors for the 384x480x128 grid.

The bandwidth was well above 300 Mb/s for 8 processors and above 400Mb/s for 4, 32 and 64 processors while for 16 processors it was about 200 Mb/s. This led to a drop in megaflop rate of about a factor of two going from 8 to 16 processors. What causes this drop in bandwidth is unclear to us.

Since for an increasing number of processors the amount of computation per processor decreases while the amount of communication is approximately constant for this problem, it could be expected that the scaling of the code would deteriorate. On the coarse grid, this computation/communication ratio effect is counteracted by the fact that the number of cache misses decreases for smaller problems (per processor) and therefore single CPU performance increases; see Table 2. The code suffers from cache-misses on both grids. However, on the coarse grid the situation improved from 4 processors onwards. This was due to a substantial increase in the level-2 cache hit rate which increased from 0.89 on 8 processors to 0.97 on 32 processors. We did not observe this strong cache effect on the fine grid. The level-1/level-2 cache hit rate increased only mildly from 0.86/0.77 on 4 processors to 0.90/0.83 on 64 processors.

5. CONCLUSIONS

The tests clearly demonstrate that on a SGI-3800 machine the MPI version of our DNS code performs well in terms of scalability. The speedup is close to ideal. Apart from implementing MPI nothing has been done to enhance the performance of this code. The single CPU performance is at best about 9 percent of the theoretical peak performance. We found that this was mainly due to cache-misses. Perhaps some increase in performance of the MPI version could be achieved in this field.

A drawback of MPI is that programming effort is needed to implement the required message-passing instructions. This can be avoided using autoparallelization, but then

Table 2
The single CPU performance and the overall performance for the MPI version.

	240 × 192 × 32		480 × 384 × 32	
# of processors	single CPU Mflops	Mflops	single CPU Mflops	Mflops
1	48	48		
2	45	90	40	81
4	41	164	32	126
8	43	341	27	218
16	59	983	15	242
32	64	2055	30	972
64			30	1901
128			28	3608

the performance is rather poor. The automatically parallelized code has been thoroughly investigated for trouble spots, where autoparallelization does a poor job. The major performance bottlenecks have been identified and reduced by means of OpenMP compiler directives. Yet, still the performance lags behind that of MPI. Continuing along this road requires a lot of effort and whether the MPI version will ever be surpassed in performance is an open question.

ACKNOWLEDGMENT

The Dutch national computing facilities foundation NCF is acknowledged for the funding of this project. Peter Michielse (SGI) is acknowledged for his helpful advice.

REFERENCES

1. P. Holmes, J.L. Lumley and G. Berkooz, *Turbulence, coherent structures, dynamical systems and symmetry*, Cambridge University Press (1996).
2. P.R. Spalart, Strategies for turbulence modelling and simulations, *Int. J. Heat and Fluid Flow*, **21**, 252 (2000).
3. TOP500 Supercomputer List.
   See: http://www.netlib.org/benchmark/top500/top500.list.html
4. R.W.C.P. Verstappen and A.E.P. Veldman, Direct numerical simulation of turbulence at lesser costs, *J. Engng. Math.*, **32**, 143–159 (1997).
5. R.W.C.P. Verstappen and A.E.P. Veldman, Spectro-consistent discretization of the Navier–Stokes equations: a challenge to RANS and LES, *J. Engng. Math.*, **34**, 163–179 (1998).
6. R.W.C.P. Verstappen and A.E.P. Veldman, Symmetry-preserving discretization of turbulent flow, submitted to *J. Comp. Phys.*

Parallel Computational Fluid Dynamics – Practice and Theory
P. Wilders, A. Ecer, J. Periaux, N. Satofuka and P. Fox (Editors)

Parallel Shallow Water Simulation for Operational Use

Dr.ir. E.A.H. Vollebregt & dr.ir. M.R.T. Roest

VORtech Computing
P.O. Box 260, 2600 AG Delft, the Netherlands
Edwin.Vollebregt@VORtech.nl, Mark.Roest@VORtech.nl

This paper discusses the parallelization of the shallow water simulation software of the Dutch Rijkswaterstaat, the governmental organization that is responsible for the maintenance of the Dutch coast and waterways. This software is used operationally for a wide range of purposes such as design studies, environmental impact studies and storm surge forecasting.

The parallelization of this software is particularly interesting because of the operational use by a variety of users for largely varying purposes, which poses special requirements on the parallel version. Further a flexible approach has been adopted for the parallelization, which proves to be extendible towards domain decomposition and model-coupling.

The parallelization of the 3D model TRIWAQ was started already in 1992, and parallel computing has been used operationally for several years now. Currently we work on domain decomposition with horizontal grid refinement, and on the parallelization of a Kalman filtering module for the shallow water models.

1. Introduction and overview

VORtech Computing has a long history in developing parallel software for the Dutch National Institute for Coastal and Marine Management (Rijkswaterstaat/RIKZ). This institute develops and maintains the simulation system SIMONA, which is used for simulation of a.o. water movement and transport processes in coastal and river water systems. SIMONA is used by many of the branches of the Dutch Ministry of Public Works and Transport, for a wide range of activities such as design studies, environmental impact studies and operational storm surge forecasting (see Figure 1).

Over the years, the models that are simulated with SIMONA have become ever larger and have included more and more features. To keep computation times within practical limits, parallelism has been introduced in the two most important modules: the 2D flow simulation module WAQUA and the 3D flow simulation module TRIWAQ. In the development of parallel versions of these modules a large number of requirements has been taken into account, which are due to the operational use of the system at many different locations, by users with varying degree of expertise, and to the size and complexity of the system.

An essential role in fulfilling the requirements is played by the communications library that was developed. This library has also opened the way towards domain decomposition

Figure 1. Examples of applications of shallow water simulation models: providing information for ship-guidance, predicting waterlevels for safety-level assessment, and investigating consequences of (a.o.) new waterworks.

and on-line model coupling, thus providing more modeling flexibility to the users. The basic concepts of the communications library have remained valid even though the current version is far more powerful than the one that was initially developed eight years ago.

In Section 2 we describe the operational use of WAQUA/TRIWAQ and the requirements posed on the parallel version. Section 3 describes the parallelization approach, the principles and practical aspects of the communication library and shows why it has been so successful. Finally Section 4 presents our conclusions.

2. Requirements on the parallel software

The examples in Figure 1 illustrate that WAQUA and TRIWAQ are used in a large number of different settings. This has the following consequences:

1. there is a large variety in the physical effects that are dominating or relevant in the different situations;
2. there is a large variety in the amount of computations required for different applications, and thus in the platforms used for the simulation;
3. there is a large number of users and a large variation between the different users of the program.

The variety of dominating aspects to be modelled manifests itself through a large number of features that is incorporated in WAQUA/TRIWAQ [1,2], characterized by:

- simulating the unsteady shallow water equations in two (depth-averaged) or three dimensions;
- using orthogonal curvilinear or spherical grids;
- allowing for complex geometries (harbours etc.), time-varying due to drying and flooding;

- supporting transport processes of conservative constituents, density effects, and including a $k - \epsilon$ turbulence model;
- allowing for 3D moving barrier constructions;
- including energy-losses due to weirs, and special discharge boundary conditions for river-applications;
- providing additional facilities for (comparison with) observed data, and for data assimilation through Kalman filtering and an adjoint model;

The WAQUA/TRIWAQ model is embedded in the SIMONA framework, which provides memory management and I/O subsystems, and a generic pre-processing mechanism.

The amount of computations required differs widely among the various applications of the program. Standard (2D) waterlevel predictions with a continental shelf model of 173×200 grid points can be carried out on a regular PC, larger grid schematizations require more advanced workstations or clusters of Linux PC's, whereas the top-range applications require supercomputing power.

For example the newest "Zeedelta"-model that is being developed consists of 153.400 active grid points in a full matrix of 471×1539 (fill-ratio 22.5%), requires at least 10 layers in the vertical direction in the transition region between fresh and salt water, requires inclusion of density effects and the $k - \epsilon$ turbulence model. For one environmental impact study different scenarios must be simulated with this model with a duration of several days or weeks and with a time step of 30 seconds (20-50.000 time steps). These simulations currently require about one day computing time per day that is simulated on a modern PC (1000 MHz) or workstation.

Thirdly the number and variety of users is illustrated by distinguishing different types of users:

- program engineers: civil engineering/numerical/software experts that further develop the simulation program. Found at a few central offices of Rijkswaterstaat and in $\pm$ 5 companies such as VORtech Computing;
- model engineers: expert users of the simulation model that develop and improve grid schematizations for regions of interest. Found at a few central offices of Rijkswaterstaat and about 10-15 (hydraulic) engineering firms;
- model users: end-users of the simulation program and grid schematizations, that want to perform simulations for different scenarios. Found at about 10 offices of Rijkswaterstaat and various (hydraulic) engineering firms.

From the setting in which WAQUA and TRIWAQ are used different requirements may be derived on the parallelization, that will now be elaborated.

In the development of the parallel versions of WAQUA and TRIWAQ a strong emphasis has been put on the requirements related to portability. These requirements are primarily due to the fact that the software is used at many different locations and on a wide variety of platforms. Most users of the simulation software do not have or need parallel computers, whereas others use networks of single- or multiprocessor workstations or PC's, and still others use large scale parallel computers at the Academic Supercomputer Center SARA in Amsterdam. The parallel software should run and deliver good performance on all these platforms.

It is not an option to maintain different versions of the code for different platforms; the maintenance of a single high quality code is expensive enough. Also, having to introduce changes into a number of different versions would readily increase the time needed before a well tested update of the simulation system can be released, thus hampering the development of the system. This is the more true because all versions should produce exactly the same results, as major policy decisions may be based on these results and confusion over the validity of predictions made by WAQUA and TRIWAQ is unacceptable.

Besides portability, the most important other system requirements concern the extendibility and the interoperability with other software.

With an operational system like SIMONA, that is constantly used for real life problems, there is always a demand for extensions and improvements. Better models of the physics of flow in seas and estuaries are introduced as the old models reveal their limitations and are improved upon. Programming such improvements should be possible for developers without extensive knowledge of parallel computing and of the way in which parallelism is implemented in the simulation software. In fact, most of the programming on the WAQUA and TRIWAQ modules is done by experts in the field of computational fluid dynamics and civil engineering, rather than by experts in the field of parallel computing or information technology in general. The way in which parallelism is introduced in WAQUA and TRIWAQ should be easy to understand, so that any experienced programmer can deal with it when making changes to these modules.

An extension that was considered important already at the time of parallelization is the development of domain decomposition functionality. Domain decomposition allows modelers to combine separate grids for different areas into a single overall model. This allows for using fine resolution only in those areas where it is really needed, and avoids some of the complications of matching a region as a whole onto a single structured grid.

Interoperability is needed because the flow simulation modules WAQUA and TRIWAQ are used to produce input for various other modules in SIMONA or even for simulation programs that are not in SIMONA. For example, the flow simulation modules are used with models of morphology changes and for particle tracing for simulation of transport of pollution. Usually, such a combined simulation is done by first running the flow simulation module, writing flow fields for all time-instances of interest to file, and then running the other simulation model, which reads its input from the file produced by the flow simulation module. But this line of working can be problematic. On the one hand, the number of time-instances for which data needs to be transferred from the flow simulation module to the other model may be too large, leading to excessively large transfer-files. On the other hand, the two models may be influencing each other (e.g. a change in morphology may lead to a change in flow), so that a one-way coupling is insufficient. In these cases the on-line coupling of different simulation models is needed.

3. The communications library

The demands listed above could be realized in the simulation system relatively easily because of the basic concepts that were used in the design of the parallelization. First of all this concerns the overall parallelization strategy, using separate computing processes and a communications library (see Section 3.1). In addition, an essential role is played by the abstract concepts behind the communications library. In this paper we mainly concentrate on the primary concepts of *"index sets"* (Sections 3.2-3.4) and *"avail and obtain operations"* (Section 3.5).

3.1. Overall parallelization strategy

Basically, the parallelization is considered as coupling several instances of a program rather than splitting a program into subprograms [3]. The program is extended with the capability to cooperate with other programs (possibly other instances of the same program), and a *partitioner* program is defined for splitting a global model input file into separate parts for the subproblems [7]. The viewpoint of cooperating processes proved to be very practical for the parallelization, because it leads to a single code to be maintained for sequential and parallel computing and the code stays familiar for other developers on WAQUA and TRIWAQ. The viewpoint greatly simplified the programming of the parallel version as well, because it avoids administration of data for both the global domain and the subdomains in the WAQUA/TRIWAQ program. Finally the viewpoint has a natural extension towards the implementation of domain decomposition and on-line couplings.

The communication between the computing processes is done by calling routines from a communication library. These routines are designed in such a way that they are meaningful and easy to understand for someone who is used to program numerical algorithms. All details regarding process numbering, the actual sending and receiving of data, synchronization, data reordering etc., and in case of domain decomposition and model coupling: data interpolation and conversion, are hidden inside the library routines. This first of all guarantees that the usual programmers working on SIMONA can easily use these routines. But at the same time it puts all system-dependent communication issues into a small number of communication routines, so that porting to a new hardware platform becomes relatively easy.

3.2. Abstractions for communication: index sets

The basic concepts of the communications library which enable all these benefits are *"index sets"* [3,4] and *"avail and obtain operations"* [5]. These are high level abstractions of what an application programmer actually wants to know about the communications.

The concept of index sets is central to the communications library. The programmer can define an arbitrary set of points at which his data is located. For example, in grid-based applications an obvious example of an index set is the grid. Another example for WAQUA and TRIWAQ concerns the locations at which special discharges are taken into account, the so-called source points; these too form an index set. These index sets are used to describe to the communications library how data is stored in data structures.

In a parallel run, each process holds a part of each of the global index sets. For example, each process has only a part of the grid and only a part of all source points. When defining the index sets, the programmer provides the global coordinates of the points, or another suitable numbering of the points about which all the cooperating processes can agree. These global numbers allow the communications library to relate data elements in different processes to each other. Also, manipulations are made possible such as locating all grid points neighboring to a process' own grid points with respect to an arbitrary stencil. This allows the handling of irregular grid partitionings while still tailoring the actual communications to precisely what is needed.

3.3. Practical use of index sets in communication

The configuration of an index set is accomplished in the program source code via an array with "coordinates" of the "indices" and via the ownership of the indices, e.g.:

```
... icoord(ix,1:2) = (m,n)  ...  iowner(ix) = p
call cocidi('fullbox', length, ndims, icoord, iowner)
```

This description is given per process for the indices of the process itself as well as for the guard band of the process. A communication-interface on an index set is defined via a stencil, which is nothing more than an array of coordinate-offsets. For a five-point stencil:

```
... istenc(1:5,1:2)=[ 0,-1; 0,0; 0,1; 1,0; -1,0 ],  noffs = 5
call cocitf('fullbox', 'stc1', noffs, istenc, mask)
```

With these definitions the communications library can determine which array values must be sent to and received from neighbouring subdomains. After this the central communication operation for parallel WAQUA/TRIWAQ *"update"* can be used, which exchanges information between neigbhouring subdomains at subdomain boundaries:

```
call cocupd(up, 'fullbox', 'stc1')
```

This example call shows how a velocity field "up" with data structure "`fullbox`" is communicated at subdomain boundaries for the standard five-point stencil "`stc1`".

The example illustrates how communication is specified using entities that are understandable and meaningful for application programmers. All awkward details w.r.t. process numbering, sending/receiving, synchronization, data reordering are hidden inside the library routines. Also the partitioning of the computational grid is hidden entirely: all kinds of irregular partitionings are allowed, and are needed because of the complex geometries used.

3.4. Extension towards domain decomposition

Domain decomposition is realized for WAQUA and TRIWAQ by just a slight extension of the strategy for parallelization [6]. Two aspects that are different w.r.t. parallel computing are:

- the subdomains are no longer determined automatically, but have become of interest for the user; however the partitioner program is still used for generating the subdomain input data.
- the communication of values between different computing processes now also requires interpolation between the different grids.

The interpolation is incorporated into the update-operation, whose goal is reformulated as "to exchange information among different computing processes". The subroutine call is extended with an optional *conversion method*:

```
call cocupd(up, 'fullbox', 'stc1', 'bilin-u')
```

A number of base conversion methods such as "bilinear interpolation" are defined, which are instantiated into actual conversion methods by configuring the coefficient (coordinate) values, in a manner similar to the definition of index sets and stencils above.

3.5. Abstractions for communication: avail and obtain

An alternative way to view communications using index sets is by considering index sets essentially as a sort of global address spaces for the data involved. Computing processes work on local copies of the data in this global data space, and communication is viewed conceptually as to put data in the global space or retrieve data from there. The avail and obtain operations can be interpreted as a consequence of this viewpoint, with particular relevance for model-coupling, coupling of different models using a functional decomposition rather than similar models using an SPMD approach.

Execution of the avail operation by a computing process states that the data provided in the call to this routine is available for other processes to use, whether any of them needs it or not. An obtain operation in a program specifies that certain information is needed at a certain point in the numerical algorithm. The obtain operation waits until all required information is provided, i.e. is blocking. An avail-point in a program can be connected to zero or more obtain points in other programs and is non-blocking. For this coupling, the subroutine call contains a name for the communication point:

```
call cocava(up, 'fullbox', mask, 'avail_up')
```

The avail and obtain communications operations have a high abstraction level because they specify only *which* data is communicated rather than *how* this is done. Further each process specifies data in its own terms, and data conversion (e.g. interpolation) may be used during the communication. Finally note that programs do not specify where the data must be sent to (in case of avail) or where the data must come from (in case of obtain). This makes sure that programmers do not make implicit or explicit assumptions on the context in which the computation program will run, and thus enhances interoperability.

Obviously, an important aspect of coupling programs is to make sure that data that is to be obtained is actually retrieved from another process at the moment when that other process avails the data. This is achieved by providing so-called *coupling algorithms*, the sequence of avail/obtain operations that are performed by a program, in an external file. This file serves as the externally visible interface to other programs. Further a coupled run requires a configuration file that lists the processes to be used and the connections between avail/obtain operations in the corresponding programs. This coupling configuration file allows for extensive checking, e.g. deadlock detection, and ensures that the communications library knows for each operation which data must be sent to or received from which other processes.

These mechanisms for model-coupling were largely developed by Delft University of Technology, in research towards the parallelization of a Kalman filter for WAQUA and TRIWAQ [5].

4. Conclusions

In this paper we have discussed special considerations for the parallelization of the shallow water simulation models WAQUA and TRIWAQ that arise as a consequence of the operational environment in which these models are used:

- portable to a wide range of platforms, using a single version of the code, and delivering good performance on all platforms used;

- hiding aspects of the parallelisation for end-users of the program, and allowing extension by non-experts in the field of parallel computing;
- applicable within a larger environment: interoperability with existing pre- and post-processing programs, and extendible towards domain decomposition and on-line model-coupling.

These requirements on the parallel system have inspired us to view the parallelization as *coupling* of different programs rather than *breaking up* a computation, and to the development of a powerful and efficient communications library. The communications library has an abstract interface (i.e. hides irrelevant aspects for the programmer) which is based on generic principles:

- index sets, to characterize data structures in a program;
- stencils, to characterize interaction patterns, neighbouring grid points;
- the update operation, to exchange information among similar computing processes;
- the avail and obtain operations, to provide information to or retrieve information from the outside world.

These concepts have allowed for extension towards domain decomposition with horizontal and vertical grid refinement and to on-line model-coupling. Thereby these concepts have proven to be very flexible so that new and unforeseen situations can be handled elegantly.

REFERENCES

1. Rijkswaterstaat/RIKZ, Users guide WAQUA, Tech. Rep. SIMONA 92-10, National Institute for Coastal and Marine Management, the Hague, the Netherlands (2001).
2. M. Zijlema, Technical documentation TRIWAQ, Tech. Rep. SIMONA 99-01, National Institute for Coastal and Marine Management, the Hague, the Netherlands (1999).
3. E. Vollebregt, Parallel software development techniques for shallow water models, Ph.D. thesis, Delft University of Technology (1997).
4. E. Vollebregt, Abstract level parallelization of finite difference methods, Scientific Programming 6 (1997) 331–344.
5. M. Roest, E. Vollebregt, Parallel kalman filtering for a shallow water flow model, in: P. Wilders, A. Ecer, J. Periaux, N. Satofuka (Eds.), ParCFD Conference 2001, Egmond aan Zee, Elsevier Science B.V., Amsterdam, The Netherlands, 2001.
6. L. Riemens, H. ten Cate, B. van 't Hof, M. Roest, Domain decomposition with vertical refinement in TRIWAQ, in: Proceedings of the 4th International Hydroinformatics Conference, 2000, cd-rom.
7. M. Roest, Partitioning for parallel finite difference computations in coastal water simulation, Ph.D. thesis, Delft University of Technology (1997).

Parallel Computational Fluid Dynamics – Practice and Theory
P. Wilders, A. Ecer, J. Periaux, N. Satofuka and P. Fox (Editors)

Parallel Deflated Krylov methods for incompressible flow

C. Vuik [a] * †, J. Frank [b] and F.J. Vermolen [a]

[a]Delft University of Technology, Department of Applied Mathematical Analysis, P.O. Box 5031, 2600 GA Delft, The Netherlands

[b]CWI, P.O. Box 94079, 1090 GB Amsterdam, The Netherlands

Efficient parallel algorithms are required to simulate incompressible turbulent flows in complex two- and three-dimensional domains. The incompressible Navier-Stokes equations are discretized in general coordinates on a structured grid. For a flow on a general domain we use an unstructured decomposition of the domain into subdomains of simple shape, with a structured grid inside each subdomain. We have developed parallel block-preconditioned Krylov methods to solve the resulting systems of linear equations. The parallel methods are further accelerated by a Deflation technique. Numerical experiments illustrate the performance of the methods.

1. Introduction

Efficient parallel algorithms are required to simulate incompressible turbulent flows in complex two- and three-dimensional domains. We consider the incompressible Navier-Stokes equations:

$$\frac{\partial \mathbf{u}}{\partial t} - \frac{1}{R_e}\Delta \mathbf{u} + \mathbf{u}(\nabla \cdot \mathbf{u}) + \nabla p = \mathbf{f}, \quad \nabla \cdot \mathbf{u} = 0, \quad \vec{x} \in \Omega,\ t \in (0,T],$$

where R_e is the Reynolds number. These equations are discretized in general coordinates using a staggered finite volume method on a structured grid [14], see Figure 1 for the placement of the unknowns. For a flow on a general domain we use an unstructured decomposition of the domain into subdomains of simple shape, with a structured grid inside each subdomain [3].

Let V^k and P^k represent the algebraic vectors containing velocity, $\mathbf{u}$, and pressure, p, unknowns at time t^k, respectively. A prediction of the velocity V^* is computed from the momentum equations:

$$\frac{V^* - V^k}{\Delta t} = F(V^k)V^* - GP^k, \tag{1}$$

*e-mail: *c.vuik@math.tudelft.nl*

†The authors thank HPαC for providing computing facilities on the Cray T3E

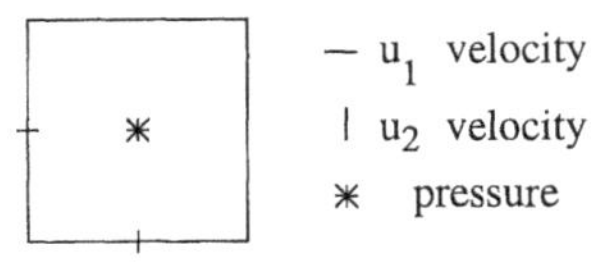

Figure 1. The staggered grid arrangement of the unknowns

where F is a nonlinear operator and G is the discrete gradient operator. The pressure correction follows from the system

$$DG\Delta P = \frac{DV^*}{\Delta t}, \tag{2}$$

where $\Delta P = P^{k+1} - P^k$ and D is the discrete divergence operator. After the pressure correction ΔP has been computed from (2), V^* is corrected by $V^{k+1} = V^* - \Delta t G \Delta P$. Above algorithm is commonly referred to as the pressure-correction method [10]. The linear systems (1) and (2) are solved with CG or GCR [9] using a block-diagonal preconditioner based on a nonoverlapping domain decomposition[15].

Efficient parallel implementation of GCR method requires, in addition to the preconditioner, a proper handling of the matrix vector multiplication and inner products. For a matrix vector product only nearest neighbor communications are required, which is efficient on most parallel computers. Inner products, on the other hand, require global communications. On present day parallel computers this is not a big issue because communication is reasonably fast [4]. This paper is a continuation of our work presented in [2,3,13,4].

2. Deflated Krylov methods

We use preconditioners based on an incomplete block LU decomposition [6]. Another preconditioning strategy that has proven successful when there are a few isolated extremal eigenvalues is Deflation [7]. We first consider a symmetric matrix $A \in \mathbb{R}^{n\times n}$ with linear system $Au = f$, $f \in \mathbb{R}^n$ where $u \in \mathbb{R}^n$ is to be determined. Let us define the projection P by

$$P = I - AZ(Z^TAZ)^{-1}Z^T, \quad Z \in \mathbb{R}^{n\times m}, \tag{3}$$

where Z is the deflation subspace, i.e. the space to be projected out of the residual. We assume that $m \ll n$ and that Z has rank m. Since $u = (I - P^T)u + P^Tu$ and because

$$(I - P^T)u = Z(Z^TAZ)^{-1}Z^TAu = Z(Z^TAZ)^{-1}Z^Tf \tag{4}$$

can be immediately computed, we need only compute P^Tu. In light of the identity $AP^T = PA$, we solve the deflated system $PA\tilde{u} = Pf$ for $\tilde{u}$ using the conjugate gradient method and premultiply this by P^T. The result is then substituted into $u = (I - P^T)u + P^Tu$

to obtain the solution. The algorithm of the Deflated ICCG method is first presented in [12].

As an example consider the case in which Z is the invariant subspace of A corresponding to the smallest eigenvalues. Note that $PAZ = 0$, so that PA has m zero-eigenvalues and the effective condition number is: $\kappa_{\text{eff}}(PA) = \frac{\lambda_n(A)}{\lambda_{m+1}(A)}$. In summary, deflation of an invariant subspace cancels the corresponding eigenvalues, leaving the rest of the spectrum untouched.

Now we consider a generalization of the projection P for a nonsymmetric matrix $A \in \mathbb{R}^{n\times n}$. In this case there is somewhat more freedom in selecting the projection subspaces. Let P and Q be given by

$$P = I - AZ(Y^TAZ)^{-1}Y^T, \quad Q = I - Z(Y^TAZ)^{-1}Y^TA.$$

where Z and Y are suitable subspaces of dimension $n \times m$. We solve the system $Au = f$ using deflation. Note that u can be written as $u = (I - Q)u + Qu$ and that $(I - Q)u = Z(Y^TAZ)^{-1}Y^TAu = Z(Y^TAZ)^{-1}Y^Tf$ can be computed immediately (cf. (4)). Furthermore Qu can be obtained by solving the deflated system, using $PA = AQ$,

$$PA\tilde{u} = Pf \tag{5}$$

for $\tilde{u}$ and premultiplying the result with Q.

Also in the nonsymmetric case, deflation can be combined with preconditioning. Suppose K is a suitable preconditioner of A, then (5) can be replaced by: solve $\tilde{u}$ from $K^{-1}PA\tilde{u} = K^{-1}Pf$, and form $Q\tilde{u}$, or solve $\tilde{v}$ from $PAK^{-1}\tilde{v} = Pf$, and form $QK^{-1}\tilde{v}$. Both systems can be solved by one's favorite Krylov subspace solver.

3. Choice of the deflation vectors

Initially the deflation vectors are chosen equal to the eigenvectors corresponding to small eigenvalues [12]. Drawbacks are: it is expensive to approximate these eigenvectors and the extra work due to the deflation increases considerably, because the eigenvectors are not sparse. It appears that it is possible to approximate the space spanned by the eigenvectors corresponding to the 'small' eigenvalues by choosing the deflation vectors equal to 1 on one subdomain and equal to 0 on all the other subdomains. In our previous work we always use non-overlapping deflation vectors. For overlapping deflation vectors the question is: how to choose the value of the deflation vectors at interface points in order to obtain an efficient, robust and parallelizable black-box deflation method.

We assume that the domain Ω consists of a number of disjoint sets $\Omega_j, j = 1, ..., m$ such that $\bigcup_{j=1}^{m} \bar{\Omega}_j = \bar{\Omega}$. The division in subdomains is motivated by the data distribution used to parallelize the solver. For the construction of the deflation vectors it is important which type of discretization is used: cell centered or vertex centered.

Cell centered

For this discretization the unknowns are located in the interior of the finite volume (element). The deflation vectors z_j are uniquely defined as:

$$z_j(\vec{x}_i) = \begin{cases} 1, & \text{for } \vec{x}_i \in \Omega_j, \\ 0, & \text{for } \vec{x}_i \in \Omega \setminus \Omega_j. \end{cases}$$

The pressure equation is discretized with a cell centered discretization.

Vertex centered

If a vertex centered discretization is used the unknowns are located at the boundary of the finite volume (element). Two different ways for the data distribution are known [8]:

- Element oriented decomposition: each finite element (volume) of the mesh is contained in a unique subdomain. In this case interface nodes occur.
- Vertex oriented decomposition: each node of the mesh is an element of a unique subdomain. Now some finite elements are part of two or more subdomains.

Note that the vertex oriented decomposition is not well suited to combine with a finite element method. Furthermore for interface points it is not uniquely defined to which subdomain they belong. Therefore we restrict ourselves to the element oriented decomposition. As a consequence of this the deflation vectors overlap at interfaces. The momentum equations are in one direction cell centered and in the other direction vertex centered, so a combination of choices for the deflation vectors is used.

In the vertex centered case the deflation vectors are defined as:

$$z_j(\vec{x}_i) = \begin{cases} 1, & \text{for } \vec{x}_i \in \Omega_j \cup (\partial\Omega_j \cap \partial\Omega), \\ \in [0,1] & \text{for } \vec{x}_i \in \partial\Omega_j \cap \Omega, \\ 0, & \text{for } \vec{x}_i \in \Omega \setminus \bar{\Omega}_j. \end{cases}$$

At the interfaces we investigate the following choices:

1. **no overlapping**

   The set S_j is defined by $S_j = (\partial\Omega_j \cap \Omega) \setminus \{(\partial\Omega_j \cap \Omega) \cap \bigcup_{i=1}^{j-1} \bar{\Omega}_i\}$.

$$z_j(\vec{x}_i) = \begin{cases} 1, & \text{for } \vec{x}_i \in S_j, \\ 0, & \text{for } \vec{x}_i \in (\partial\Omega_j \cap \Omega) \setminus S_j. \end{cases}$$

2. **complete overlapping**

$$z_j(\vec{x}_i) = 1, \text{ for } \vec{x}_i \in \partial\Omega_j \cap \Omega.$$

3. **average overlapping**

   Suppose $n_{neighbors}(i)$ is equal to the number of subdomains Ω_j such that $\vec{x}_i \in \bar{\Omega}_j$, then

$$z_j(\vec{x}_i) = \frac{1}{n_{neighbors}(i)}, \text{ for } \vec{x}_i \in \partial\Omega_j \cap \Omega.$$

4. Parallel implementation

In this section we describe an efficient parallel implementation of the subdomain deflation method with Z as defined in Section 3. We distribute the unknowns according to subdomain across available processors. For the discussion we will assume one subdomain per processor. The coupling with neighboring domains is realized by the use of virtual cells added to the local grids. In this way, a block-row of $Au = f$ corresponding to the subdomain ordering

$$A = \begin{bmatrix} A_{11} & \cdots & A_{1m} \\ \vdots & \vdots & \vdots \\ A_{m1} & \cdots & A_{mm} \end{bmatrix}, \tag{6}$$

can be represented locally on one processor: the diagonal block A_{ii} represents coupling between local unknowns of subdomain i, and the off-diagonal blocks of block-row i represent coupling between local unknowns and the virtual cells.

In parallel, we first compute and store $(Z^TAZ)^{-1}$ in factored form on each processor. Then to compute PAp we first perform the matrix-vector multiplication $w = Ap$, requiring nearest neighbor communications. Then we compute the local contribution to the restriction $\tilde{w} = Z^Tw$ and distribute this to all processors. With this done, we can solve $\tilde{e} = (Z^TAZ)^{-1}\tilde{w}$ and compute $(AZ)^T\tilde{e}$ locally. The total communications involved in the matrix-vector multiplication and deflation are a nearest neighbor communication and a global gather-broadcast of dimension m.

5. Numerical experiments

It appears that the solution of the pressure equation is the most time-consuming part. Furthermore the convergence of Krylov methods applied to the pressure equation resembles that of the Poisson equation. Therefore we consider as test example the Poisson equation on a square domain. In the following subsections we give the results for cell centered and vertex centered discretizations.

5.1. Cell centered discretization

Since the pressure matrix is nonsymmetric [11] we do not exploit the symmetry of the Poisson matrix in these experiments.

The domain is composed of a $\sqrt{p} \times \sqrt{p}$ array of subdomains, each with an $n \times n$ grid. With $h = \Delta x = \Delta y = 1.0/(n\sqrt{p})$ the cell-centered discretization is

$$4u_{i,j} - u_{i+1,j} - u_{i-1,j} - u_{i,j-1} - u_{i,j+1} = h^2 f_{i,j}.$$

The right hand side function is $f_{i,j} = f(ih, jh)$, where $f(x,y) = -32(x(1-x) + y(1-y))$. Homogeneous Dirichlet boundary conditions $u = 0$ are defined on $\partial\Omega$, implemented by adding a row of ghost cells around the domain, and enforcing the condition, for example, $u_{0,j} = -u_{1,j}$ on boundaries.

For the tests, GCR is restarted after 30 iterations, and modified Gram-Schmidt was used as the orthogonalization method for all computations. The solution was computed to a fixed tolerance of 10^{-6}. Block RILU is used as preconditioner [1]. We compare results for a fixed problem size on the 300×300 grid using 4, 9, 16 and 25 blocks. In Table 1 the iteration counts and wall clock times on a Cray T3E are given. Note that without Deflation the number of iterations increases when the number of blocks grows. This implies that the parallel efficiency decreases when one uses more processors.

Table 1
Results for various number of blocks without Deflation

	$p = 4$	$p = 9$	$p = 16$	$p = 25$
Iterations	341	291	439	437
Wall clock time	65	26	22	15

We also present timing results on the Cray T3E for a problem on a 480×480 grid using Deflation. The results are given in Table 2. In this experiment we use Deflated Block RILU preconditioned GCR. Note that the number of iterations decreases when the number of blocks increases. This leads to an efficiency larger than 1. The decrease in iterations is partly due to the improved approximation of the RILU preconditioner for smaller subdomains. On the other hand when the number of blocks increases, more small eigenvalues are projected to zero which also accelerates the convergence (see [5]). We expect that there is some optimal value for the number of subdomains, because at the extreme limit there is only one point per subdomain and the coarse grid problem ($\tilde{e} = (Z^T AZ)^{-1}\tilde{w}$) is identical to the original problem so there is no speedup at all.

Table 2
Speedup of the Deflated GCR method using a 480×480 grid

p	iterations	time	speedup	efficiency
1	485	710	-	-
4	322	120	5	1.2
9	352	59	12	1.3
16	379	36	20	1.2
25	317	20	36	1.4
36	410	18	39	1.1
64	318	8	89	1.4

5.2. Vertex centered discretization

For the vertex centered discretization we have only results for the Deflated ICCG method on a sequential computer. At this moment we are working on a parallel version. In our first experiment we use a 41×41 grid and 7 subdomains. The subdomains are layers parallel to the x-axis. In Figure 2 the results are given for different choices of the deflation vectors at the interfaces. Note that the results of the average and no overlap of the deflation vectors are more or less the same, whereas complete overlap leads to worse results. We prefer the average overlap technique, due to its black box nature.

We also varied the number of subdomains. The results for average overlap are given in Figure 3. We see again (compare Table 2) that the convergence is more or less independent of the number of subdomains.

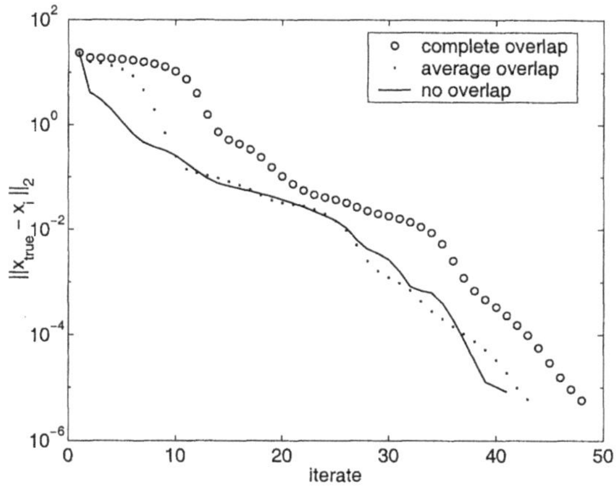

Figure 2. Convergence of the Deflated ICCG method for 41 x 41 internal nodes. The computations are done with average, complete overlap and without overlap.

6. Conclusions

Deflation of Krylov methods can easily be used in combination with existing software. For vertex centered discretizations the average overlap deflation vectors are optimal. Furthermore it appears that for these methods the number of iterations decreases when the number of processors increases. This leads to efficiencies larger than 1. So we conclude that Deflation is a very efficient technique to accelerate parallel block preconditioners.

REFERENCES

1. O. Axelsson and G. Lindskog. On the eigenvalue distribution of a class of preconditioning methods. *Numer. Math.*, 48:479–498, 1986.
2. E. Brakkee, A. Segal, and C.G.M. Kassels. A parallel domain decomposition algorithm for the incompressible Navier-Stokes equations. *Simulation Practice and Theory*, 3:185–205, 1995.
3. E. Brakkee, C. Vuik, and P. Wesseling. Domain decomposition for the incompressible Navier-Stokes equations: solving subdomain problems accurately and inaccurately. *Int. J. for Num. Meth. Fluids*, 26:1217–1237, 1998.
4. J. Frank and C. Vuik. Parallel implementation of a multiblock method with approximate subdomain solution. *Appl. Num. Math.*, 30:403–423, 1999.
5. J. Frank and C. Vuik. On the construction of deflation-based preconditioners. *SIAM Journal on Scientific Computing*, 23:442–462, 2001.
6. J.A. Meijerink and H.A. van der Vorst. An iterative solution method for linear systems of which the coefficient matrix is a symmetric M-matrix. *Math. Comp.*, 31:148–162, 1977.

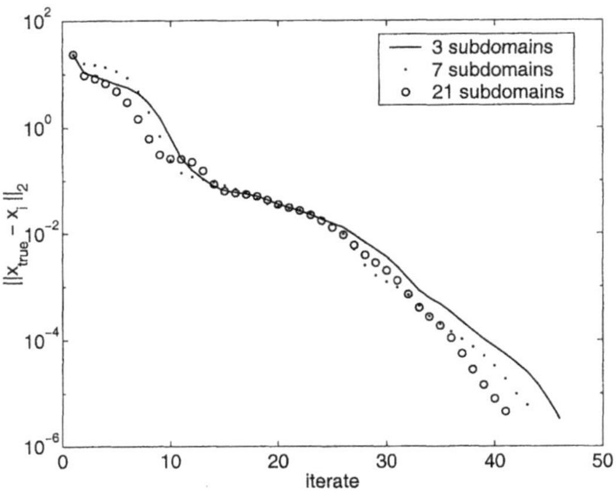

Figure 3. Convergence of the Deflated ICCG method for 41 x 41 internal nodes. The number of subdomains is varied.

7. R. A. Nicolaides. Deflation of conjugate gradients with applications to boundary value problems. *SIAM J. Numer. Anal.*, 24(2):355–365, 1987.
8. E. Perchat, L. Fourment, and T. Coupez. Parallel incomplete factorisations for generalised Stokes problems: application to hot metal forging simulation. Report, EPFL, Lausanne, 2001.
9. H.A. van der Vorst and C. Vuik. GMRESR: a family of nested GMRES methods. *Num. Lin. Alg. Appl.*, 1:369–386, 1994.
10. J. van Kan. A second-order accurate pressure-correction scheme for viscous incompressible flow. *SIAM J. Sci. Stat. Comput.*, 7:870–891, 1986.
11. C. Vuik. Solution of the discretized incompressible Navier-Stokes equations with the GMRES method. *Int. J. for Num. Meth. Fluids*, 16:507–523, 1993.
12. C. Vuik, A. Segal, and J.A. Meijerink. An efficient preconditioned CG method for the solution of a class of layered problems with extreme contrasts in the coefficients. *J. Comp. Phys.*, 152:385–403, 1999.
13. C. Vuik, R.R.P. van Nooyen, and P. Wesseling. Parallelism in ILU-preconditioned GMRES. *Parallel Computing*, 24:1927–1946, 1998.
14. P. Wesseling, A. Segal, and C.G.M. Kassels. Computing flows on general tree-dimensional nonsmooth staggered grids. *J. Comp. Phys.*, 149:333–362, 1999.
15. J. Xu and J. Zou. Some nonoverlapping domain decomposition methods. *SIAM Review*, 40:857–914, 1998.

Parallel Computational Fluid Dynamics – Practice and Theory
P. Wilders, A. Ecer, J. Periaux, N. Satofuka and P. Fox (Editors)

Parallel CFD Applications Under DLB Environment

E.Yilmaz and A. Ecer

Computational Fluid Dynamics Laboratory
Department of Mechanical Engineering
Indiana University-Purdue University Indianapolis (IUPUI)
Indianapolis, IN 46202, USA

1. INTRODUCTION

Since the introduction of parallel computers almost two decades ago, parallel computing environments became a reasonable choice for fast and efficient computing needs in research centers and academia. Main stimulus that drive parallel environment is its cost advantage over supercomputers.

In recent years, a global computing approach has emerged to bring powers of all parallel computers in a common pool to make resources available to users in remote locations. The new concept, named as Computational Grid, [1], is a large collection of computers linked via the Internet so that their combined processing power can be harnessed to work on time-consuming problems. Yet, there is no widely accepted software environment for grid applications to handle inter-site communication, transfer of data and balance of loads between remote sites except some research projects like Globus project, [2], from Argonne National Laboratory to make basis for new environment. It is worth to note that there is a successful application of such grid concept, but in a different way, at personal computer level with thousands of computers to analyze space observation data for the Search for Extraterrestrial Intelligence [3], SETI, leaded by University of California Berkeley.

Either in Grid environment or individually distributed computing environments, it is necessary to have a kind of load control and balance on computers. In such distributed computing environments, multiple users can access such resources at any time, which in turn causes random loading of compute-nodes. This requires a kind of load leveler or balancer to keep up efficiency of computing in favor of all users.

Recently, a software tool was developed for efficient load balancing in heterogeneous computing environment in CFD laboratory in IUPUI, [4], based on earlier load balancing studies, [5], in the same laboratory. This tool provides an environment that allows each parallel job to do its own load balancing while ensures that the load balancing of one parallel job does not affect the load balancing of other parallel jobs. If each user does dynamic load balancing without cooperation with other users, there will be no global optimal parallel load distribution. The objective of the present study is to demonstrate the use and benefit of the Dynamic Load-Balancing environment, [4], for parallel CFD applications on distributed computers connected in local and wide area networks. Demonstrations are accomplished with three different parallel CFD programs. Communications in parallel computing is established via a parallel library [6,7], (GPAR), developed in CFD laboratory in IUPUI.

2. PARALLEL ENVIRONMENT

In the present parallel application programs, GPAR (A Grid-based Database System for Parallel Computing) is used to achieve communication between the compute nodes and blocks. GPAR is a database management system for the interface and block combinations and relations between them. It provides an upper-level simplified parallel computing environment. Information related to these data groups are made available throughout the computation. Each block may have several interfaces. Each interface defines the partial boundary of the block connected to one of its neighbor. Each interface has a twin interface belonging to the neighbor block. Block solver contains no communication and it utilizes the database for data storage and update. Interface solver includes communication between neighbors. More details can be obtained in references given above.

3. DYNAMIC LOAD BALANCING ENVIRONMENT

Dynamic Load Balancing (DLB) is a software tool that allows each parallel job to do its application level load balancing while ensuring that system load is balanced. The DLB supports load balancing of parallel jobs that run on computers having different architecture and different operating systems as long as the parallel job has an executable code on these machines. There are two main components of DLB: one is at DLB system level and the other is at DLB users level. At the system level, computers are added to the pool that is available to the users provided that the users have accounts at all or some of those computers. Computer loads, communication speed, and all other status are recorded at system level such that the users can access these records for the purpose of load balancing and monitoring. At the users level, parallel jobs are initiated and submitted to DLB environment via a graphical users interface. Some control parameters such as number of DLB cycles, I/O file locations etc. can be specified by the users. The users can form a subset of all available computers in the pool for a specific parallel job. Study on using graphical user interface as a web service is under way. Figure 1 shows general structure of the DLB tool.

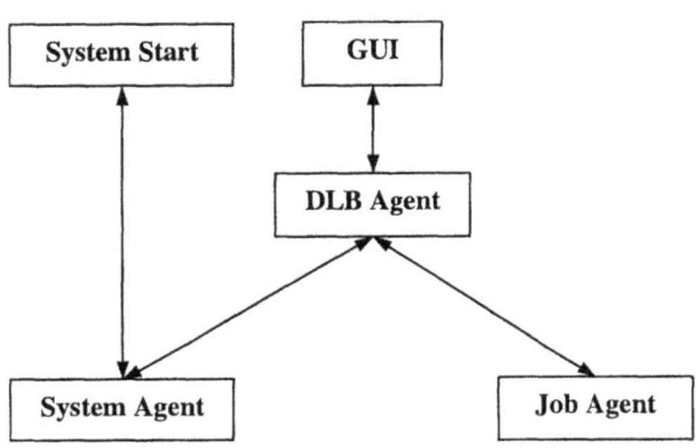

Figure 1. General structure of DLB tool

Graphical User Interface, (GUI), performs the job for users to start and stop DLB Agent conveniently. Presently, GUI is running as a user process. A typical graphical user interface is given in Figure 2.

The System Agent is a program that runs on every computer. It accepts the registration of all jobs from all users through DLB agents, records computer loads and communication speeds between computers, and tracks jobs doing DLB. System Agent information is accessible by the users.

DLB Agent is responsible for dynamic load balancing of parallel jobs. DLB Agent collects computer and application related information from System Agent and Job Agent for load balancing calculation. Currently, DLB Agent is running as a user process and each user needs to execute one DLB Agent.

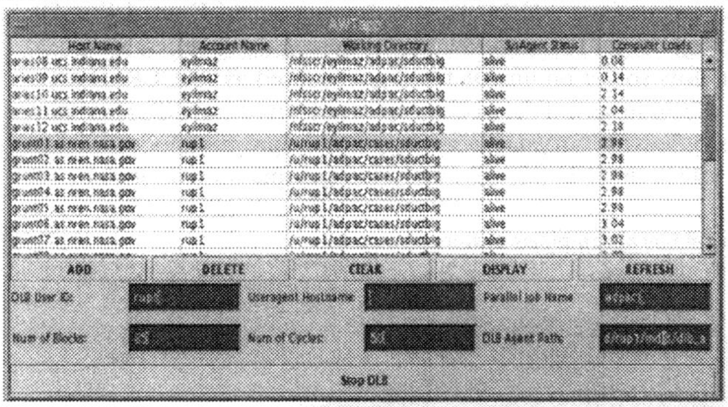

Figure 2. A Typical Graphical User Interface for DLB User Agent

Job Agent is responsible to start and stop user's parallel job and gather the execution timing information for the parallel job. The timing information is sent to the DLB Agent for load balancing.

Software requirements for DLB environment is Java run-time environment (JRE): JRE 1.2.2 or above, message passing libraries for parallel computing, that is PVM or MPICH, a library to get timing from application program, if timing data is not provided by other means. Details of the DLB environment are given in Ref. [4].

4. COMPUTING ENVIRONMENT

In the present study, three clusters, formed by Unix, Linux, and Windows 2000 operating systems, are used. The first cluster is located at the CFD laboratory in IUPUI and consists of six Unix-based IBM RS6000 workstations, 14 Windows 2000 based PII PCs and six PIII based Linux PCs. Second cluster consists of 129 compute nodes (564 CPUs) IBM SP2 parallel computer system in Indiana University in Bloomington-IN. The third cluster is a Beowulf system, which is Linux-based system, with 64 nodes (128 CPUs) PIII PCs in NASA Glenn Research Center in Cleveland-OH. In dynamic load balancing applications, some of these nodes available to us are used to test the environment. Connection between IUPUI and IU is a fast connection different than the connection with other sites, which use regular Internet connections. Figure 2 gives computer network used in the present study.

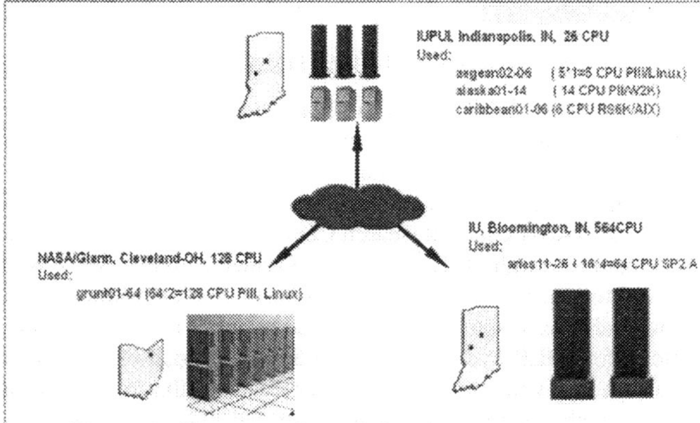

Figure 3. Computer network used in the present study

5. RESULTS

The current research involves parallel processing of three CFD programs for DLB applications. First one is Euler version of the ADPAC code (An Advanced Ducted Propfan Analysis Code) [6,8], second program is the Parallel CFD Test Case that solves heat equations, and the finally an Euler flow solver on unstructured grid named as PACER3D [9].

In Figure 4, ADPAC program is run for a test case on computers from three different sites. Block-distribution history was recorded for fifty DLB cycles. Total compute nodes chosen from the pool of computers is 32, which gives 69 processors for 85 computational blocks.

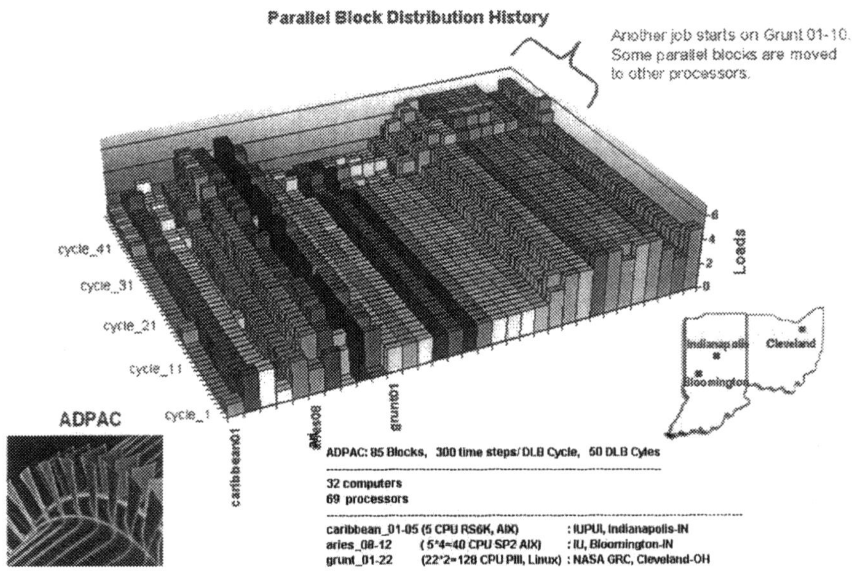

Figure 4. Parallel Block Distribution History by DLB for ADPAC Code

After a few DLB cycles loads are almost balanced and came to a steady level almost for all compute nodes. There are minor fluctuations on some of the SP2 nodes due to random access from the users of that system. Another parallel job, not submitted through DLB environment, is started at 37th cycle of DLB for ADPAC on Grunt 01-10, which operates on Linux at NASA Glenn Research Center, to see the effect of load balancing. Loads that are not submitted through DLB is seen as extraneous loads. Some of the blocks of ADPAC case are moved to other Grunt nodes with almost even distribution on these nodes. Introduction of this new load caused some minor fluctuation at the beginning, however it came almost steady case after a few DLB cycles. In Figure 5, overall load-distribution on the computers used for this problem is given. As can be observed from that result, overall loads are evenly distributed within each cluster of computers for the benefit of all users whether they run DLB with parallel jobs or not. Total loads are sum of the parallel blocks submitted through the DLB plus other loads not submitted through the DLB, which includes the serial and parallel loads submitted by other means.

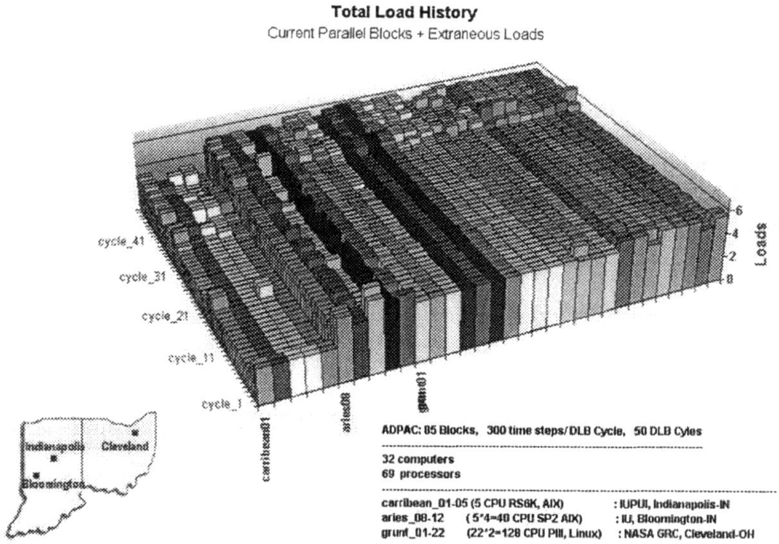

Figure 5. Parallel Block Distribution History by DLB for ADPAC Code

In the next application, we used Parallel CFD (PCFD) test case with Pacer3D code as external loads to observe the effect of two parallel jobs submitted through the DLB environment. Total compute nodes chosen from the pool of computers is 56, which gives 131 processors for 50 computational blocks for the PCFD test case. Computers chosen from three different locations have Unix, Linux, and Windows 2000 operating systems. As for the Pacer3D, 50 block partitions is used for unstructured grid test case; five from RS6K workstations and ten from SP2 with total of 45 CPUs. Two DLB cycles were performed for the Pacer3D during 24 DLB cycles of the PCFD test case. One cycle for the Pacer3D case took almost 5-6 cycles of the DLB for the PCFD test case. Average elapsed time for the PCFD test case was recorded as given in Figure 6. Block distribution for the PCFD test case and overall load distribution at different the DLB cycles of the PCFD test case are given in Figure 7. Note that blocks are equally distributed at the beginning as in Figure 7a then it is balanced in the next cycles. The even distribution has resulted in very high elapsed time, as given in Figure 6,

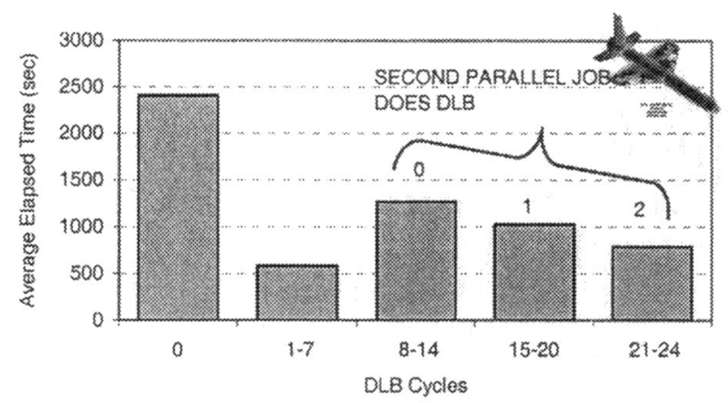

Figure 6. Average Elapsed Time for Parallel CFD Test Case with another parallel job

when compared with balanced distribution of the loads between cycles 1-7. At the middle of 8th DLB cycle for the PCFD test case, the other parallel job is started. Its effect and its block distribution are shown in Figure 7b. Average elapsed time for this PCFD test case is increased from the balanced distribution. However, as the load balancing is performed for the second job at 15th cycle, it resulted more decrease in the average elapsed time for the PCFD test case. Final loads and parallel blocks distribution are given in Figure 7c. This case shows that two DLB even helps each other more for efficient distribution of loads on compute nodes used together.

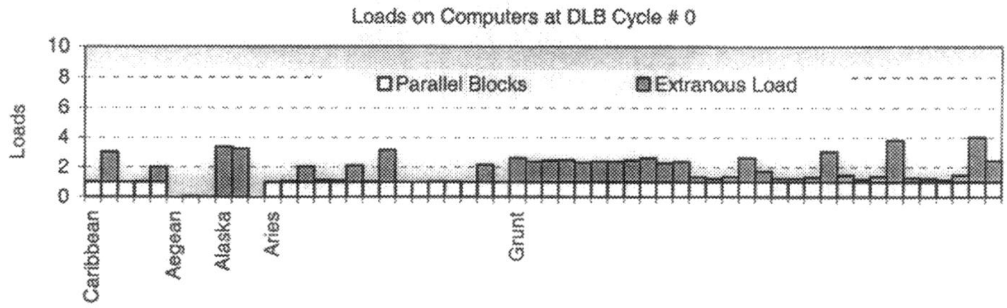

(a)

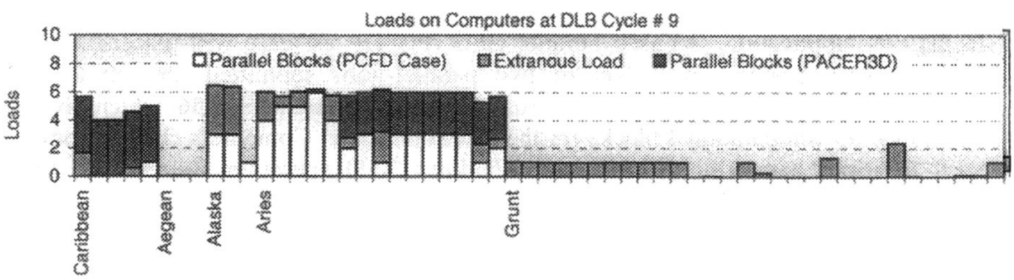

(b)

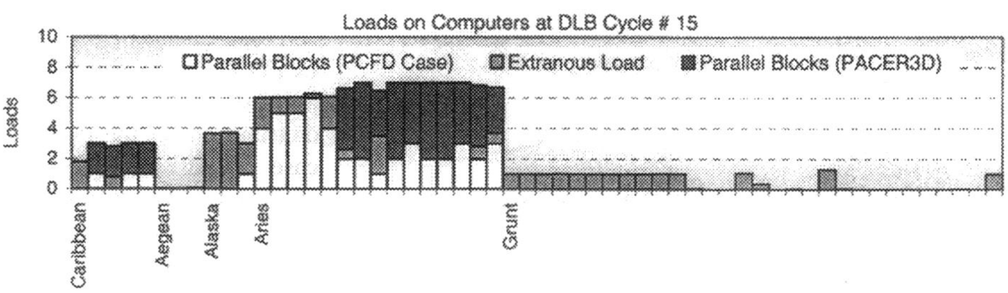

(c)

Figure 7. Block distribution for PCFD test case and Pacer3D case and Load distribution on all computers at different DLB cycles

6. CONCLUSIONS

We have demonstrated application of the dynamic load-balancing environment on several CFD programs. It has been run on different operating systems on several computers. When multiple parallel jobs are run through the DLB environment, there is no conflict or inefficiency in synchronization of the overall loads distribution on computers. Therefore, the DLB environment can be used in multiple job submission for the parallel and serial programs as well. The environment itself is user friendly with the Graphical User Interface form and takes the advantages of Java programming language for most general portability. To demonstrate such applicability, it has been run on Unix, Linux and Windows NT platforms combined all together for parallel execution and load balancing. Application programs have basics to more sophisticated CFD algorithms from simple heat equation to turbomachinery problems.

It has been observed that load balancing improves computation time of the CFD programs by redistributing the blocks on available and fast computers. When a new load is introduced on to the computer system, the DLB records the change and consider that load in the next load balancing cycle for efficient distribution of blocks. There might be huge number of computers in the resource pool of the DLB environment. But user can use only some part of those computers. Even among them DLB can choose some of them for maximum efficiency based on loads on the computers, their CPU speed, network communication speed, and block-to-block communication cost. From the users point it can be said that present tool is sufficient to handle efficient distribution of the loads in a multi-users environment.

As future improvements, web-based login to the environment and submitting job through it would make this tool very accessible from any remote site. Post-processing features of the timing results on the same graphical environment would be very helpful. Algorithms behind block moving works well, however, to come to steady state loads distribution might take a few cycles of the DLB depending on the computer environment and loads on it. This might result from very random access to computers by other users.

ACKNOWLEDGEMENTS

We would like to thank to the NASA Glenn Research Center and the Indiana University Computer Center to let us use their computer resources. We are also grateful to the staffs in the CFD laboratory in IUPUI for their help in using resources and their help on setting software environments.

REFERENCES

1. I. Foster, C. Kesselman, and S. Tuecke, "The Anatomy of the Grid: Enabling Scalable Virtual Organizations," (to be published in *Intl. J. Supercomputer Applications*, 2001).
2. I. Foster, and C. Kesselman, "Globus: A Metacomputing Infrastructure Toolkit, " *Intl J. Supercomputer Applications*, 11 (2): 115-128, 1997.
3. D. Anderson et al., "Internet Computing for SETI," *Bioastronomy 99: A New Era in Bioastronomy*, ASP Conference Series No. 213 (Astronomical Society of the Pacific: San Francisco), p. 511, 2000.
4. Y.P. Chien, J.D. Chen, A. Ecer, H.U. Akay, and J.Zhou "DLB 2.0: A Distributed Environment Tool for Supporting Balanced Execution of Multiple Jobs on Networked

Computers," *Proceeding of Parallel CFD 2001*, May 21-23, 2001, Elsevier Science, Amsterdam, The Netherlands, (in print).

5. Y.P. Chien, A. Ecer, H.U. Akay, and F. Carpenter, "Dynamic Load Balancing on Network of Workstations for Solving Computational Fluid Dynamics Problems," *Computer Methods in Applied Mechanics and Engineering*, Vol. 119, 1994, pp. 17-33.
6. A. Ecer, H.U. Akay, W.B. Kemle, H. Wang, D. Ercoskun, and E.J. Hall, "Parallel Computation of Fluid Dynamics Problems," *Computer Methods in Applied Mechanics and Engineering*, Vol. 112, 1994, pp. 91-108.
7. "*A Data-Parallel Software Environment, GPAR*," CFD Laboratory Indiana University Purdue University Indianapolis, Indianapolis IN, USA (in preparation).
8. E.J. Hall, R.A. Delaney, and J.L. Bettner, *Investigation of Advanced Counterrotation Blade Configuration Concepts for High Speed Turboprop Systems*, NASA Contractor Report CR-187106, May 1991.
9. E. Yilmaz, M.S. Kavsaoglu, H.U. Akay, and I.S. Akmandor, "Cell-vertex Based Parallel and Adaptive Explicit 3D Flow Solution on Unstructured Grids," *International Journal of CFD*, 2001, Vol. 14, pp. 271-286

Parallel Computational Fluid Dynamics – Practice and Theory
P. Wilders, A. Ecer, J. Periaux, N. Satofuka and P. Fox (Editors)

Parallel Performance of a CFD Code on SMP Nodes

Mitsuo Yokokawa[a], Yoshinori Tsuda[a], Minoru Saito[a], and Kenji Suehiro[b]

[a]Earth Simulator Research and Development Center
Japan Atomic Energy Research Institute
3173-25, Showa-machi, Kanazawa-ku, Yokohama, Kanagawa 236-0001, Japan.

[b]NEC Corporation
1-10, Nisshincho, Fuchu, Tokyo 183-8501, Japan

Three different programming interfaces of microtasking, MPI and HPF, which are the parallel programming interfaces for a cluster of shared memory symmetric multiprocessor (SMP) nodes, are used in parallelizing a CFD code. We have evaluated performance of the parallelized code written by these interfaces on a single-node and multi-node SX-5 system. Microtasking achieves the best performance for the single-node execution of all interfaces, and the performance of HPF programming is almost the same as that of MPI programming when the problem size is large enough. In the two-node execution, homogeneous MPI programming achieves the highest speedup of 6.0 by 8 MPI processes over the two nodes.

1. INTRODUCTION

It is quite essential to employ parallel processing in carrying out large-scale computational fluid dynamics simulations, but parallel programming on distributed memory parallel computers is known to be difficult because parallel task assignment and data mapping of a program onto such parallel computers are highly skillful and time-consuming jobs. Moreover, programming models on various kinds of parallel computers are not unique, and a lot of researchers and engineers not in computer science field struggle to make their application programs run efficiently on a variety of parallel computer architectures.

Parallel programming interfaces such as MPI (Message Passing Interface)[1], HPF (High Performance Fortran)[2], and OpenMP[3] have been proposed for parallel programming models. OpenMP is supposed to be a favorable interface for a shared memory symmetric multiprocessor (SMP) system, because it treats nothing but task parallelization by using compiler directives and it does not need to take care of partitions of data mapped onto the shared memory. It is, however, known to be difficult to extend its capability of parallel programming to distributed memory parallel systems. On the other hand, MPI, as well as HPF, is considered as one of the major parallel programming interfaces for distributed memory parallel computers. MPI is a library which can describe explicitly data transfer between parallel tasks. Users should be well aware of data assignment to parallel tasks in MPI programming. HPF is a high-level data parallel language designed to provide a clear and easy programming interface to understand. Users can parallelize their sequential programs by mainly inserting directives specifying data mapping on distributed memory. MPI and HPF can be also considered as the programming models on shared memory system.

Recently, it has been recognized that hybrid parallel programming models are extremely important to create a program by using the above interfaces on a cluster of SMP nodes, which has a memory hierarchy of shared memory system and distributed memory system. Users should consider data mapping to the memory hierarchy in making a program so that they can obtain high performance. However, there are several implementations for hybrid programming according to the combination of the parallel programming interfaces, e.g. OpenMP on shared memory system and MPI on distributed memory system. Of course, we can choose a homogeneous programming model in which only a interface suited for the distributed parallel programming model is used on both shared and distributed memory systems, e.g. MPI on the both systems. Since performance of a program on different computer systems is usually changed even if the same programming model is taken, a performance evaluation should be carried out to demonstrate which interface is better to use on a target computer system.

In this paper, we have applied three different parallel programming interfaces such as microtasking, MPI and HPF to a CFD code, and evaluated their performance on SX-5 SMP nodes. A comparison between hybrid parallel programming models and homogeneous programming models is also presented.

2. PROGRAMMING MODELS ON SX-5 SMP NODES

A hybrid programming model is important on SX-5 SMP nodes which provide two-level parallel programming environment of shared and distributed memory systems; parallel processing in an SMP node, and parallel processing among nodes via internode communication network[4]. Moreover, vector processing in a processor can be taken into account. Users should make a program by considering the environment to obtain the best performance on the SX-5 nodes.

Some programming interfaces are provided to help users make a parallel program on the SX-5 nodes (Fig. 1). The most fundamental interface is an automatic vectorization by a compiler. Users can immediately obtain high performance by vector calculations in a processor for continuum models like CFD simulations. An automatic parallelization by loop-level microtasking is also prepared as a proper parallel programming interface on the shared memory of the SX-5 SMP node. Moreover, the parallel programming interfaces such as MPI and HPF are also supplied as parallel programming models in a node. The program with HPF directives would be transformed into a program with MPI libraries.

MPI and HPF can be also employed for the parallel programming interfaces on two or more nodes. A hybrid programming model of microtasking within a node and MPI or HPF among nodes is available in addition to a homogeneous programming model by using MPI or HPF without microtasking over the nodes.

If we take the homogeneous programming model, message transfer between parallel processes in a node is realized by a memory copy and message transfer between parallel processors resided on different nodes is made by using an internode crossbar switch (IXS).

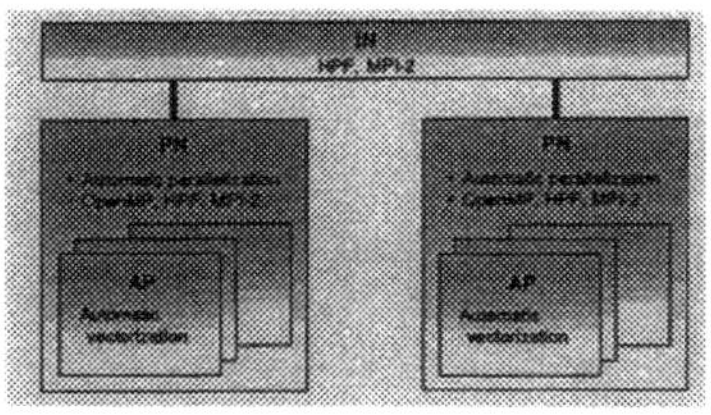

Figure 1. Hierarchical programming interfaces available on SX-5

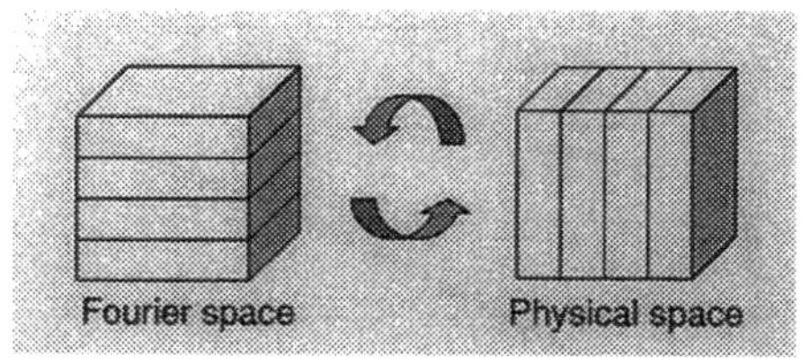

Figure 2. Transposition of data in three-dimensional FFT

Latency and throughput of message transfer in the node are different from those between nodes. It seems that the hybrid programming model by microtasking in a node and MPI or HPF as internode parallel implementations is appropriate because microtasking does not use any additional works for the memory copy. Vector processing should be always used in any implementation. We have applied these models to parallel programming models for a CFD code.

3. OUTLINE OF A CFD CODE

A computational fluid dynamics code or Trans6, which is a pseudospectral code for homogeneous isotropic turbulent flows, is used in this evaluation[5]. The Navier-Stokes equations and the continuity equation are employed for the simulation of three-dimensional incompressible viscous fluid in a cube. We assume a cyclic boundary condition on the wall of the cube for the isotropic flows. Fourier expansions are used for the discretization of physical space and the 4-th order Runge-Kutta method is used for the time integration. The problem size N^3 is defined as the number of Fourier coefficients.

The pseudospectral method is used to decrease the huge computational works in calculating a convolution of the nonlinear terms of the Navier-Stokes equations[6]. The convolution of two variables in spectral space is calculated as follows; Firstly two Fourier coefficients in spectral space are transformed into physical space. Then the products of those two variables at all discretized grids in physical space are calculated. Finally the products are transformed back into Fourier spectral space. Aliasing errors are removed by phase shifts and the spherical truncation is applied for the modes $k > \frac{\sqrt{2}N}{3}$. The three-dimensional FFT (3D-FFT) is executed 72 times during a time advancement of the Runge-Kutta method.

In parallel implementation of Trans6, the assignment of its calculation to the parallel processors is a great concern to be considered, because the 3D-FFT has a global data dependency due to its integral feature. Considering that the 3D-FFT can be carried out efficiently in any direction in vector operations, it is appropriate that the calculations on several two-dimensional planes are assigned to a node by domain decomposition method and the two-dimensional FFT is applied to the divided slab on each node.The FFT in the rest direction should be performed after the 3-dimensional array data are transposed

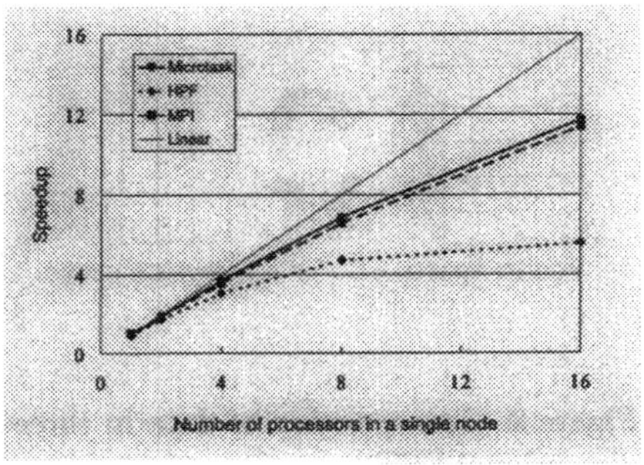

Figure 3. Speedup for the size of 128^3 on a node; Solid, broken, and dotted lines denote the results by microtasking, HPF, and MPI, respectively

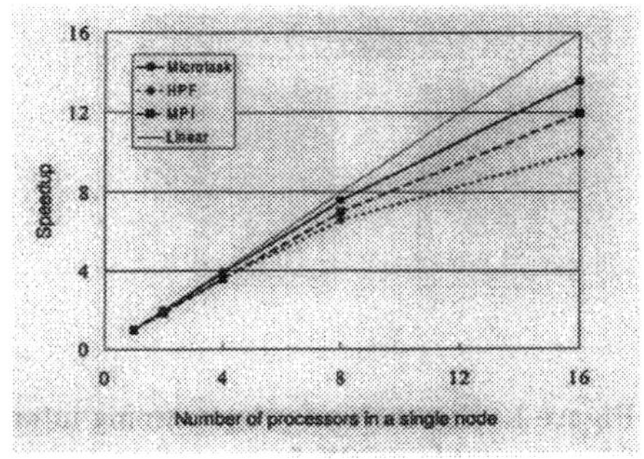

Figure 4. Speedup for the size of 256^3 on a node; Solid, broken, and dotted lines denote the results by microtasking, HPF, and MPI, respectively

(Fig. 2). The transposition of 3-dimensional array data need all-to-all communications in parallel implementation on the distributed memory parallel system, and the transposition makes computational works of data transfer very expensive.

4. PARALLEL PERFORMANCE OF THE CODE

We have evaluated parallel efficiency of several parallel implementations both on a single node and on two nodes[7].

The code is compiled by FORTRAN90/SX for microtasking, which has a capability of automatic loop parallelization on the SMP node. For the MPI implementation, MPI/SX, which is an MPI-2 libraries for SX-5, is used for the parallelization. Some HPF directives are inserted to the program and compiled by an HPF/SX V2 compiler for the HPF implementation.

As for the reference, the execution time of the sequential version of Trans6 on the single processor is measured by changing the problem size of 128^3, 256^3, and 512^3. The time is 3.055sec, 17.182sec, and 156.431sec, respectively. Sustained performance of 2.9Gflops, 4.66Gflops, and 4.56Gflops is obtained for each problem size. Since the peak speed of a processor of SX-5 is 8 Gflops, more than half speed of the peak is achieved for large problem size.

4.1. Parallel performance on a node

Three parallel implementations, which are microtasking, MPI, and HPF implementations, are considered in the measurement of execution time within the single node. The execution time is measured by changing the number of processors in the node as 1, 2, 4, 8, and 16 for each problem size. Each MPI or HPF process is assigned to the different processor. For example, 16 MPI processes or 16 HPF processes are invoked on the 16 processors.

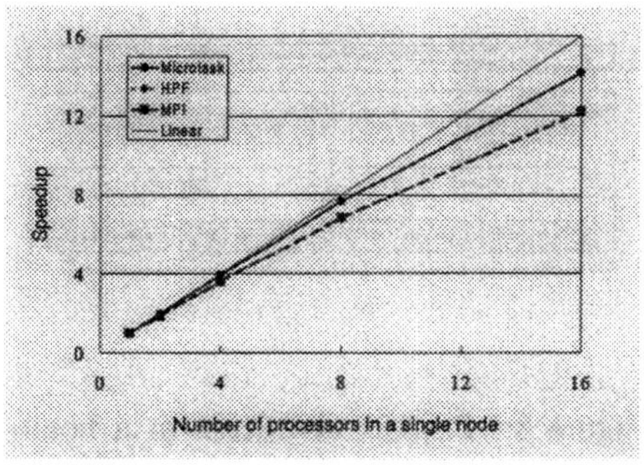

Figure 5. Speedup for the size of 512^3 on a node; Solid, broken, and dotted lines denote the results by microtasking, HPF, and MPI, respectively

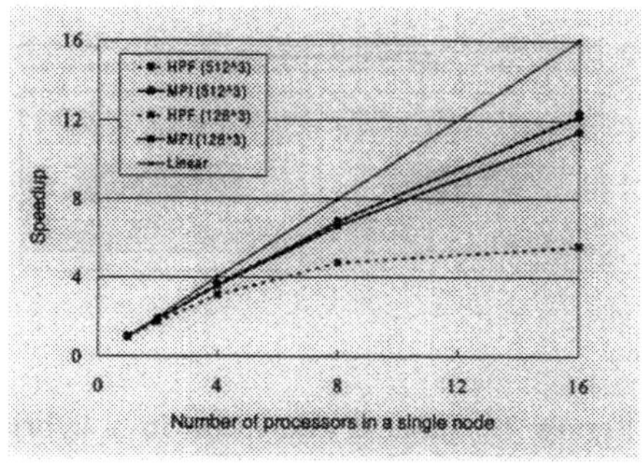

Figure 6. Comparison between MPI and HPF for the size of 128^3 and 512^3

Figures 3, 4, and 5 denote speedup which is the ratio of execution time of the parallel implementation to the one of the sequential verison executed on the single processor for the probelm size of 128^3, 256^3, and 512^3, respectively. Microtasking exhibits the highest performance of all parallel implementations. The speedup of 7.67 is obtained by 8 microtasks for the size of 512^3, and speedup of 14.14 by 16 microtasks. Because it is not necessary to transpose the data in microtasking since all the data reside on the shared memory and can be accessed by any processors without extra overhead. On the other hand, the MPI and HPF implementations need actual data copy in the memory when the transposition of the data occurs, because the data are separated logically on the shared memory.

The execution time of the HPF implementation is higher than that of the MPI implementation for the problem size of 128^3 and 256^3. The execution time of the 16 HPF processes is about two times larger than that of 16 MPI processes for the size of 128^3. However the difference between the execution time of HPF and the one of MPI decreases as the problem size becomes larger (Fig. 6). For the size of 512^3, the execution time of HPF is almost the same as the one of MPI, because the ratio of the cost of data transfer to the calculation cost becomes smaller.

4.2. Parallel performance on two nodes

A hybrid parallel implementation and a homogeneous parallel implementation are compared in execution time with two nodes. The execution time for four cases were measured for the problem size of 256^3 by using up to 8 processors across the two nodes. The processes are assigned to the two nodes equally so that load balance is taken. Therefore the number of processes in every node is equal.

Cases 1 and 3 are the hybrid implementations in which either MPI or HPF is taken for the inter-node parallel programming model between two nodes and microtasking is used in a node (Fig. 7). For example, two MPI or HPF processes and 4 microtasks on each

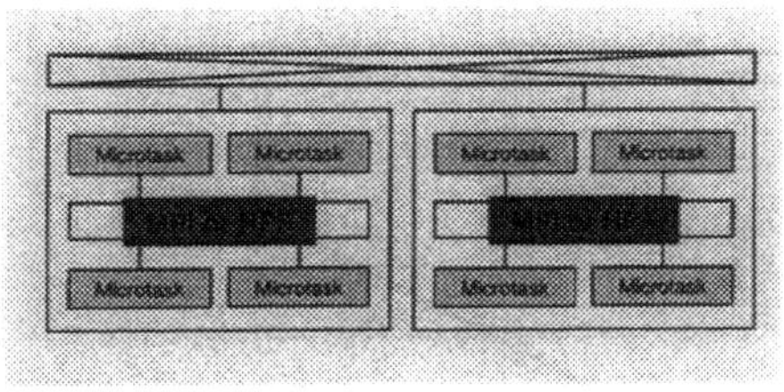

Figure 7. Process allocation in a hybrid parallel implementation (Cases 1 and 3)

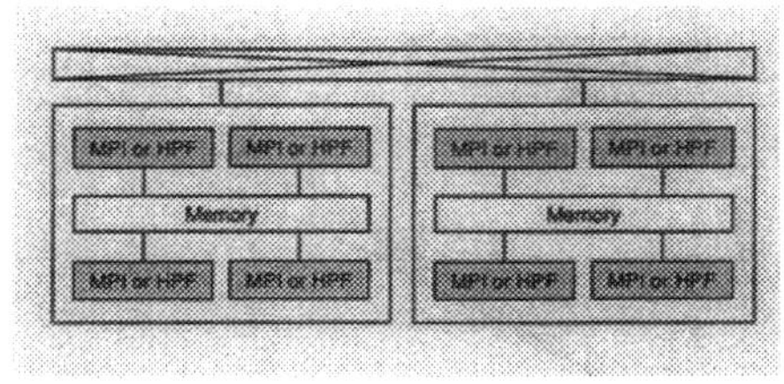

Figure 8. Process allocation in a homogeneous parallel implementation (Cases 2 and 4)

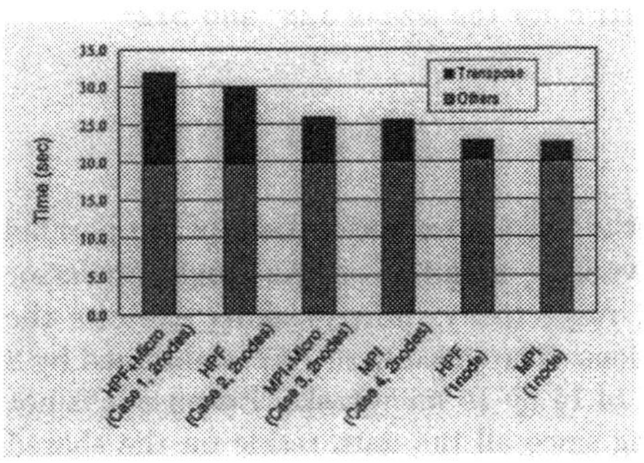

Figure 9. CPU time consumed by data transposition

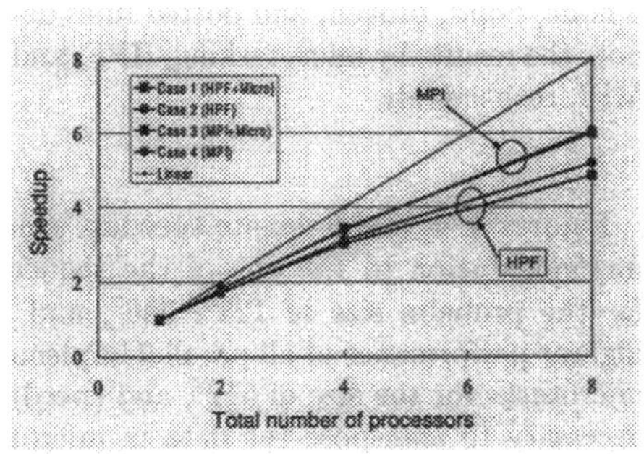

Figure 10. Comparison of speedup of one-level and hybrid implementations

node are invoked when the total number of processes is 8. Just one specified processor treats data transfer between two nodes via inter-node crossbar switch.

Cases 2 and 4 are the homogeneous implementations in which each process on the processor is either an HPF process or an MPI process and no microtasking is used (Fig. 8). Data transfer is realized by a memory copy in a node and by message passing between two nodes.

Figure 9 shows speedup of four cases. It is found that the case 4 achieves the highest performance in this experiment; Its speedup is about 6.0 by 8 MPI processes. Performance of the case 1 is less than that of the case 2, and performance of the case 3 is less than that of the case 4. These results denote that the hybrid implementation is not necessary to achieve good performance compared to the homogeneous implementation. It is clear that other processors but a processor, which is in charge of data transfer in the hybrid implementation, are idle during the data communication between two nodes. Only one processor can handle data transposition and the message size is not changed so long as the number of nodes is not changed. This gives the hybrid implementation with two nodes

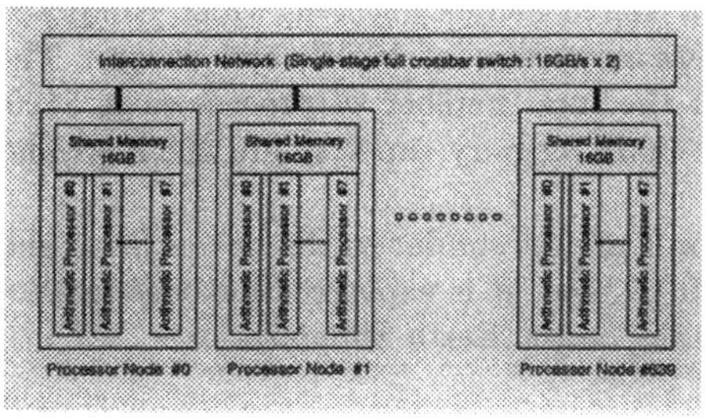

Figure 11. Configuration of the Earth Simulator

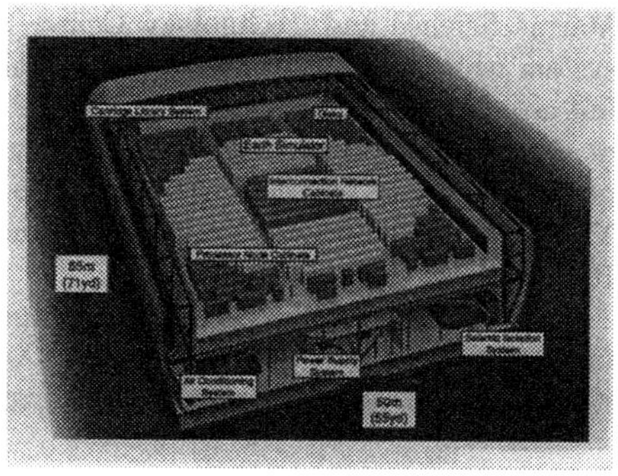

Figure 12. Artist's Representation of the Earth Simulator

lower performance.

Figure 9 shows the total CPU time for each case; Transposition time and the others are denoted by the different color in the figure. For the reference, the results for MPI and HPF implementations on a single node are also depicted. Though the time but transposition time is almost the same in all implementations, the transposition time is different in each implementation. The transposition time for the hybrid implementations by HPF is large compared to other implementations, unrelated to either hybrid or homogeneous implementations. HPF compiler cannnot make an efficeint object for data transfer between the nodes at present. It will be improved near future. It is also clear that the data transfer between nodes is expensive in parallelization, though the transposition time required for the single node execution is small.

5. CONCLUDING REMARKS

We have presented parallel performance of a CFD code on the SX-5 SMP cluster. Performance of the HPF implementation is almost the same as one of the MPI implementation as the problem size becomes larger, although it is rather low compared to performance of the microtasking implementation. HPF performance using the two nodes system is worse than MPI performance for the problem size of 256^3.

In the case that the problem size is large and the memory size required for the code is beyond the memory capacity of a node, a parallel computation over several nodes must be necessary and cannot be avoided. Therefore, hybrid parallel programming models are quite important and its efficiency should be improved and evaluated more precisely.

Here the Earth Simulator should be mentioned briefly in the end of this paper. Because it will be operational in March 2002, and many presearchers are very interested in the programming models available on the Earth Simulator in making their programs.

The Earth Simulator is also an SMP cluster which is being developed by National Space Development Agency of Japan, Japan Atomic Energy Research Institute, and Japan

Marine Science and Technology Center(Fig. 12). It is a distributed memory parallel system which consists of 640 processor nodes (PN) connected by a 640 × 640 single-stage full crossbar switch (Fig. 11)[8]. Each PN is a shared memory system which composed of eight vector processors (AP), a shared memory system of 16GB, and so on. The peak performance of each AP is 8Gflops and the total number of processors is 5120. Therefore the total peak performance and the main memory capacity are 40Tflops and 10TB, respectively.

Since the architecture of the Earth Simulator is almost similar to the one of SX-5 and the most capabilities of compilers for SX-5 will be used, it is expected that this kind of study would be a good guide for the programming on the Earth Simulator.

ACKNOWLEDGMENT

The authors would like to thank Mr. S. Kitawaki for his valuable comments to the paper. They would also like to thank all members at Earth Simulator Research and Development Centerfor for their valuable discussions.

REFERENCES

1. "MPI-2: Extensions to the Message-Passing Interface," Message Passing Interface Forum, July (1997).
2. "High Performance Fortran Language Specification Version 2.0," High Performance Fortran Forum, January (1997).
3. "OpenMP Fortran Application Program Interface Version1.0," OpenMP Architecture Review Board, October (1997).
4. K. Kinoshita, "Hardware System of the SX Series," NEC RESEARCH & DEVELOPMENT, Vol.39, No.4, pp.362-368 (1998).
5. M. Yokokawa, et al., "Parallelization of A Fourier Pseudospectral CFD Code," Proc. of the PERMEAN'95, pp.54-59, Japan (1995).
6. C. Canuto et al., Spectral Methods in Fluid Dynamics, Springer-Verlag, Berlin Heidelberg, 1988.
7. M. Takahashi, et al., "An Evaluation of HPF Implementation on Cenju-4," IPSJ SIG Notes, Vol.99, No.103, pp.49-54 (1999) (*in Japanese*).
8. M. Yokokawa, et al., "Basic Design of the Earth Simulator," High Performance Computing, LNCS 1615, Springer, pp.269-280 (1999).

www.ingramcontent.com/pod-product-compliance
Ingram Content Group UK Ltd.
Pitfield, Milton Keynes, MK11 3LW, UK
UKHW050948290926
15470UKWH00005B/1711

9 780444 506726